Benchmark Papers in Electrical Engineering and Computer Science

Editor: John B. Thomas—Princeton University

A selection from the published volumes in this series

SYSTEM SENSITIVITY ANALYSIS / *José B. Cruz, Jr.*
RANDOM PROCESSES: Multiplicity Theory and Canonical Decompositions / *Anthony Ephremides and John B. Thomas*
ALGEBRAIC CODING THEORY: History and Development / *Ian F. Blake*
COMPUTER-AIDED CIRCUIT DESIGN: Simulation and Optimization / *S. W. Director*
ENVIRONMENTAL MODELING: Analysis and Management / *Douglas Daetz and Richard H. Pantell*
CIRCUIT THEORY: Foundations and Classical Contributions / *M. E. Van Valkenburg*
TWO-DIMENSIONAL DIGITAL SIGNAL PROCESSING / *Sanjit K. Mitra and Michael P. Ekstrom*
COMPUTER ARITHMETIC / *Earl E. Swartzlander, Jr.*
ARRAY PROCESSING: Applications to Radar / *Simon Haykin*
NONPARAMETRIC DETECTION: Theory and Applications / *Saleem A. Kassam and John B. Thomas*
DECONVOLUTION OF SEISMIC DATA / *V. K. Arya and J. K. Aggarwal*
REPRODUCING KERNEL HILBERT SPACES: Applications in Statistical Signal Processing / *Howard L. Weinert*
DISTRIBUTED PARAMETER SYSTEMS THEORY, PART I: Control / *Peter Stavroulakis*
DISTRIBUTED PARAMETER SYSTEMS THEORY, PART II: Estimation / *Peter Stavroulakis*

DISCRETE TRANSFORMS AND THEIR APPLICATIONS

Edited by

K. RAMAMOHAN RAO
University of Texas at Arlington

A HUTCHINSON ROSS BENCHMARK® BOOK

To my wife, Karuna Rao, and to my children, Ramesh Rao and Rekha Rao McGee

Benchmark Papers in Electrical Engineering and Computer Science, Volume 28
Library of Congress Catalog Card Number: 84-20832
ISBN: 0-442-27669-9

Manufactured in the United States of America.

Published by Van Nostrand Reinhold Company Inc.
135 West 50th Street
New York, New York 10020

Van Nostrand Reinhold Company Limited
Molly Millars Lane
Wokingham, Berkshire RG11 2PY, England

Van Nostrand Reinhold
480 Latrobe Street
Melbourne, Victoria 3000, Australia

Macmillan of Canada
Division of Gage Publishing Limited
164 Commander Boulevard
Agincourt, Ontario MIS 3C7, Canada

15 14 13 12 11 10 9 8 7 6 5 4 3 2 1

Library of Congress Cataloging in Publication Data
Main entry under title:
Discrete transforms and their applications.
(Benchmark papers in electrical engineering and computer science; v. 28)
Reprint of papers originally published 1969-1983. With editor's comments
"A Hutchinson Ross Benchmark book."
Includes indexes.
1. Electric engineering—Mathematics—Addresses, essays, lectures.
2. Transformations (Mathematics—Addresses, essays, lectures. I. Rao, K. Ramamohan (Kamisetty Ramamohan) II. Series.
TK153.D57 1984 515.7'23'0246213 84-20832
ISBN 0-442-27669-9

CONTENTS

PART II: APPLICATIONS

SERIES EDITOR'S FOREWORD

This Benchmark Series in Electrical Engineering and Computer Science is aimed at sifting, organizing, and making readily accessible to the reader the vast literature that has accumulated. Although the series is not intended as a complete substitute for a study of this literature, it will serve at least three major critical purposes. In the first place, it provides a practical point of entry into a given area of research. Each volume offers an expert's selection of the critical papers on a given topic as well as his views on its structure, development, and present status. In the second place, the series provides a convenient and time-saving means for study in areas related to but not contiguous with one's principal interests. Last, but by no means least, the series allows the collection, in a particularly compact and convenient form, of the major works on which present research activities and interests are based.

Each volume in the series has been collected, organized, and edited by an authority in the area to which it pertains. In order to present a unified view of the area, the volume editor has prepared an introduction to the subject, has included his comments on each article, and has provided a subject index to facilitate access to the papers.

We believe that this series will provide a manageable working library of the most important technical articles in electrical engineering and computer science. We hope it will be equally valuable to students, teachers, and researchers.

This volume, *Discrete Transforms and Their Applications,* has been edited by K. R. Rao of the University of Texas at Arlington. It contains 42 papers on the definitions, applications, and limitations of discrete transforms. Although the fast Fourier transform has received the most attention in the literature, this volume collects, organizes, and references the large amount of research on the many other discrete transforms that lend themselves to modern applications.

JOHN B. THOMAS

PREFACE

Discrete transforms is a relatively new field and has evolved rapidly over the last 20 years. This field has attracted the attention of various researchers resulting not only in generating and developing various transforms, but also in applying them to different disciplines. The field has emerged from a scientific curiosity to actual hardware realization and commercial application. *Discrete Transforms and Their Applications* is divided into two parts: the first part deals with various discrete transforms; the second part deals with specific applications. These papers are aimed at both the novice and the expert. The former receives an introduction to the transforms and their advantages, limitations, and possible applications. The latter can review the discipline and probe further into the field. Thus engineers, researchers, and academia with varying degrees of familiarity with discrete transforms can benefit from this volume.

The editor would like to acknowledge the help and assistance of the faculty, graduate students, and staff at the University of Texas at Arlington. Also, the understanding and encouragement of his wife Karuna and children, Ramesh and Sasirekha, is deeply appreciated.

K. RAMAMOHAN RAO

CONTENTS BY AUTHOR

DISCRETE TRANSFORMS AND THEIR APPLICATIONS

INTRODUCTION

The introduction of the fast Fourier transform, (FFT), an efficient computational algorithm for the discrete Fourier transform (DFT) by Cooley and Tukey (1965), has brought to the limelight various other discrete transforms (Ahmed and Rao, 1975; Elliott and Rao, 1982; Rao and Jalali, 1982). Some of the analog functions from which these transforms have been derived date back to the early 1920s, for example: Walsh functions (Walsh, 1923) and Haar functions (Haar, 1910, 1912). Other transforms, however, have been developed fairly recently, for example: discrete cosine (DCT) in 1974 (Paper 1), rapid (RT) in 1969 (Paper 7), slant (ST) in 1974 (Paper 22), slant-Haar (SHT) in 1974 (Paper 26), Hadamard-Haar (HHT) in 1975 (Paper 27), discrete Legendre (DLT) in 1982 (Paper 14), and discrete sine (DST) in 1976 (Paper 24). Except for the RT, all the others are orthogonal. Hence energy is preserved in both the data and transform domains. Also, mapping from the data domain to the transform domain is unique. Fast algorithms developed for the forward transform are equally applicable, except for minor changes, to the inverse transform. Also, the transforms discussed in this volume, including the RT, are separable. A consequence of this property is that the multidimensional transforms can be implemented by a sequence of 1D-(one dimensional) transforms. Fast algorithms and other properties of the 1D-transform are preserved for the multidimensional case.

Fast algorithms, in general, reduce the computational complexity of the 1D-transform from $O(N^2)$ arithmetic operations (addition and/or multiplication) to $O\ (N \log_2 N)$ operations. Here N is the size of the data sequence. Other benefits are savings in memory and reduced finite word-length arithmetic error. Also, most of the transforms can be implemented with real arithmetic. These developments coupled with improvements in the semiconductor technology (logic, memory, computers, size, cost, etc.) have led to the application of discrete transforms to digital signal processing, initially by simulation followed by actual hardware realization. Books relating to these transforms have been published recently (Ahmed and Rao, 1975; Elliott and Rao, 1982; Beauchamp, 1975; Harmuth, 1972, 1977).

The papers in this volume have been divided into two parts. Part I describes various discrete transforms, properties, and fast algorithms and their application in digital signal processing. Both simulation and hardware realization are outlined and the disadvantages, if any, pointed out. Papers 1 through 6 deal with the discrete cosine transform (DCT) which, apart from the DFT, has been utilized mostly in signal processing because it compares very favorably with the statistically optimal Karhunen-Loeve transform (KLT). This set of papers discusses the origin and derivation of the DCT, fast recursive algorithm involving real arithmetic only, implementation via the Walsh-Hadamard transform (WHT), and comparative performance and hardware design with specific application to speech processing.

All aspects of the RT, in addition to its application in pattern recognition, are described in Papers 7, 8, and 9. In spite of its nonorthogonality, it has strong potential in character recognition and scene matching because of some very interesting properties. Papers 10 through 13 relate to the discrete sine transform (DST). Paper 10 in particular describes the sinusoidal family of transforms, including the DCT and DST. The remaining papers describe the properties and fast algorithms of the DST. It is interesting to note that the DST, although sinusoidal, falls short in performance compared to the DCT. A more recent transform called the discrete Legendre (DLT) has been developed in Paper 14. Although it appears to perform better than the WHT, its properties need to be compared with other transforms. Also, larger size transforms ($N = 16, 32, \ldots$) need to be developed.

A very simple transform called the Walsh-Hadamard (WHT) is treated in Papers 15 through 20. Among its various properties, the one dealing with the power spectra invariant to various shifts or groupings is particularly attractive. Because of its simplicity (need only add/subtract), it has been utilized in data compression involving both image transmission and storage. Another simple transform is the Haar (HT), which is described in detail in Paper 21. A modification to avoid the irrational numbers is also proposed. The slant transform, which is tailored for image coding, is described in Papers 22 and 23. While the former describes its development and properties in detail with application in image coding, the latter shows how it can be developed progressively through the WHT.

The Karhunen-Loeve transform (KLT) and its properties are described in Papers 24 and 25. The emphasis has been how, under certain boundary conditions, the KLT reduces to the DST, which has a fast algorithm. Part I concludes with Papers 26 through 29, which deal with hybrid transforms such as the slant-Haar and Hadamard-Haar. Also, an integer version (approximation) of the DCT and the generation of some of the transforms via the WHT are treated in Papers 28 and 29.

The papers in Part II deal with the applications of the discrete transforms and are divided into three sections, that is, Papers 30, 31, and 32: pattern recognition; Papers 33 through 37: speech coding; and Papers 38 through 42: image coding. In these papers the emphasis has been on why these transforms are specifically applicable to these areas and how they can be applied. Their realization in hardware is also discussed as well as the actual simulation of these transforms in these areas and their performance evaluation. This has led to the exploration of other applications and the investigation of different architectural designs for the hardware implementation of the transforms.

As in any other volume, logistics require that the selection of papers be limited in number. Therefore, a number of papers are not reprinted here that are equally important. Some of these papers are listed at the end of the editor's comments in each section. The papers selected in this volume provide sufficient background and generate enough interest and enthusiasm to excite the reader to further explore the general field of discrete transforms.

REFERENCES

Ahmed, N., and K. R. Rao, 1975, *Orthogonal Transforms for Digital Signal Processing,* Springer, Berlin, 263p.

Beauchamp, K. G., 1975, *Walsh Functions and their Applications,* Academic Press, New York, 236p.

Cooley, J. W., and J. W. Turkey, 1965, An Algorithm for the Machine Calculation of Complex Fourier Series, *Math. Comput.* **19:**297-301.

Elliott, D. F., and K. R. Rao, 1982, *Fast Transforms: Algorithms, Analyses and Applications,* Academic Press, New York, 488p.

Haar, A., 1910, Zur Theorie der Orthogonalen Funktionensysteme, *Math. Ann.* **69:**331-371.

Haar, A., 1912, Zur Theorie der Orthogonalen Funktionensysteme, *Math. Ann.* **71:**38-53.

Harmuth, H. F., 1972, *Transmission of Information by Orthogonal Functions,* 2nd ed., Springer, Berlin, 393p.

Harmuth, H. F., 1977, *Sequency Theory,* Academic Press, New York.

Rao, K. R., and A. Jalali, 1982, *Discrete Transforms,* IEEE 16th Asilomar Conference on Circuits, Systems, and Computers, Pacific Grove, Calif., pp. 290-294.

Walsh, J. L., 1923, A Closed Set of Orthogonal Functions, *Am. J. Math.* **45:**5-24.

Part I

DISCRETE TRANSFORMS

Editor's Comments on Papers 1 Through 6

DISCRETE COSINE TRANSFORM

Apart from the discrete Fourier transform (DFT), the discrete cosine transform (DCT) has been most widely used in signal and image processing. Several variations and versions of the DCT have been developed. It has been realized in hardware, leading to commercial and industrial applications. These realizations are based on different designs (ex.: Ward and Stanier, 1983) aimed at various real-time implementations. Both its properties and the fast algorithms have contributed to its popularity.

DCT was originally developed in 1974 (Paper 1). It is shown that the DCT basis vectors form a class of discrete Chebyshev polynomials and that it performs very close to the statistically optimal transform, the Karhunen-Loeve transform (KLT) (Paper 24). Also, an N-point DCT can be computed through a $2N$-point fast Fourier transform (FFT) (Paper 1). Subsequently others (Haralick, 1976; Makhoul, 1980; Narasimha and Peterson, 1978; Tseng and Miller, 1978) improved this algorithm. Algorithms involving only real arithmetic for efficient implementation of a $2N$-point DCT ($2^n = N$) have also been developed (Paper 2; Corrington, 1978; Belt, Keele, and Murray, 1977). This is later extended to DCT of an arbitrary number of points (Wagh and Ganesh, 1980; Paper 35). The DCT can also be computed via the Walsh-Hadamard transform (WHT) (Paper 3) and also via the arcsine transform (Dyer, Ahmed, and Hummels, 1980). As this transform is separable, multidimensional DCT can be implemented by applying the ID-DCT algorithms along each dimension. Other techniques such as polynomial transform (Nussbaumer, 1981) and matrix factorization (Kamangar and Rao, 1982) resulting in fast algorithms for the 2D-DCT have also been developed.

Of the various algorithms just cited the one developed by Chen, Smith, and Fralick (Paper 2) appears to be most heavily utilized both for simulation and in hardware (Jalali and Rao, 1982). This algorithm involves real arithmetic, has a recursive structure, and requires a reduced number of computations. DCT processors have been built for applications in teleconferencing (Chen, 1981), RPV image date compression (Bessette and Schaming, 1980; Chan and Whiteman, 1983), and transmultiplexers (Paper 33). Breadboard implementation of the DCT using SSI and MSI TTL for single speech-channel encoding was also carried out (Paper 4). DCT has been applied to least mean square (LMS) adaptive filtering with applications in the speech-related areas such as spectral analysis, echo cancellers, and adaptive line enhancer (Narayan, Peterson, and Narasimha, 1983). For the transform domain LMS adaptive filtering, in general, the DCT performs better than the DFT.

The reasons for the universal utilization of the DCT can be further substantiated by the decorrelation property of the DCT as reported in Paper 5. DCT is not only asymptotically equivalent to the KLT of Markov-1 signals (so also the DFT) but also is a better approximation to KLT than the DFT (Paper 5). Kitajima, Saito, and Kurobe (1977) also show that the DCT has a better decorrelation even for a small block size and is relatively immune to the statistical changes compared to the DFT. Clarke (1981) has shown that the DCT is the limiting case of the KLT of a first order Markov random process as the adjacent correlation coefficient tends to unity. This is further substantiated by Flickner and Ahmed (Paper 6). Ahmed and Flickner (1982) also show that the DCT is much more effective in reducing the block-edge effects than the other discrete transforms.

REFERENCES

Ahmed, N., and M. D. Flickner, 1982, *Some Considerations of the Discrete Cosine Transform,* IEEE 16th Asilomar Conference on Circuits, Systems, and Computers, Pacific Grove, Calif., pp. 295–299.

Belt, R. A., R. V. Keele, and G. G. Murray, 1977, Digital TV Microprocessor System, *Natl. Telecommun. Conf., Los Angeles, Calif. Proc.* **10:**6-1–6-6.

Bessette, O. E., and W. B. Schaming, 1980, *A Two Dimensional Discrete Cosine Transform Video Bandwidth Compression System,* IEEE National Aerospace Electronics Conference (NAECON) Proceedings 1–7, Dayton, Ohio.

Chan, L. C., and P. Whiteman, 1983, Hardware Constrained Hybrid Coding of Video Imagery, *IEEE Aerosp. Electron. Syst. Trans.* **19:**71–83.

Chen, W. H., 1981, Scene Adaptive Coder, *Int. Conf. Commun. Proc.* **22:**5.1–5.6.

Clarke, R. J., 1981, Relation Between the Karhunen-Loeve and Cosine Transforms, *IEE Proc.* **128:**359–360.

Corrington, M. S., 1978, *Implementation of Fast Cosine Transform Using Real Arithmetic,* IEEE National Aerospace Electronics Conference (NAECON) Proceedings, Dayton, Ohio.

Dyer, S. A., N. Ahmed, and D. R. Hummels, 1980, *Computation of the Discrete Cosine Transform Via the Arcsine Transform,* IEEE International Conference on Acoustics, Speech, and Signal Processing Proceedings, Denver, Colo., pp. 231–234.

Haralick, R. M., 1976, A Storage Efficient Way to Implement the Discrete Cosine Transform, *IEEE Comput. Trans.* **C-25:**764–765.

Jalali, A., and K. R. Rao, 1982, A High Speed FDCT Processor for Real Time Processing of NTSC Color TV Signal, *IEEE Electromagn. Compat. Trans.* **EMC-24:**278–286.

Kamangar, F. A., and K. R. Rao, 1982, Fast Algorithms for the 2-D Discrete Cosine Transform, *IEEE Comput. Trans.* **C-31:**899–906.

Kitajima, H., T. Saito, and T. Kurobe, 1977, Comparison of the Discrete Cosine and Fourier Transforms as Possible Substitutes for the Karhunen-Loeve Transform, *Inst. Electr. Commun. Eng. Japan Trans.* **E-60:**279–283.

Makhoul, J., 1980, A Fast Cosine Transform in One and Two Dimensions, *IEEE Acoust., Speech, Signal Process. Trans.* **ASSP-28:**27–34.

Narasimha, M. J., and A. M. Peterson, 1978, On the Computation of the Discrete Cosine Transform, *IEEE Commun. Trans.* **COM-26:**934–936.

Narayan, S. S., A. M. Peterson, and M. J. Narasimha, 1983, Transform Domain LMS Algorithm, *IEEE Acoust., Speech, Signal Process. Trans.* **ASSP-31:**609–615.

Nussbaumer, H. J., 1981, *Fast Polynomial Transform Computation of the 2-D DCT,* Proceedings of the International Conference on Digital Signal Processing., Florence, Italy, pp. 276–283.

Tseng, B. D., and W. C. Miller, 1978, On Computing the Discrete Cosine Transform, *IEEE Comput. Trans.* **C-27:**966–968.

Wagh, M. D., and H. Ganesh, 1980, A New Algorithm for the Discrete Cosine Transform of Arbitrary Number of Points, *IEEE Comput. Trans.* **C-26:**269–277.

Ward, J. S., and B. J. Stanier, 1983, Fast Discrete Cosine Transform Algorithm for Systolic Arrays, *Electron. Lett.* **19:**58–60.

Discrete Cosine Transform

N. AHMED, T. NATARAJAN, AND K. R. RAO

***Abstract*—A discrete cosine transform (DCT) is defined and an algorithm to compute it using the fast Fourier transform is developed. It is shown that the discrete cosine transform can be used in the area of digital processing for the purposes of pattern recognition and Wiener filtering. Its performance is compared with that of a class of orthogonal transforms and is found to compare closely to that of the Karhunen–Loève transform, which is known to be optimal. The performances of the Karhunen–Loève and discrete cosine transforms are also found to compare closely with respect to the rate-distortion criterion.**

***Index Terms*—Discrete cosine transform, discrete Fourier transform, feature selection, Haar transform, Karhunen–Loève transform, rate distortion, Walsh–Hadamard transform, Wiener vector and scalar filtering.**

INTRODUCTION

In recent years there has been an increasing interest with respect to using a class of orthogonal transforms in the general area of digital signal processing. This correspondence addresses itself towards two problems associated with image processing, namely, pattern recognition [1] and Wiener filtering [2].

In pattern recognition, orthogonal transforms enable a noninvertible transformation from the pattern space to a reduced dimensionality feature space. This allows a classification scheme to be implemented with substantially less features, with only a small increase in classification error.

In discrete Wiener filtering applications, the filter is represented by an $(M \times M)$ matrix G. The estimate $\hat{X}$ of data vector X is given by GZ, where $Z = X + N$ and N is the noise vector. This implies that approximately $2M^2$ arithmetic operations are required to compute $\hat{X}$. Use of orthogonal transforms yields a G in which a substantial number of elements are relatively small in magnitude, and hence can be set equal to zero. Thus a significant reduction in computation load is realized at the expense of a small increase in the mean-square estimation error.

The Walsh–Hadamard transform (WHT), discrete Fourier transform (DFT), the Haar transform (HT), and the slant transform (ST), have been considered for various applications [1], [2], [4]–[9] since these are orthogonal transforms that can be computed using fast algorithms. The performance of these transforms is generally compared with that of the Karhunen–Loève transform (KLT) which is known to be optimal with respect to the following performance measures: variance distribution [1], estimation using the mean-square error criterion [2], [4], and the rate-distortion function [5]. Although the KLT is optimal, there is no general algorithm that enables its fast computation [1].

In this correspondence, a discrete cosine transform (DCT) is introduced along with an algorithm that enables its fast computation. It is shown that the performance of the DCT compares more closely to that of the KLT relative to the performances of the DFT, WHT, and HT.

Manuscript received January 29, 1973; revised June 7, 1973.
N. Ahmed is with the Departments of Electrical Engineering and Computer Science, Kansas State University, Manhattan, Kans.
T. Natarajan is with the Department of Electrical Engineering, Kansas State University, Manhattan, Kans.
K. R. Rao is with the Department of Electrical Engineering, University of Texas at Arlington, Arlington, Tex. 76010.

DISCRETE COSINE TRANSFORM

The DCT of a data sequence $X(m)$, $m = 0, 1, \cdots, (M-1)$ is defined as

$$G_x(0) = \frac{\sqrt{2}}{M} \sum_{m=0}^{M-1} X(m)$$

$$G_x(k) = \frac{2}{M} \sum_{m=0}^{M-1} X(m) \cos \frac{(2m+1)k\Pi}{2M}, \qquad k = 1, 2, \cdots, (M-1) \quad (1)$$

where $G_x(k)$ is the kth DCT coefficient. It is worthwhile noting that the set of basis vectors $\{1/\sqrt{2}, \cos((2m+1)k\Pi)/(2M)\}$ is actually a class of discrete Chebyshev polynomials. This can be seen by recalling that Chebyshev polynomials can be defined as [3]

$$\hat{T}_0(\xi_p) = \frac{1}{\sqrt{2}}$$

$$\hat{T}_k(\xi_p) = \cos(k \cos^{-1} \xi_p), \qquad k, p = 1, 2, \cdots, M \quad (2)$$

where $\hat{T}_k(\xi_p)$ is the kth Chebyshev polynomial.

Now, in (2), ξ_p is chosen to be the pth zero of $\hat{T}_M(\xi)$, which is given by [3]

$$\xi_p = \cos \frac{(2p-1)\Pi}{2M}, \qquad p = 1, 2, \cdots, M. \quad (3)$$

Substituting (3) in (2), one obtains the set of Chebyshev polynomials

$$\hat{T}_0(p) = \frac{1}{\sqrt{2}}$$

$$\hat{T}_k(p) = \cos \frac{(2p-1)k\Pi}{2M}, \qquad k, p = 1, 2, \cdots, M. \quad (4)$$

From (4) it follows that the $\hat{T}_k(p)$ can equivalently be defined as

$$T_0(m) = \frac{1}{\sqrt{2}}$$

$$T_k(m) = \cos \frac{(2m+1)k\Pi}{2M}; \qquad k = 1, 2, \cdots, (M-1),$$

$$m = 0, 1, \cdots, M-1. \quad (5)$$

Comparing (5) with (1) we conclude that the basis member cos ((2*m* +

$1)k\Pi)/(2M)$ is the kth Chebyshev polynomial $T_k(\xi)$ evaluated at the mth zero of $T_M(\xi)$.

Again, the inverse cosine discrete transform (ICDT) is defined as

$$X(m) = \frac{1}{\sqrt{2}} G_x(0) + \sum_{k=1}^{M-1} G_x(k) \cos \frac{(2m+1)k\Pi}{2M}, \quad m = 0, 1, \cdots, (M-1). \quad (6)$$

We note that applying the orthogonal property [3]

$$\sum_{m=0}^{M-1} T_k(m)\, T_l(m) = \begin{cases} M/2, & k = l = 0 \\ M/2, & k = l \neq 0 \\ 0, & k \neq l \end{cases} \quad (7)$$

to (6) yields the DCT in (1).

If (6) is written in matrix form and Λ is the $(M \times M)$ matrix that denotes the cosine transformation, then the orthogonal property can be expressed as

$$\Lambda^T \Lambda = \frac{M}{2} [I] \quad (8)$$

where Λ^T is the transpose of Λ and $[I]$ is the $(M \times M)$ identity matrix.

MOTIVATION

The motivation for defining the DCT is that it can be demonstrated that its basis set provides a good approximation to the eigenvectors of the class of Toeplitz matrices defined as

$$\psi = \begin{bmatrix} 1 & \rho & \rho^2 & \cdots & \rho^{M-1} \\ \rho & 1 & \rho & \cdots & \rho^{M-2} \\ \cdot & & \cdot & & \cdot \\ \cdot & & & \cdot & \cdot \\ \cdot & & & & \cdot \\ \rho^{M-1} & \rho^{M-2} & & \cdots & 1 \end{bmatrix}, \quad 0 < \rho < 1. \quad (9)$$

For the purposes of illustration, the eigenvectors of ψ for $M = 8$ and $\rho = 0.9$ are plotted (see Fig. 1) against

$$\left\{ \frac{1}{\sqrt{2}}, \cos \frac{(2m+1)k\Pi}{16}, \quad k = 1, 2, \cdots, 7, m = 0, 1, \cdots, 7 \right\} \quad (10)$$

which constitute the basis set for the DCT. The close resemblance (aside from the 180° phase shift) between the eigenvectors and the set defined in (10) is apparent.

ALGORITHMS

It can be shown that (1) can be expressed as

$$G_x(0) = \frac{\sqrt{2}}{M} \sum_{m=0}^{M-1} X(m)$$

$$G_x(k) = \frac{2}{M} \mathrm{Re} \left\{ e^{(-ik\Pi)/(2M)} \sum_{m=0}^{2M-1} X(m)\, W^{km} \right\}, \quad k = 1, 2, \cdots, (M-1) \quad (11)$$

where

$$W = e^{-i2\Pi/2M}, \quad i = \sqrt{-1},$$

$$X(m) = 0, \quad m = M, (M+1), \cdots, (2M-1)$$

and Re $\{\cdot\}$ implies the real part of the term enclosed. From (11) it follows that all the M DCT coefficients can be computed using a $2M$-point fast Fourier transform (FFT). Since (1) and (6) are of the same form, FFT can also be used to compute the IDCT. Similarly, if a discrete sine transform were defined, then the Re $\{\cdot\}$ in (11) would be replaced by Im $\{\cdot\}$, which denotes the imaginary part of the term enclosed.

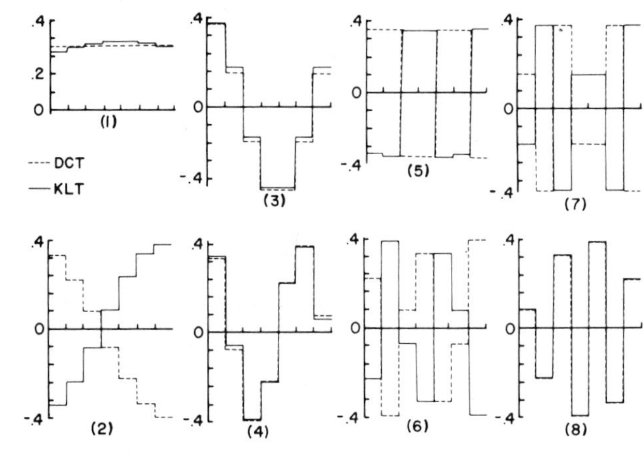

Fig. 1. Eigenvectors of (8 × 8) Toeplitz matrix (ρ = 0.9) and basis vectors of the DCT.

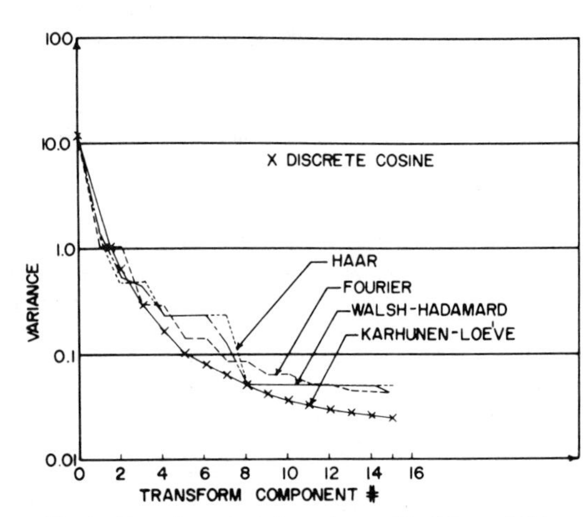

Fig. 2. Transform domain variance; $M = 16$, $\rho = 0.95$.

COMPUTATIONAL RESULTS

In image processing applications, ψ in (9) provides a useful model for the data covariance matrix corresponding to the rows and columns of an image matrix [6], [7]. The covariance matrix in the transform domain is denoted by Ψ and is given by

$$\Psi = \Lambda \psi \Lambda^{*T} \quad (12)$$

where Λ is the matrix representation of an orthogonal transformation and Λ^* is the complex conjugate of Λ. From (12) it follows that Ψ can be computed as a two-dimensional transform of ψ.

Feature Selection

A criterion for eliminating features (i.e., components of a transform vector), which are least useful for classification purposes, was developed by Andrews [1]. It states that features whose variances (i.e., main diagonal elements of Ψ) are relatively large should be retained. (Fig. 2 should be retained.) Fig. 2 shows the various variances ranked in decreasing order of magnitude. From the information in Fig. 2 it is apparent that relative to the set of orthogonal transforms shown, the DCT compares most closely to the KLT.

Wiener Filtering

The role of orthogonal transforms played in filtering applications is illustrated in Fig. 3 [2]. Z is an $(M \times 1)$ vector which is the sum of a

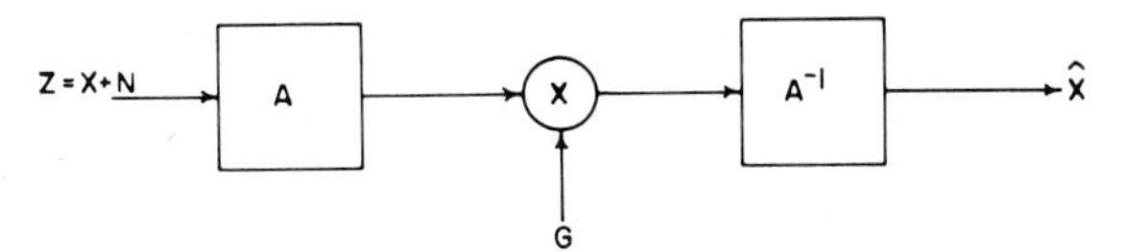

Fig. 3. Wiener filtering model.

TABLE I
MEAN-SQUARE ERROR PEFORMANCE OF VARIOUS TRANSFORMS FOR SCALAR WIENER FILTERING; $\rho = 0.9$

Transform \ M	2	4	8	16	32	64
Karhunen–Loève	0.3730	0.2915	0.2533	0.2356	0.2268	0.2224
Discrete cosine	0.3730	0.2920	0.2546	0.2374	0.2282	0.2232
Discrete Fourier	0.3730	0.2964	0.2706	0.2592	0.2441	0.2320
Walsh–Hadamard	0.3730	0.2942	0.2649	0.2582	0.2582	0.2559
Haar	0.3730	0.2942	0.2650	0.2589	0.2582	0.2581

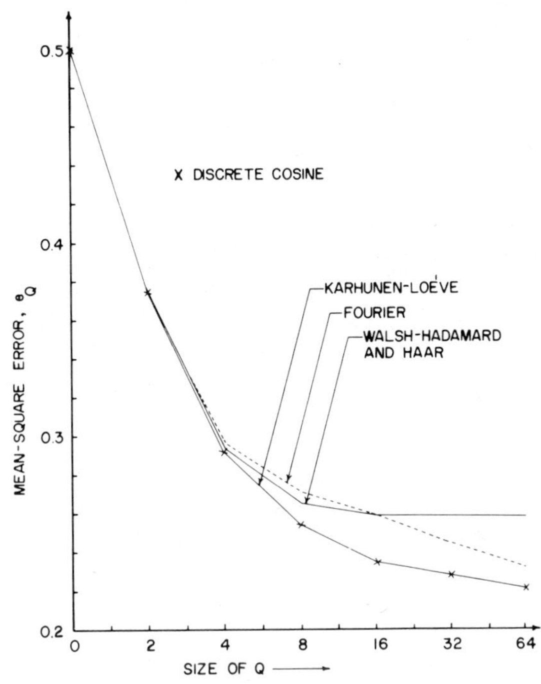

Fig. 4. Mean-square error performance of various transforms for scalar Wiener filtering; $\rho = 0.9$.

vector X and a noise vector N. X is considered to belong to a random process whose covariance matrix is given by ψ which is defined in (9). The Wiener filter G is in the form of an $(M \times M)$ matrix. A and A^{-1} represent an orthonormal transform and its inverse, respectively, while $\hat{X}$ denotes the estimate of X, using the mean-square error criterion.

We restrict our attention to the case when G is constrained to be a diagonal matrix Q. This class of Wiener filters is referred to as *scalar* filters while the more general class (denoted by G) is referred to as vector filters. The additive noise (see Fig. 3) is considered to be white, zero mean, and uncorrelated with the data. If the mean-square estimation error due to scalar filtering is denoted by e_Q, then e_Q can be expressed as [4]

$$e_Q = 1 - \frac{1}{M}\sum_{s=1}^{M} \frac{\Psi_x^2(s, s)}{\Psi_x(s, s) + \Psi_n(s, s)} \tag{13}$$

where Ψ_x and Ψ_n denote the transform domain covariance matrices of the data and noise, respectively. Table I lists the values of e_Q for different values of M for the case $\rho = 0.9$ and a signal-to-noise ratio of unity. From Table I it is evident that the DCT comes closest to the KLT which is optimal. This information is presented in terms of a set of performance curves in Fig. 4.

ADDITIONAL CONSIDERATIONS

In conclusion, we compare the performance of the DCT with KLT, DFT, and the identity transforms, using the rate-distortion criterion [5]. This performance criterion provides a measure of the information rate R that can be achieved while still maintaining a fixed distortion D, for encoding purposes. Considering Gaussian sources along with the mean-square error criterion, the rate-distortion performance measure is given by [5]

$$R(A, D) = \frac{1}{2M}\sum_{j=1}^{M} \max\left\{0, \ln\left(\frac{\sigma_j}{\theta}\right)\right\} \tag{14a}$$

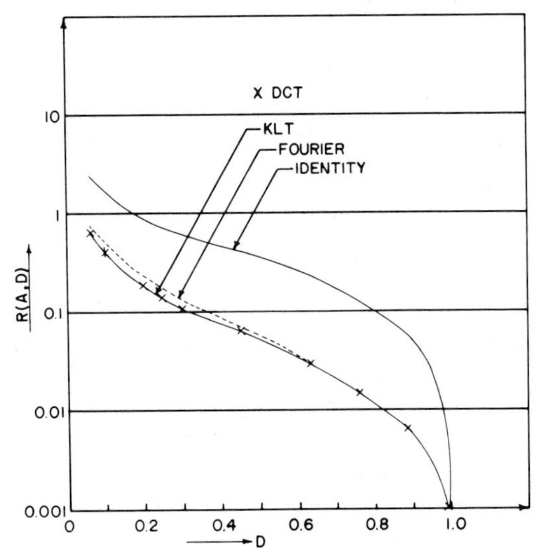

Fig. 5. Rate versus distortion for $M = 16$ and $\rho = 0.9$.

$$D = \frac{1}{M} \sum_{j=1}^{M} \min(\theta, \sigma_j) \tag{14b}$$

where A denotes the orthogonal transformation and the σ_j are the main diagonal terms of the transform domain covariance matrix Ψ in (12).

The rate-distortion pertaining to $M = 16$ and $\rho = 0.9$ is shown in Fig. 5, from which it is evident that the KLT and DCT compare more closely than the KLT and DFT.

Summary

It has been shown that the DCT can be used in the area of image processing for the purposes of feature selection in pattern recognition; and scalar-type Wiener filtering. Its performance compares closely with that of the KLT, which is considered to be optimal. The performances of the KLT and DCT are also found to compare closely, with respect to the rate-distortion criterion.

Acknowledgment

The authors wish to thank Dr. H. C. Andrews, University of Southern California, for his valuable suggestions pertaining to the rate-distortion computation.

References

[1] H. C. Andrews, "Multidimensional rotations in feature selection," *IEEE Trans. Comput.*, vol. C-20, pp. 1045–1051, Sept. 1971.
[2] W. K. Pratt, "Generalized Wiener filtering computation techniques," *IEEE Trans. Comput.*, vol. C-21, pp. 636–641, July 1972.
[3] C. T. Fike, *Computer Evaluation of Mathematical Functions.* Englewood Cliffs, N. J.: Prentice-Hall, 1968, ch. 6 and 7.
[4] J. Pearl, "Walsh processing of random signals," *IEEE Trans. Electromag. Compat.*, vol. EMC-13, pp. 137–141, Aug. 1971.
[5] J. Pearl, H. C. Andrews, and W. K. Pratt, "Performance measures for transform data coding," *IEEE Trans. Commun. Technol.*, vol. COM-20, pp. 411–415, June 1972.
[6] W. K. Pratt, "Walsh functions in image processing and two-dimensional filtering," in *1972 Proc. Symp. Applications of Walsh Functions*, AD-744 650, pp. 14–22.
[7] W. K. Pratt, L. R. Welch, and W. Chen, "Slant transforms in image coding," in *1972 Proc. Symp. Applications of Walsh Functions*, pp. 229–234.
[8] W. H. Chen and W. K. Pratt, "Color image coding with the slant transform," in *1973 Proc. Symp. Applications of Walsh Functions*, AD-763000, pp. 155–161.
[9] W. K. Pratt, "Spatial transform coding of color images," *IEEE Trans. Commun. Technol.*, vol. COM-19, pp. 980–992, Dec. 1971.

A Fast Computational Algorithm for the Discrete Cosine Transform

WEN-HSIUNG CHEN, C. HARRISON SMITH, AND S. C. FRALICK

Abstract—**A Fast Discrete Cosine Transform algorithm has been developed which provides a factor of six improvement in computational complexity when compared to conventional Discrete Cosine Transform algorithms using the Fast Fourier Transform. The algorithm is derived in the form of matrices and illustrated by a signal-flow graph, which may be readily translated to hardware or software implementations.**

INTRODUCTION

The Discrete Cosine Transform (DCT) has been successfully applied to the coding of high resolution imagery [1-5]. The conventional method of implementing the DCT utilized a double size Fast Fourier Transform (FFT) algorithm employing complex arithmetic throughout the computation [1]. Use of the DCT in a wide variety of applications has not been as extensive as its properties would imply due to the lack of an efficient algorithm. This report describes a more efficient algorithm involving only real operations for computing the Fast Discrete Cosine Transform (FDCT) of a set of N points. The algorithm can be extended to any desired value of $N = 2^m$, $m \geqslant 2$. The generalization consists of alternating cosine/sine butterfly matrices with binary matrices to reorder the matrix elements to a form which preserves a recognizable bit-reversed pattern at every other node. The generalization is not unique—several alternate methods have been discovered—but the method described herein appears to be the simplest to interpret. It is not necessarily the most efficient FDCT which could be constructed but represents one technique for methodical extension. The method takes $(3N/2)(\log_2 N - 1) + 2$ real additions and $N \log_2 N - 3N/2 + 4$ real multiplications: this is approximately six times as fast as the conventional approach using a double size FFT.

Paper approved by the Editor for Communication Theory of the IEEE Communications Society for publication without oral presentation. Manuscript received January 24, 1977.

W.-H. Chen was with the Western Development Laboratories Division, Ford Aerospace and Communications Corporation, Palo Alto, CA 94303. He is now with Compression Laboratories, Inc., Campbell, CA 95008.

C. H. Smith is with the Western Development Laboratories Division, Ford Aerospace and Communications Corporation, Palo Alto, CA 94303.

S. C. Fralick was with the Western Development Laboratories Division, Ford Aerospace and Communications Corporation, Palo Alto, CA 94303. He is now with Comtech Advanced Systems, Inc., Sunnyvale, CA 94086.

DISCRETE COSINE TRANSFORM

The discrete cosine transform of a discrete function $f(j)$, $j = 0, 1, \cdots, N-1$ is defined as [1]

$$F(k) = \frac{2c(k)}{N} \sum_{j=0}^{N-1} f(j) \cos\left[\frac{(2j+1)k\pi}{2N}\right]; \quad k = 0, 1, \cdots, N-1 \tag{1}$$

and the inverse transform is

$$f(j) = \sum_{k=0}^{N-1} c(k)F(k) \cos\left[\frac{(2j+1)k\pi}{2N}\right]; \quad j = 0, 1, \cdots, N-1 \tag{2}$$

where

$$c(k) = \frac{1}{\sqrt{2}} \quad \text{for } k = 0$$

$$= 1 \quad \text{for } k = 1, 2, \cdots, N-1.$$

The transform possesses a high energy compaction property which is superior to any known transform with a fast computational algorithm. [1-5] The transform also possesses a circular convolution-multiplication relationship which can readily be used in linear system theory. [6]

A FAST COMPUTATIONAL ALGORITHM

The discrete cosine transform of an $N \times 1$ data vector $[f]$ can be expressed in a matrix form as

$$[F] = \frac{2}{N}[A_N][f] \tag{3}$$

where $[A_N] = [c(k)\cos(2j+1)k\pi/2N]$; $j, k = 0, 1, \cdots, (N-1)$ as defined in equation (1) and $[F]$ is the $N \times 1$ transformed vector. The fast computational algorithm to be presented here is based upon the matrix decomposition of the $[A_N]$ matrix. As shown below, this matrix can first be written into the following recursive form:

$$[A_N] = [P_N]\left[\begin{array}{c|c} A_{N/2} & 0 \\ \hline 0 & R_{N/2} \end{array}\right][B_N] \tag{4}$$

where $[B_N]$ is defined in equation (7),

$$[R_{N/2}] = \left[c(k)\cos\frac{(2j+1)(2k+1)\pi}{2N}\right]; \quad j, k = 0, 1, \cdots, \frac{N}{2}-1$$

and $[P_N]$ is an $N \times N$ permutation matrix which permutes the transformed vector from a bit reversed order to a natural order. As in all unitary transforms the 2×2 DCT can be written as

$$[A_2] = \frac{1}{\sqrt{2}}\begin{bmatrix} 1 & 1 \\ 1 & -1 \end{bmatrix}. \tag{5}$$

It can be seen from the recursive nature of equation (4) that $[A_2]$ can be extended into higher order matrices as long as there is a generalized method of decomposing the $[R_{N/2}]$ matrix.

The following discussion presents one systematic way of decomposing the $[R_{N/2}]$ matrix. It is emphasized that this method of decomposition is not unique and is not optimum. Several methods have been found which require fewer computational steps but with no apparent generalization to larger sizes.

The $[R_{N/2}]$ matrix is decomposed into $(2\log_2 N - 3)$ matrices in the following manner:

$$[R_{N/2}] = [M1][M2][M3][M4]\cdots[M(2\log_2 N - 3)]. \tag{6}$$

The matrices are of four distinct types.

- Type 1: $[M1]$, the first matrix
- Type 2: $[M(2\log_2 N - 3)]$, the last matrix
- Type 3: $[Mq]$, the remaining odd numbered matrices $[M3]$, $[M5]$, etc.
- Type 4: $[Mp]$, the even numbered matrices $[M2]$, $[M4]$, etc.

Before describing the four types of matrices in detail, the following definitions are provided for notational efficiency:

$$B_N = \begin{bmatrix} I_{N/2} & \bar{I}_{N/2} \\ \bar{I}_{N/2} & -I_{N/2} \end{bmatrix}$$

$$B_N{}^* = \begin{bmatrix} -I_{N/2} & \bar{I}_{N/2} \\ \bar{I}_{N/2} & I_{N/2} \end{bmatrix} \tag{7}$$

where

$[I_{N/2}]$ is an identity matrix of order $\frac{N}{2}$

$[\bar{I}_{N/2}]$ is the opposite diagonal identity matrix

$$[S_i^k] = \sin\frac{k\pi}{i}[I_{N/2i}]$$

$$[\bar{S}_i^k] = \sin\frac{k\pi}{i}[\bar{I}_{N/2i}]$$

$$[C_i^k] = \cos\frac{k\pi}{i}[I_{N/2i}]$$

$$[\bar{C}_i^k] = \cos\frac{k\pi}{i}[\bar{I}_{N/2i}]\,. \tag{8}$$

In equations (8) above, the identity matrices $[I_{N/2i}]$ specify the order of the diagonal sine or cosine matrices $[S_i^k]$, $[\bar{S}_i^k]$, $[C_i^k]$, $[\bar{C}_i^k]$ with the condition that $[I_{N/2i}] \equiv 1$ for $i > N/2$.

The four types of matrices may now be described in detail with reference to the right hand side of equation (9).

TYPE 1

The first matrix $[M1]$ is formed by concatenating $S_{2N}{}^{a_j}$ matrices (of order 1) along the upper left to middle of the main diagonal and $C_{2N}{}^{a_j}$ matrices along the middle to lower right. The opposite diagonal is formed by $\bar{C}_{2N}{}^{a_j}$ matrices along the upper right to middle and $\bar{S}_{2N}{}^{a_j}$ matrices along the middle to lower left. For this type matrix the values of a_j are the binary bit-reversed representation of $N/2 + j - 1$ for $j = 1, 2, \cdots, N/2$.

TYPE 2

The last matrix $[M(2\log_2 N - 3)]$ is formed by concatenating $I_{N/8}$, $-C_4{}^1$, $C_4{}^1$, $I_{N/8}$ matrices along the upper left to lower right of the main diagonal and concatenating $O_{N/8}$, $\bar{C}_4{}^1$, $\bar{C}_4{}^1$, $O_{N/8}$ matrices along the upper right to lower left of the opposite diagonal.*

TYPE 3

The remaining odd matrices $[Mq]$ are formed by repeated concatenation of the matrix sequence $I_{N/2i}$, $-C_i{}^{k_j}$, $-S_i{}^{k_j}$ and $I_{N/2i}$ where $i = N/(2^{(q-1)/2})$ for $j = 1, 2, \cdots, i/8$ along the upper left to middle of the main diagonal and the matrix sequence $I_{N/2i}$, $C_i{}^{k_j}$, $S_i{}^{k_j}$ and $I_{N/2i}$ for $j = i/8 + 1, \cdots, i/4$ along the middle to lower right. The opposite diagonal is formed similarly, using the matrix sequence $O_{N/2i}$, $\bar{S}_i{}^{k_j}$, $-\bar{C}_i{}^{k_j}$, $O_{N/2i}$ along the upper right to middle and the matrix sequence $O_{N/2i}$, $-\bar{S}_i{}^{k_j}$, $\bar{C}_i{}^{k_j}$, $O_{N/2i}$ along the middle to lower left. Repeated concatenation of a matrix sequence along a diagonal is clearly illustrated in equation (9), where for clarity the k_j's have been replaced by b_j's and c_j's, etc., because the value of k_j depends on the matrix index q. For this type matrix, the values of the k_j are the binary bit-reversed variables $(i/4) + j - 1$.

* 0_j is a null matrix of order j.

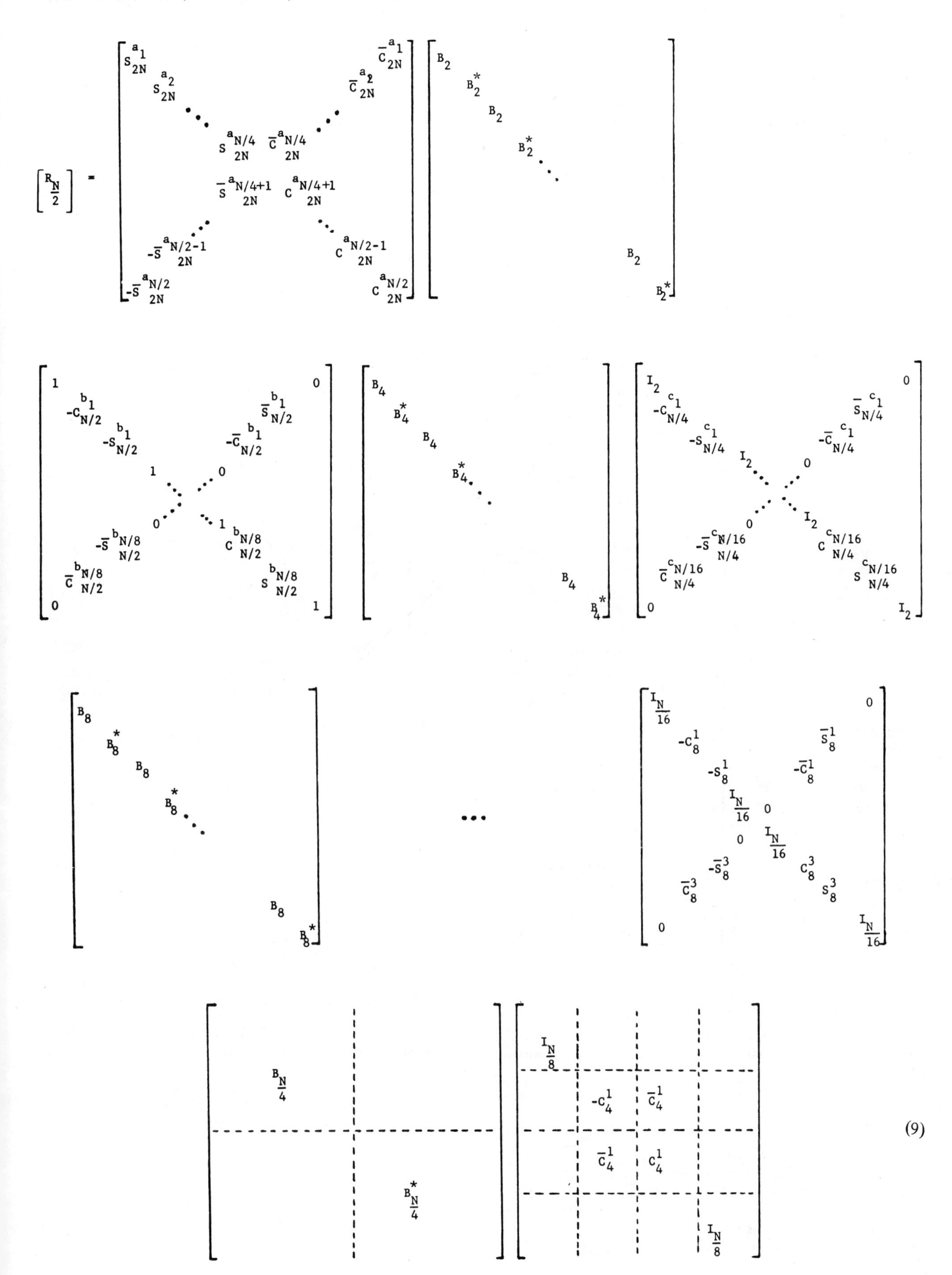

(9)

TYPE 4

The even numbered matrices $[Mp]$ are binary matrices formed by alternating B_l and B_l^* matrices along the upper left to lower right of the main diagonal. The subscript l indicates the order of the B or B^* matrix, and takes on the value of $2^{P/2}$.

A specific example of $[R_{N/2}]$ for $N = 16$ is shown in Eq. (10).

$$\left[R_{\frac{16}{2}}\right] = \begin{bmatrix} \sin\frac{\pi}{32} & & & & & & & \cos\frac{\pi}{32} \\ & \sin\frac{9\pi}{32} & & & & & \cos\frac{9\pi}{32} & \\ & & \sin\frac{5\pi}{32} & & & \cos\frac{5\pi}{32} & & \\ & & & \sin\frac{13\pi}{32} & \cos\frac{13\pi}{32} & & & \\ & & & -\sin\frac{3\pi}{32} & \cos\frac{3\pi}{32} & & & \\ & & -\sin\frac{11\pi}{32} & & & \cos\frac{11\pi}{32} & & \\ & -\sin\frac{7\pi}{32} & & & & & \cos\frac{7\pi}{32} & \\ -\sin\frac{15\pi}{32} & & & & & & & \cos\frac{15\pi}{32} \end{bmatrix} \begin{bmatrix} 1 & 1 & & & & & & \\ 1 & -1 & & & & & & \\ & & -1 & 1 & & & & \\ & & 1 & 1 & & & & \\ & & & & 1 & 1 & & \\ & & & & 1 & -1 & & \\ & & & & & & -1 & 1 \\ & & & & & & 1 & 1 \end{bmatrix}$$

$$\begin{bmatrix} 1 & & & & & & & 0 \\ & -\cos\frac{\pi}{8} & & & & & \sin\frac{\pi}{8} & \\ & & -\sin\frac{\pi}{8} & & & -\cos\frac{\pi}{8} & & \\ & & & 1 & 0 & & & \\ & & & 0 & 1 & & & \\ & & -\sin\frac{3\pi}{8} & & & \cos\frac{3\pi}{8} & & \\ & \cos\frac{3\pi}{8} & & & & & \cos\frac{3\pi}{8} & \\ 0 & & & & & & & 1 \end{bmatrix} \begin{bmatrix} 1 & & & 1 & & & & \\ & 1 & 1 & & & & & \\ & 1 & -1 & & & & & \\ 1 & & & -1 & & & & \\ & & & & -1 & & & 1 \\ & & & & & -1 & 1 & \\ & & & & & 1 & 1 & \\ & & & & 1 & & & 1 \end{bmatrix}$$

$$\begin{bmatrix} 1 & & & & & & & 0 \\ & 1 & & & & & 0 & \\ & & -\cos\frac{\pi}{4} & & & \cos\frac{\pi}{4} & & \\ & & & -\cos\frac{\pi}{4} & \cos\frac{\pi}{4} & & & \\ & & & \cos\frac{\pi}{4} & \cos\frac{\pi}{4} & & & \\ & & \cos\frac{\pi}{4} & & & \cos\frac{\pi}{4} & & \\ & 0 & & & & & 1 & \\ 0 & & & & & & & 1 \end{bmatrix} \tag{10}$$

The computational steps required for $[F]$ of equation (3) can be found from equation (4) with the following recursive relations:

$$K_{A_N} = N + K_{A_{N/2}} + K_{R_{N/2}} \tag{11a}$$

$$K_{A_N}' = K_{A_{N/2}}' + K_{R_{N/2}}' \tag{11b}$$

where K_{A_i} and K_{A_i}' are the number of additions and multiplications for $[A_i]$, and K_{R_i} and K_{R_i}' are the number of additions and multiplications for $[R_i]$. The number of additions for $[R_{N/2}]$ can easily be determined from equation (9) by noting that $\log_2 N - 2$ stages of decomposed matrices are half occupied on the opposite diagonals and $\log_2 N - 1$ stages are fully occupied on both diagonals. Therefore

$$K_{R_{N/2}} = \frac{N}{4}(\log_2 N - 2) + \frac{N}{2}(\log_2 N - 1)$$

$$= \frac{3N}{4}\log_2 N - N \qquad N \geqslant 4. \tag{12a}$$

As for the number of multiplications, only the odd matrices consist of multiplicative terms. In these matrices the first matrix consists of N multipliers, the last matrix consists of $N/4$ multipliers, and the rest of the $\log_2 N - 3$ matrices each consists of $N/2$ multipliers. Thus

$$K_{R_{N/2}}' = N + \frac{N}{4} - \frac{N}{2}(\log_2 N - 3)$$

$$= \frac{N}{2}\log_2 N - \frac{N}{4}; \qquad N \geqslant 8. \tag{12b}$$

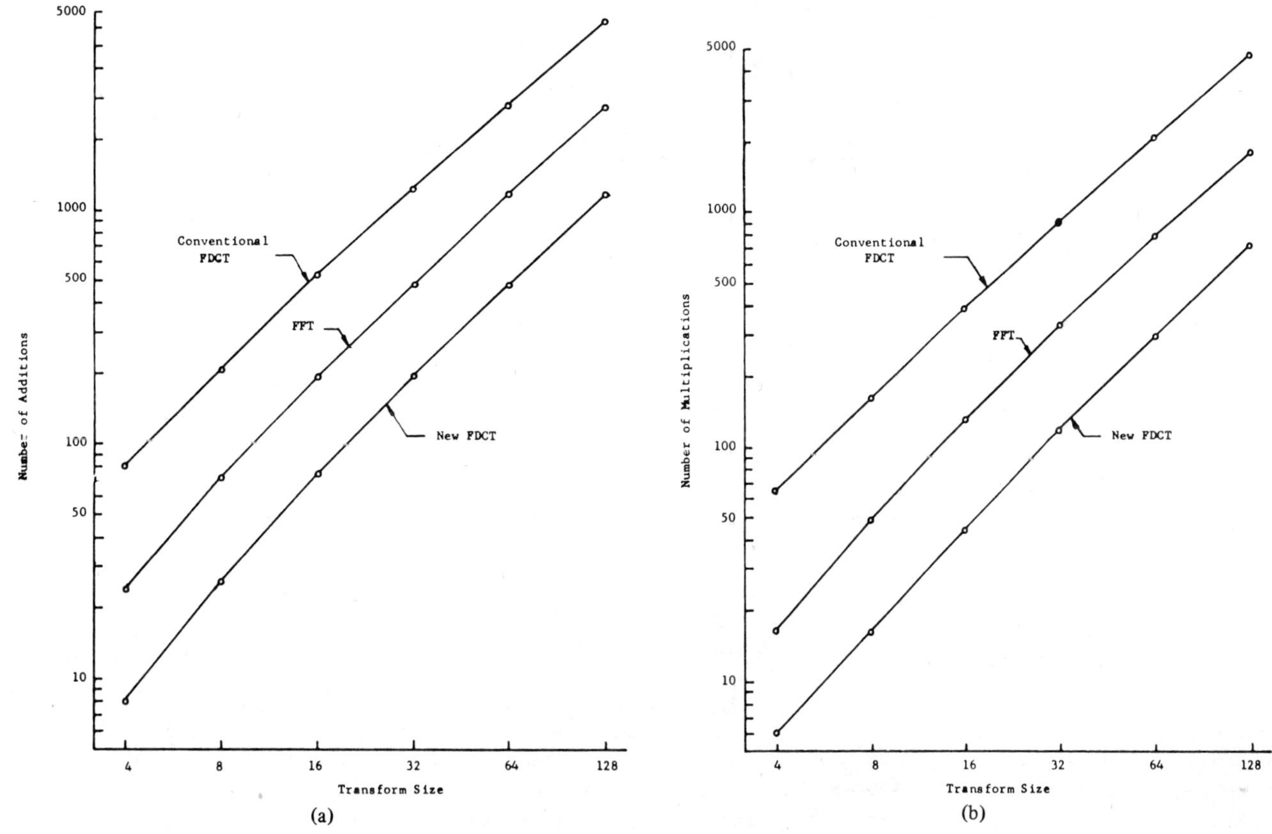

Figure 1. Comparison of Computational Steps for Conventional FDCT, FFT and FDCT. (a) Additions. (b) Multiplications.

For $N = 4$, $K_{R_2}' = 4$ since the only matrix involved is the first matrix. Now substituting equations (12a) and (12b) into equations (11a) and (11b), knowing (from equation 5) that

$$K_{A_2} = 2 \tag{13a}$$

$$K_{A_2}' = 2 \tag{13b}$$

one can obtain

$$K_{A_N} = \frac{3N}{2} (\log_2 N - 1) + 2 \tag{14a}$$

$$K_{A_N}' = N \log_2 N - \frac{3N}{2} + 4 \qquad N \geqslant 4. \tag{14b}$$

For purposes of comparison the conventional approach of computing the DCT utilizing an FFT takes $2N \log 2N$ complex additions and $N(\log 2N + 1)$ complex multiplications (which are equivalent to $6N \log 2N + 2N$ additions and $4N(\log 2N + 1)$ real multiplications). Figure 1 plots the number of computational steps versus the transform sizes for the conventional algorithm and the algorithm presented in this paper. It can easily be seen that the algorithm presented here takes less than 1/6 as many steps as the conventional algorithm. Also plotted in the figure are the computational steps for FFT. It can be seen that the new algorithm takes only 1/3 as many steps as the FFT.

Figure 2 is a signal-flow graph for $N = 4, 8, 16, 32$ arranged in the fashion described. Note that the input samples are in natural order from top to bottom. For every N, the output transform coefficients are in bit-reversed order.

Note that as N increases the even coefficients of each successive transform are obtained directly from the coefficients of the prior transform by doubling the subscript of the prior coefficients.

It can be seen that extension of the signal-flow graph to the next power of 2 merely involves adding a set of ±1 butterflies to accommodate the new set of input samples and a series of alternating cosine/sine butterflies and ±1 butterflies to yield the new set of odd transform coefficients.

In Figure 2, the coefficients have not been normalized. To obtain the normalized N-point DCFT coefficients, the appropriate terminal points of the flow graph of Figure 2 should each be multiplied by $2/N$. This signal-flow graph represents the forward transform matrix $[A_N]$. The inverse transform matrix $[A_N]^{-1}$ is simply $(N/2)[A_N]^T$.

Thus, except for normalization factor these FDCT signal-flow graphs are bidirectional, i.e., the inverse transform may be computed by introducing the vector $[F]$ at the output and recovering the vector $[f]$ at the input. This follows from the fact that every butterfly pair in the signal-flow graph is a unitary matrix except for a normalization factor.

SUMMARY

A Fast Discrete Cosine Transform (FDCT) algorithm has been developed which may be extended to any desired value of $N = 2^m \geqslant 2$. The algorithm has been interpreted in the form of matrices and illustrated by a signal-flow graph. The

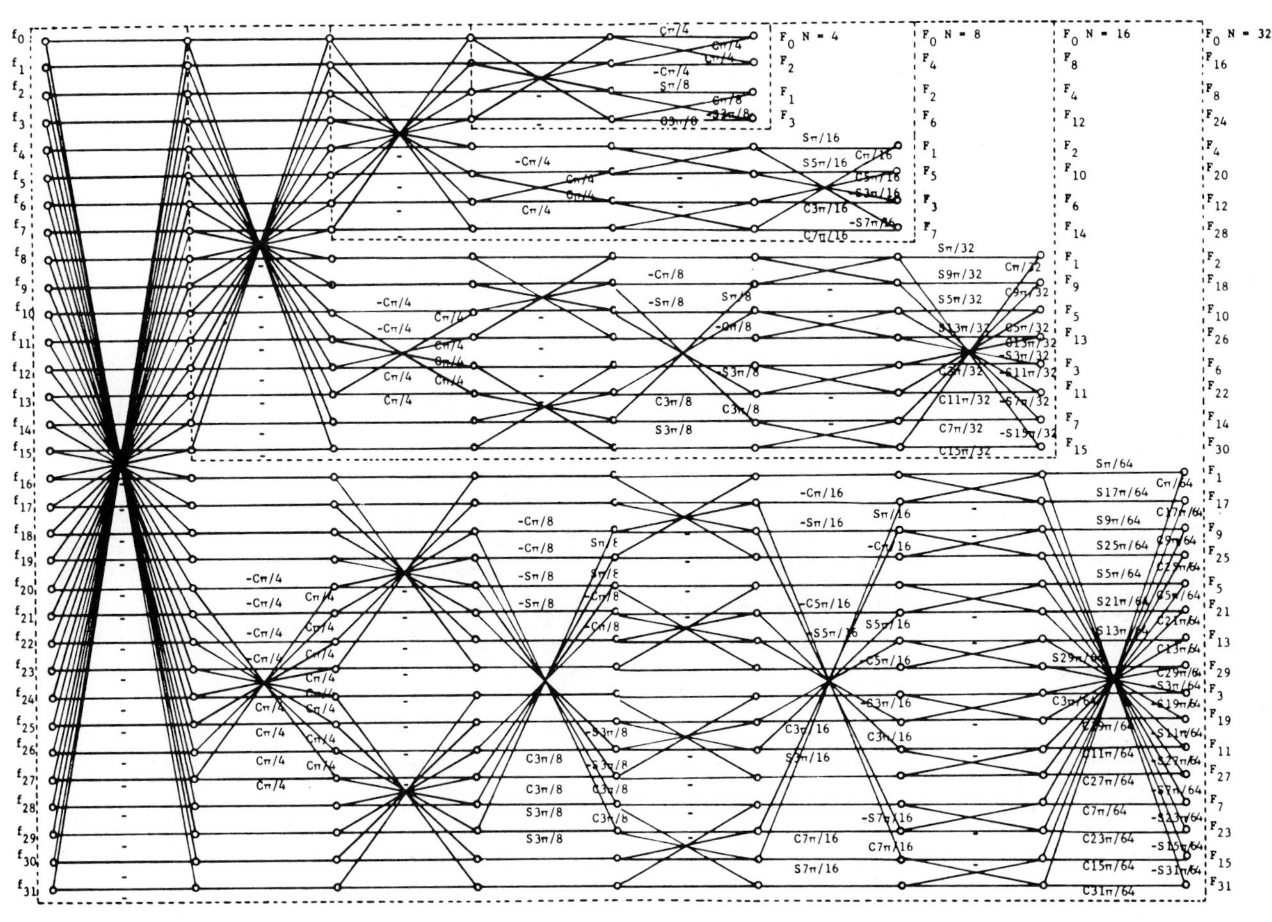

Figure 2. FDCT Flow Graph for $N = 4$, $N = 8$, $N = 16$ and $N = 32$; $Ci = \cos i$, $Si = \sin i$.

signal-flow graph may be readily translated to hardware (or software) implementation. The number of computational steps has been shown to be less than 1/6 of the conventional DCT algorithm employing a 2-sided FFT.

ACKNOWLEDGMENT

The authors are deeply indebted to Miss J. Del Mastro for her patience and skill in typing this paper.

REFERENCES

1. N. Ahmed, T. Natarjan, and K. R. Rao, "Discrete Cosine Transform," *IEEE Transactions on Computer,* January 1974, pp. 90-93.
2. A. Habibi, "Hybrid Coding of Pictorial Data," *IEEE Transactions on Communications,* Vol. COM-22, No. 5, May 1974, pp. 614-624.
3. J. A. Rose, W. K. Pratt, G. S. Robinson, and A. Habibi, "Interframe Transform Coding and Predictive Coding Methods," *Proceedings of ICC 75,* IEEE Catalog No. 75CH0971-2GSCB, pp. 23.17-23.21.
4. W. Chen and C. H. Smith, "Adaptive Coding of Color Images Using Cosine Transform," *Proceedings of ICC 76,* IEEE Catalog No. 76CH1085-0 CSCB, pp. 47-7 to 47-13.
5. J. R. Persons and A. G. Tescher, "An Investigation of MSE Contributions in Transform Image Coding Schemes," *Proceedings of SPIE,* August 1975, pp. 196-206.
6. W. Chen and S. C. Fralick, "Image Enhancement Using Cosine Transform Filtering," *Proceedings of the Symposium on Current Mathematical Problems in Image Science,* Montery, Ca., November 1976, pp. 186-192.

3

On a Real-Time Walsh–Hadamard/Cosine Transform Image Processor

D. HEIN AND NASIR AHMED, SENIOR MEMBER, IEEE

***Abstract*–A real-time image processor which is capable of video compression using either the sequency-ordered Walsh–Hadamard transform $(WHT)_w$, or the discrete cosine transform (DCT), is considered. The processing is done on an intraframe basis in (8 × 8) data**

Manuscript received March 30, 1978. This work was supported by the NASA-Ames Research Center, Moffett Field, CA, under Interchange NCA2-OR363-702.
The authors are with the Department of Electrical Engineering, Kansas State University, Manhattan, KS 66506. (913) 532-5600.

blocks. The $(WHT)_w$ coefficients are computed directly, and then used to obtain the DCT coefficients. This is achieved via an (8 × 8) transformation matrix which is orthonormal, and has a block-diagonal structure. As such, it results in substantial savings in the number of multiplications and additions required to obtain the DCT, relative to its direct computation. Some aspects of a hardware implementation of the processor are also included.

Key Words: Image processor, real time, Walsh-Hadamard/cosine transform.

I. INTRODUCTION

In recent years, there has been a growing interest in using the Walsh-Hadamard transform $(WHT)_w$ and the discrete cosine transform (DCT) for image processing, relative to other orthogonal transforms, e.g., the Haar transform, and the slant transform. Real-time hardware systems that enable video bandwidth compressions for either the $(WHT)_w$ or the DCT have been designed and implemented [1]-[3]. Our objective in this paper is to consider a *single* processor that is capable of video compression using the $(WHT)_w$ or the DCT. The processing is done on an intraframe basis in (8 X 8) data blocks. A novel feature of the processor is that it computes the DCT via the $(WHT)_w$, by means of a transformation matrix which is orthonormal, and has a block-diagonal structure. The result is a substantial savings in the number of multiplications and additions required to obtain the DCT, relative to its direct computation, i.e., without computing the $(WHT)_w$ first. A computer simulation has shown that a hardware implementation is feasible. Some aspects of the proposed hardware implementation are also included.

The input to the proposed real-time image processor is a National Television System Commission (NTSC) signal, low-pass filtered to 4 MHz. The resulting signal is sampled at 8×10^6 samples per second. The image is then buffered and segmented into (8 X 8) blocks, and a two-dimensional transform is performed via simultaneous one-dimensional transforms of size 8, in the horizontal and vertical directions. Thus the number of one-dimensional transforms of size 8 this requires is approximately 10^6 per second. An important factor in the design of the processor we seek is the ability to secure video bandwidth compression with either the $(WHT)_w$ or the DCT. System cost and circuit complexity are also of importance.

II. MOTIVATION

For the purposes of discussion, we consider the case $N = 8$. Fig. 1 shows the corresponding (8 X 8) transform matrices which are denoted by $[H_w(3)]$ and $[\Lambda(3)]$ for the Walsh and discrete cosine transforms, respectively [4]. In Fig. 1(a), the Sal and Cal functions associated with the $(WHT)_w$ are identified with respect to the rows of $[H_w(3)]$.

Examination of $[H_w(3)]$ and $[\Lambda(3)]$ in Fig. 1 shows that there is a one-to-one correspondence between the sequencies of the rows of these transform matrices. This implies that the Sal–Cal (i.e., odd–even) structure of the $(WHT)_w$ is preserved in the DCT. In other words, the basis vectors of the DCT are essentially "amplitude-modulated" versions of the basis vectors of the $(WHT)_w$. Hence the motivation for seeking an algorithm to compute the DCT via the $(WHT)_w$, as it relates to the image processor discussed in Section I.

$$[H_w(3)] = \frac{1}{\sqrt{8}}\begin{bmatrix} 1 & 1 & 1 & 1 & 1 & 1 & 1 & 1 \\ 1 & 1 & 1 & 1 & -1 & -1 & -1 & -1 \\ 1 & 1 & -1 & -1 & -1 & -1 & 1 & 1 \\ 1 & 1 & -1 & -1 & 1 & 1 & -1 & -1 \\ 1 & -1 & -1 & 1 & 1 & -1 & -1 & 1 \\ 1 & -1 & -1 & 1 & -1 & 1 & 1 & -1 \\ 1 & -1 & 1 & -1 & -1 & 1 & -1 & 1 \\ 1 & -1 & 1 & -1 & 1 & -1 & 1 & -1 \end{bmatrix} \begin{matrix} \mathrm{Wal}(0,t) \\ \mathrm{Sal}(1,t) \\ \mathrm{Cal}(1,t) \\ \mathrm{Sal}(2,t) \\ \mathrm{Cal}(2,t) \\ \mathrm{Sal}(3,t) \\ \mathrm{Cal}(3,t) \\ \mathrm{Sal}(4,t) \end{matrix}$$

(a)

$$[\Lambda(3)] = \begin{bmatrix} 0.354 & 0.354 & 0.354 & 0.354 & 0.354 & 0.354 & 0.354 & 0.354 \\ 0.490 & 0.416 & 0.278 & 0.098 & -0.098 & -0.278 & -0.416 & -0.490 \\ 0.462 & 0.191 & -0.191 & -0.462 & -0.462 & -0.191 & 0.191 & 0.462 \\ 0.416 & -0.098 & -0.490 & -0.278 & 0.278 & 0.490 & 0.098 & -0.416 \\ 0.354 & -0.354 & -0.354 & 0.354 & 0.354 & -0.354 & -0.354 & 0.354 \\ 0.278 & -0.490 & 0.098 & 0.416 & -0.416 & -0.098 & 0.490 & -0.278 \\ 0.191 & -0.462 & 0.462 & -0.191 & -0.191 & 0.462 & -0.462 & 0.191 \\ 0.098 & -0.278 & 0.416 & -0.490 & 0.490 & -0.416 & 0.278 & -0.098 \end{bmatrix}$$

(b)

Fig. 1. $(WHT)_w$ and DCT matrices for transformation size 8.

III. COMPUTATIONAL ALGORITHM

Let $\{C(3)\}$ and $\{W(3)\}$ be (8 X 1) column vectors which denote the DCT and $(WHT)_w$ transforms of size 8; i.e.,

$$\{C(3)\} = [\Lambda(3)]\{X(3)\} \tag{1}$$

and

$$\{W(3)\} = [H_w(3)]\{X(3)\} \tag{2}$$

where $\{X(3)\}$ denotes the (8 X 1) data vector.

We now rearrange the rows of $\{C(3)\}$, $[\Lambda(3)]$, $\{W(3)\}$, and $[H_w(3)]$ in bit-reversed order. If this rearrangement is indicated by the symbol "$\hat{}$", then corresponding to (1) and (2) we have

$$\{\hat{C}(3)\} = [\hat{\Lambda}(3)]\{X(3)\} \tag{3}$$

and

$$\{\hat{W}(3)\} = [\hat{H}_w(3)]\{X(3)\}. \tag{4}$$

Since $[\hat{H}_w(3)]$ is an orthonormal matrix, we can rewrite (3) in the form

$$\{\hat{C}(3)\} = [\hat{\Lambda}(3)][\hat{H}_w(3)]^T[\hat{H}_w(3)]\{X(3)\} \tag{5}$$

where $[\hat{H}_w(3)]^T$ denotes the transpose of $[\hat{H}_w(3)]$.

Substituting (4) in (5), we obtain

$$\{\hat{C}(3)\} = [T(3)]\{\hat{W}(3)\} \tag{6}$$

where

$$[T(3)] = [\hat{\Lambda}(3)][\hat{H}_w(3)]^T.$$

Equation (6) yields the DCT in terms of the $(WHT)_w$ via the transformation matrix $[T(3)]$. Evaluation of $[T(3)]$ leads to

$$[T(3)] = \begin{bmatrix} 1.0 & & & & & & & \\ & 1.0 & & & & 0 & & \\ & & 0.923 & 0.383 & & & & \\ & & -0.383 & 0.923 & & & & \\ & & & & 0.907 & -0.075 & 0.375 & 0.180 \\ & 0 & & & 0.214 & 0.768 & -0.513 & 0.318 \\ & & & & -0.318 & 0.513 & 0.768 & 0.214 \\ & & & & -0.180 & -0.375 & -0.075 & 0.907 \end{bmatrix} \qquad (7)$$

Examination of $[T(3)]$ in (7) shows that it has two important properties: i) it is orthonormal, being the product of two orthonormal matrices $[\hat{\Lambda}(3)]$ and $[\hat{H}_w(3)]^T$ (see (6)), and ii) it has a block diagonal structure.

The orthonormal property of $[T(3)]$ implies that the transformation in (6) is unique, while its block-diagonal structure results in a reduction in the number of arithmetic operations to compute the DCT. Specifically, we observe that 20 multiplications are required to compute the DCT via (6), as opposed to 64 multiplications required for a direct computation via (1). Again, the number of additions required is reduced from 56 to 38, where 14 of them are done following the $(\text{WHT})_w$ computation in (4). Since 10^6 transforms of size 8 are to be computed each second, the savings in the number of multiplications and additions (to compute the DCT) is approximately 44 and 22 million per second, respectively. It is this savings that makes a hardware implementation feasible (some aspects of this are discussed in the next section).

The preceeding method of computing the DCT may be generalized for N other than 8. A similar matrix decomposition has been reported in [5] and demonstrates that computation time is reduced for N less than or equal to 64, relative to the $2N$-point FFT method.

IV. HARDWARE IMPLEMENTATION CONSIDERATIONS

A software simulation was first carried out to determine the precision requirements for the multipliers and intermediate results. It was determined that sufficient accuracy could be achieved by using 12 bits for the arithmetic computations. This was done assuming that the digitized video consisted of 6 bits per picture element (pel).

Instead of using multipliers, actual implementation will be done via READ ONLY memories (ROM's). These ROM's will be preprogrammed with the correct product for each of the multiplying elements belonging to $[T(3)]$ in (7). Then, a direct table lookup using 12-bit input words requires a ROM of size

$$S_1 = 20 \times 2^{12} \times 12 = 983040 \text{ bits} \cong 120 \text{ Kbytes} \qquad (8)$$

where the factor of "20" is the number of multiplications associated with $[T(3)]$ in (7).

It is evident that the size of the ROM needed is much too large to be practical. However, there are techniques for reducing the memory requirements for multiplication [6]. Let A be the 12-bit $(\text{WHT})_w$ coefficient. This may be represented as a sum of two 6-bit numbers

$$A = A_1 \times 64 + A_0 \qquad (9)$$

where A_1 and A_0 are the most and least significant halves of A. Multiplication by a constant α results in

$$A \times \alpha = \alpha \times A_1 \times 64 + \alpha \times A_0. \qquad (10)$$

Each half can then be multiplied by using a 64×12 bit table in the ROM for the most significant half, and a 64×6 bit table for the least significant half. The final 12-bit product is then obtained by addition of these two results. This approach requires a ROM of size

$$S_2 = 20 \times 2^6 \times (12 + 6) = 23040 \text{ bits} \cong 3\text{K bytes}. \qquad (11)$$

From (8) and (11) it is apparent that this approach leads to a substantial reduction in the size of the ROM required.

The use of ROM lookup tables, as opposed to using discrete multipliers, also results in a significant reduction in system cost and complexity. In addition, execution speed is also increased. The interested reader may refer to the Appendix for a more detailed discussion of this technique.

The proposed system is shown in Fig. 2, where the input to the multiplying array is the $(\text{WHT})_w$ coefficients. Details pertaining to a hardware implementation for computing the $(\text{WHT})_w$ are available elsewhere, e.g., see [1], [7], [8]. The circuit requires four clock cycles to compute the DCT coefficients. The ROM's used have eight address lines, six of which are used by the $(\text{WHT})_w$ input and the other two select the proper multiplier for each clock cycle. One of the four DCT coefficients that requires a combination of four of the $(\text{WHT})_w$ coefficients is computed during each cycle. The result is then switched to the appropriate output register. The two coefficients that require two $(\text{WHT})_w$ coefficients are computed during every other clock cycle by storing intermediate results in a temporary register on odd clock cycles. The other two coefficients are passed through unchanged. Provisions will also be made to pass all of the $(\text{WHT})_w$ coefficients through unaltered, so that either the $(\text{WHT})_w$ or the DCT coefficients are obtained for further processing.

Fig. 2. A hardware implementation.

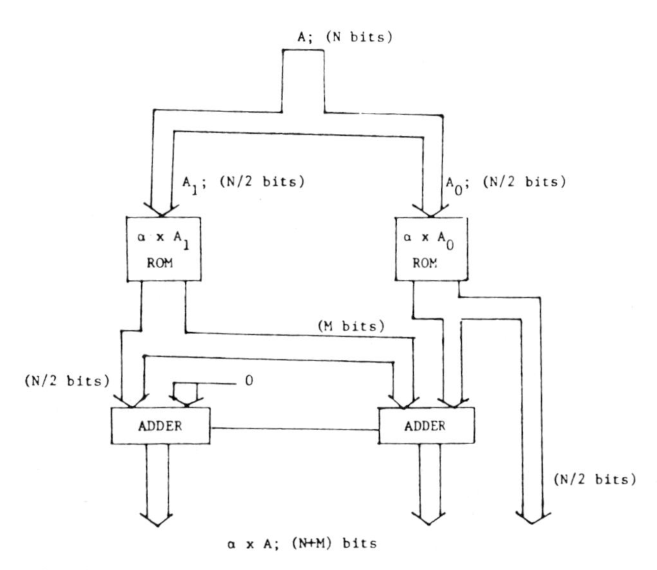

Fig. 3. Computation of $\alpha \times A$ defined in (A3) using two ROMS.

V. CONCLUSIONS

The image processor described in this paper is unique in that it can process image data using either the $(\mathrm{WHT})_w$ or the DCT. The processing is done on an intraframe basis in (8 X 8) blocks. The $(\mathrm{WHT})_w$ is computed directly, while the DCT is computed via the $(\mathrm{WHT})_w$ using a transformation matrix which has a block diagonal structure. A hardware implementation which avoids the use of discrete multipliers is proposed. Simulation results indicate that an implementation involving a table lookup approach with a 3K byte ROM is feasible.

APPENDIX

Consider the N-bit integer A. Assuming N is even, we may represent A as

$$A = A_1 \times 2^{N/2} + A_0 \tag{A1}$$

where A_1 and A_0 are the $(N/2)$-bit most and least significant halves of the number A, respectively.

Let α be an M-bit fractional binary number less than 1. We form the product

$$\alpha \times A = \alpha \times A_1 \times 2^{N/2} + \alpha \times A_0. \tag{A2}$$

In general, $\alpha \times A_0$ and $\alpha \times A_1 \times 2^{N/2}$ will be $(N/2 + M)$-bit numbers with M and $(M - N/2)$ fractional bits, respectively. Forming the product, $\alpha \times A$, by addition of the two halves results in an $(M + N)$-bit number with M fractional bits, as illustrated in Fig. 3.

Thus multiplication by a constant may be performed by using two ROM's with $2^{N/2}$ entries, rather than one ROM with 2^N entries. The corresponding ROM sizes required are summarized below for values of N from 2 through 16.

Word Length	Memory Required	
	One ROM	Two ROMS
N	2^N	$2(2^{N/2})$
2	4	4
4	16	8
6	64	16
8	256	32
10	1024	64
12	4096	128
14	16384	256
16	65536	512

ACKNOWLEDGMENT

The authors are grateful to Dr. D. R. Lumb, Dr. S. C. Knauer, Dr. H. W. Jones, and Mr. L. R. Hofman of the NASA-Ames Research Center, Moffett Field, CA, for their assistance and encouragement, throughout the duration of the work reported in this paper.

REFERENCES

[1] S. C. Knauer, "Real-time video compression algorithm for Hadamard transform processing," *IEEE Trans. Electromagn. Compat.*, vol. EMC-18, pp. 28–36, Feb. 1976.

[2] H. J. Whitehouse, R. W. Means, and E. H. Wrench, "Real time television image bandwidth reduction using charge transfer

devices," *Proc. Soc. Photo-Optical Instrumentation Engineers*, pp. 36–47, 1975.

[3] H. J. Whitehouse *et al.*, "A digital real time intraframe video bandwidth compression system," in *Proc. Soc. Photo-Optical Instrumentation Engineers*, pp. 64–78, 1977.

[4] N. Ahmed and K. R. Rao, *Orthogonal Transforms for Digital Signal Processing.* New York/Berlin/Heidelberg: Springer-Verlag, 1975, chaps. 6, 7.

[5] C. Chen, "Development and implementation of a class of matrix production algorithms," Ph.D. dissertation, Kansas State University, Manhattan, KS, 1975.

[6] A. Barna and D. I. Porat, *Integrated Circuits in Digital Electronics.* New York: Wiley, 1973, chap. 11.

[7] H. F. Harmuth, *Sequency Theory–Foundations and Applications.* New York: Academic Press, 1977, chap. 2.

[8] S. C. Noble, "A comparison of hardware implementations of the Hadamard transform for real-time image coding," *Proc. Soc. Photo-Optical Instrumentation Engineers*, pp. 207–211, 1975.

An Approach to the Implementation of a Discrete Cosine Transform

GUIDO BERTOCCI, MEMBER, IEEE, BRIAN W. SCHOENHERR, AND DAVID G. MESSERSCHMITT, SENIOR MEMBER, IEEE

Abstract—**An approach to the implementation of a discrete cosine transform (DCT) for application to coding speech is described. The approach is oriented toward single speech channel encoding. In addition, a detailed computer simulation of an adaptive transform coder is described.**

The purpose of the computer simulation is to determine the internal precision at various points in the implementation required to avoid subjective degradation. Specific recommmendations are made on the required internal precision in the implementation of the discrete cosine transform.

A breadboard implementation of the DCT using SSI and MSI TTL logic based on the results of the computer simulation is reported.

I. INTRODUCTION

ADAPTIVE transform coding is, together with subband coding, a promising method of encoding speech at bit rates below 16 kbits/s [1]. A significant obstacle to the widespread application of transform coding is, however, its great complexity. A computationally intensive portion of the transform coder is the front end discrete cosine transform (DCT). In this paper an architecture for the implementation of a DCT is recommended. A detailed computer simulation of a transform coder, including the bit allocation algorithm as well as DCT, was performed for the purpose of determining the required internal finite wordlength precisions.

This is a critical problem, since choosing too high a precision complicates the implementation, and insufficient precision will result in degradation of speech quality beyond that inherent in the encoding technique. The simulation was carefully designed to accurately reflect these finite precision effects. The simulation was run on actual speech followed by informal listening tests to determine the effects of insufficient precision.

This paper does not consider the implementation of the bit allocation algorithm in a transform coder. The bit allocation method used in the simulator was that recommended in [1]. The design of the bit allocation portion of the coder is the most challenging part, particularly from an algorithmic standpoint.

Manuscript received August 11, 1981; revised October 20, 1981. This work was supported in part by the National Science Foundation under Grant 78-16966 and by GTE Lenkurt. The work of G. Bertocci and B. W. Schoenherr was performed in partial fulfillment of the M.S. degree from the University of California, Berkeley, CA.

G. Bertocci is with Bell Laboratories, Holmdel, NJ 07733.

B. W. Schoenherr is with Bell Laboratories, North Andover, MA 08145.

D. G. Messerschmitt is with the Department of Electrical Engineering and Computer Science, University of California, Berkeley, CA 94720.

As a ground rule in this study, it was assumed that a single channel was being encoded, as opposed to the encoding of a large group of channels simultaneously. The available circuit techniques and device speeds were assumed to be constrained by those available in MOS-LSI. The DCT was actually implemented using SSI and MSI TTL parts using the architecture recommended here.

The proposed implementation architecture is described in Section II, and the simulation and results are described in Section III. A brief description of the breadboarded DCT is given in Section IV.

II. AN APPROACH TO IMPLEMENTATION OF THE DCT CODER

It has been recommended that a discrete cosine transform (DCT) [2] is the most appropriate fixed (nonadaptive) transform for speech signals. It is given by

$$G_x(0) = \frac{2^{1/2}}{N} \sum_{m=0}^{N-1} X(m) \tag{2.1}$$

$$G_x(k) = \frac{2}{N} \sum_{m=0}^{N-1} X(m) \cos\left(2\pi \frac{(2m+1)k}{4N}\right)$$

$$k = 1, 2, \cdots, (N-1)$$

where $X(m)$, $0 \leq m < N$, are the N speech samples in the block, and $G_x(k)$ are the transform coefficients. The inverse DCT is given by an analogous equation.

Several approaches to the implementation of the DCT were considered, including the straightforward implementation of (2.1) using either an entirely digital or a partially analog approach using switched-capacitor techniques. "Fast DCT" algorithms which have been proposed [3] were also considered. At a sampling rate of 8 kHz, the multiply rate for a straightforward implementation of (2.1) is only a modest one million per second for the recommended value of $N = 128$ [1]. Thus, the added control complexity of a fast algorithm is obviously not justified for a single channel transform coder (although it would be valuable in a multichannel application). For a single channel coder, the switched capacitor techniques were not estimated to offer an appreciable die area advantage over an all-digital implementation, but would offer a significantly more difficult design. As a result we settled on a straightforward digital implementation of (2.1).

There are two possible methods of calculating the N transform coefficients from N successive speech samples.

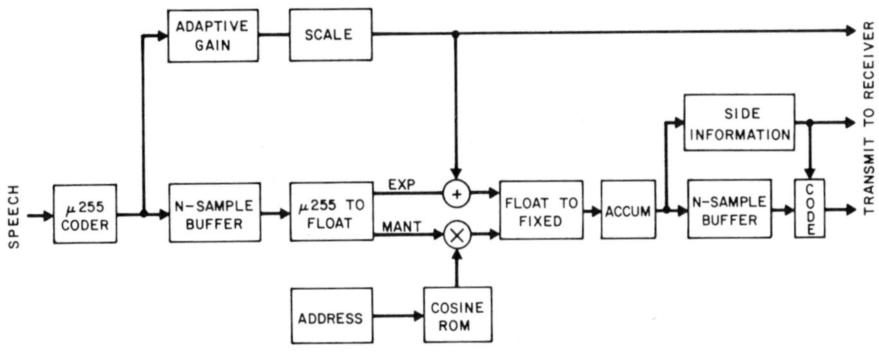

Fig. 1. Block diagram of a per-channel transform coder.

1) Calculate all N transform coefficients simultaneously, keeping, at any time, N partial accumulations. As each speech sample arrives, update each of the N partial accumulations.

2) Store in memory a block of N speech samples as they arrive. Simultaneously, calculate the transform coefficients of the previously stored block. This is done by calculating one transform coefficient per sampling interval, using all N stored speech samples.

The considerations in choosing one of these methods are as follows.

1) Method 2 requires storage locations for $2N$ speech samples, including those currently being received, as well as the last block on which the transform coefficients are currently being calculated. Method 1 requires no storage of speech samples.

2) Method 1 requires the storage of N partial accumulations, while method 2 has no partial accumulations. Both methods require memory for N transform coefficients for storage while the adaptive quantization side information is being calculated.

3) Both methods results in $2N$ sample periods of delay. In the case of method 1, the first block is used to calculate the transform coefficients, and the second to calculate the side information. For method 2, the first block is used to store the N speech samples, and the second is used to calculate both the transform coefficients and the side information.

4) Method 2 is compatible with adaptive feedforward quantization [1] in which a quantization scaling factor to be sent to the receiver as side information is calculated for the block of speech samples before the transform coefficients are calculated. This adaptive quantization results in a normalization of the transform coefficients, resulting in a reduction in the number of bits of precision in the accumulation and storage of transform coefficients and in the subsequent adaptive quantization algorithms which operate on a per-transform coefficient (as opposed to block) basis. In method 1, on the other hand, the transform coefficients are calculated prior to reception of the entire block of speech samples. Thus the accumulation requires greater precision.

In view of these considerations, the fourth point was considered overriding, and method 2 was chosen. This choice also results in fewer bits of required memory, in view of the fewer number of bits of precision required for speech samples as compared to partial accumulations and transform coefficients which have not been adaptively quantized.

This choice results in the configuration of Fig. 1. It is assumed that the speech is first encoded using the $\mu = 255$ encoding law commonly used in telephone networks. This type of coder is widely available in monolithic form and has adequate precision and dynamic range for this application. As the speech samples are read into an N-sample buffer, the adaptive feedforward quantization algorithm (the box labeled "adaptive gain") calculates some measure of the average speech power. This measure is then mapped into a scale factor in the box labeled "scale," which is applied before accumulation of the transform coefficients and is also transmitted as side information. As the samples are read from the N-sample buffer, they are converted into a true floating point representation (which is very similar to $\mu = 255$). Prior to accumulation, we must multiply by the values of the cosine as in (2.1). Since the cosine has a well-defined level, it can be stored in ROM in a fixed point representation. The multiply with the speech sample can be performed only on the mantissa of the speech sample floating point representation. The resulting values are then normalized by adding the previously determined scale factor to the exponent, and the resulting value is converted to fixed point prior to accumulation. The accumulation of the N values then determines one transform coefficient as in (2.1). The accumulation of these N values occurs during one speech sampling interval, and the N-sample buffer is read N times in order to determine the N transform coefficients.

As the N transform coefficients are calculated, they are stored in an N-sample buffer. Simultaneously, the adaptive quantization side information is calculated. Then, as the N transform coefficients are read from the buffer, they are coded using the appropriate step size and number of bits of precision, as determined by the side information.

Several aspects of the transform coder implementation deserve to be discussed in more detail. In the following sections we discuss the adaptive gain algorithm, the $\mu = 255$ to floating point conversion, and the generation of the cosine.

A. $\mu = 255$ to Floating Point Conversion

The conversion of a $\mu = 255$ sample to floating point to expedite a subsequent multiplication has been used in [4]. In this section the details of this conversion are developed. The $\mu = 255$ output level is given by [5]

$$X = 2^L(V + 16.5) - 16.5 \tag{2.2}$$

where X is the output (analog) level corresponding to $\mu = 255$

code (L, V), L is the segment number, $0 \leqslant L < 8$, and V is the level on a segment, $0 \leqslant V < 16$. The effect of the sign bit has not been incorporated into (2.2). We can find a floating point representation for X by setting it equal to $2^L V'$, where L is the exponent (the same as the segment number) and V' is the mantissa. Solving for V':

$$V' = V + 16.5(1 - 2^{-L}). \tag{2.3}$$

It is straightforward to see that $0 \leqslant V' < 32$, and that a full precision representation of V' would require 13 bits. However, if we used only 5 bits, the accuracy in V' would be ± 0.5, which would result in an accuracy in X of $\pm 2^{L-1}$. This is exactly half a $\mu = 255$ step size, since on segment L the step size is 2^L. Thus, a 5 bit precision on V' results in a maximum error of half a step size in the $\mu = 255$ law, and adding bits to V' reduces the error correspondingly.

The actual conversion from V to V' can be done by simple combinatorial logic. It should also be apparent that if the original speech should be encoded in fixed point rather than $\mu = 255$, the conversion to floating point would also be advantageous, although slightly more complicated.

B. Adaptive Gain

It is recommended in [1] that the input speech samples be normalized by their sample variance prior to calculation of the transform coefficients. However, once the speech samples have been expressed in floating point form, it should be adequate (and much simpler) to normalize them in terms of the exponents only. In particular, if we calculate the sample mean of the exponents in the block of N speech samples

$$L = \frac{1}{N} \sum_{i=0}^{N-1} L_i \tag{2.4}$$

where L_i is the exponent of sample i, and then subtract L from the exponent of each speech sample, the block will be normalized such that the average exponent is zero. This is roughly equivalent to normalizing by the geometric mean of the block of speech samples. Also note that L_i is the same as the segment number of the $\mu = 255$ sample, so that this normalization factor can be calculated directly from the $\mu = 255$ samples as they are being stored in the buffer (as shown in Fig. 1).

C. Generation of the Cosine

From (2.1), we see that the cosine must be generated at the $4N$ points uniformly spaced on a circle shown in Fig. 2. Obviously, only $N + 1$ values in one quadrant need be stored in the ROM, and the others can easily be inferred by symmetry. In addition, since N will undoubtedly be a power of two, it is convenient to store only N values. Fortuitously, the three angles marked with a question mark can never occur for N a power of two[1] (this is suggested by (2.1) and has been verified by a computer program for $N = 128$). This implies that the ROM need only have N addresses. In addition, angle zero is

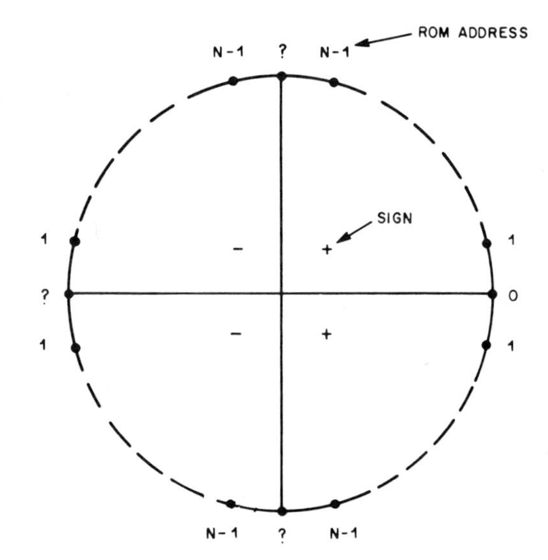

Fig. 2. $4N$ angles for which cosine must be generated.

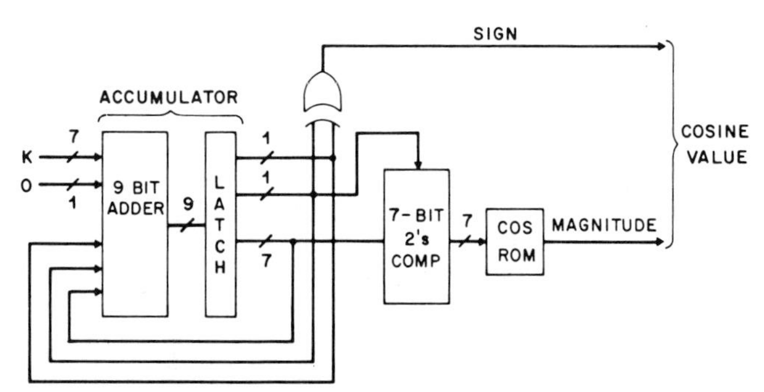

Fig. 3. Cosine value with address generator (shown for $N = 128$).

only addressed in calculation of the $k = 0$ transform coefficient, and hence ROM address zero can contain the value $2^{-1/2}$ rather than unity. The ROM addresses are then shown on the diagram for all four quadrants, and the required signs are also shown.

From (2.1), the ROM address can easily be derived from the quantity

$$I(m, k) = (2m + 1)k \text{ modulo } 4N \tag{2.5}$$

where m is the sample number in the block, and k is the transform coefficient number. If N is a power of two, the modulo operation is simply accomplished by using finite precision arithmetic in the address generation and ignoring any overflow. Of course, k is simply generated by a counter. The $I(m, k)$ can be generated sequentially from the relation

$$I(m + 1, k) = I(m, k) + 2k \tag{2.6}$$

as shown in Fig. 3. The accumulator simply adds $2k$ to the last value of $I(m, k)$ at each new speech sample. The resulting 9 bit address (for $N = 128$) specifies which of the $4N$ points is desired on Fig. 2. The two most significant bits specify the quadrant. The exclusive-or gate then determines the correct sign bit. In quadrants one and three, the seven least significant bits are used as the ROM address, and in quadrants two and four the two's complement of the 7 bit address is used.

[1] This was pointed out to the authors by H.-H. Lu.

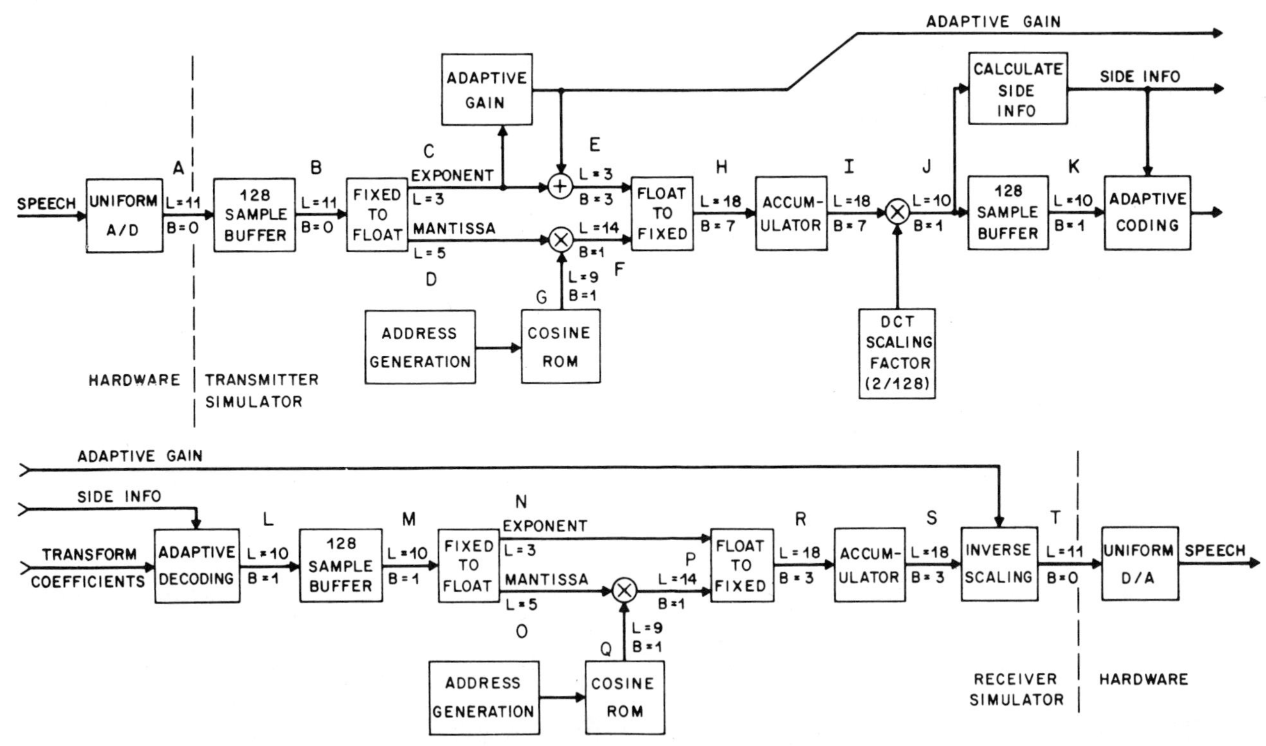

Fig. 4. Transform coder simulator architecture (L = bit length, B = binary point location).

III. GOAL OF COMPUTER SIMULATION

The quality of the speech processed by a transform coder should depend only on the bit rate of the transmitted speech as determined by the bit-allocation algorithm. Determining the minimum precision required to implement a transform coder is important in order to reduce to a minimum the memory requirements and complexity of the transform coder.

The goal of the computer simulation is to determine experimentally the required precision at each point of the transform coder. In particular, it is important to determine the required precision to implement the discrete cosine transform since most of the memory needed for a transform coder lies in implementing the DCT.

Since the simulator was not a goal in itself but a tool to determine required wordlength precisions, one compromise was made to accelerate the implementation of the transform-coder simulator. The compromise lies in the use of a 12 bit uniform coder instead of a μ = 255 coder. A 12 bit uniform coder interfaced to a PDP 11/40 computer was available, greatly simplifying the collection of digitized voice samples for processing through the simulator.

A. Architecture of Computer Simulator

Fig. 4 shows the architecture of the computer simulator. As mentioned previously, there is one basic difference, the use of a 12 bit uniform coder. This substitution causes another small change. The adaptive gain is calculated using the exponent instead of using the μ = 255 segment number and scaling. The result of these two processes is, however, identical.

1) DCT Scaling Factor: The DCT scaling factor is defined to be

$$\frac{2}{N} \quad \text{where } N = \text{block length.} \tag{3.1}$$

The $2/N$ factor comes from (2.1). By introducing the $2/N$ factor after all the partial coefficients have been accumulated, leading zeros are not carried through the entire discrete cosine transform.

Since N is a power of 2, multiplying by $2/N$ simply shifts the binary point. For the simulator, the location of the binary point is a very important parameter and must be maintained properly for each coefficient throughout the simulation (Section III-B discusses number representation). In actual hardware the location of the binary point is implicit, thus eliminating the necessity to multiply by $2/N$.

Thus, even though there appears to be an additional change in architecture from Fig. 1 (at point I in Fig. 4) this change is necessary only for the simulation and not in a hardware implementation.

B. Number Representation

The heart of the finite precision computer simulator is the ability to specify the number of bits in a coefficient and the location of the binary point. The location of the binary point is very important since it determines the tradeoff between the range of numbers that can be represented and the precision. Adding a bit to the left of a binary point increases the range and reduces the probability of an overflow. Adding a bit to the right of a binary point increases the precision.

For each coefficient, the binary point and bit length must

be specified. The binary point is defined relative to the most significant bit. For example, the coefficient 1.01 would be represented as $L = 3$ and $B = 1$, which means a wordlength of 3 bits and 1 bit to the left of the binary point. In addition, a sign bit is required. In Fig. 4 at points A–T the minimum precision needed to implement a one-channel transform coder is shown. The required wordlengths as indicated in Fig. 4 are given by

$$\text{total wordlength in bits} = L + 1. \tag{3.2}$$

This precision was determined from the simulation as described in Section III-D.

The simulator uses signed magnitude arithmetic for ease of implementation. Signed magnitude arithmetic creates one problem that does not exist with two's complement arithmetic: there is a duplicate representation of zero. For coefficients with bit lengths less than 4 there is a significant loss of information. For example, a coefficient with two bits can only represent three distinct values with signed magnitude arithmetic as opposed to four distinct values with two's complement arithmetic.

Since most of the transmitted coefficients are short, an average length of 2 for 16 kbit/s speech, the redundancy factor is important. To determine the minimum required precision for the DCT for 16 kbit/s speech, a higher bit rate is necessary to compensate for the redundant zero.

C. Overflows

One of the critical problems is overflowing a transform coefficient during accumulation. An overflow is defined to be a loss of most significant bits due to an insufficient number of allocated bits to the left of the binary point. An overflow can occur at two critical points. One is during the calculation of the DCT or IDCT and the other is during the scaling of the transform coefficients before truncating to a specified bit length for transmission.

Overflows during the calculation of the DCT or IDCT can be easily avoided by allocating bits to the left of the binary point. One subtle point to consider is that it is possible for a transform coefficient to exceed its final value during the accumulation of partial transforms. This is due to the possibility that several positive partial transforms are accumulated before negative partial transforms or vice versa.

Overflows during the scaling of the transform coefficients are very critical. The purpose of scaling the coefficients before truncating to an assigned bit length is to remove all leading zeros. However, the algorithm to determine the proper scaling factor is based on linearly interpolating between the averaged side information coefficients [1]. As a result, the scaling factor for a particular coefficient may cause a transform coefficient to overflow.

In an attempt to reduce the overflows during scaling, the simulator counts the number of overflows for each block of transform coefficients after each coefficient is scaled. If the number of overflows exceeds a predetermined tolerance, the transform coefficients are rescaled with a smaller scaling factor

TABLE I
SIMULATION RESULTS ("WHY DO I OWE")

Label	Simul 1	Simul 2[a]	Simul 3
D	15	5	4
F	30	14	12
G	15	9	8
I	30	18	15
J	30	10	10
L	30	10	10
O	30	5	4
Q	15	9	8
P	30	14	12
R	30	18	15
SNR (dB)	16.57	16.16	14.87

[a] Values recommended for minimum required precision as shown in Fig. 4.

(each coefficient has one less leading zero truncated). This process is repeated until the number of overflows is below the tolerance. The number of fewer leading zeros truncated is sent to the receiver as side information.

In [1], the $\log_2$ of the variance of the transform coefficients is proposed as the side information. In an attempt to simplify the algorithm, the simulator uses the number of leading zeros for each coefficient instead of squaring the entire transform coefficient. This is very similar to using the exponent of a floating point number.

D. Simulation Results

Three phrases low-pass filtered at 3500 Hz were used for processing. The three phrases are "Why do I owe," "Why not be louder" and the letter "e."

The strategy for determining the required precision was to first set all bit lengths to the maximum allowable by the simulator. The number of overflows allowed during scaling of the transform coefficients was varied until the SNR was maximized. Three overflows per block of 128 appeared to be close to optimal for all three samples.

The next step was to reduce the precision at each point in the coder, starting from the maximum wordlength of 30 or 15 depending on the coefficient, until there was noticeable degradation in the SNR and quality of the speech (the latter determined by informal listening tests comparing the original uncoded speech with the processed speech).

Table I shows the resulting SNR (not segmental SNR) for three simulations using different wordlengths. Each of the three simulations were conducted at 16 kbits/s. Simulation 1 uses the maximum precision allowed by the simulator. Simulations 2 and 3 show a threshold at which performance, as measured by SNR, drops dramatically with a small reduction in coefficient precision.

The labels in Table I correspond to the labels (A–T) in Fig. 4. Only the values that changed from simulation to simulation are shown. All nonlisted values are those shown in Fig. 4. In Fig. 4 "L" denotes the length of each word in bits and "B" denotes the location of the binary point relative to the most significant bit.

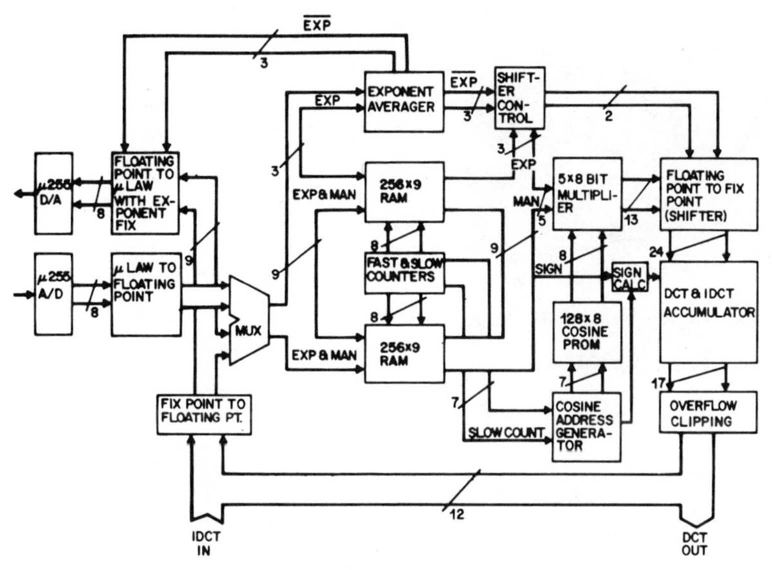

Fig. 5. DCT/IDCT hardware architecture.

The transform coder output bit rate used for the simulations was 16 kbits/s. This bit rate does not take into account the side information or the inefficiencies due to the use of signed magnitude notation. These two factors have an opposing effect. Including the side information would increase the bit rate for the same SNR while using a more efficient representation of coefficients would decrease the bit rate for the same SNR.

Once the minimum required precision for 16 kbits/s was determined, simulations were conducted to determine if the same precision was also adequate for 24 kbits/s. The SNR for 24 kbits/s using the same precision as shown in Fig. 4 was only 0.7 dB lower than the SNR obtained with the maximum precision allowable by the simulator at 24 kbits/s. This result indicates that the precision as shown in Fig. 4 is sufficiently conservative to allow for improvements in the coding algorithm at 16 kbits/s without being limited by the DCT implementation.

IV. BREADBOARD IMPLEMENTATION

The DCT-AQF section of the transform coder was implemented with SSI and MSI TTL hardware. Because similar functions are required by both the DCT and IDCT, the hardware that performed these functions was operated at twice its normal frequency so that data from both the DCT and IDCT could be interleaved. Hardware that is shared by both the DCT and IDCT includes the memories, the cosine value generator, the multiplier, the floating point to fixed point converter, the accumulator, and the overflow detection circuitry.

The hardware implementation architecture is shown in Fig. 5. Fig. 5 clearly shows the shared structure for the DCT and IDCT. The word widths recommended in Fig. 4, derived by the software simulation, were used for the implementation except for three modifications.

The first modification involved the cosine value generator PROM. The PROM has 8 bit word widths representing values less than unity. Each of these values had its least significant bit rounded to account for lesser significant bits, except for the second through sixth values. These should have been rounded to 1.00000000 but, because the multiplier used only numbers less than unity, these values were truncated to 0.1111111. This error should be insignificant because, at most, this represents 5/128 of an accumulation where the least significant seven bits are truncated.

The second modification involved the floating point to fixed point converter. Because of the time required to input a value to the accumulator, only shifts up to six positions were allowed. This is also of minor consequence because the adaptive quantization of the input makes the occurrence of seven shifts highly unlikely.

The third modification involved the accumulator. It was determined that an additional five chips would be needed to implement a full precision accumulator with 24 bits instead of 18 as suggested by the simulations. Therefore, the additional cost was determined to be worthwhile in light of the fact that the hardware could be used for other purposes than 16 kbit/s transform coding.

The circuit implementation required 139 assorted SSI and MSI TTL integrated circuits, as well as seven MOS integrated circuits (codec, filters, and memories). Three power supplies were required (–5 V, +5 V, and +12 V). A 16.384 MHz clock was used.

The performance in looped operation, the DCT output connected to the IDCT input, was found to be limited only to that of the input and output coder-decoder, with a 2 block (32 ms) delay. Fig. 6 shows a mainly 920 Hz input spectrum and the resulting output spectrum. The SNR as calculated from the output spectrum, using a 10 Hz window with a spectrum analyzer, is approximately 38 dB which is characteristic of a codec.

From the implementation of the DCT and IDCT it became clear that most of the cost was in the multiplications and format conversions and not in the accumulation of the coefficients. This result is partially due to the fact that semiconductor memories are available in units that are powers of two (i.e., a 16 bit accumulator would have saved significantly more chips than an 18 bit accumulator).

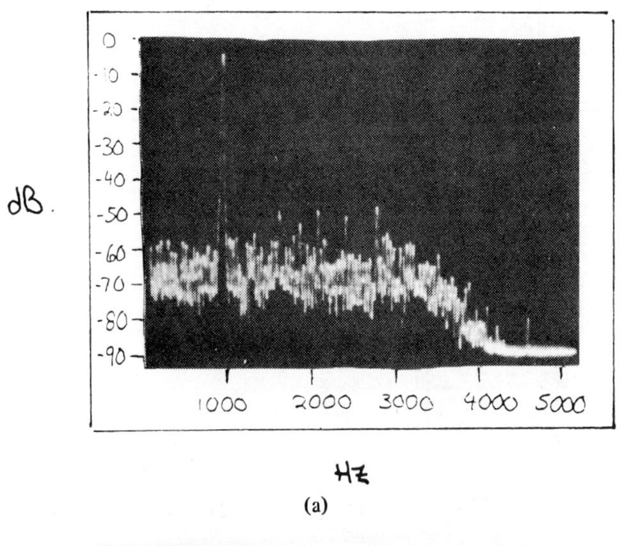

(a)

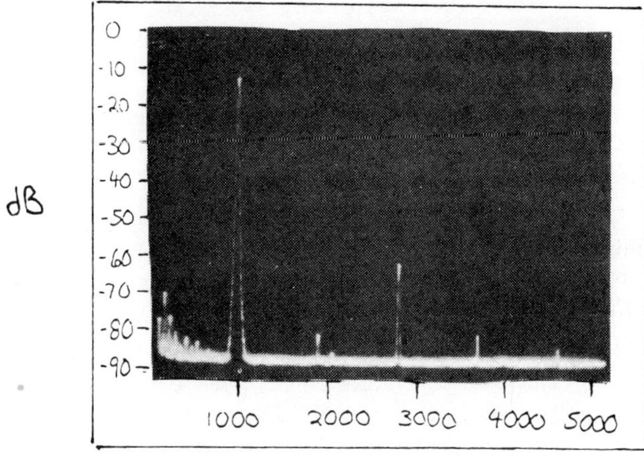

(b)

Fig. 6. (a) Output spectrum (10 Hz window). (b) Input spectrum (10 Hz window).

REFERENCES

[1] R. Zelinski and P. Noll, "Adaptive transform coding of speech signals," *IEEE Trans. Acoust., Speech, Signal Processing*, vol. ASSP-25, Aug. 1977.

[2] N. Ahmed, T. Natarajan, and K. Rao, "Discrete cosine transform," *IEEE Trans. Comput.*, vol. C-23, Jan. 1974.

[3] J. Makhoul, "A fast cosine transform in one and two dimensions," *IEEE Trans. Acoust., Speech, Signal Processing*, vol. ASSP-28, Feb. 1980.

[4] D. L. Duttweiler, "A twelve-channel digital echo canceler," *IEEE Trans. Commun.*, vol. COM-26, May 1978.

[5] H. Kaneko, "A unified formulation of segment companding laws and synthesis of codecs and digital companders," *Bell Syst. Tech. J.*, vol. 49, Sept. 1970.

[6] J. Tribolet and R. Crochiere, "Frequency domain coding of speech," *IEEE Trans. Acoust., Speech, Signal Processing*,, vol. ASSP-27, pp. 512–530, Oct. 1979.

[7] ——, "An analysis/synthesis framework for transform coding of speech," in *Proc. IEEE Conf. Acoust., Speech, Signal Processing*, 1979, pp. 81–84.

[8] J. Huang and P. Schultheiss, "Block quantization of correlated Gaussian random variables," *IEEE Trans. Commun. Syst.*, vol. CS-11, pp. 289–296, Sept. 1963.

[9] S. Campanella and G. Robinson, "A comparison of orthogonal transformations for digital speech processing," *IEEE Trans. Commun. Technol.*, vol. COM-19, pp. 1045–1049, Dec. 1971.

[10] R. Zelinski and P. Noll, "Approaches to adaptive transform speech coding at low bit rates," *IEEE Trans. Acoust., Speech, Signal Processing*, vol. ASSP-27, Feb. 1979.

[11] B. Kernighan and D. Ritchie, *The C Programming Language*. Englewood Cliffs, NJ: Prentice-Hall, 1978.

5

ASSP-24:428–429 (1976)

Comparison of the Cosine and Fourier Transforms of Markov-1 Signals

MASSIH HAMIDI AND JUDEA PEARL

***Abstract*—This correspondence compares the effectiveness of the discrete cosine and Fourier transforms in decorrelating sampled signals with Markov-1 statistics. It is shown that the discrete cosine transform (DCT) offers a higher (or equal) effectiveness than the discrete Fourier transform (DFT) for all values of the correlation coefficient. The mean residual correlation is shown to vanish as the inverse square root of the sample size.**

I. INTRODUCTION

In a recent paper Ahmed *et al.* [1] proposed a new transform called discrete cosine transform (DCT) and, based on empirical evidence, conjecture that its performance is closer to the optimal Karhunen–Lóeve transform (KLT) than the other commonly used transforms (i.e., discrete Fourier, Walsh–Hadamard, Haar). Means *et al.* [2] actually used the DCT for encoding TV pictures in real time.

Pearl showed [3] that for a signal statistic characterized by a covariance matrix T, $|T - T_U|^2$ (to be defined later) constitutes a measure of performance for a transform U, in the sense that the error bounds (in coding and filtering) are increasing functions of $|T - T_U|^2$.

The purpose of this investigation is to determine the relation between $|T - T_C|^2$ (the norm obtained using the DCT) and $|T - T_F|^2$ (the norm obtained using the discrete Fourier transform (DFT)), thus testing the conjecture of Ahmed and his collaborators.

II. DEFINITIONS AND NOMENCLATURE

Let T be a Töeplitz matrix and U an orthogonal transform. Let $T' = UTU^{-1}$ be the representation of T in the new basis, and $T'_U = \text{diag}\,(T'_{11}, T'_{22}, \cdots, T'_{ii}, \cdots, T'_{MM})$. We define T_U to be the representation of T'_U in the first basis, i.e.,

$$T_U = U^{-1} T'_U U$$

and $|T - T_U|^2$ the Hilbert–Schmidt norm of $T - T_U$, i.e.,

$$|T - T_U|^2 = \frac{1}{M}\left(\sum_{m,n=0}^{M-1} |(T - T_U)_{mn}|^2\right).$$

The cosine transform representation of a Töeplitz matrix T is given by CTC^{-1} where C is an $M \times M$ matrix defined by

$$\begin{cases} C_{oj} = \dfrac{2}{M}\dfrac{1}{\sqrt{2}} & j = 0, 1, \cdots, M-1 \\[2ex] C_{kj} = \dfrac{2}{M}\cos\dfrac{k\pi}{2M}(2j+1) & k = 1, 2, \cdots, M-1 \\ & j = 0, 1, \cdots, M-1. \end{cases}$$

A simple algebraic manipulation shows

$$CC^* = C^*C = \frac{2}{M} I$$

(where X^* indicates the complex conjugate transpose of X). Hence

$$C^{-1} = \frac{M}{2} C^*.$$

In contrast, the DFT is defined by a unitary matrix F where

$$F_{kj} = M^{-1/2} \exp\left(i\frac{2\pi}{M}kj\right) \qquad k, j = 0, 1, \cdots, M-1.$$

III. COMPARISON OF $|T - T_C|^2$ AND $|T - T_F|^2$

For any orthogonal matrix U (e.g., $U = C$ or $U = F$) we have

$$|T - T_U|^2 = |T' - T'_U|^2 = |T'|^2 - \frac{1}{M}\sum_{m=0}^{M-1} |(UTU^{-1})_{mm}|^2$$

$$= |T|^2 - \frac{1}{M}\sum_{m=0}^{M-1} |(UTU^{-1})_{mm}|^2,$$

i.e., the higher the norm of the diagonal vector of the transformed matrix, the lower $|T - T_U|$ and the better the transform. Hence, to compare $|T - T_C|^2$ and $|T - T_F|^2$, it suffices to compare

$$\sum_{m=0}^{M-1} |(CTC^{-1})_{mm}|^2 \text{ and } \sum_{m=0}^{M-1} |(FTF^{-1})_{mm}|^2.$$

We consider matrices of the form

$$T = \begin{pmatrix} 1 & \rho & \rho^2 \cdots \rho^{M-1} \\ \rho & 1 & \cdots \rho^{M-2} \\ \vdots & & \ddots \quad \vdots \\ \rho^{M-1} & & \cdots \quad 1 \end{pmatrix}$$

which represent covariance matrices of Markov-1 signals, with $0 \leqslant \rho \leqslant 1$ being the covariance coefficient between adjacent samples.

Clearly, for $\rho = 0$ and $\rho = 1$ the cosine and Fourier transforms are equivalent, since T is diagonal in both representations. For an intermediate value of ρ we obtained

$$(CTC^{-1})_{00} = \frac{1+\rho}{1-\rho} - \frac{2}{M}\frac{\rho(1-\rho^M)}{(1-\rho)^2}$$

and

$$(CTC^{-1})_{mm} = \frac{1}{2}\left(\frac{e^{i\alpha}+\rho}{e^{i\alpha}-\rho} + \frac{e^{-i\alpha}+\rho}{e^{-i\alpha}+\rho}\right) - \frac{\rho(1-(-1)^m\rho^M)}{M} \cdot \left(\frac{e^{-i(\alpha/2)}}{e^{-i\alpha}-\rho} + \frac{e^{i(\alpha/2)}}{e^{i\alpha}-\rho}\right)^2$$

for $m \neq 0$, where $\alpha = m\pi/M$.

An elementary (but tedious) computation leads to

$$\sum_{m=0}^{M-1} |(CTC^{-1})_{mm}|^2 = \frac{M(1+\rho^2)}{1-\rho^2} - \frac{4\rho^2}{(1-\rho^2)^2} + \frac{2(1-\rho^{2M})\rho^2}{M(1-\rho^2)^3}[3(1+\rho^2)+4\rho] - \frac{4\rho^2(1-\rho^M)^2}{M^2(1-\rho)^4}$$

for $M = 2k$, $k > 1$ (i.e., M even $\geqslant 4$).

Combined with

$$M|T|^2 = M\frac{1+\rho^2}{1-\rho^2} - \frac{2\rho^2(1-\rho^{2M})}{(1-\rho^2)^2},$$

we finally obtain the desired norm

$$M|T - T_c|^2 = \frac{2\rho^2(1+\rho^{2M})}{(1-\rho^2)^2} - \frac{2\rho^2(1-\rho^{2M})}{M(1-\rho^2)^3}(3+4\rho+3\rho^2) + \frac{4\rho^2(1-\rho^M)^2}{M^2(1-\rho)^4}.$$

Manuscript received December 17, 1975; revised March 5, 1976. This work was supported in part by the Naval Undersea Center, San Diego, CA, under Contract N66001-75-0-226-MJE.

The authors are with the School of Engineering and Applied Science, University of California, Los Angeles, CA 90024.

Note that

$$\lim_{M \to \infty} |T - T_c| = 0$$

implying that the DCT is asymptotically equivalent [4] to the Karhunen–Lóeve transform (KLT) of Markov-1 processes. Moreover, since for large M and $\rho \neq 1$, we have

$$|T - T_c| = \sqrt{2}\, \frac{\rho}{1 - \rho^2}\, O(M^{-1/2})$$

we conclude that the degradation in performance in filtering and coding [3] vanishes like $M^{-1/2}$.

An asymptotic equivalence between the KLT and the DCT was also argued by Shanmugam [5] using a circulant extension of T. His argument, however, remains incomplete as the relation between the DCT and the circulant matrices used in [5] is rather unclear.

In order to calculate $|T - T_F|^2$, recall [3] that for $T_{ij} = t(|i - j|)$ we have

$$(T - T_F)_{ij} = \frac{|i - j|}{M}\,[t(|i - j|) - t(M - |i - j|)]$$

and substituting $t(|i - j|) = \rho^{|i-j|}$ we obtain

$$M|T - T_F|^2 = \frac{2\rho^2(1 + \rho^{2M})}{(1 - \rho^2)^2} - \frac{2(1 + \rho^2)\rho^2(1 - \rho^{2M})}{M(1 - \rho^2)^3} - \frac{\rho^M(M^2 - 1)}{3}.$$

It shows that the asymptotic behavior of $|T - T_F|$ for large M is identical to that of $|T - T_c|$. Thus, the performance difference between the DCT and the DFT must vanish like M^{-1} Indeed, for large M one obtains the positive difference

$$|T - T_F|^2 - |T - T_c|^2 \cong \frac{4\rho^2}{M^2(1 - \rho^2)(1 + \rho)^2} \qquad \rho < 1,$$

indicating that the cosine transform is closer to optimal than the Fourier transform over the entire range of $0 < \rho < 1$.

For moderate values of M we should examine the expressions for $|T - T_F|^2$ and $|T - T_C|^2$ over the range $0 \leq \rho \leq 1$. The two are plotted, in a normalized form, in Fig. 1. We chose $|T - I|^2$ as a common normalizing factor, where I is the identity matrix, and so

$$|T - I|^2 = \frac{2\rho^2}{M(1 - \rho^2)^2}\,[M - 1 - M\rho^2 + \rho^{2M}].$$

It measures the degree of cross correlation contained in the unprocessed signal, and, therefore, the maximum amount of decorrelation that can be accomplished by any transform (i.e., the KLT). The ratio

$$\frac{|T - T_U|^2}{|T - I|^2}$$

represents the fractional correlation left "undone" by a transformation U.

Fig. 1 shows that for $M = 8$, 16, 64, and for the entire range of $0 < \rho < 1$, $|T - T_F|^2$ is higher than $|T - T_C|^2$. The difference between the two are quite noticeable, occasionally reaching a ratio of 2 : 1.

Conclusions

We established that the DCT is asymptotically equivalent to the KLT of Markov-1 signals and demonstrated that the rate of convergence is on the order of $M^{-1/2}$. $|T - T_C|^2$ is shown to be smaller than $|T - T_F|^2$ for all values of M and ρ, i.e., the discrete cosine transform offers a better approximation to the KLT of Markov-1 signals than the DFT.

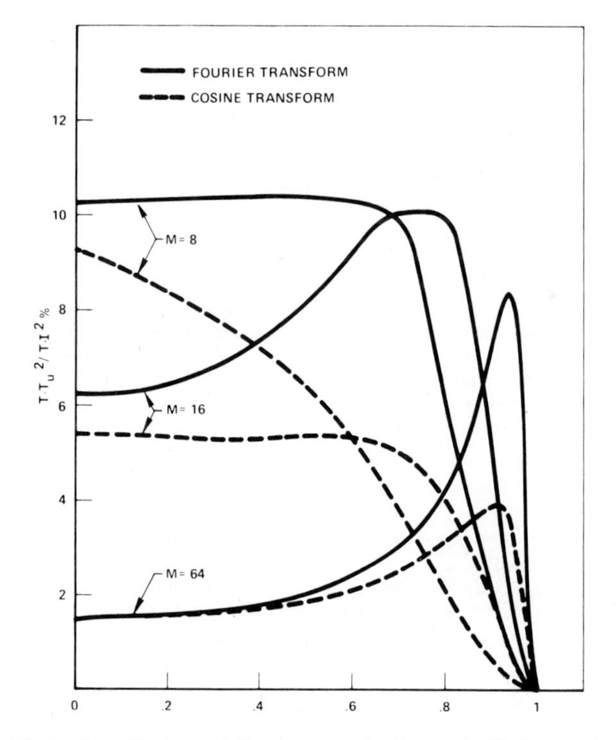

Fig. 1. Normalized correlation measures for Fourier (solid lines) and cosine (broken lines) transforms.

References

[1] N. Ahmed, T. Natarajan, and K. R. Rao, "Discrete cosine transform," *IEEE Trans. Comput.* (Corresp.), vol. C-23, pp. 90–93, Jan. 1974.

[2] R. W. Means, H. J. Whitehouse, and J. M. Speiser, "Television encoding using a hybrid discrete cosine transform and a differential pulse code modulator in real time," in *Proc. IEEE Nat. Telecommunications Conf.*, San Diego, CA, Dec. 1974.

[3] J. Pearl, "On coding and filtering stationary signals by discrete Fourier transforms," *IEEE Trans. Inform. Theory* (Corresp.), vol. IT-19, pp. 229–232, Mar. 1973.

[4] —, "Asymptotic equivalence of spectral representations," *IEEE Trans. Acoust., Speech, Signal Processing*, vol. ASSP-23, pp. 547–551, Dec. 1975.

[5] K. S. Shanmugam, "Comments on discrete cosine transform," *IEEE Trans. Comput.* (Corresp.), vol. C-24, p. 759, July 1975.

6

A Derivation for the Discrete Cosine Transform

MYRON D. FLICKNER AND NASIR AHMED

***Abstract*—The purpose of this letter is to derive the discrete cosine transform (DCT) as a limiting case of the Karhunen-Loève transform (KLT) of a first-order Markov process, as the correlation coefficient approaches 1.**

I. Introduction

The discrete cosine transform (DCT) was introduced in 1974 by Ahmed, Natarajan, and Rao [1], [2]. Since then it has been widely used to secure data compression via transform coding techniques, e.g., image [3] and speech compression [4]. The superior performance of the DCT in such applications is attributed to the basic property that its basis vectors "closely approximate" those of the Karhunen-Loève transform (KLT) of a first-order Markov process, for values of the correlation coefficient that are close to 1. In this letter we show that the DCT can be readily derived as the limiting case of the KLT of first-order Markov processes, as the correlation coefficient approaches 1.

II. First-Order Markov Process

Let $\{x_t\}$ be a wide-sense stationary random sequence, defined on the set of integers $T = \{1, 2, \cdots, N\}$ for some positive integer N, and having the statistical properties

$$E\{x_t\} = 0$$
$$E\{x_t^2\} = 1$$
$$E\{x_t x_s\} = r_{|t-s|}, \qquad s, t = 1, 2, \cdots, N \tag{1}$$

Manuscript received March 8, 1982.
M. D. Flickner was with the Department of Electrical Engineering, Kansas State University, Manhattan, KS. He is now with IBM Research, Computer Science Department, San Jose, CA 95193.
N. Ahmed is with the Department of Electrical Engineering, Kansas State University, Manhattan, KS 66506.

where E denotes statistical expectation, and $-1 < r_{|t-s|} < 1$ is the correlation coefficient. It follows that if

$$X' = (x_1, x_2, \cdots, x_N) \tag{2}$$

where the x_i are random variables, the corresponding data covariance matrix Σ_x is given by

$$\Sigma_x = E(XX') \tag{3}$$

where Σ_x is an $N \times N$ matrix. It is known that Σ_x is a symmetric positive definite matrix which has a Toeplitz form, i.e., all its elements along the northwest–southeast diagonals are equal.

We restrict our attention to the case

$$r_{|t-s|} = \rho^{|t-s|}, \qquad |\rho| < 1. \tag{4}$$

Then (1), (3), and (4) yield

$$\Sigma_x = \begin{bmatrix} 1 & \rho & \rho^2 & \cdots & \rho^{N-1} \\ \rho & 1 & \rho & \cdots & \rho^{N-2} \\ \cdot & & \ddots & & \\ \cdot & & & & \\ \rho^{N-1} & & \cdots & & 1 \end{bmatrix}, \qquad |\rho| < 1 \tag{5}$$

which is defined as the covariance matrix of a *first-order Markov process*.

Ray and Driver [5][1] have shown that the Karhunen–Loève expansion of first-order Markov processes for N even can be expressed in closed form as follows:

$$x(m) = \sum_{k=0}^{N-1} a(k) K_x(k) \sin\left\{\omega_k\left[m - \left(\frac{N-1}{2}\right)\right] + (k+1)\frac{\pi}{2}\right\},$$

$$0 \leq m, \quad k \leq N-1 \tag{6a}$$

where $K_x(k)$ is the kth KLT coefficient; $\{\omega_k\}$ is the positive roots of the transcendental equation

$$\tan(N\omega_k) = -\frac{(1-\rho^2)\sin\omega_k}{\cos\omega_k - 2\rho + \rho^2\cos\omega_k}. \tag{6b}$$

$$b(m,k) = \sin\left\{\omega_k\left[m - \left(\frac{N-1}{2}\right)\right] + (k+1)\frac{\pi}{2}\right\} \tag{6c}$$

are the orthogonal KLT basis functions whose normalization factors $a(k)$ are

$$a(k) = \sqrt{\frac{2}{N + \lambda_k}}, \qquad 0 \leq k \leq N-1 \tag{6d}$$

where λ_k is the kth eigenvalue of Σ_x in (5), given by

$$\lambda_k = \frac{1-\rho^2}{1 - 2\rho\cos\omega_k + \rho^2}, \qquad 0 \leq k \leq N-1. \tag{7}$$

III. DCT Derivation

We consider the limiting case when $\rho \to 1$. Then from (6b) it follows that the ω_k are roots of the simplified equation

$$\tan(N\omega_k) = 0, \qquad 1 \leq k \leq N-1 \tag{8}$$

where $0 < \omega_k < \pi$.

Equation (8) implies that

$$\omega_k = k\pi/N, \qquad 1 \leq k \leq N-1. \tag{9}$$

Again, (7) implies that the condition $\rho \to 1$ yields a set of $N-1$ eigenvalues that are zero, i.e.,

$$\lambda_k = 0, \qquad 1 \leq k \leq N-1. \tag{10}$$

Thus from (1) it is apparent that the normalization factors $a(k)$ in (6d) also simplify to yield

$$a(k) = \sqrt{\frac{2}{N}}, \qquad 1 \leq k \leq N-1. \tag{11}$$

Now, (9) implies that $\omega_k = k\pi/N$ for $1 \leq k \leq N-1$ are the roots of (6b). To show that $\omega_0 = 0$ is also a root of (6b) is difficult analytically, but is easily verified by resorting to numerical techniques such as the Newton iteration method. To illustrate, we plot the 8 roots of (6b) as ρ is varied from 0 to 1 in Fig. 1. We observe that ω_0 indeed moves to 0 as $\rho \to 1$. Hence the N roots of (6b) as $\rho \to 1$ are given by

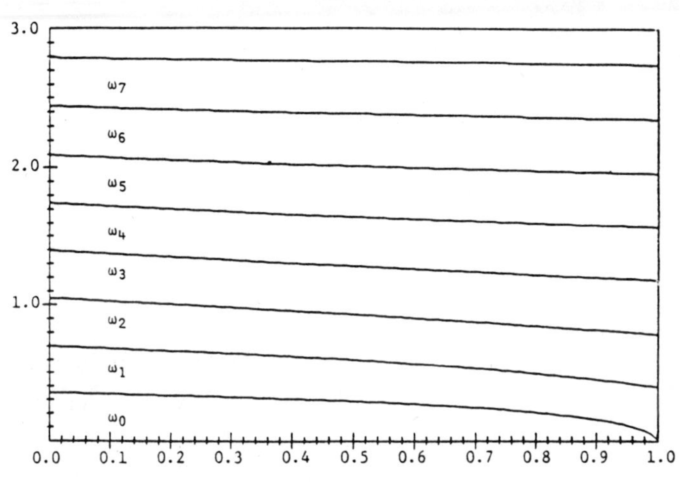

Fig. 1. The roots ω_k for $N = 8$.

$$\omega_k = k\pi/N, \qquad 0 \leq k \leq N-1. \tag{12}$$

Next, let us examine the KLT basis functions $b(m,k)$ in (6c) as $\rho \to 1$. To this end, substitution of the ω_k in (12) in (6c) leads to the following orthogonal simplified basis set:

$$b(m,k) = \cos\left[\frac{(2m+1)k\pi}{2N}\right], \qquad 0 \leq m, \quad k \leq N-1. \tag{13}$$

The corresponding normalization factors $a(k)$ are given by (11) for $1 \leq k \leq N-1$. To find $a(0)$, we use the property

$$\sum_{i=0}^{N-1} \lambda_i = \operatorname{tr}(\Sigma_x) = N \tag{14}$$

since the KLT is an orthonormal transformation. Equations (10) and (14) imply that

$$\lambda_0 = N \tag{15}$$

and hence (6d) yields

$$a_0 = \frac{1}{\sqrt{N}}. \tag{16}$$

Thus substitution of (11), (13), and (16) in (6a) leads to

$$x(m) = \frac{1}{\sqrt{N}} K_x(0) + \sqrt{\frac{2}{N}} \sum_{k=1}^{N} K_x(k) \cos\left[\frac{(2m+1)k\pi}{2N}\right],$$

$$\text{for } 0 \leq m, \quad k \leq N-1. \tag{17}$$

It is clear that (17) is exactly the definition of the inverse DCT as defined in [1], [2] with $K_x(k)$ being the kth DCT coefficient. Since the basis set $b(m,k)$ in (13) is orthogonal with normalization factors $a(k)$ in (11) and (16), the DCT is given by (17) to be

$$K_x(0) = \frac{1}{\sqrt{N}} \sum_{m=0}^{N-1} x(m)$$

and

$$K_x(k) = \sqrt{\frac{2}{N}} \sum_{m=0}^{N-1} x(m) \cos\left[\frac{(2m+1)k\pi}{2N}\right], \qquad 1 \leq k \leq N-1 \tag{18}$$

which is the desired result.

IV. Concluding Remarks

The DCT can be derived as the limiting case of the KLT for first-order Markov processes as $\rho \to 1$. Further, plots of ω_k versus ρ (e.g., Fig. 1) for different values of N show that the ω_k are reasonably constant. As such, the DCT provides an excellent approximation to the KLT for

[1] In particular, see [5, example 1, p. 665].

values of ρ close to 1. Other values of ρ lead to a family of orthonormal transforms, as shown in [6]. Since the ω_k in (12) are equally spaced on the unit circle, the DCT can be computed via the fast Fourier transform (FFT), e.g., see [7].

ACKNOWLEDGMENT

The authors wish to thank Dr. D. Curtis of the Department of Mathematics, and Dr. P. Kirmser and D. Haran of the Electrical Engineering Department, Kansas State University, Manhattan, KS, for their assistance and valuable suggestions.

REFERENCES

[1] N. Ahmed, T. Natarajan, and K. R. Rao, "Discrete cosine transform," *IEEE Trans. Comput.*, vol. C-23, pp. 90-93, Jan. 1974.

[2] N. Ahmed and K. R. Rao, *Orthogonal Transforms for Digital Signal Processing*. New York: Springer, 1975, pp. 169-171.

[3] W. H. Chen and C. H. Smith, "Adaptive coding of monochrome and color images," *IEEE Trans. Commun.*, vol. COM-25, pp. 1285-1292, Nov. 1977.

[4] J. M. Tribolet and R. E. Crochiere, "Frequency domain coding of speech," *IEEE Trans. Acoust. Speech, Signal Processing*, vol. ASSP-27, pp. 512-530, Oct. 1979.

[5] W. D. Ray and R. M. Driver, "Further decomposition of the Karhunen-Loeve series representation of a stationary random process," *IEEE Trans. Informat. Theory*, vol. IT-16, pp. 663-668, Nov. 1970.

[6] A. K. Jain, "A sinusoidal family of unitary transforms," *IEEE Trans. Pattern Analysis Mach. Intell.*, vol. PAMI-1, pp. 356-365, Oct. 1979.

[7] J. Makhoul, "A fast cosine transform in one and two dimensions," *IEEE Trans. Acoust. Speech, Signal Processing*, vol. ASSP-28, pp. 27-34, Feb. 1980.

Editor's Comments on Papers 7, 8, and 9

RAPID TRANSFORM

Whereas all the other discrete transforms described in this volume are orthogonal, the rapid transform (RT) is not orthogonal (Paper 7). It has no inverse and the recovery of the original data requires additional information (Vlasenko and Rao, 1979). In spite of this, RT has been used in machine print and handwritten character recognition (Paper 7; Wang and Shiau, 1973; Milson and Rao, 1976; Paper 31). This can be attributed to some of its interesting properties, such as the invariance to circular shift and/or reflection of the data sequence and slight rotation of a 2-D pattern (Paper 7). This invariance property has been utilized by Schutte, Frydrychowicz, and Schroder (1980) for scene matching. Other variant sets have been developed by Burkhardt and Muller (1980). Various properties including generalization of RT have been outlined by Wagh and Kanetkar (1975, 1977) and Wagh (1975, 1977).

RT has a fast algorithm and its flow-graph is similar to that of the Walsh-Hadamard transform (WHT) (Paper 15). For an N-point data sequence, the fast algorithm requires $N \log_2 N$ additions/subtractions and has an "in-place" structure. Improved algorithms for efficient implementation of RT have also been developed by Ulman (Paper 8) and Kunt (Paper 9).

REFERENCES

Burkhardt, H., and X. Muller, 1980, On Invariant Sets of Certain Class of Fast Translation-Invariant Transforms, *IEEE Acoust., Speech, Signal Process. Trans.* **ASSP-28:**517–523.

Milson, T. E., and K. R. Rao, 1976, A Statistical Model for Machine Print Recognition, *IEEE Syst., Man, Cybern. Trans.* **SMC-6:**671–678.

Schutte, H., S. Frydrychowicz, and J. Schroder, 1980, *Scene Matching with Translation*

Invariant Transforms, IEEE 5th International Conference on Pattern Recognition Proceedings, Miami Beach, Florida, pp. 195–198.

Vlasenko, V., and K. R. Rao, 1979, Unified Matrix Treatment of Discrete Transforms, *IEEE Comput. Trans.* **C-28:**934–938.

Wagh, M. D., 1975, Periodicity in *R*-Transformation, *Inst. Electron. Telecommun. Eng. J.* **21:**560–561.

Wagh, M. D., 1977, An Extension of *R*-Transform to Patterns of Binary Lengths, *Int. J. Comput. Math.*, sec. B., **7:**1–12.

Wagh, M. D., and S. V. Kanetkar, 1975, A Multiplexing Theorem and Generalization of R-Transform, *Int. J. Comput. Math.*, sec. A, **5:**163-171.

Wagh, M. D., and S. V. Kanetkar, 1977, A Class of Translation Invariant Transforms, *IEEE Acoust., Speech, Signal Process. Trans.* **ASSP-25:**203–205.

Wang, P. P., and R. C. Shiau, 1973, Machine Recognition of Printed Chinese Characters Via Transformation Algorithms, *Pattern Recognition* **5:**303–321.

7

Reprinted from *Inf. Control* **15:**130–154 (1969)

A Transformation with Invariance under Cyclic Permutation for Applications in Pattern Recognition

H. REITBOECK AND T. P. BRODY

Westinghouse Electric Corp., Research and Development Center, Churchill Boro., Pittsburgh, Pennsylvania 15235

The paper describes a transformation that can be used to characterize patterns independent of their position. Examples of the application of the transform for the machine recognition of letters are discussed. The program succeeded in a recognition rate of 80–100% for letters having different position, distortions, inclination, rotation up to 15° and size variation up to 1:3 relative to a reference set of 10 letters. Results with a program for the autonomous learning of new varieties of a pattern (using a learning matrix as an adaptive classifier) are given. When executed on a digital computer, this transform is 10-100 times faster than the fast Fourier transform (depending on the number of sampling points).

LIST OF SYMBOLS

$\mathbf{W}_j$ = Stored pattern vector (pattern class j)
d_{ij} = Euclidean distance between vectors
$\mathbf{Z}_i$ = Pattern vector (pattern class i)
Z_{ij} = Component j of pattern vector i
$\mathbf{X}$ = A selected pattern vector
X_i = Component i of pattern vector $\mathbf{X}$
$X_i^{(R)}$ = Component i of transformed pattern vector $\mathbf{X}$ after R transformation steps, represented by column (R) in the diagrams
$X_{m,n}^{(R)}$ = Component (m) in group (n) of column (R) (for one dimensional transform)
$X_{i,j}^{(R)}$ = Element in row (i), column (j) of layer (R) (for two dimensional transform)
N = 2^M = Dimensionality of pattern vector, resp. number of variables in a column

$W(l)$ = Complex factor, $W(l) = \exp\left(j\frac{2\pi l}{N}\right)$

P = Permutation operator, defined by Eq. C.2.

1. INTRODUCTION

Recent success in several areas of pattern recognition due to the application of Fourier transforms (performed either optically or digitally with the algorithm of the fast Fourier transform (Cooley and Tukey, 1965) led us to search for simpler (i.e., faster) algorithms for various special problems in the machine recognition of patterns. An algorithm has been developed that has been successfully used for the recognition of letters. When executed on a digital computer, this algorithm needs only a fraction of the computation time needed for the fast Fourier transform (FFT). In spite of its simplicity, this algorithm is practical for the machine recognition of letters and digits; it even seems to cope better with certain distortions that are characteristic of handprinted letters and digits (inclination, small rotation, etc.) than does the standard two-dimensional Fourier transform.

2. DISCUSSION

This section describes a new transform with invariance under cyclic permutation that can be used to characterize patterns independent of their position in the pattern space. Several properties of the transform are discussed and a proof for the shift invariance is given (see Section 2.4 and Appendix A). Its applicability for classifying varieties of handprinted characters of various size, inclination, distortion, and small rotation is demonstrated.

2.1 PROPERTIES REQUIRED OF THE TRANSFORM

One basic requirement in the machine recognition of patterns is the ability to operate on the input function (i.e., the pattern to be recognized) in such a manner that a characteristic output function is generated that is independent of the position of the pattern in its (normally two-dimensional) pattern space.

Such operation T should, therefore, transform functions $f(x, y)$ into functions $T\{f\} = g(u, v)$, defined over a certain range, with the following properties:

If $T\{f\} = g(u, v)$, and if $f(x + h, y + k) = f_1(x, y)$, and $T\{f_1\} = g_1(u, v)$, then $g_1(u, v) = g(u, v) \cdot m(h, k)$ where either $m(h, k) \equiv 1$ or

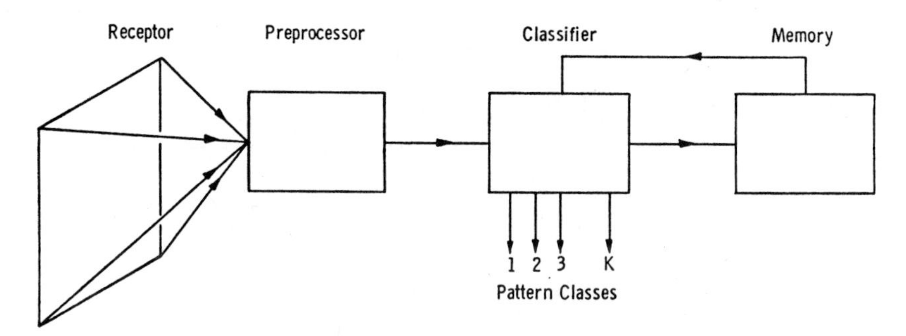

FIG. 1. Basic diagram of the receptor

$m(h, k)$ is a function of modulus 1 so that

$$| T\{f_1(x, y)\} | = | g(u, v) | = | T\{f(x, y)\} |.$$

A transformation of particular interest having these properties is the Fourier transform; it has been widely used in pattern recognition since economic methods for its execution became available, such as optical methods, and the use of a fast algorithm for the machine calculation of the discrete Fourier transform: the fast Fourier transform.

Besides shift invariance, the transform should have other properties. The transformed function $g(u, v)$ should be immune to (a) a limited amount of noise and distortion of the input pattern, (b) a limited angle of rotation of the input pattern with respect to some reference axis, and (c) shearing of the input patterns (required for the recognition of hand-printed characters). The transform should also be able to classify patterns independent of their size.

A new transformation which we have successfully used for the recognition of handprinted characters is denoted as "rapid transformation" or "R-transform". The R-transform fulfills the above requirements.

2.2 The Perceptor

The applicability of the R-transform for pattern recognition purposes has been tested with a computer simulation of a "perceptor" as shown in Fig. 1. This perceptor consists of a "receptor", a preprocessor, and a classifier/memory system. In most of the experiments, the receptor had a resolution of 16×16 elements. The input patterns are written on cards with zeroes and ones.[1]

[1] The transform is also applicable when the variables assume continuous values (patterns having grey values); however, no such patterns have been tested yet.

In the next stage, the preprocessor calculates the R-transform of the input pattern. The result is a square matrix of order N; a part of these N^2 values is selected (in our examples it was $N^2/4$ resp. $N^2/16$, i.e., 64, resp. 16 out of 256 values) and passed for further processing to the classifier/memory system.

The classifier/memory system is a computer simulation of the Steinbuch learning matrix, which calculates the Euclidean distance between the unknown "pattern vector" $\mathbf{Z}_0$ (i.e., a multidimensional vector having the selected matrix elements as its coordinates) and all stored pattern vectors $\mathbf{W}_i \,|_{i=1}^{k}$ that represent the k pattern classes.

The distance between $\mathbf{Z}_0$ and $\mathbf{W}_i$ is given by

$$|d_{0,i}| = |\mathbf{Z}_0 - \mathbf{W}_i| = \sum_{j=0}^{N-1} Z_{0,j}^2 + \sum_{j=0}^{N-1} W_{i,j}^2 - 2\sum_{j=0}^{N-1} Z_{0,j} W_{i,j} \qquad (1)$$

and a minimum evaluator selects that pattern vector $\mathbf{W}_m$ which has the smallest distance, $|d_{0,m}|$, from $\mathbf{Z}_0$. The $\mathbf{Z}_0$ is then classified as belonging to pattern class m, represented by the pattern vector $\mathbf{W}_m$.

In the operation mode of a "closed learning loop," the classified vector $\mathbf{Z}_0$ influences the stored representative vector $\mathbf{W}_m$. When the learning loop is open, $\mathbf{W}_m$ remains unchanged.

For evaluation of the pattern vector $\mathbf{W}_m$ that has minimum distance to $\mathbf{Z}_0$, it is unnecessary to calculate $|d_{0,i}|$ according to Eq. 1. Since $\sum_{j=0}^{N-1} Z_{0,j}^2$ is a constant term in all $d_{0,i}\,|_{i=1}^{k}$, it is irrelevant for the evaluation of the minimum and can therefore be omitted.

In the following examples, the maximum of the values

$$C_{0,i}\Big|_{i=1}^{k} = \left(\sum_{j=0}^{N-1} Z_{0,j} W_{i,j}\right) - \left(\frac{1}{2}\sum_{j=0}^{N-1} W_{i,j}^2\right), \qquad (2)$$

has been used for the classification, giving identical results with the minimum Euclidean distance (MED) classification.

The $\sum_{j=0}^{N-1} Z_{0,j} W_{i,j}$ represents the scalar product of the pattern vectors $\mathbf{Z}_0$ and $\mathbf{W}_i$ or the cross correlation between the sets $\{Z_{0,0}, Z_{0,1} \cdots Z_{0,N-1}\}$ and $\{W_{i,0}, W_{i,1} \cdots W_{i,N-1}\}$

2.3 The Transform

2.3.1 Algorithm A

The principle of the transform for a one-dimensional array of eight input variables, representing the components of an 8 dimensional pattern vector **X**, is shown in Fig. 2.

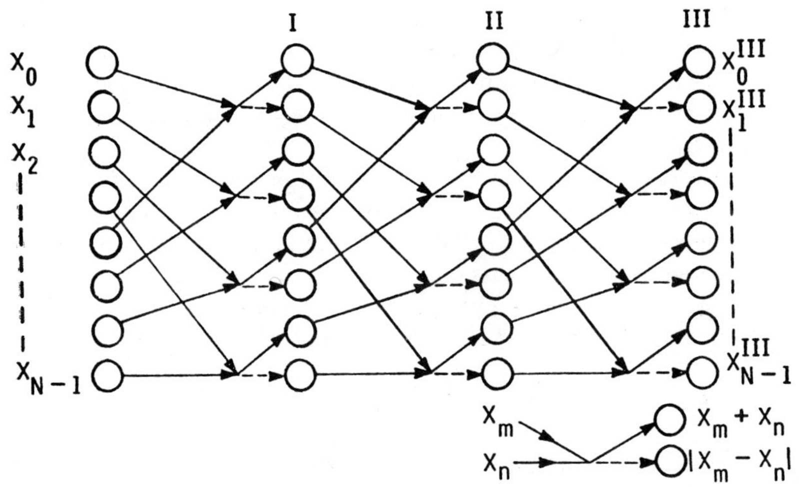

FIG. 3. Tree-graph of R-transform, Algorithm B for 8 input variables

For an array of $N = 2^M$ input variables (numbered from 0 to $N - 1$), M transformation steps are required.

In the first transformation step the input variables are divided into two groups, numbered from 0 to $(N/2) - 1$ and from $N/2$ to $N - 1$. The variables of the first transform column (1) are then calculated by

$$
\begin{aligned}
X_i^{(1)} &= X_i + X_{i+N/2} \quad \Big|_{i=0}^{N/2-1} \\
X_{i+N/2}^{(1)} &= |X_i - X_{i+N/2}| \quad \Big|_{i=0}^{N/2-1} .
\end{aligned}
\tag{3.1}
$$

In the second transformation step the two groups of the variables in column (1) are again divided into two subgroups each, giving the variables of column (2) by

$$
\begin{aligned}
X_i^{(2)} &= X_i^{(1)} + X_{i+N/4}^{(1)} \quad \Big|_{i=0}^{N/4-1} \\
X_{i+N/4}^{(2)} &= |X_i^{(1)} - X_{i+N/4}^{(1)}| \quad \Big|_{i=0}^{N/4-1} \\
X_{i+N/2}^{(2)} &= X_{i+N/2}^{(1)} + X_{i+3N/4}^{(1)} \quad \Big|_{i=0}^{N/4-1} \\
X_{i+3N/4}^{(2)} &= |X_{i+N/2}^{(1)} - X_{i+3N/4}^{(1)}| \quad \Big|_{i=0}^{N/4-1} .
\end{aligned}
\tag{3.2}
$$

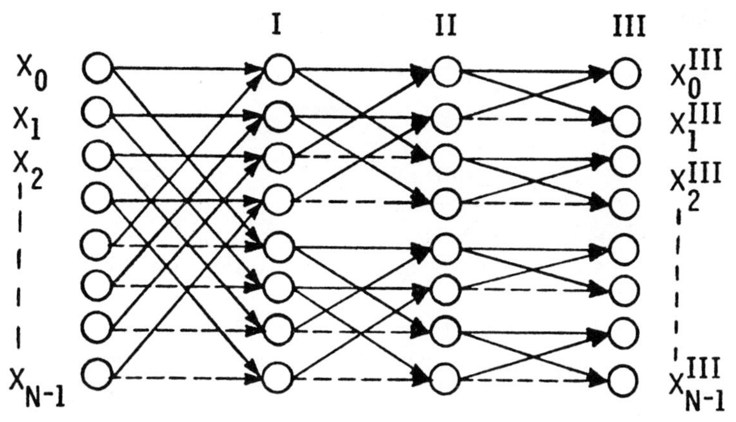

FIG. 2 Tree-graph of R-transform, Algorithm A, for 8 input variables

This procedure is repeated $M(= \log_2 N)$ times.

In general, the variables in any column (R) are calculated from the variables in the preceding column $(R - 1)$ by

$$X_{m+2ns}^{(R)} = |X_{m+2ns}^{(R-1)} + X_{m+(2n+1)s}^{(R-1)}| \qquad 4.1$$

$$X_{m+(2n+1)s}^{(R)} = |X_{m+2ns}^{(R-1)} - X_{m+(2n+1)s}^{(R-1)}| \Bigg|_{m=0}^{s-1} \Bigg|_{n=0}^{t-1} \qquad 4.2$$

with $s = 2^{M-R}$ and $t = 2^{R-1}$. 4.3

These operations are illustrated in Fig. 2 for $N = 8$. The dotted arrows indicate subtraction, while the solid arrows indicate addition. The absolute value is then taken at each cell.

2.3.2. Algorithm B[2]

A second algorithm, giving the same results[3] is shown in Fig. 3. The number of the transform steps required for N input variables $X_0 \cdots X_{N-1}$ is again $M = \log_2 N$.

In every transform step the variables of the new column (R) are

[2] The tree graphs for algorithm A and B of the R-transform are identical with two (out of several possible) tree graphs for the execution of the Hadamard transform (Hadamard, 1893). The R-transform and the Hadamard transform differ, however, in the arithmetic operations in the nodal points and in their general properties.

[3] See Appendix C.

calculated from the variables of the preceding column $(R - 1)$ by

$$\left. \begin{aligned} X_{2i}^{(R)} &= |X_i^{(R-1)} + X_{i+N/2}^{(R-1)}| \\ X_{2i+1}^{(R)} &= |X_i^{(R-1)} - X_{i+N/2}^{(R-1)}| \end{aligned} \right|_{i=0}^{N/2-1} . \tag{5}$$

The advantage of this algorithm is that the same operations are performed in every transform step.

For two-dimensional patterns one can use either two one-dimensional transforms in sequence for x and y, or a two-dimensional transform as described in 2.3.3.

If two one-dimensional transforms are used, either the rows or the columns of the input field will be "floating," depending on whether the transform in x or y is carried out first, i.e., if the transform in y is done first, the output remains unchanged not only when the whole pattern is shifted but also when a relative shift along single lines exists.

This can be of advantage for the recognition of handwritten patterns that have a varying angle of inclination.

2.3.3 Algorithm AA

An algorithm for a two-dimensional transform on the basis of Algorithm **A**, however, with every transform column (R) replaced by a two dimensional transform layer (R), is given below. Instead of two variables of every column (R), as in the one dimensional case, this algorithm uses four variables of every layer (R) for the calculation of every variable in the subsequent layer. Assume the variables of the input field (layer (0)) form the array

$$\begin{pmatrix} X_{0,0} & \cdots & X_{0,j} & \cdots & X_{0,N-1} \\ | & & | & & | \\ | & & | & & | \\ | & & | & & | \\ X_{i,0} & \cdots & X_{i,j} & \cdots & X_{i,N-1} \\ | & & | & & | \\ | & & | & & | \\ | & & | & & | \\ X_{N-1,0} & \cdots & X_{N-1,j} & \cdots & X_{N-1,N-1} \end{pmatrix} .$$

The corresponding variables of the subsequent layers are again denoted by $X_{i,j}^{(1)}$ for the first transform layer, $X_{i,j}^{(2)}$ for the second layer, etc.

The variables of the first transform layer are then given by

$$X^{(1)}_{i,j} = X_{i,j} + X_{i+N/2,j} + X_{i,j+N/2} + X_{i+N/2,j+N/2} \qquad \Big|_{i,j=0}^{N/2-1}$$

$$X^{(1)}_{i+N/2,j} = |X_{i,j} - X_{i+N/2,j}| + |X_{i,j+N/2} - X_{i+N/2,j+N/2}| \qquad \Big|_{i,j=0}^{N/2-1}$$

$$X^{(1)}_{i,j+N/2} = |(X_{i,j} + X_{i+N/2,j}) - (X_{i,j+N/2} + X_{i+N/2,j+N/2})| \qquad \Big|_{i,j=0}^{N/2-1}$$

$$X^{(1)}_{i+N/2,j+N/2} = \big|\,|X_{i,j} - X_{i+N/2,j}| - |X_{i,j+N/2} - X_{i+N/2,j+N/2}|\,\big| \qquad \Big|_{i,j=0}^{N/2-1}$$

The variables of layer (2) are calculated by dividing the variables of layer (1) into subgroups of variables that are $N/4$ elements apart, in analogy to Eq. 2, and so forth, until $M = \log_2 N$ transformation steps have been executed. Algorithm AA for a two dimensional array of 8×8 input variables is shown in Fig. 4. A similar two-dimensional transform can be derived on the basis of Algorithm B.

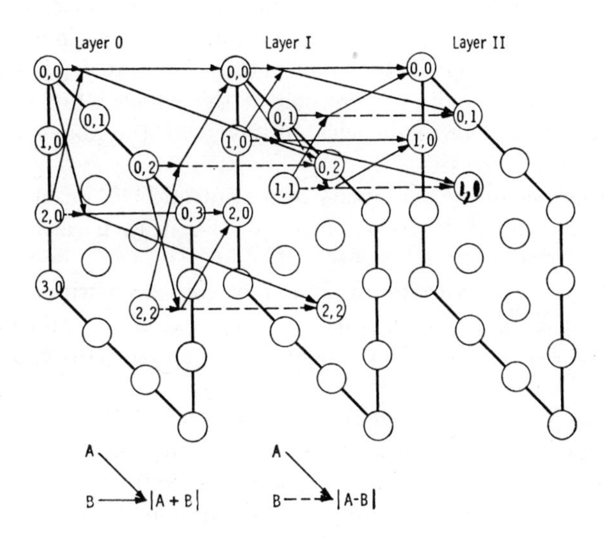

FIG. 4. Two dimensional R-transform for field of 4×4 input variables

2.4 Properties of the Transform

2.4.1 Invariance

2.4.1.1 *Cyclic permutation* The transform is invariant under cyclic permutation, i.e., $T\{f(X_0, X_1, X_2 \cdots X_i \cdots X_{N-1})\} = T\{f(X_h, X_{h+1}, X_{h+2} \cdots X_{h+i} \cdots X_{h+N-1}\}$ with h being any integer and $X_{N+k} \equiv X_k$. The transformed data field does not, therefore, depend on shifts of the input pattern in x (and y) (see Figs. 5a, b, and c). For a general proof, see Appendix A.

2.4.1.2 *Reflection* The transform is invariant under reflection,[4] i.e.

$$T\{f(X_0, X_1, X_2 \cdots X_i \cdots X_{N-1})\} = T\{f(X_{N-1}, X_{N-2}, \cdots X_{N-i-1} \cdots X_0)\}.$$

2.4.2 Periodicity of the Output Function

The transform of a one-dimensional array of N input variables is periodic over its basic interval if $N/2$ or more consecutive input variables are zero (Note: Since $X_i \equiv X_{N+i} \rightarrow X_0 \equiv X_N$; hence X_{N-1} and X_0 are consecutive variables.) If the input function contains Z variables that are zero, so that $2^{M-1} \geqq 2^K \geqq N - Z > 2^{K-1}$, then the transformed function is periodic over its basic interval with a frequency of 2^{M-K}.

For two-dimensional patterns the output field is periodic in i and/or j if the linear dimension of the input pattern in i and/or j is smaller than half of the corresponding dimension of the whole input field. This is shown in Fig. 5. The frequencies are given by the above equation for one-dimensional arrays.

Via this check of the periodicity of the output field, input patterns can be identified independently of powers of two in their size. This is done as follows:

The value of the variable $X_{0,0}^{(M)}$ is first compared with the variables $X_{0,N/2}^{(M)}$, and $X_{N/2,0}^{(M)}$. The equality of $X_{0,0}^{(M)}$ and $X_{0,N/2}^{(M)}$ and/or $X_{N/2,0}^{(M)}$ is taken as a criterion that the periodicity (in i and/or j) is two or a power of two.

Next, $X_{0,0}^{(M)}$ is compared with $X_{0,N/4}^{(M)}$ and $X_{N/4,0}^{(M)}$. If these are different, then a periodicity of two is assumed, and for the calculation of the cross correlation function every variable of one basic interval of the periodic field is then compared with every second variable of the unperiodic field. A corresponding procedure is used for higher periodicities.

[4] See Appendix B.

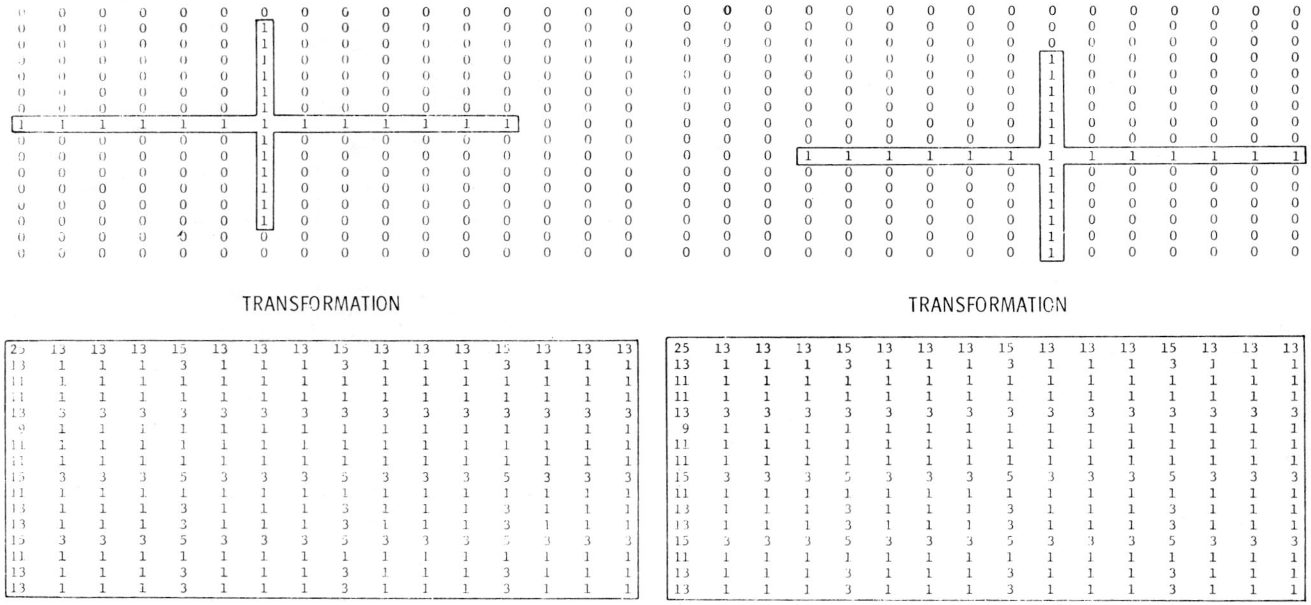

0	0	0	0	0	0	0	0	0	0	0	0	0	0	0	0
0	0	0	0	0	0	1	0	0	0	0	0	0	0	0	0
0	0	0	0	0	0	1	0	0	0	0	0	0	0	0	0
0	0	0	0	0	0	1	0	0	0	0	0	0	0	0	0
0	0	0	0	0	0	1	0	0	0	0	0	0	0	0	0
0	0	0	0	0	0	1	0	0	0	0	0	0	0	0	0
0	0	0	0	0	0	1	0	0	0	0	0	0	0	0	0
1	1	1	1	1	1	1	1	1	1	1	1	1	0	0	0
0	0	0	0	0	0	1	0	0	0	0	0	0	0	0	0
0	0	0	0	0	0	1	0	0	0	0	0	0	0	0	0
0	0	0	0	0	0	1	0	0	0	0	0	0	0	0	0
0	0	0	0	0	0	1	0	0	0	0	0	0	0	0	0
0	0	0	0	0	0	1	0	0	0	0	0	0	0	0	0
0	0	0	0	0	0	1	0	0	0	0	0	0	0	0	0
0	0	0	0	0	0	0	0	0	0	0	0	0	0	0	0
0	0	0	0	0	0	0	0	0	0	0	0	0	0	0	0

TRANSFORMATION

25	13	13	13	15	13	13	13	15	13	13	13	15	13	13	13
13	1	1	1	3	1	1	1	3	1	1	1	3	1	1	1
11	1	1	1	1	1	1	1	1	1	1	1	1	1	1	1
11	1	1	1	1	1	1	1	1	1	1	1	1	1	1	1
13	3	3	3	3	3	3	3	3	3	3	3	3	3	3	3
9	1	1	1	1	1	1	1	1	1	1	1	1	1	1	1
11	1	1	1	1	1	1	1	1	1	1	1	1	1	1	1
11	1	1	1	1	1	1	1	1	1	1	1	1	1	1	1
15	3	3	3	5	3	3	3	5	3	3	3	5	3	3	3
11	1	1	1	1	1	1	1	1	1	1	1	1	1	1	1
13	1	1	1	3	1	1	1	3	1	1	1	3	1	1	1
13	1	1	1	3	1	1	1	3	1	1	1	3	1	1	1
15	3	3	3	5	3	3	3	5	3	3	3	5	3	3	3
11	1	1	1	1	1	1	1	1	1	1	1	1	1	1	1
13	1	1	1	3	1	1	1	3	1	1	1	3	1	1	1
13	1	1	1	3	1	1	1	3	1	1	1	3	1	1	1

0	0	0	0	0	0	0	0	0	0	0	0	0	0	0	0
0	0	0	0	0	0	0	0	0	0	0	0	0	0	0	0
0	0	0	0	0	0	0	0	0	0	0	0	0	0	0	0
0	0	0	0	0	0	0	0	0	1	0	0	0	0	0	0
0	0	0	0	0	0	0	0	0	1	0	0	0	0	0	0
0	0	0	0	0	0	0	0	0	1	0	0	0	0	0	0
0	0	0	0	0	0	0	0	0	1	0	0	0	0	0	0
0	0	0	0	0	0	0	0	0	1	0	0	0	0	0	0
0	0	0	0	0	0	0	0	0	1	0	0	0	0	0	0
0	0	0	1	1	1	1	1	1	1	1	1	1	1	1	1
0	0	0	0	0	0	0	0	0	1	0	0	0	0	0	0
0	0	0	0	0	0	0	0	0	1	0	0	0	0	0	0
0	0	0	0	0	0	0	0	0	1	0	0	0	0	0	0
0	0	0	0	0	0	0	0	0	1	0	0	0	0	0	0
0	0	0	0	0	0	0	0	0	1	0	0	0	0	0	0
0	0	0	0	0	0	0	0	0	1	0	0	0	0	0	0

TRANSFORMATION

25	13	13	13	15	13	13	13	15	13	13	13	15	13	13	13
13	1	1	1	3	1	1	1	3	1	1	1	3	1	1	1
11	1	1	1	1	1	1	1	1	1	1	1	1	1	1	1
11	1	1	1	1	1	1	1	1	1	1	1	1	1	1	1
13	3	3	3	3	3	3	3	3	3	3	3	3	3	3	3
9	1	1	1	1	1	1	1	1	1	1	1	1	1	1	1
11	1	1	1	1	1	1	1	1	1	1	1	1	1	1	1
11	1	1	1	1	1	1	1	1	1	1	1	1	1	1	1
15	3	3	3	5	3	3	3	5	3	3	3	5	3	3	3
11	1	1	1	1	1	1	1	1	1	1	1	1	1	1	1
13	1	1	1	3	1	1	1	3	1	1	1	3	1	1	1
13	1	1	1	3	1	1	1	3	1	1	1	3	1	1	1
15	3	3	3	5	3	3	3	5	3	3	3	5	3	3	3
11	1	1	1	1	1	1	1	1	1	1	1	1	1	1	1
13	1	1	1	3	1	1	1	3	1	1	1	3	1	1	1
13	1	1	1	3	1	1	1	3	1	1	1	3	1	1	1

(a)

Fig. 5. Properties of the transform: (a) output unaffected by shift of input pattern (b) periodic output for small input patterns (c) transforms of patterns having a size ratio of 1:2.

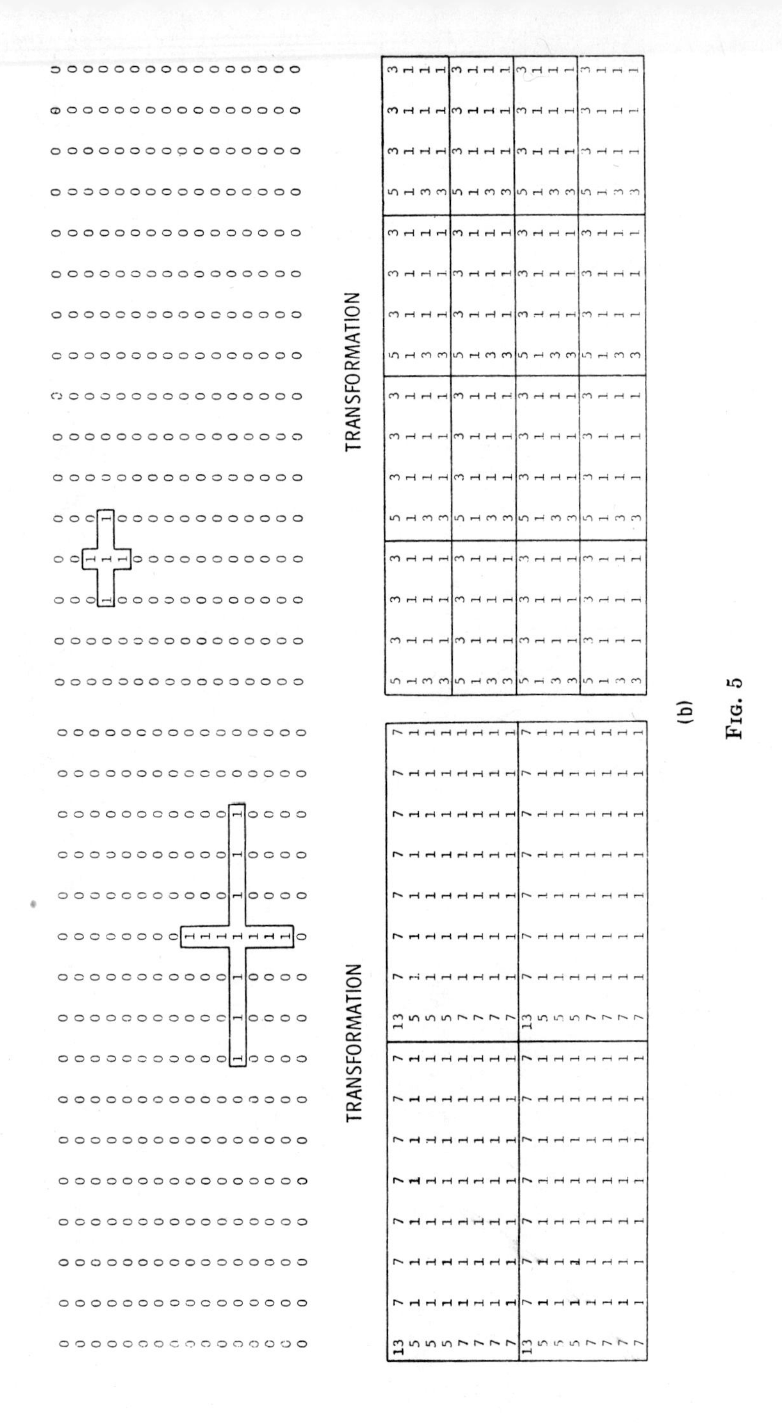

(b)

FIG. 5

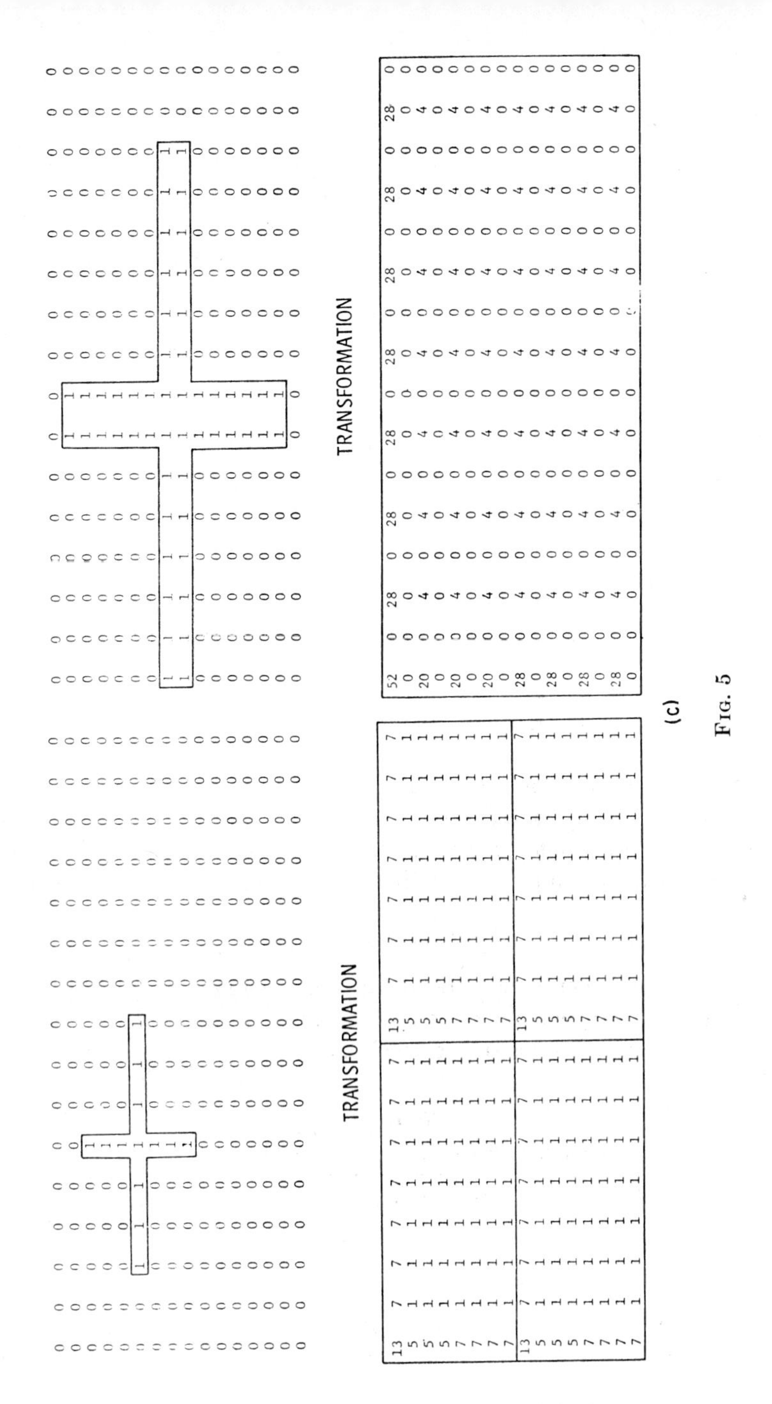

(c)

Fig. 5

Note that the equivalence of the variables in the above tests is a necessary, but insufficient criterion for the periodicity, and that the periodicity is a necessary but insufficient criterion for the size of the input pattern. Degenerate cases do exist where the test fails, but these are highly improbable for input patterns such as letters and digits.

2.4.3 Applicability

The transform is applicable for binary and for analog inputs.

2.4.4 Input Variables

The number of input variables must be a power of two $N = 2^M$; only a fraction of the N output variables is, however, sufficient to characterize the input function.

2.4.5 Binary Functions

For binary inputs ($X_i = {}^0_1$) the N input variables can represent 2^N input patterns. The corresponding number of output patterns is found to be

$$(N - 3) \log_2 N + \frac{N^2}{4} - \frac{5}{2} N + 10 \quad \text{for} \quad N > 2.$$

The number of input and output patterns for a one-dimensional array of $N = 4, 8, 16$, and 32 input variables is given in Table 1.

Table 1 shows that for increasing N the number of possible input patterns increases much faster than the number of output patterns.

2.4.6 Sensitivity

The output function can be made immune to certain distortions; for example, when two one dimensional transforms are used in sequence for x and y, it is insensitive to inclination and small rotation of the

TABLE 1

INPUT AND OUTPUT PATTERNS FOR A ONE-DIMENSIONAL ARRAY

Input Variables (N)	Possible input patterns (2^N)	Possible output patterns
4	16	6
8	256	21
16	65,536	86
32	$\sim 4.3 \times 10^9$	331

input pattern as long as the general shape of the pattern is conserved. This makes the transform especially suited for the recognition of hand-printed letters and digits, where it allows a correct classification, even in cases where the standard two dimensional Fourier transform fails.

A two-dimensional transform that uses four variables for the calculation of every new variable in the next layer (Algorithm AA) has the advantage of relative insensitivity to changes in the linewidth; however, it does not cope as well with inclination and rotation as do two one-dimensional transforms.

2.4.7 Limitations

1. The transform cannot discriminate between mirror images (see 2.4.1.2). This discrimination can be performed, however, (e.g. between 6 and 9) by utilizing additional information from one or two simple scans in the visual field.

2. The transform cannot be used to filter out details from complex images.

3. The transformation is relatively sensitive to changes in the width of patterns that consist predominantly of two parallel lines such as *H* and *U*. This is, however, also the case for the Fourier transform, since the spatial frequencies change drastically when the width of these letters is changed. In order to cope with this problem, we used two word lines in the learning matrix for the classification of these letters.

2.4.8 Speed

A main advantage of the transform in comparison with the fast Fourier transform is the short computation time, when executed on a digital computer. This is illustrated in Figs. 6 and 7.

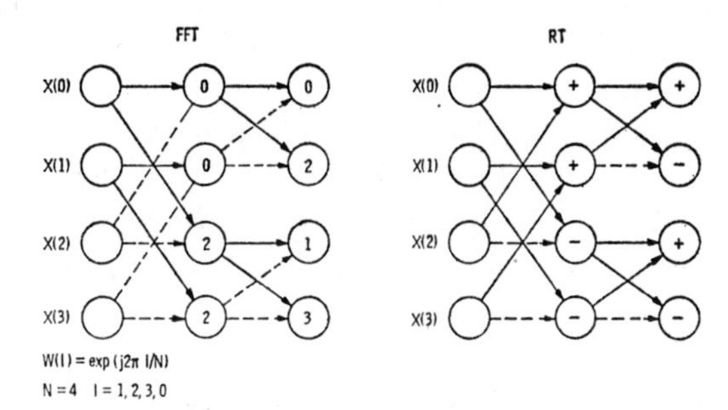

FIG. 6. Comparison of R-transform and fast Fourier transform

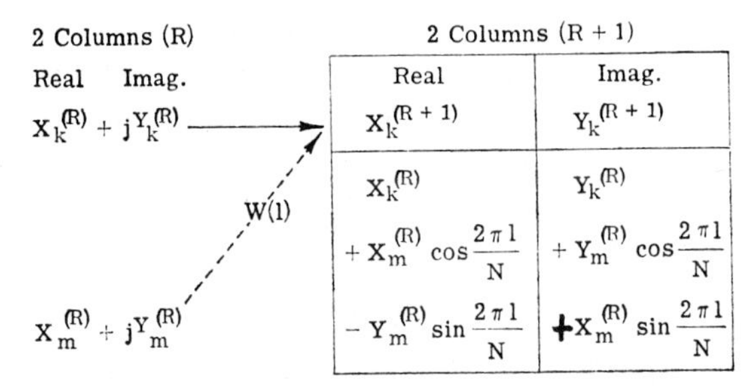

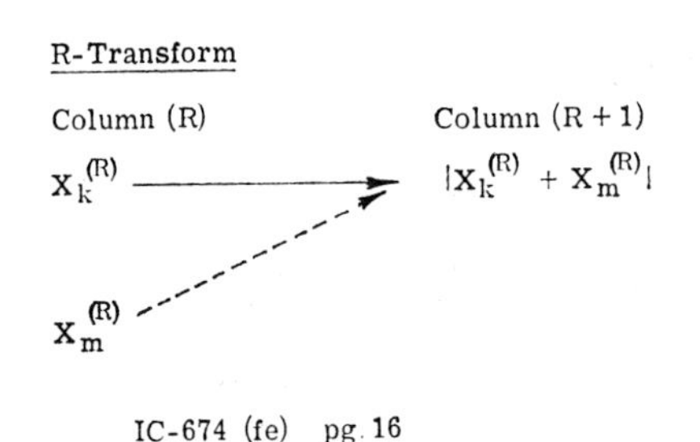

FIG. 7. Comparison of fast Fourier transform and R-transform operations

In the case of the FFT, the variables are complex, whereas in the R-transform all the variables have real values. The FFT therefore requires twice as much storage capacity as the R-transform.

In Fig. 6, the dotted line in the flow chart for the computation of the FFT indicate that the (complex) number that is shifted along this line must be multiplied by a complex factor $W(1)$, where 1 is the number in the corresponding cell.

For the calculation of every new variable, therefore, four multiplications and four additions are necessary. When one assumes $\sim$25 additions per multiplication,[5] then the calculation of every variable needs approximately 100 additions. In the case of the R-transform, the full drawn lines again indicate addition, while the dotted lines indicate subtraction.

[5] This, of course, depends on the size of the numbers to be multiplied, which in turn depends on the accuracy to which the phase factors have to be calculated, i.e., on the size of the input field. Our figure is based on an assumed accuracy of 5 (decimal) significant figures.

The R-transform, therefore, needs only one addition (or subtraction) for the calculation of every variable (See Fig. 7).

2.5 Results

Fig. 8 shows 10 standard patterns (the letters A, B, C, D, E, F, G, H, K, and M) that have been trained. The classification of distorted versions

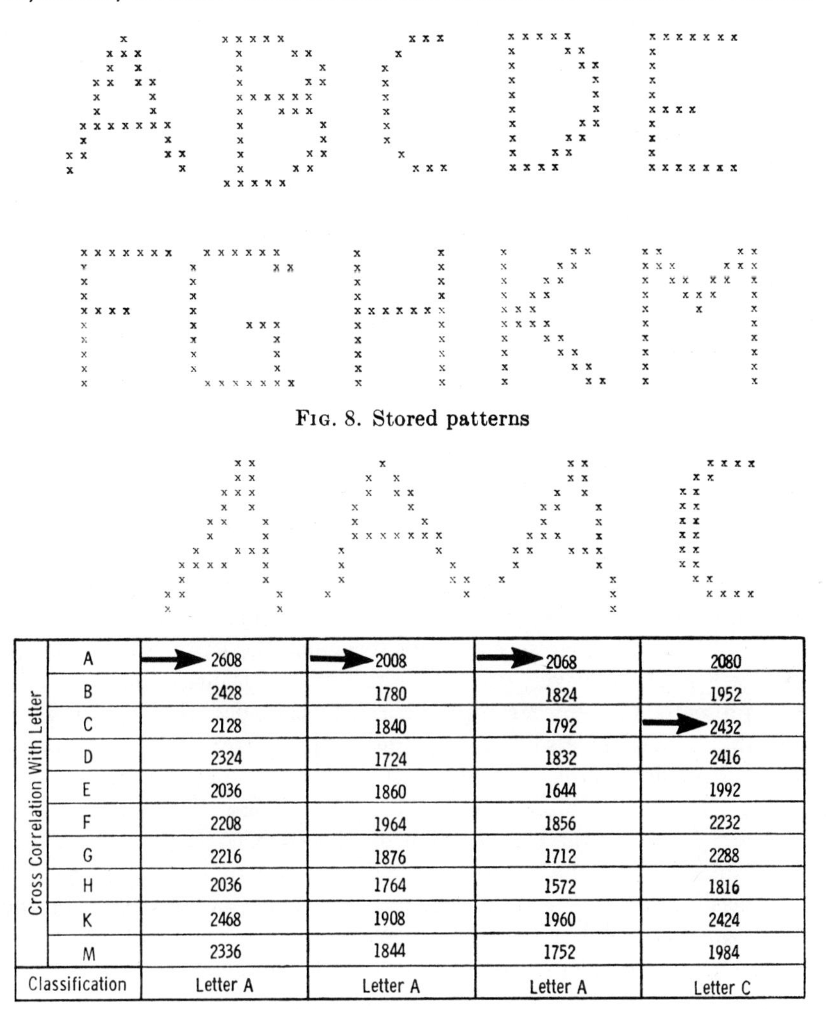

Fig. 8. Stored patterns

Cross Correlation With Letter				
A	→ 2608	→ 2008	→ 2068	2080
B	2428	1780	1824	1952
C	2128	1840	1792	→ 2432
D	2324	1724	1832	2416
E	2036	1860	1644	1992
F	2208	1964	1856	2232
G	2216	1876	1712	2288
H	2036	1764	1572	1816
K	2468	1908	1960	2424
M	2336	1844	1752	1984
Classification	Letter A	Letter A	Letter A	Letter C

Fig. 9. Classification of distorted patterns

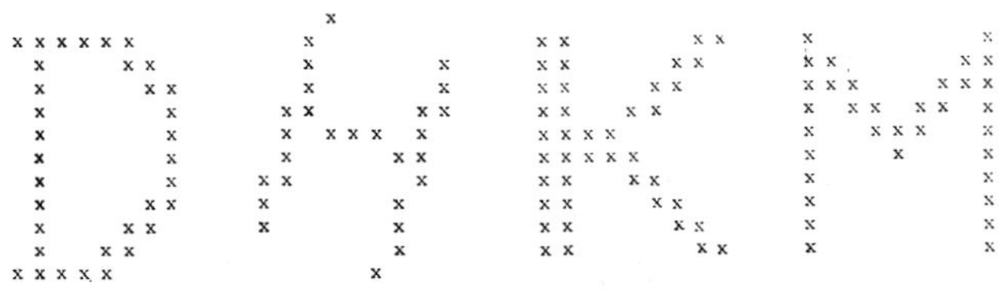

Cross Correlation With Letter				
A	2700	1902	3654	2554
B	2880	1800	3852	2208
C	2248	1760	2816	2012
D	→ 2982	1800	3488	2176
E	2372	1660	3364	1962
F	2388	1872	3164	2058
G	2620	1672	3488	2094
H	2324	→ 2012	3312	1974
K	2644	1912	→ 4040	2186
M	2380	1596	3160	→ 2900
Classification	Letter D	Letter H	Letter K	Letter M

Fig.10-Classification of distorted patterns

FIG. 10. Further examples of distorted patterns correctly classified

of these letters is shown in Figs. 9 and 10. The program succeeded in a correct classification.[6]

Figure 11 shows the process of autonomous learning of new varieties of a pattern. With only "knowledge" of the original version of a hand-printed "D", the program could not recognize the handwritten cursive "D" of Fig. 11. However, when several "D's" with increasing deviation were presented to the program, and correctly classified, then the program was able to recognize several varieties of that letter. Subsequent tests for the classification of other letters show that these classifications were undisturbed by the "broadening" of the pattern class "D".

Figure 12 shows the classification of 10 letters (A to M) that are of approximately half the size of the standard pattern set of Fig. 8. (For the classification of the small letters, the periodicity properties of the transform were used, as described in Section 2.4.) The small letters are considerably distorted in comparison with the large standard patterns,

[6] With a program using the fast Fourier transform, the rotated H of Fig. 10 was not recognized; it was classified as an A.

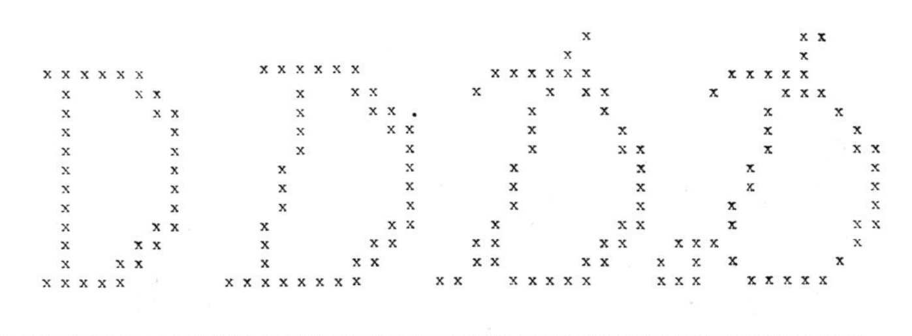

Cross Correlation With Letter					
	A	2700	3736	3788	3512
	B	2880	3996	4144	3812
	C	2248	3032	3116	2968
	D	→ 2982	→ 4253	→ 4363	→ 4096
	E	2372	3492	3370	3108
	F	2388	3224	3274	3068
	G	2620	3864	3738	3468
	H	2324	2984	3042	2768
	K	2644	3812	4006	3660
	M	2380	3588	3756	3604
Classification		Letter D	Letter D	Letter D	Letter D

FIG. 11. Illustration of autonomous learning behavior

when they are magnified two times, since the magnification also causes a corresponding increase in the line width.

In the first test run, when the small letters were directly matched with the standard pattern set of Fig. 8, 60% of the small letters were correctly classified. However, when the mean value of the transforms of full-size and half-size letters was stored as a representative, 100% correct classification was achieved for both full-size and half-size letters of that set. This indicates that the transforms of full-size and half-size letters are sufficiently similar (when the rules of Section 2.4 are observed), so that they can be summed up in the same memory cells, i.e., they can be treated like variations of a certain letter of approximately the same size. Figure 12 shows the results of the classification of half-size letters in the second run.

3. CONCLUSIONS

The R-transform, when used in conjunction with a simple, minimum Euclidean distance classifier of the output vectors, has given a good

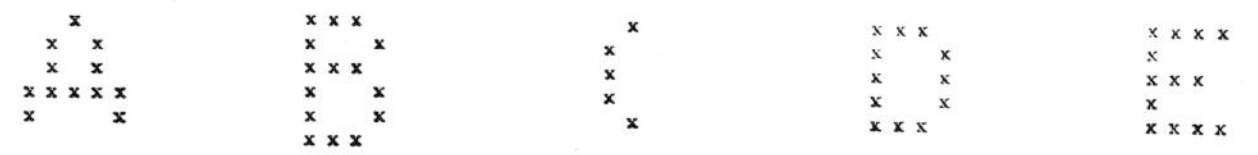

A

Cross Correlation With Letter					
A	→ 5812	5864	1400	4068	4472
B	4504	→ 7408	1024	5312	5024
C	3712	4688	→ 2448	3568	3696
D	3980	6508	1212	→ 5820	5116
E	4166	5686	1014	4926	→ 6262
F	4398	5406	1774	4414	4490
G	4740	6748	1212	5244	5340
H	3150	5726	558	4646	4222
K	4358	5998	1438	4446	4478
M	4808	5208	1200	3424	3560
Classification	Letter A	Letter B	Letter C	Letter D	Letter E

B

Cross Correlation With Letter					
A	3748	5700	3508	3532	4972
B	3616	6480	4592	3696	4040
C	3248	4240	2800	3024	3832
D	3740	6060	4284	3452	3668
E	4206	5310	3758	3342	3450
F	→ 4734	5038	3550	3278	3586
G	3996	→ 6748	3868	3500	4056
H	2966	4614	→ 5750	3350	2682
K	3358	5166	3998	→ 4898	3842
M	2568	4816	2576	2600	→ 6336
Classification	Letter F	Letter G	Letter H	Letter K	Letter M

FIG. 12. (a), (b). Classification of half-size letters

performance as a character-recognition algorithm. The principle of relating every input element to every other input element, inherent in the tree-graphs of Figs. 2 and 3, is here combined with a simplified and hence fast computation at each nodal point. The resulting gain in speed (by a factor of 10 or more) with respect to the fast Fourier transform is, of course, traded against the loss of phase information. This loss, however, seems not to impair the performance in the particular problem investigated.

The simplicity of the *RT* computation has the added advantage of being well suited to its realization in a high-speed, special purpose computer. A further significant increase of speed could be expected from such a computer.

The advantages and limitations of the R-transform with respect to other pattern-recognition tasks remain to be explored. At the present time it apears that it will be more useful in the high-speed processing of simple patterns then in the analysis of more complex patterns.

ACKNOWLEDGMENT

We would like to acknowledge the contribution of D. K. McLain and D. H. Shaffer, who have developed an alternative proof of the shift-invariance property.

APPENDIX A: PROOF OF SHIFT INVARIANCE FOR ALGORITHM A

The tree graph of Algorithm A consists of a column of $N = 2^M$ input variables, followed by M columns of operations, performed on pairs of the variables of the preceding column.

For a given column $(R - 1)$ the operations for the calculation of the N variables in the next column (R) are defined by Eq. (4). From the tree graph Fig. 2 and from Eq. (4) it is evident that the operations in the columns follow symmetrical patterns, so that every column $(R > 0)$ can be divided into 2^R groups with 2^{M-R} variables each.

In the following proof for the shift invariance we shall use a shorthand notation for Eq. (4) which emphasizes this symmetry:

$$X_{m,2n}^{(R)} = |X_{m,n}^{(R-1)} + X_{m+s,n}^{(R-1)}| \qquad \Big|_{m=0}^{s-1} \quad \Big|_{n=0}^{t-1} \qquad \text{A.1.1}$$

$$X_{m,2n+1}^{(R)} = |X_{m,n}^{(R-1)} - X_{m+s,n}^{(R-1)}| \qquad \text{A.1.2}$$

with

$$s = 2^{M-R}, \qquad t = 2^{R-1}. \qquad \text{A.1.3}$$

Equation A.1 is converted to the original notation of Eq. 4 when s times the second subscript is added to the first subscript of the variables in column (R), e.g. $X_{m,2n+1}^{(R)} \leftrightarrow X_{m+(2n+1)s}^{(R)}$, and $2s$ times the second subscript added to the first subscript of the variables in column $(R - 1)$, e.g. $X_{m,n}^{(R-1)} \leftrightarrow X_{m+2ns}^{(R-1)}$.

In the new notation, the second subscript represents the group number within a column, as illustrated in Fig. 13 for $M = 3$. The variables in column (0) are defined as being cyclic with a period of $N = 2^M$, so that

$$X_{m,0}^{(0)} \equiv X_{m+2^M,0}^{(0)}. \qquad \text{A.2}$$

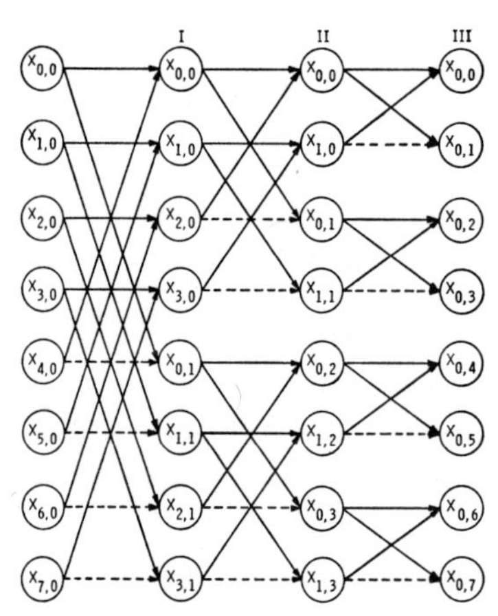

FIG. 13. Tree-graph of R-transform, Algorithm A, in group notation (for 8 input variables).

We consider now a 2^{M-1}-fold cyclic permutation of the variables in column (0),

$$Y_{i,0}^{(0)} = X_{i+2^{M-1},0}^{(0)}$$

which corresponds to a shift of the input pattern by 2^{M-1} places, that is, to an interchange of the top and bottom halves of the variables of column (0). From Eqs. (3.1) and A.2 we obtain then for the variables in column (1):

$$\begin{aligned} Y_{i,0}^{(1)} &= |Y_{i,0}^{(0)} + Y_{i+2^{M-1},0}^{(0)}| \\ &= |X_{i+2^{M-1},0}^{(0)} + X_{i+2^{M},0}^{(0)}| = X_{i,0}^{(1)} \Big|_{i=0}^{2^{M-1}-1} \end{aligned} \qquad \text{A.3.1}$$

and

$$\begin{aligned} Y_{i,1}^{(1)} &= |Y_{1,0}^{(0)} - Y_{i+2^{M-1},0}^{(0)}| \\ &= |X_{i+2^{M-1},0}^{(0)} - X_{i+2^{M},0}^{(0)}| = X_{i,1}^{(1)} \Big|_{i=0}^{2^{M-1}-1} \end{aligned} \qquad \text{A.3.2}$$

i.e. the variables in column (1) remain unchanged.

By a similar argument we see that the variables in column (2) are invariant to a 2^{M-2}-fold cyclic permutation of the variables in column (0), and in general, we state the following

THEOREM. *The variables in any group* (n) *of a given column* (R) *are cyclic with a period of* 2^{M-R} *and, therefore, invariant to a* 2^{M-R}*-fold cyclic permutation of the variables in column (0),*

$$\text{i.e.} \qquad Y^{(R)}_{m,n} = X^{(R)}_{m+2^{M-R},n} = X^{(R)}_{m,n}\Bigg|_{n=0}^{2^R-1}, \qquad \text{A.4}$$

$$\text{for} \qquad Y^{(0)}_{m,0} = X^{(0)}_{m+2^{M-R},0}.$$

Proof. We assume the truth of the proposition for group (n) in column ($R-1$),
i.e.

$$Y^{(R-1)}_{m,n} = X^{(R-1)}_{m+2^{M-(R-1)},n} = X^{(R-1)}_{m,n}, \qquad \text{A.5}$$

for

$$Y^{(0)}_{m,0} = X^{(0)}_{m+2^{M-(R-1)},0},$$

and calculate the values of the variables in column (R) for a 2^{M-R} fold cyclic permutation of the variables in column (0),

$$Y^{(0)}_{m,0} = X^{(0)}_{m+2^{M-R},0}.$$

From A.1 we obtain for the variables in group ($2n$) of column (R):

$$Y^{(R)}_{m,2n} = \left| Y^{(R-1)}_{m,n} + Y^{(R-1)}_{m+s,n} \right| \Bigg|_{m=0}^{s-1}$$

$$= \left| X^{(R-1)}_{m+2^{M-R},n} + X^{(R-1)}_{m+2^{M-R+1},n} \right| \Bigg|_{m=0}^{s-1}$$

and with A.5

$$= \left| X^{(R-1)}_{m+2^{M-R},n} + X^{(R-1)}_{m,n} \right| \Bigg|_{m=0}^{s-1} = X^{(R)}_{m,2n} \Bigg|_{m=0}^{s-1} \qquad \text{A.6.1}$$

A similar argument shows the truth of the proposition for the variables in group ($2n+1$) of column (R):

$$Y^{(R)}_{m,2n+1} = X^{(R)}_{m,2n+1}. \qquad \text{A.6.2}$$

From the assumption that the theorem is true for $R-1$ it follows, therefore, that it is also true for R, and since it is true for $R = 1$ it is true for all $1 \leqq R \leqq M$.

For the last column ($R = M$) we obtain that the variables $X^{(M)}_{0,n}$ remain unchanged if the input is shifted by

$$s = 2^{M-M} = 1 \text{ place}$$

i.e.,

$$Y_{0,n}^{(M)} = X_{0,n}^{(M)},$$

for

$$Y_{m,0}^{(0)} = X_{m+1,0}^{(0)},$$

and, therefore, by any integer number of places. The invariance under cyclic permutation of Algorithm A is proven.

APPENDIX B: INVARIANCE UNDER REFLECTION

Another property of the transform is that it does not discriminate between mirror images (2.4.7.1).

A formal proof of this property is as follows:

A mirror image of the input function $X_{m,0}^{(0)}\Big|_{m=0}^{2^M-1}$ is defined by $Y_{m,0}^{(0)}$ where

$$Y_{m,0}^{(0)} = X_{2^M-1-m,0}^{(0)}. \qquad \text{B.1}$$

Using A1.1 we get for group (0) in column (1)

$$Y_{m,0}^{(1)} = \left| Y_{m,0}^{(0)} + Y_{m+2^{M-1},0}^{(0)} \right| \Bigg|_{m=0}^{2^{M-1}-1},$$

and with Eq. A.2 and B.1 we obtain

$$Y_{m,0}^{(1)} = \left| X_{2^M-1-m,0}^{(0)} + X_{2^{M-1}-1-m,0}^{(0)} \right|,$$

but

$$\left| X_{2^M-1-m,0}^{(0)} + X_{2^{M-1}-1-m,0}^{(0)} \right| = X_{2^{M-1}-1-m,0}^{(1)},$$

hence, $Y_{m,0}^{(1)}\Big|_{m=0}^{s-1}$ is the mirror image of $X_{m,0}^{(1)}\Big|_{m=0}^{s-1}$, with $s = 2^{M-1}$.

A similar argument shows that the second group of column (1)

$$Y_{m,1}^{(1)}\Big|_{m=1}^{s-1} \text{ is the mirror image of } X_{m,1}^{(1)}\Big|_{m=0}^{s-1}.$$

With an induction proof analogous to that used in Appendix A it can be shown that in any column (R) each group is the mirror image of that group for the original input.

In the penultimate column ($M - 1$) the groups, finally, consist only of two variables where the operation of mirroring becomes equivalent to a unit shift. This, by previous argument, leaves the output invariant.

We have recently defined a general rule for all those permutations of the input which leave the output invariant. This rule of course, includes the shift- and mirror-invariances as special cases. We will discuss the rule and its implications for the pattern-separation capability of the transform in a subsequent paper.

APPENDIX C: EQUIVALENCE OF ALGORITHM A AND B

THEOREM. *The variables in any column $1 \leqq R \leqq M$ of Algorithm A are identical to the variables of corresponding index in the same column R of Algorithm B if the permutation operator P, defined in Eq. (C.2) is applied R times to the indices of the variables $X_i^{(R,A)}$ in Eq. (4)*

$$X_i^{(R,B)} = X_{P^R(i)}^{(R,A)}, \tag{C.1}$$

with

$$P^R(i) = i \cdot 2^R \text{ div } 2^M + i \cdot 2^R \text{ mod } 2^M, \tag{C.2}$$

where "div" and "mod" are arithmetic operators that respectively produce the integer quotient and the remainder.[7]

Proof. Applying the permutation operator P^R to the indices of the variables

$$X_{m+2ns}^{(R,A)} \Big|_{m=0}^{s-1} \Big|_{n=0}^{t-1} \tag{C.3}$$

in column (R) of Eq. (4) with $s = 2^{M-R}$ and $t = 2^{R-1}$ we obtain under consideration of the limitations for m and n

$$P^R(2ns) = (2ns \cdot 2t) \text{ div } (2st) = 2n$$

$$P^R(m) = (m \cdot 2t) \text{ mod } (2st) = 2mt,$$

and

$$P^R(m + 2ns) \Big|_{m=0}^{s-1} \Big|_{n=0}^{t-1} = 2tm + 2n \Big|_{m=0}^{s-1} \Big|_{n=0}^{t-1},$$

and with

$$K = tm + n, \tag{C.3.1}$$

we obtain

$$P^R(m + 2ns) = 2K \Big|_{K=0}^{st-1}. \tag{C.4.1}$$

[7] The effect of the operator P becomes more transparent when the index is written in binary notation $(i) = (u_{M-1}, u_{M-2} \cdots u_0)$ binary. The $P^R(i)$ is then $P^R(u_{M-1}, u_{M-2}, \cdots u_0) = (u_{M-R-1}), \cdots u_0, \cdots u_{M-R})$.

A similar argument shows that the indices of the variables

$$X^{(R,A)}_{m+(2n+1)s} \Big|_{m=0}^{s-1} \Big|_{n=0}^{t-1},$$

of Eq. (4) can be rewritten, after permutation, as

$$P^R[m + (2n + 1)s] \Big|_{m=0}^{s-1} \Big|_{n=0}^{t-1} = (2K + 1) \Big|_{K=0}^{st-1}, \qquad \text{C.4.2}$$

$(R - 1)$-fold application of the permutation operator to the indices of the variables in column $(R - 1)$ of Eq. (4) yields

$$P^{(R-1)}(m + 2ns) \Big|_{m=0}^{s-1} \Big|_{n=0}^{t-1} = K \Big|_{K=0}^{st-1}, \qquad \text{C.4.3}$$

and

$$P^{(R-1)}[m + (2n + 1)s] \Big|_{m=0}^{s-1} \Big|_{n=1}^{t-1} = K + st \Big|_{K=0}^{st-1}. \qquad \text{C.4.4}$$

Comparing Eq. (4) with Eq. (5) under consideration of the Eqs. (C.4) shows that the permutation operator Eq. (C.2) transforms Algorithm A into Algorithm B.

For any input set $X_i^{(0)} \Big|_{i=0}^{2M-1}$ the variables in any column (R) of Algorithm B assume, therefore, the same values as the permuted variables in column (R) of Algorithm A.

For the last column $(R = M)$ we obtain from Eq. (C.2)

$$P^M(i) = i,$$

i.e., the variables in the last column of both Algorithm A and B are identical for any given input set.

Received: January 13, 1969; revised: May 15, 1969

REFERENCES

Cooley, J. W. and Tukey, J. W. (1965), An algorithm for the machine calculation of complex Fourier series. *Math. Comput.* **19,** 297-301.

Hadamard, J. (1893), Resolution d'une question relative aux determinants. *Bull. Sci. Math.*, ser. 2, Vol. 17, pt. I, pp. 240–246.

ERRATUM

Figures 2 and 3 appear in the wrong order.

8

COMPUTATION OF THE HADAMARD TRANSFORM AND THE *R*-TRANSFORM IN ORDERED FORM

L. J. Ulman

***Abstract*—This correspondence describes a method of computing the Hadamard transform and the *R*-transform. This method produces the transform components in the order of the increasing sequency of the Walsh functions represented by the rows of a symmetric Hadamard matrix for which the sequency of each row is larger than the sequency of the preceding row.**

***Index Terms*—Hadamard transform, invariance with cyclic data shifts, pattern recognition, *R*-transform, Walsh–Fourier transform.**

Walsh functions and the Walsh–Fourier transform have been discussed in recent articles [1]–[3]. Algorithms for computing discrete Walsh–Fourier or Hadamard transforms are described in [2] and [3]. The *R*-transform described in [4] may also be considered in terms of a Hadamard matrix. Its major property is that transformed data is independent of cyclic shifts of the input pattern. One form of computation of a one-dimensional *R*-transform is based on a flow graph which is similar to the flow graph used for the Cooley–Tukey FFT algorithm. This computation is the same as that shown for the fast Walsh–Fourier transform in [3] (Fig. 2), except that the absolute value is taken for each computation which involves both $+1$ and -1 multiplying factors. The fast Hadamard transform described in [2] provides the transform components in "ordered" form, where these components are in the order of the sequency of the Walsh functions represented by the rows of a symmetric Hadamard matrix for which the sequency of each row is larger than the sequency of the preceding row. The algorithms described in [3] and [4] provide the components in a "natural" form (not in order of sequency). The new form of computation described here can be used to compute the Hadamard transform and the *R*-transform in "ordered" form. It is for use with sampled data where the number of samples considered is an integral power of 2. The fast Hadamard transform of [2] is not suitable for use with the *R*-transform since it is not based on a computation starting with pairs of the N input samples which are separated by $N/2$ samples as required for invariance of the *R*-transform under cyclic permutation of the input samples.

Manuscript received October 2, 1969; revised November 18, 1969.

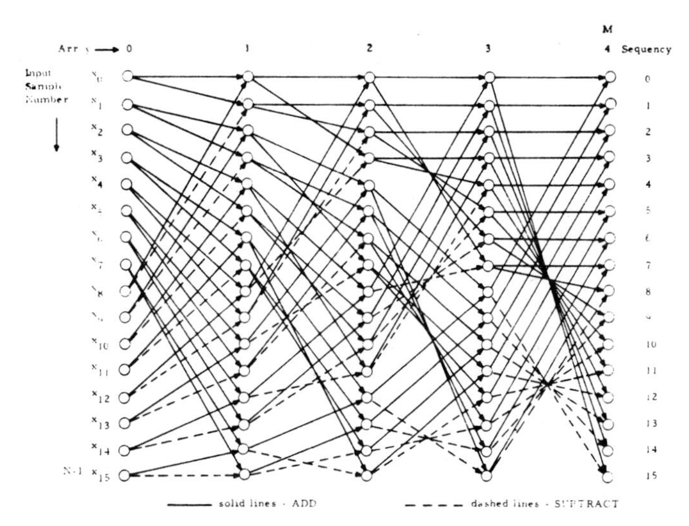

Fig. 1. Ordered Hadamard transform computation.

```
N = 2^M = Number of sample points.
L = Index of array used to compute values for array LP.
NZN = Number of computation groups for array being computed.
J = Index of point in array being computed

Reserve storage for an (M+1) by N array for XR.
Read input data into XR (1, J).
Output data appears in XR (M2, J) where M2 = M + 1.

Program
      NH = N/2
      DO 120 L = 1, M
      LP = L + 1
      LM = L - 1
      NY = 0
      NZ = 2**LM
      NZI = 2*NZ
      NZN = N/NZI
      DO 110 I = 1, NZN
      NX = NY + 1
      NY = NY + NZ
      JS = (I-1) *NZI
      JD = JS + NZI + 1
      DO 100 J = NX, NY
      JS = JS + 1
      J2 = J + NH
      XR (LP, JS) = XR (L, J) + XR (L, J2)
      JD = JD - 1
      XR (LP, JD) = XR (L, J) - XR (L, J2)  See Note
100 CONTINUE
110 CONTINUE
120 CONTINUE

NOTE: For R-transform use XR (LP, JD) = ABS (XR(L, J) - XR (L, J2))
```

Fig. 2. FORTRAN program for ordered Hadamard transform and ordered R-transform.

Fig. 1 is a diagram showing the computational steps for finding the Hadamard transform or the R-transform of an input function defined by 16 samples. Subscripts of the rows in this diagram are numbered from 0 to $(N-1)$, where $N=2^M$ (M an integer) is the number of input samples, and the columns are numbered from 0 to M. The arrays (columns) represent the N input values and the N results of each of M transformation steps. The value at each node of arrays 1 through M is found by calculating the sum or difference of two values in the preceding array from a row i and a row $(i+N/2)$. In the diagram the solid arrows represent addition and the dashed arrows represent subtraction so that when a subtraction is involved, the value connected to the dashed arrow is subtracted from the value connected by the solid arrow. If the R-transform is being computed, absolute values are used for the results of each computation which is indicated as a subtraction. The values in array M are the Hadamard transform (or R-transform) values in order of sequency.

The method of computation which has been described was developed during a study of the identification of radars by means of video waveforms. The ordered transform was desired to observe the sequency characteristics of the video waveforms and to determine the effects of filtering and other operations on the correlation properties of the transforms of sampled waveforms.

Fig. 2 is a FORTRAN program for computing the ordered Hadamard transform and the ordered R-transform as described above. Note that in the program the rows of Fig. 1 are numbered from 1 to N, and the columns are numbered from 1 to $(M+1)$ to avoid zero subscript values. This program was used for a relatively small number of input samples where the number of storage locations required was not a problem. For applications where the available storage is a limiting factor, it has been suggested that the storage efficiency could be increased by alternating the computations between two arrays of length N. As the computation cannot be performed "in place," the number of storage locations required is at least twice the number of input samples.

REFERENCES

[1] H. F. Harmuth, "A generalized concept of frequency and some applications," *IEEE Trans. Information Theory*, vol. IT-14, pp. 375–382, May 1968.
[2] W. K. Pratt, J. Kane, and H. C. Andrews, "Hadamard transform image coding," *Proc. IEEE*, vol. 57, pp. 58–68, January 1969.
[3] J. L. Shanks, "Computation of the fast Walsh-Fourier transform," *IEEE Trans. Computers* (Short Notes), vol. C-18, pp. 457–459, May 1969.
[4] H. Reitboeck and T. P. Brody, "A transformation with invariance under cyclic permutation for applications in pattern recognition," Westinghouse Research Labs, Scientific Paper 68-1F1-ADAPT-P1, September 17, 1968.

9

On Computation of the Hadamard Transform and the R Transform in Ordered Form

MURAT KUNT

***Abstract*—This correspondence describes an improved computational algorithm for the Hadamard transform and the R transform. By performing the computation "in place", the number of storage locations is minimized and the speed is increased. The transformed coefficients are in the order of increasing sequency.**

***Index Terms*—Gray code, Hadamard transform, R transform, sequency ordering.**

Various types of computational algorithm for the Hadamard transform have been reported recently [1]–[3]. The algorithm described in [1] provides the transformed coefficients in "ordered" form which corresponds to the order of increasing sequency [4]. However, it requires $N/2$ auxiliary storage locations where $N = 2^m$ is the number of input samples, and it cannot be used for the computation of the R transform [3].

The method proposed by Shanks [2] performs the computation "in place" which means that intermediate results are stored in the locations of the input samples without any need for auxiliary storage. However, the transformed coefficients are in the "natural" form (not in the order of increasing sequency).

The computational algorithm for both the Hadamard and the R transforms in ordered form is given by Ulman [3]. This method requires N auxiliary storage locations for intermediate results.

The algorithm proposed here can be performed in place, it gives the transformed coefficients in ordered form and it is faster than Ulman's algorithm.

Let us first consider the Cooley–Tukey fast Fourier transform (FFT) algorithm. Generally in this algorithm a dummy parameter having the value $+1$ or -1 is placed on the exponent in order to perform the direct $(+1)$ or the inverse (-1) transform. If this parameter is set to 0, all multiplicative factors are reduced to ± 1 and the FFT algorithm computes the Hadamard transform in the "bit-reversed order". It is therefore an easy task to exploit the structure of the FFT in order to establish an algorithm for the fast Hadamard transform (FHT) in the bit-reversed order without using complex numbers.

Walsh functions represented by the rows of a symmetric Hadamard matrix are characters of the dyadic group. Their reordering in the order of increasing sequency can be carried out with Gray code. The corresponding points in the dyadic space are then ordered into a path of minimum length [5].

The R transform introduced by Reitboeck and Brody [6] has the interesting property of invariance under cyclical permutation of the input data. This transform may be considered in terms of a Hadamard transform matrix in which absolute value is taken after each subtraction.

Fig. 1 is an eight-point flow graph showing the computational steps (numbered from 0 to 5) for Hadamard and R transforms.

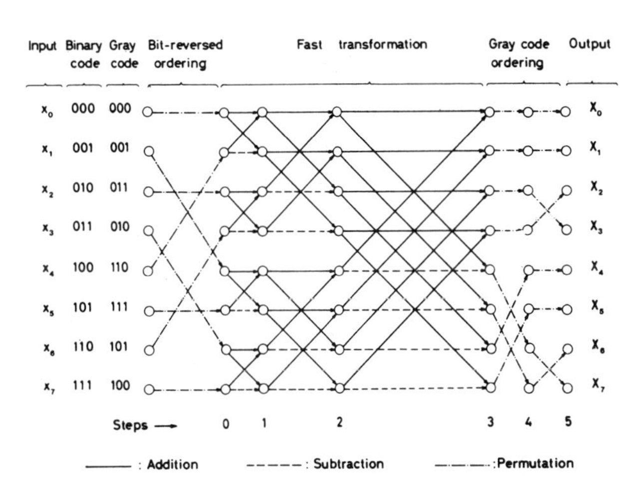

Fig. 1. Eight-point flow graph for efficient Hadamard and R transforms computation.

```
      SUBROUTINE FHT(X,N,NU)
C THIS SUBROUTINE COMPUTES THE HADAMARD TRANSFORM OR
C THE R-TRANSFORM OF A SAMPLED SIGNAL. THE TRANSFORMED
C COEFFICIENTS ARE IN THE ORDER OF INCREASING SEQUENCY.
C X = REAL ARRAY CONTAINING THE SIGNAL'S SAMPLES BEFORE
C     THE CALL AND THE COEFFICIENTS AFTER THE CALL.
C N = 2**NU = NUMBER OF SAMPLES.
C RESERVE STORAGE FOR X(N) IN THE CALLING PROGRAM.
      DIMENSION X(N)
      N1=N-1 $N2=N/2 $NU1=NU-1 $N4=N/4 $I=1
      DO 1 J=1,N1
      IF(I.LE.J)  GO TO 2
      Z=X(I) $X(I)=X(J) $X(J)=Z
2     K=N2
3     IF(I.LE.K) GO TO 1
      I=I-K $K=K/2 $GO TO 3
1     I=I+K
      DO 4 I=1,NU
      IA=2**I $IB=IA/2
      DO 4 J=1,IB
      DO 4 K=J,N,IA
      IC=K+IB $Z=X(IC) $X(IC)=X(K)-Z
C FOR THE R-TRANSFORM USE X(IC)=ABS(X(K)-Z)
4     X(K)=X(K)+Z
      FAC=SQRT(2.)**NU
      DO 5 I=1,N
5     X(I)=X(I)/FAC
      DO 6 I=1,NU1
      IGA=2**(I-1)
      DO 7 L=1,IGA
      DO 7 J=1,N4
      IY=N2+J+(L-1)*N/IGA $IX=IY+N4
      Z=X(IX) $X(IX)=X(IY)
7     X(IY)=Z
      N2=N2/2
6     N4=N4/2
      RETURN
      END
```

Fig. 2. Fortran subroutine for efficient Hadamard and R transforms computation.

Each column corresponds to a particular step, starting with the input samples at left and ending with the transformed coefficients at right. Binary and Gray codes equivalent of the subscripts of the input samples are indicated for reordering purposes. A bit-reversed reordering based on binary code is first carried out, changing for example x_1 at address 001 with x_4 at address 100 (step 0). Then, a fast transformation is performed which is quite similar to the

Manuscript received June 26, 1973; revised April 17, 1975.
The author is with the Research Laboratory of Electronics, Massachusetts Institute of Technology, Cambridge, Mass. 02139, on leave from Ecole Polytechnique Federale de Lausanne, Lausanne, Switzerland.

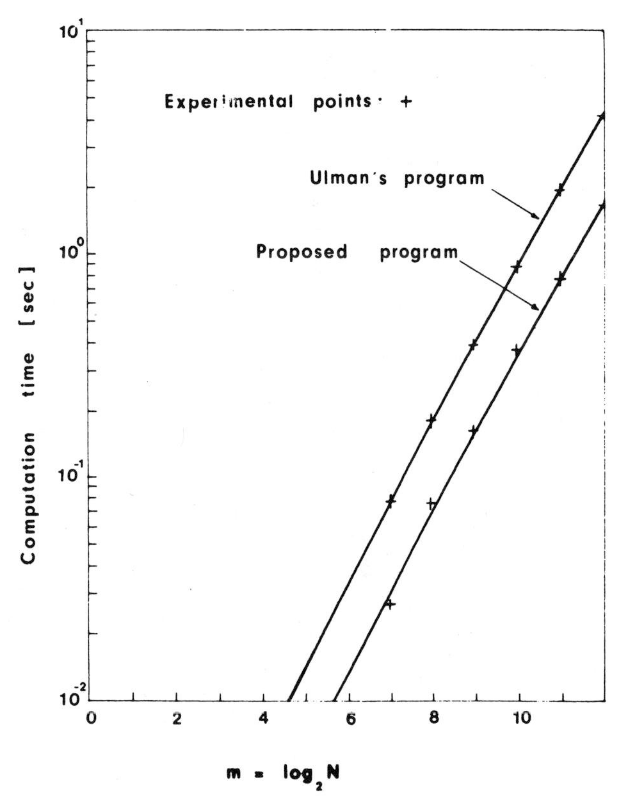

Fig. 3. Comparison of the computation time for Ulman's program and our subroutine.

Cooley–Tukey algorithm (Steps 1, 2, and 3). In this transformation the value at a node of the step j is found by computing the sum or difference of two values in the preceding step $j - 1$ from rows differing by 2^{j-1}. This structure allows to store intermediate results in the storage locations of the input data. In this flow graph solid lines represent addition and the dashed lines represent substraction. If the R-transform is being performed, absolute value must be taken at each node reached by a dashed line.

Finally, at Steps 4 and 5 Gray code reordering is carried out by placing the value of a node with a given binary coded address to the node which has the same code word in Gray code. The listing of a Fortran subroutine performing these computations is shown on Fig. 2. Note that the results are normalized in the loop "DO 5." Two successive calls of this subroutine lead to the initial input data. From a computation time point of view, we can compare the performances of this subroutine and Ulman's program. It is well known that the computation time of this kind of fast transformation is given by

$$TN \log_2 N \tag{1}$$

where T is a constant depending on the structure of the algorithm and the access and cycle time of the computer used. We have estimated the values of T for Ulman's program and our subroutine running on a CDC 6500 computer operating under Scope 3.4 with a RUN Fortran compiler. For $N = 4096 = 2^{12}$ we find $T_u = 85.4\ \mu s$ for Ulman's program and $T_0 = 33.7\ \mu s$ for ours. Theoretical curves obtained from (1) and experimental points for different values of N are shown on Fig. 3. These curves show that our program is faster by a factor 2.5.

ACKNOWLEDGMENT

The author wishes to thank Prof. F. de Coulon for his constant and enthusiastic support.

REFERENCES

[1] W. K. Pratt, J. Kane, and H. C. Andrews, "Hadamard transform image coding," *Proc. IEEE*, vol. 57, pp. 58–68, Jan. 1969.
[2] J. L. Shanks, "Computation of the fast Walsh–Fourier transform," *IEEE Trans. Comput.* (Short Notes), vol. C-18, pp. 457–459, May 1969.
[3] L. J. Ulman, "Computation of the Hadamard transform and the *R*-transform in ordered form," *IEEE Trans. Comput.* (Corresp.), vol. C-19, pp. 359–360, Apr. 1970.
[4] H. F. Harmuth, "A generalized concept of frequency and some applications," *IEEE Trans. Inform. Theory*, vol. IT-14, pp. 375–382, May 1968.
[5] C.-K. Yuen, "Remarks on the ordering of Walsh functions," *IEEE Trans. Comput.* (Corresp.), vol. C-21, p. 1452, Dec. 1972.
[6] H. Reitboeck and T. P. Brody, "A transformation with invariance under cyclic permutation for applications in pattern recognition," *Inform. Contr.*, vol. 15, pp. 130–154, Aug. 1969.

Editor's Comments on Papers 10 Through 13

DISCRETE SINE TRANSFORM

Discrete sine transform (DST), which was originally developed by Jain (Paper 24), belongs to the family of unitary transforms (Paper 10). This family includes the DST, DCT, and DFT. Minor variations of these transforms such as even DCT, odd DCT, DST-I, and DST-II (Kekre and Solanki, 1978) also belong to this family. Jain (Paper 10) has shown that all these transforms are asymptotically equivalent (Paper 5) to the KLT.

Jain (Paper 24) has shown that, for a first order Markov process, the KLT reduces to the DST if the boundary values of a sequence are fixed or known. He has also shown that the basis functions of the DST are the eigenvectors of a symmetric tridiagonal Toeplitz matrix and that an M-point DST can be implemented via a $2(M+1)$-point FFT. The effectiveness of the DST in signal processing is shown by Yip and Rao (Paper 11). As in the case of the DCT, with suitable changes, the DST can also be computed via a $(M+1)$-point FFT. Since both the DCT and DST have sinusoidal functions as their basis vectors, it is reasonable to expect a fast recursive algorithm involving only real arithmetic for the DST as in the case for the DCT (Paper 2). Such an algorithm has been developed by Yip and Rao (Paper 12). Wang (Paper 13) has shown that the DST can also be implemented via the DCT. As the DST is an even-odd transform (EOT) it can also be obtained from the WHT using a conversion matrix that has a sparse block-diagonal structure (Paper 11). An EOT is one that has one-half even vectors (Ex. v_1, v_2, v_3, v_4, v_4, v_3, v_2, v_1) and one-half odd vectors (Ex. v_1, v_2, v_3, v_4, $-v_4$, $-v_3$, $-v_2$, $-v_1$). Other techniques for implementation of the DST have also been developed (Srinivasan and Rao, 1981).

Yip and Rao (Paper 11) have compared the performance of DST with the

DCT based on decorrelation (Paper 5) and rate distortion (Paper 1) and have shown that for small correlation (adjacent correlation coefficient $\rho < 0.6$) DST performs as well as the DCT. This is further substantiated by Ahmed and Flickner (1982), who also show that the DST is a limiting case of the KLT of a first-order Markov process as ρ approaches zero. As the DST has no constant basis vector, its performance can be improved considerably if the mean or average of the sequence is processed separately and the DST is applied to a zero-mean sequence.

REFERENCES

Ahmed, N., and M. Flickner, 1982, *Some Considerations of the Discrete Cosine Transform,* IEEE 16th Asilomar Conference on Circuits, Systems and Computers, Pacific Grove, Calif., pp. 295–299.

Kekre, H. B., and J. K. Solanki, 1978, Comparative Performance of Various Trigonometric Unitary Transforms for Transform Image Coding, *Int. J. Electron.* **44:**305–315.

Srinivasan, R., and K. R. Rao, 1981, *Fast Algorithms for the Discrete Sine Transform,* 24th Midwest Symposium on Circuits and Systems, Albuquerque, New Mexico, Western Periodicals, North Hollywood, Calif., pp. 230–233.

10

A Sinusoidal Family of Unitary Transforms

ANIL K. JAIN, MEMBER, IEEE

Abstract—A new family of unitary transforms is introduced. It is shown that the well-known discrete Fourier, cosine, sine, and the Karhunen-Loeve (KL) (for first-order stationary Markov processes) transforms are members of this family. All the member transforms of this family are sinusoidal sequences that are asymptotically equivalent. For finite-length data, these transforms provide different approximations to the KL transform of the said data. From the theory of these transforms some well-known facts about orthogonal transforms are easily explained and some widely misunderstood concepts are brought to light. For example, the near-optimal behavior of the even discrete cosine transform to the KL transform of first-order Markov processes is explained and, at the same time, it is shown that this transform is not always such a good (or near-optimal) approximation to the above-mentioned KL transform. It is also shown that each member of the sinusoidal family is the KL transform of a unique, first-order, non-stationary (in general), Markov process. Asymptotic equivalence and other interesting properties of these transforms can be studied by analyzing the underlying Markov processes.

Index Terms—Fast transforms, image transforms, Karhunen-Loeve transform, orthogonal transforms, unitary matrices.

I. Introduction

SINCE the discovery of fast Fourier transform (FFT) algorithms [1], [2], interest in unitary and orthogonal transforms, which yield similar fast computational algorithms, has increased rapidly. Many different transforms, e.g., Walsh-Hadamard, Haar, slant, cosine, and sine [3], [4], have been discovered.[1] To a certain extent, the discovery of many unitary transforms has been triggered by attempts to find fast, FFT-type or other, computational algorithms for the family of Karhunen-Loeve (KL) transforms. Since a KL transform is constructed from the eigenvectors of a covariance matrix (of the data that are to be transformed), there is no single unique KL transform for all random processes. Much of the effort in studying KL transforms has been directed at the KL transform of first-order stationary Markov processes whose covariance matrix is given by the Toeplitz matrix R defined as

$$R_{i,j} = \rho^{|i-j|}, \quad |\rho| < 1 \tag{1}$$

where $1 \leqslant i, j \leqslant N$. Although the eigenvectors of R are known analytically [5], they are expensive to generate computationally and have no known fast algorithm to transform a vector of data. Intuitively, the reason is that these eigenvectors, being of the sinusoidal form

$$\phi_m(k) = a_m \sin(w_m k + \theta_m), \quad 1 \leqslant k, \ m \leqslant N \tag{2}$$

where $\{w_m\}$ = solutions of a transcendental equation, are nonharmonic (i.e., the frequency w_m is a nonlinear function of m), which prohibits the existence of an FFT-type algorithm. For higher order Markov processes, closed-form solution of the KL transform is not known in general, and the possibility of a fast algorithm seems even more remote. Recently, Ahmed *et al.* [6] have shown empirically that the discrete cosine transform performs very close to the KL transform of (2) for values of ρ [see (1)] near 0.9. Hamidi and Pearl [7] have extended these results and have shown that the cosine transform yields better performance than the discrete Fourier transform (DFT) for all positive values of ρ. Asymptotic results pertaining to cosine and KL transforms for all finite-order Markov processes have been discussed by Yemini and Pearl [8]. These and other experimental results have led to a widely extrapolated belief that the cosine transform is the best substitute (among the currently known fast unitary transforms) for the KL transform for all first-order, stationary Markov processes. The results of this paper will show that such is not the case. In fact, we shall see that even for first-order Markov processes, there is not unique fast transform within the class examined that will always yield the best approximation to its KL transform. For a higher order Markov

Manuscript received December 30, 1977; revised December 15, 1978. This work was supported in part by the Naval Ocean Systems Center, San Diego, CA, under Contract N00953-77-C-0033MJE and in part by Army Research Office Grant DAAG29-78-G-0206.

The author is with the Department of Electrical Engineering, University of California, Davis, CA 95616.

[1] For a unified treatment of these transforms, see Fino and Algazi [22].

TABLE I
SOME MEMBERS OF THE SINUSOIDAL TRANSFORM FAMILY

No.	J MATRIX PARAMETERS	TRANSFORM	EIGENVECTORS $\phi_m(k)$, $1 \le m,k \le N$	EIGENVALUES λ_m	Δ	Δ_c COMMUTING DISTANCE
1	$k_1=\rho=k_2$ $k_3=k_4=0$	KLT for 1st order stationary Markov Process	$a_m \sin(\omega_m k+\theta_m)$	Solution of a Transcendental Equation	0	0
2	$k_1=k_2=0$ $k_3=k_4=-1$	DFT	$\frac{1}{\sqrt{N}} \exp \pm [\frac{i2\pi(m-1)(k-1)}{N}]$	$1-2\alpha \cos\frac{2\pi(m-1)}{N}$	$2\alpha^2(1+\rho^2)$	$8\alpha^4(1+\rho^2)$
3	$k_1=k_2=0$ $k_3=k_4=0$	EVEN SINE -1 (EDST-1)	$\sqrt{\frac{2}{N+1}} \sin \frac{mk\pi}{N+1}$	$1-2\alpha \cos\frac{m\pi}{N+1}$	$2\rho^2\alpha^2$	$4\rho^2\alpha^4$
4	$k_1=k_2=1$ $k_3=k_4=0$	EVEN COSINE -1 (EDCT-1)	$\frac{1}{\sqrt{N}}$ $m = 1$, $1 \le k \le N$ $\sqrt{\frac{2}{N}} \cos \frac{(2k-1)(m-1)\pi}{2N}$, $m = 2, \ldots N$	$1-2\alpha \cos\frac{(m-1)\pi}{N}$	$2(1-\rho)^2\alpha^2$	$4(1-\rho)^2\alpha^4$
5	$k_1=k_2=-1$ $k_3=k_4=0$	EVEN SINE -2 (EDST-2)	$\sqrt{\frac{2}{N}} \sin \frac{(2k-1)m\pi}{2N}$, $m \ne N$ $\frac{1}{\sqrt{N}}$, $m = N$	$1-2\alpha \cos\frac{m\pi}{N}$	$2(1+\rho)^2\alpha^2$	$4(1+\rho)^2\alpha^4$
6	$k_1=0$, $k_2=+1$ $k_3=k_4=0$	ODD SINE -1 (ODST-1)	$\frac{2}{\sqrt{2N+1}} \sin \frac{(2m-1)k\pi}{2N+1}$	$1-2\alpha \cos\frac{(2m-1)\pi}{2N+1}$	$(1-\rho)^2\alpha^2+\rho^2\alpha^2$	$2(1-\rho)^2\alpha^4+2\rho^2\alpha^4$
7	$k_1=0$, $k_2=-1$ $k_3=k_4=0$	ODD SINE -2 (ODST-2)	$\frac{2}{\sqrt{2N+1}} \sin \frac{2mk\pi}{2N+1}$	$1-2\alpha \cos\frac{2m\pi}{2N+1}$	$(1+\rho)^2\alpha^2+\rho^2\alpha^2$	$2(1+\rho)^2\alpha^4+2\rho^2\alpha^4$
8	$k_1=+1$, $k_2=0$ $k_3=k_4=0$	ODD COSINE -1 (ODCT-1)	$\frac{2}{\sqrt{2N+1}} \cos \frac{(2k-1)(2m-1)\pi}{2(2N+1)}$	$1-2\alpha \cos\frac{2m\pi}{2N+1}$	$(1-\rho)^2\alpha^2+\rho^2\alpha^2$	$2(1-\rho)^2\alpha^4+2\rho^2\alpha^4$
9	$k_1=-1$, $k_1=0$ $k_3=k_4=0$	ODD SINE -3 (ODST-3)	$\frac{2}{\sqrt{2N+1}} \sin \frac{(2k-1)m\pi}{2N+1}$	$1-2\alpha \cos\frac{2m\pi}{2N+1}$	$(1+\rho)^2\alpha^2+\alpha^2\rho^2$	$2(1+\rho)^2\alpha^4+2\rho^2\alpha^4$
10	$k_1=+1$, $k_2=-1$ $k_3=k_4=0$	EVEN COSINE -2 (EDCT-2)	$\sqrt{\frac{2}{N}} \cos \frac{(2k-1)(2m-1)\pi}{4N}$	$1-2\alpha \cos\frac{(2m-1)\pi}{2N}$	$2\alpha^2(1+\rho^2)$	$4\alpha^4(1+\rho^2)$
11	$k_1=-1$, $k_2=+1$ $k_3=k_4=0$	EVEN SINE -3 (EDST-3)	$\sqrt{\frac{2}{N}} \sin \frac{(2k-1)(2m-1)\pi}{4N}$	$1-2\alpha \cos\frac{(2m-1)\pi}{2N}$	$2\alpha^2(1+\rho^2)$	$4\alpha^4(1+\rho^2)$

process, therefore, one may not even expect to find a single best substitute for its KL transform. Instead, one may find the best substitute for a KL transform only for a certain set of values for the statistical parameters of the underlying random process. Alternatively, one may model the random process in such a way that its KL transform is a fast-transform member of the sinusoidal family.

The sinusoidal transforms introduced here, and specifically the sine transform members (see Table I), are different from the previously introduced "fast KL transform algorithm" [9]-[12], which has also been called the "pinned KL transform" [13]. This algorithm is based on the fact that certain random processes can be decomposed as a sum of two mutually orthogonal processes and can be written as

$$u = u^o + u^b \tag{3}$$

where u^b, called the boundary response, is completely determined by the boundary values of the random process sample functions, and the KL transform of the residual process $u - u^b = u^o$ is a sine transform. This sine transform has a fast computational algorithm associated with it and is a member of the sinusoidal family introduced here. This sine transform is not the KL transform of the original random process u, but is the KL transform of the modified process u^o. Rate distortion calculations of the fast KL transform algorithm for data compression of first-order stationary Markov processes have shown superior performance than the cosine transform. However, this algorithm and related considerations are not the subject of this paper. Details may be found in [9]-[12] and [23].

II. THE SINUSOIDAL TRANSFORM FAMILY

Consider the parametric family of matrices

$$J = J(k_1, k_2, k_3, k_4) = \begin{bmatrix} 1-k_1\alpha & -\alpha & & k_3\alpha \\ -\alpha & 1 & \ddots & \\ & \ddots & 1 & -\alpha \\ k_4\alpha & & -\alpha & 1-k_2\alpha \end{bmatrix}. \tag{4}$$

This is a variation of the well-known tridiagonal Jacobi matrix.[2] For $k_3 = k_4$, J is a symmetric matrix, and for suitably

[2] A matrix J is called a Jacobi matrix if $J_{m,n} = 0$, for $(m - n) \geq 2$ [14]. Hence, for $k_3 = k_4 = 0$, (4) is a Jacobi matrix.

chosen α it would be admissable as a positive definite covariance matrix. For $k_3 = k_4 = 0$, and

$$k_1 = \rho, \quad |\rho| < 1$$
$$\alpha \triangleq \rho/(1+\rho^2)$$
$$\beta^2 = (1-\rho^2)/(1+\rho^2) \tag{5}$$

it can be shown that

$$J(\rho,\rho,0,0) = \beta^2 R^{-1} \tag{6}$$

where R is the covariance matrix of the stationary, first-order Markov process defined in (1). Since the eigenvectors of a matrix are invariant under all commuting transformations, the eigenvectors of R^{-1} [and hence of $J(\rho,\rho,0,0)$] and R are identical. Since $J(\rho,\rho,0,0)$ is also a covariance matrix, the KL transform of its underlying random process is the same as the KL transform of the stationary first-order Markov process, which can be written as

$$u_{k+1} = \rho u_k + \epsilon_k, \qquad k = 1, \cdots, N-1$$
$$E[u_1^2] = 1 \quad E[u_1 \epsilon_1] = 0$$
$$E[\epsilon_k] = 0 \quad E[\epsilon_k \epsilon_l] = (1-\rho^2)\delta_{k,l}. \tag{7}$$

Now consider all those J matrices in (4) that are admissable as covariance matrices, i.e., for those values of $k_1, k_2, k_3 = k_4$, and α for which J is positive definite. *The sinusoidal family of unitary transforms is the class of complete orthonormal sets of eigenvectors generated by these J matrices.* Note that each covariance matrix guarantees an associated complete orthonormal set of eigenvectors. Table I summarizes some of the sinusoidal transforms obtained by solving for orthonormal vector solutions of

$$J\phi_m = \lambda_m \phi_m, \qquad 1 \leqslant m \leqslant N \tag{8}$$

for different sets of k_i, $1 \leqslant i \leqslant 4$. In this table ϕ_m represents the mth column of the sinusoidal transform matrix Φ_m. All of the eleven transforms listed in Table I are different in the sense that they are eigenvectors of different noncommuting matrices. It is clear that the well-known transforms such as the KLT (number 1), the DFT (number 2), the sine transform (number 3) [10], and the cosine transform (number 4) [6] are members of this family. From (4), (8), and the fact that $k_3 = k_4$, all the sinusoidal transforms are solutions of the homogeneous second-order difference equation

$$\phi_m(k) - \alpha[\phi_m(k-1) + \phi_m(k+1)] = \lambda_m \phi_m(k), \qquad 2 \leqslant k \leqslant N-1 \tag{9a}$$

subject to the parametric family of boundary conditions

$$(1-k_1\alpha)\phi_m(1) - \alpha\phi_m(2) + k_3\alpha\phi_m(N) = \lambda_m\phi_m(1)$$
$$k_3\alpha\phi_m(1) - \alpha\phi_m(N-1) + (1-k_2\alpha)\phi_m(N) = \lambda_m\phi_m(N). \tag{9b}$$

Thus all the sinusoidal transforms satisfy the same difference equation [i.e., 9(a)] and differ only in the boundary conditions (9b). Table I also lists the eigenvalues associated with the different transforms. These are, of course, the eigenvalues of the J matrices. Interestingly, two different sinusoidal transforms may yield identical eigenvalues, e.g., the J matrices corresponding to ODST-2 (number 7) and ODCT-1 (number 8) have identical sets of eigenvalues, although their eigenvectors are different. The fact that sinusoidal transforms are solutions of (9a) and (9b) can be checked by direct substitution. The unitary property of these transforms can also be verified easily by showing that $\Phi\Phi^{*T} = \Phi^{*T}\Phi = I$. The proofs of these results are straightforward and are obtained by assuming the general solution of (9a) to be of the form

$$\phi_m(k) = A \exp\{ik\theta\} + B \exp\{-ik\theta\}$$

where A, B, and θ are complex in general, and depend on the value of m. These are determined by applying boundary conditions of (9b).

III. Properties of the Sinusoidal Transforms

The following properties of the sinusoidal family can either be proved or are evident from Table I.

A. Orthonormal Properties

The vectors $\{\phi_m(k)\}$ of each family member form a complete orthonormal set of basis vectors in an N-dimensional vector space for $1 \leqslant m, k \leqslant N$. If Φ is the matrix whose mth column is the eigenvector $\{\phi_m(k), 1 \leqslant k \leqslant N\}$, then

$$J\Phi = \Phi\Lambda$$

or

$$\Phi^{*T} J \Phi = \Lambda, \qquad \Lambda = \text{diagonal}\{\lambda_i, 1 \leqslant i \leqslant N\} \tag{10a}$$

where * denotes the complex conjugate, T denotes the matrix transpose, and $\{\lambda_i\}$ are the eigenvalues of J. Hence, if x is an $N \times 1$ vector whose covariance matrix is $f(J)$ for an arbitrary function $f(\cdot)$, then the sinusoidal transformation

$$\hat{x} = \Phi^{*T} x \tag{10b}$$

is its *Karhunen–Loeve* (KL) transformation. Notice that if Φ is not Hermitian, i.e., $\Phi \neq \Phi^{*T}$, then $\hat{x} = \Phi x$ will not be the KL transformation of x. For example, the EDCT-1, EDST-2, ODST-1, ODST-3 are nonsymmetric, even though they are orthogonal. Hence, the transform domain variances of $\hat{x} = \Phi x$ and $\hat{x} = \Phi^T x$ will be different. This can be seen from Table I by comparing the ODST-1 (number 6) and ODST-3 (number 9). These two transform matrices are transposes of each other. Yet their J matrices as well as the eigenvalues $\{\lambda_m\}$ (which represent the variances of $\hat{x}_m$) are different. Thus, if a given transform were a good approximation to a KL transform (of some arbitrary random process), its transpose need not be so.

B. Comparison with KL Transform

Comparison among the sinusoidal family members for various applications may be made by comparing their J matrices. For example, various transforms could be compared with respect to the transform number 1 in Table I, which is the KL transform of a first-order stationary Markov process. Suppose we wish to know how close the various sinusoidal transforms are to this KL transform. Let us define the difference norm

$$\Delta = \| J(k_1, k_2, k_3, k_4) - J(\rho,\rho,0,0) \| \tag{11}$$

where $\|A\| = \Sigma_{i,j} a_{i,j}^2$ is defined as the weak norm of a matrix A. This norm is indicative of the distance between the covariance matrix corresponding to a sinusoidal transform and the covariance matrix corresponding to the KLT in question. [Note that $J(\rho, \rho, 0, 0)$ is not the covariance matrix of the above-mentioned Markov process; rather, it is proportional to the inverse of that covariance matrix; see (8).]

Table I lists the expression for Δ for different sinusoidal transforms. From this it is evident that

$$\Delta(\text{EDST-1}) \geqslant \Delta(\text{EDCT-1}) \quad \text{for} \quad 0.5 \leqslant \rho \leqslant 1 \tag{12a}$$

$$\Delta(\text{EDCT-1}) \geqslant \Delta(\text{EDST-1}) \quad \text{for} \quad -0.5 \leqslant \rho \leqslant 0.5 \tag{12b}$$

$$\Delta(\text{EDCT-1}) \geqslant \Delta(\text{EDST-1}) \geqslant \Delta(\text{EDST-2}) \quad \text{for} \quad -1 \leqslant \rho \leqslant -0.5 \tag{12c}$$

$$\Delta(\text{ODCT-1}) \geqslant \Delta(\text{EDCT-1}) \quad \text{for} \quad 0.5 \leqslant \rho \leqslant 1. \tag{12d}$$

Here, (12a) implies the even cosine-1 transform is better than the even sine-1 transform only for $0.5 \leqslant \rho \leqslant 1$; (12b) and (12c) imply the even sine-1 and even sine-2 transforms perform better than even cosine-1 for other values of ρ; (12d) implies the even cosine-1 performs better than the odd cosine-1 for $0.5 \leqslant \rho \leqslant 1$. Similar inequalities for other sinusoidal transforms can be derived quite straightforwardly.

The validity of these inequalities and hence of the difference norm of (11) as a measure of performance of different transforms can be tested by comparing the actual performances of these transforms. One method of comparing different transforms with a KL transform is to compare their data compression ability. If for any unitary transform Φ and a covariance matrix R we define

$$\sigma_k^2 = [\Phi^{*T} R \Phi]_{k,k} \tag{13}$$

then $\{\sigma_k^2\}$ represents the variances of the transform domain elements of $\hat{x} = \Phi^{*T} x$. If the elements of $\hat{x}$ are ranked in a decreasing order of their variances, i.e., $\sigma_1^2 \geqslant \sigma_2^2 \geqslant \cdots \geqslant \sigma_N^2$, then we will call the quantity

$$J_m = \sum_{k=m+1}^{N} \sigma_k^2 \Big/ \sum_{k=1}^{N} \sigma_k^2 \tag{14}$$

as the "basis restriction error" of the transform Φ for a restriction m. It represents the normalized, minimum mean square error of restricting $\hat{x}$ to m degrees of freedom (i.e., $\hat{x}$ is allowed to have only m nonzero elements). Since $\{\sigma_k^2\}$ depend on the transform Φ via (13), the basis restriction error, for any given m, will vary with Φ and will be minimum for the KL transform. In order to test the inequalities of (12a)-(12d), we let R be given by (1). Figs. 1-3 show comparisons of basis restriction errors of the various transforms for different values of ρ. In the case when a DFT is used, the transformed vector $\hat{x}$ exhibits a conjugate symmetry, i.e.,

$$\hat{x}_k = \hat{x}_{N-k+2}^*, \qquad k = 2, \cdots, N$$

$$\hat{x}_1 = \hat{x}_1^*, \qquad \hat{x}_{(N/2)+1} = \hat{x}_{(N/2)+1}^*.$$

This gives

$$\sigma_k^2 = \sigma_{N-k+2}^2, \qquad k = 2, \cdots, N$$

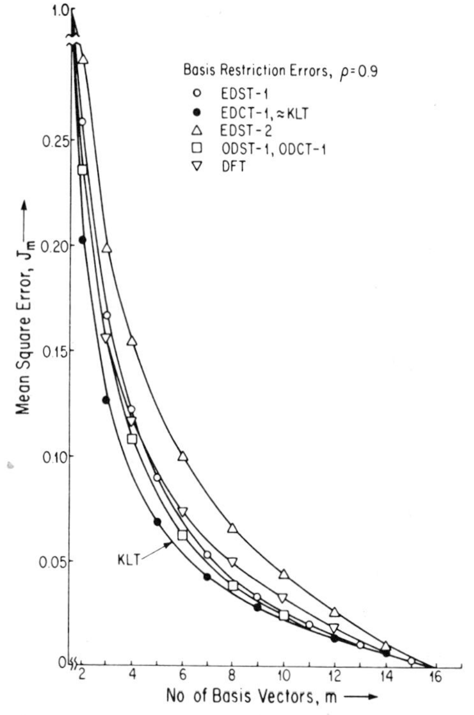

Fig. 1. Basis restriction errors of sinusoidal transforms for a first-order, stationary Markov process of duration $N = 16$, and correlation $\rho = 0.9$.

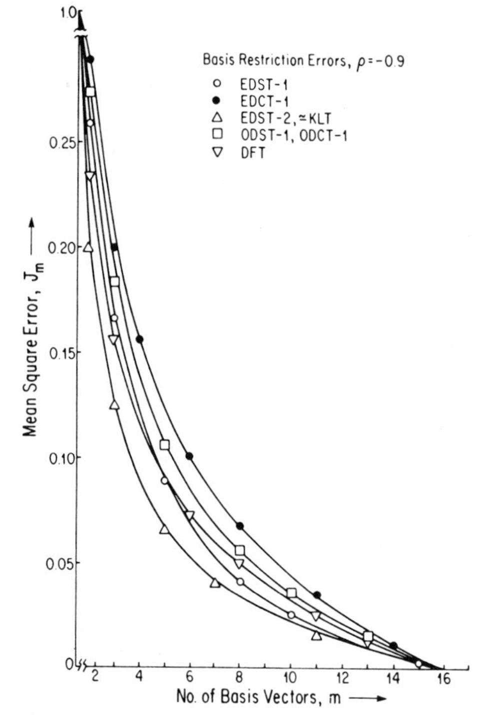

Fig. 2. Basis restriction errors of sinusoidal transforms for a first-order, stationary Markov process of duration $N = 16$, and correlation $\rho = -0.9$.

which results in relatively poor performance of the DFT. An improved ordering of σ_k^2 (with respect to basis restriction error) is obtained by ordering the real and imaginary points of $\hat{x}_k$ according to their variances. This amounts to a permutation between the real and imaginary parts of the vectors of DFT and yields an improved performance. In Figs. 1-4, we have considered this latter ordering.

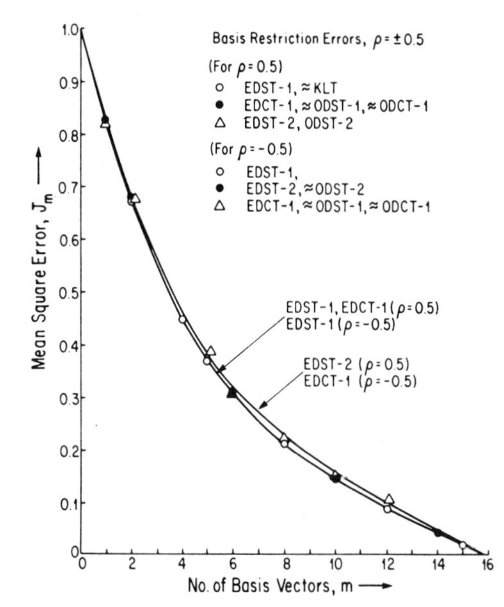

Fig. 3. Basis restriction errors of sinusoidal transforms for a first-order, stationary Markov process of duration $N = 16$, and correlation $\rho = \pm 0.5$.

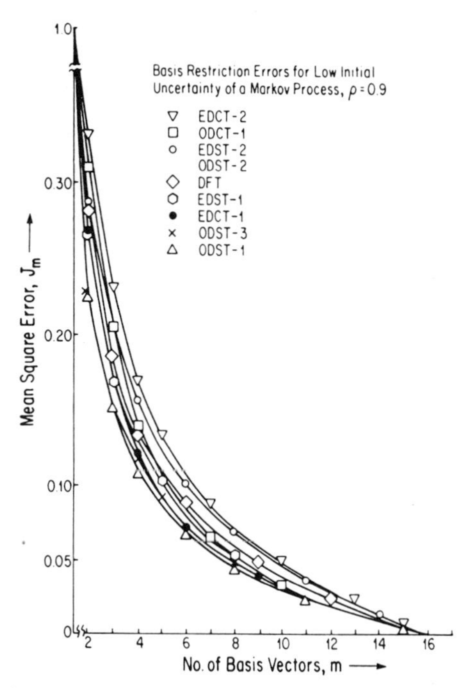

Fig. 4. Basis restriction errors of sinusoidal transforms for a first-order, stationary Markov process of duration $N = 16$, and correlation $\rho = 0.9$, and starting with zero initial condition.

For $\rho = 0.9$, we get the previously reported results of Ahmed *et al.* [6] where the EDCT-1 (generally called, simply, the cosine transform) performs very close to the KLT. This is also expected from our inequality of (12a) and from Table I by noting that Δ(EDCT-1) is minimum (except for the case of the KL transform itself) whenever $0.5 \leqslant \rho < 1$. However, when $\rho = -0.9$, the cosine transform EDCT-1 has the worst performance!

Fig. 4 shows the results for a first-order homogeneous Markov process that starts with zero initial condition and $\rho = 0.9$. Here we find that the odd sine transform (ODST-1) gives the best performance. The various transforms perform in accordance with (12c), and other similar inequalities may be concluded using the criterion of (11). For $|\rho| < 0.5$, the inequality (12b) is also satisfied. However, the performance differences among the various transforms are marginal. These results show that the performance ranking induced by Δ is validated by the performance ranked by the basis restriction error J_m. The inferior performance of the cosine transform, EDCT-1, for $\rho < 0.5$, with respect to some of the other sinusoidal transforms, contradicts the widely held belief that the former is always the best approximation to the KL transform of (2). Although the previous work on cosine transform, such as the empirical results of Ahmed *et al.* [6], who studied its properties for values of $\rho \simeq 0.9$, some asymptotic results and comparisons of cosine transform and DFT reported by Yemini and Pearl [8] and Hamidi and Pearl [7] (where a different criterion is used), and other experimental studies [15], has not specifically compared the cosine transform with all the other transforms for negative or small positive values of ρ; this widely held belief has emerged by their extrapolation. Our results, based on the simple criterion of (11) [and verified by evaluation of (14)], show that the best approximating transform varies with the value of ρ. This is not unreasonable to expect because, after all, the KL transform basis vectors of (2) also vary with the value of ρ.

C. KL Transform Approximation for Arbitrary Random Processes

From the foregoing discussion, one might wonder how the KL transform of an arbitrary random process might be substituted by a (fast) sinusoidal transform. Let A, B be any two covariance matrices. If these matrices commute (i.e., $AB = BA$), then they have an identical set of eigenvectors. Now, define *commuting distance* between any two A and B matrices as

$$\Delta_c = \|AB - BA\| \qquad \Delta_n = \Delta_c/(\|A\| \cdot \|B\|). \tag{15}$$

We conjecture that Δ_c or Δ_n may be used as a measure of proximity of the eigenvectors of A and B. Table I lists the commuting distances between the KL transform covariance matrix R and other J matrices. Comparisons with Δ values show that Δ_c satisfies the same inequalities as does Δ in (12a)–(12d). The difference norm $\Delta = \|R - J\|$ will not be very meaningful for an arbitrary covariance matrix R and the given J matrix. For example, if $R = J^2$, then R and J have the same eigenvectors, but their difference norm Δ is not zero. The above criterion also suggests that for a random process whose covariance function can be measured, if one could reasonably model $R \simeq f(J)$ for a suitably chosen J matrix (e.g., among those J matrices that yield fast transform), then its KL transform would be a (fast) sinusoidal transform. For example, the function f could be an nth-order polynomial. In general, $f(J)$ will not be a Toeplitz matrix so that many nonstationary random processes may be modeled by it. Finally, since all the J matrices are highly sparse (they are tridiagonal whenever $k_3 = k_4 = 0$), only $0(N^2)$ operations are needed in the calculation of Δ_c or Δ_n.

Example: Consider the symmetric, banded Toeplitz matrix

$$H = \{h_{i,j}\}, \qquad 1 \leqslant i,\ j \leqslant N \tag{16}$$

where $h_{i,j} = h_{i-j} = h_{j-i}$ and $h_k = 0$, $|k| > p$. Such matrices arise frequently in the modeling of stationary random processes. For instance, H could represent the covariance function of a pth-order moving average model. Alternatively, H could be the model of a point-spread function (PSF) of an imaging system [18] (note that H represents a finite impulse response (FIR) of a noncausal system). The matrix H has a decomposition

$$H = f(J) - H_b \tag{17}$$

where $f(x)$ is a pth-order polynomial in x and H_b is a matrix of rank at most $2p$. For example, if J is chosen to be the circulant matrix $J(0, 0, -1, -1)$, then H_b is a matrix given as

$$H_b = \begin{bmatrix} 0 & f^T \\ f & 0 \end{bmatrix} \tag{18}$$

where f is a $p \times p$ lower triangular, Toeplitz matrix, defined as

$$f_{i,j} = f_{i-j} = h_{i-j}, \quad i \geq j$$
$$= 0, \quad i < j. \tag{19}$$

If J is chosen to be $J(1, 1, 0, 0)$ then, following [16], it can be shown that (17) is satisfied when

$$H_b = \begin{bmatrix} \tilde{g} & 0 \\ 0 & \tilde{g} \end{bmatrix} \tag{20}$$

where $\tilde{g}$ is a $(p-1) \times (p-1)$ Hankel matrix whose elements are

$$\tilde{g}_{i,j} = h_{i+j}, \quad 1 \leq i, \; j \leq p. \tag{21}$$

Since $h_k = 0$, for $|k| > p$, (21) implies

$$\tilde{g}_{i,j} = h_{i+j}, \quad 2 \leq i + j \leq p. \tag{22}$$

Similarly, the H_b matrices, corresponding to other J matrices, could be found, and the best approximation to the KL transform of H is obtained by minimizing the error

$$\Delta_c = \| JH - HJ \| = \| JH_b - H_bJ \|$$

or

$$\Delta_n = \| JH_b - H_bJ \| / (\| H \| \cdot \| J \|).$$

To extend the previous example to, say, pth-order homogeneous Markov processes, we note that the covariance matrix of such an $N \times 1$ process can be written as [16]

$$R^{-1} = H + H_d \tag{23}$$

where H_d is a sparse matrix of rank at most $2p$. Specifically, H_d contains $p \times p$ matrix blocks in its upper left and lower right corner and is zero elsewhere. Combining (15) and (23) and defining $\tilde{H}_b = H_b - H_d$, we get

$$R = (f(J) - \tilde{H}_b)^{-1} \tag{24}$$

where $\tilde{H}_b$ is of the form

$$\tilde{H}_b = \begin{bmatrix} g_{11} & 0 & g_{12} \\ 0 & 0 & 0 \\ g_{21} & 0 & g_{22} \end{bmatrix} = \begin{bmatrix} I_p & 0 \\ 0 & 0 \\ 0 & I_p \end{bmatrix} \cdot \begin{bmatrix} g_{11} & g_{12} \\ g_{21} & g_{22} \end{bmatrix} \begin{bmatrix} I_p & 0 & 0 \\ 0 & 0 & I_p \end{bmatrix} \tag{25}$$

where $\{g_{ij}\}$ are $p \times p$ matrix blocks and I_p is a $p \times p$ identity matrix. Now using the *ABCD* lemma[3] for inversion of $(A - BCD)$, it can be shown that (24) reduces to

$$R = [f(J)]^{-1} + [f(J)]^{-1} \Delta R \, [f(J)]^{-1} \tag{26}$$

where ΔR is a sparse matrix of rank at most $2p$. For the case when J is a circulant matrix, the derivation of (26) is given in [17]. The general result of (26) for any of the J matrices follows by proceeding in a manner similar to the development of [17]. The explicit form of ΔR is not relevant to our discussion here and its derivation is left to the reader. Using (26) in (15) for $A = R, B = J$, we get

$$\Delta_c = \| [f(J)]^{-1} (J\Delta R - \Delta RJ) [f(J)]^{-1} \| \tag{27}$$

which can be used for evaluating the approximation of the KLT (for R) by different transforms. Equation (27) can be simplified considerably by observing that J and $f(J)$ are diagonalized by the sinusoidal transform Φ and Δ_c is invariant under a unitary transformation. Hence

$$\Delta_c = \sum_{i,j} \left[\frac{(\lambda_i - \lambda_j)}{f_i f_j} \mu_{ij} \right]^2 \tag{28}$$

where

$$f_i = f(\lambda_i), \quad \mu_{ij} = [\Phi \Delta R \Phi^{*T}]_{i,j} \tag{29}$$

and $\{\lambda_i\}$ are the eigenvalues of J. Since ΔR is a sparse matrix and Φ could be a fast transform, μ_{ij} can be easily computed.

D. Fast Sinusoidal Transforms

A large number of sinusoidals yield fast algorithms for the transformation $\hat{x} = \phi x$ or $\hat{x} = \phi^{*T} x$. Table II shows algorithms for several transforms via the DFT. Hence, these transforms could be implemented via a suitable FFT algorithm in $0(N \log_2 N)$ operations. In general, all the transforms that are harmonic sinusoidal functions will yield such a fast implementation. Thus it follows that all of these transforms will also yield a chirp z-transform (CZT) implementation [18] that can be performed in real time via charge-transfer and surface acoustic-wave devices. These devices allow sampled analog signals at input and output and perform computations in time duration that are proportional to N rather than $N \log N$ [19]. Equation (26) suggests that the inverse of any nearly banded, Toeplitz matrix of the form of (23) may be obtained easily via the fast sinusoidal transforms. The desired inverse R, given by (26), is obtained by calculation of $[f(J)]^{-1}$ and ΔR. Since $f(J)$ is diagonalized by Φ, $[f(J)]^{-1}$ requires only $0(N \log N)$ operations; see, e.g., [17] when Φ is the DFT and [16] when Φ is the sine transform EDST-1. Calculation of ΔR requires at most $0(p^3)$ operations. This is because ΔR depends only on $2p$ independent quantities. When H_d in (23) is zero, ΔR can be computed in only $0(p^2)$ operations (see, e.g., [17]). This is because the $2p$ equations for determining ΔR become Toeplitz in structure and could be solved in $0(p^2)$ operations via a Levinson-Trench algorithm. Once the representation of (26) is obtained, any calculation of the type $x = Ry$ (which is the solution of the nearly Toeplitz equation $R^{-1}x = y$) can be obtained in $0(N \log N) + 0(p^2)$ operations

[3] $(A - BCD)^{-1} = A^{-1} - A^{-1}B(C^{-1} - DA^{-1}B)^{-1}DA^{-1}$.

TABLE II
FAST COMPUTATION ALGORITHM FOR DIFFERENT SINUSOIDAL TRANSFORMS VIA DFT, WHICH IS IMPLEMENTED BY A SUITABLE FFT ALGORITHM

Transform	Computation algorithm for $y_m = \sum_{k=1}^{N} x_k \phi_m(k)$, $m = 1, , , ,N$. x_k is assumed zero for unspecified indices
DFT	$y_m = \mathrm{DFT}\{x_k\}_N \overset{\Delta}{=} \frac{1}{\sqrt{N}} \sum_{k=1}^{N} x_k \exp\{\frac{i2\pi(k-1)(m-1)}{N}\}$
EDST-1	$y_m = 2\mathrm{Im}[\mathrm{DFT}\{z_k\}_{2N+2}]$, $z_k = x_{k-1}$, $2 \le k \le N+1$
EDCT-1	$y_m = 2\mathrm{Re}[\mathrm{DFT}\{z_k\}_{2N} \cdot \exp\{\frac{i(m-1)\pi}{2N}\}]$; $z_k = x_k$, $1 \le k \le N$; $2 \le m \le N$ $y_1 = \frac{1}{\sqrt{N}} \sum_{k=1}^{N} x_k$
EDST-2	$y_{m-1} = 2\mathrm{Im}[\mathrm{DFT}\{z_k\}_{2N} \cdot \exp\{\frac{i(m-1)\pi}{2N}\}]$; $z_k = x_k$, $1 \le k \le N$; $2 \le m \le N$ $y_N = \frac{1}{\sqrt{N}} \sum_{k=1}^{N} x_k$
ODST-1	$y_m = 2\mathrm{Im}[\mathrm{DFT}\{z_k\}_{2N+1} \cdot \exp\{\frac{i2\pi(m-1)}{2N+1}\}]$; $z_k = x_k \exp\{\frac{ik\pi}{2N+1}\}$, $1 \le k \le N$
ODST-2	$y_{m-1} = 2\mathrm{Im}[\mathrm{DFT}\{z_k\}_{2N+1}]$; $z_k = x_{k+1}$, $1 \le k \le N$, $2 \le m \le N+1$
ODCT-1	$y_m = 2\mathrm{Re}[\mathrm{DFT}\{z_k\}_{2N+1} \cdot \exp\{\frac{i(2m-1)\pi}{2(2N+1)}\}]$; $z_k = x_k \exp\{\frac{i(k-1)\pi}{2N+1}\}$, $1 \le k \le N$
ODST-3	$y_{m-1} = 2\mathrm{Im}[\mathrm{DFT}\{z_k\}_{2N+1} \cdot \exp\{\frac{i(m-1)\pi}{2N+1}\}]$; $z_k = x_k$, $1 \le k \le N$, $2 \le m \le N+1$
EDCT-2	$y_m = 2\mathrm{Re}[\mathrm{DFT}\{z_k\}_{2N} \cdot \exp\{\frac{i(2m-1)\pi}{2N}\}]$; $z_k = x_k \exp\{\frac{i\pi(k-1)}{2N}\}$, $1 \le k \le N$
EDST-3	$y_m = 2\mathrm{Im}[\mathrm{DFT}\{z_k\}_{2N} \cdot \exp\{\frac{i(2m-1)\pi}{2N}\}]$; $z_k = x_k \exp\{\frac{i\pi(k-1)}{2N}\}$, $1 \le k \le N$

via the fast algorithm. This follows because $z \triangleq [f(J)]^{-1} y$ requires $0(N \log N)$ operations and ΔRz requires only $0(p^2)$ operations. Thus, these fast transforms could be useful in substituting the KL transform of certain random processes and also in fast and exact solutions of sparse, nearly Toeplitz, matrix inversion problems.

E. Asymptotic Equivalence

All of the sinusoidal transforms are asymptotically equivalent (see Section V). This can be seen intuitively by inspection of the J matrices. Since the J matrices differ only at boundary points (i.e., the corner elements), as the size of these matrices gets larger, the boundary effects get smaller. For (doubly) infinite J matrices, the boundary positions go to infinity and all the J matrices are the same. This equivalence can also be established by showing asymptotic equivalence of random processes underlying the J matrices and is done in the next section. It should be pointed out that asymptotic equivalence does not imply that for small (or even practical) values of N these transforms perform quite closely. For example, Fig. 1 shows that performances at compression of $2 (m = 8)$ for the $N = 16$ case can be different by about 1.6 to 2 dB when $\rho = 0.9$. These differences will be larger for higher values of ρ and will be even larger for two-dimensional signals, such as images. In fact, performance differences for images are large enough to warrant the use of the best approximating fast transform.

IV. THE UNDERLYING MARKOV PROCESSES

In general, each sinusoidal transform is the KL transform of a nonstationary, first-order Markov process. From Section II, we know that the J matrix of (6) is related to the covariance matrix of the stationary Markov process of (7). The KL transform of this process is the (slow) transform defined in (2). Now we wish to identify the Markov processes that correspond to the general J matrix. First, assume $k_3 = k_4 = 0$. Let us define a noncausal representation for a stochastic process x as

$$Jx = \nu \tag{30}$$

where ν is a zero mean random process vector whose covariance is given by

$$R_\nu = E[\nu\nu^T] = \beta^2 J. \tag{31}$$

From (30), it follows easily that

$$R_x = E[xx^T] = \beta^2 J^{-1}. \tag{32}$$

Comparison with (6) shows that $R_x = R$ when $J = J(\rho, \rho, 0, 0)$. Equations (30) and (31) can be written explicitly as

$$x_k - \alpha(x_{k-1} + x_{k+1}) = \nu_k, \quad 2 \leqslant k \leqslant N-1 \tag{33a}$$

$$(1 - k_1\alpha)x_1 - \alpha x_2 = \nu_1 \tag{33b}$$

$$(1 - k_2\alpha)x_N - \alpha x_{N-1} = \nu_N \tag{33c}$$

and

$$E[\nu_k \nu_l] = \beta^2(\delta_{k,l} - \alpha\delta_{k-1,l} - \alpha\delta_{k+1,l}). \quad (34)$$

The representation of (30) or, equivalently, (33) is a minimum variance representation of a random process $\{x_k\}$ whose covariance is $\beta^2 J^{-1}$. This means that the linear combination

$$\bar{x}_k = \alpha(x_{k-1} + x_{k+1}), \quad 2 \leq k \leq N-1$$

$$\bar{x}_1 = \alpha x_2/(1 - k_1\alpha)$$

$$\bar{x}_N = \alpha x_N/(1 - k_2\alpha)$$

constitutes the best linear mean square estimate of x_k obtained from all possible neighbors of x_k. In other words, the mean square value of the error $\nu_k = x_k - \bar{x}_k$; i.e., $E[\nu_k^2]$ is minimized with the above definitions of $\bar{x}_k$. This follows by observing that the orthogonality relation

$$E[\bar{x}_k \nu_k] = 0, \quad \text{for all } k$$

which implies $E[x_l \nu_k] = 0$, $k \neq l$, and $E[x_k \nu_k] = E[\nu_k^2] = \beta^2$, is satisfied because, from (30) and (31), we get

$$E[x\nu^T] = J^{-1}E[\nu\nu^T] = \beta^2 I.$$

If $\{x_k\}$ were to be a first-order Markov process, it must have a representation

$$x_k = r_k x_{k-1} + s_k, \quad 2 \leq k \leq N \quad (35)$$

where $\{r_k\}$ and $\{s_k\}$ are deterministic and random white-noise sequences, respectively. Repeated substitution of (35) in (33a) yields the identity

$$[(1 - \alpha r_{k+1})r_k - \alpha]\, x_{k-1} - \alpha s_{k+1} + (1 - \alpha r_{k+1}) s_k = \nu_k. \quad (36)$$

Now, we let the sequence $\{r_k\}$ be such that

$$(1 - \alpha r_{k+1}) r_k - \alpha = 0$$

or

$$r_k = (1 - \alpha r_{k+1})^{-1}\alpha, \quad k = 2, \cdots, N-1. \quad (37)$$

Then (36) reduces to

$$-\alpha s_{k+1} + (1 - \alpha r_{k+1}) s_k = \nu_k, \quad 2 \leq k \leq N. \quad (38)$$

For (38) to be a valid realization for the random process $\{\nu_k\}$, (34) must be satisfied. Defining

$$E[s_k s_l] = p_k \delta_{k,l} \quad (39)$$

and substituting (38) in (34), using (39), and solving for p_k, we get

$$p_k = \frac{\beta^2 r_k}{\alpha}, \quad k = 3, \cdots, N-1. \quad (40)$$

Using (34) in (32c), we obtain

$$r_N = \alpha/(1 - k_2\alpha), \quad |r_N| < 1 \text{ for } |k_2| < 1 \quad (41)$$

and

$$s_N = \nu_N/(1 - k_2\alpha). \quad (42)$$

For (35) to be a Markov representation we must assume x_1 and s_2 to be independent. The variances of the initial state x_1 and random variable s_2 are then obtained from (42) and (34) as

$$E[x_1^2] = \beta^2/(1 - \alpha r_2 - k_1\alpha) \quad (43a)$$

$$E[s_2^2] = (r_2 + k_1)/\alpha. \quad (43b)$$

This completely specified (35). Thus, we see that for each J matrix there is a nonstationary (in general) Markov process given by (35) whose KL transform is determined by the eigenvectors of J. The covariance of the random process x is $\beta^2 J^{-1}$. Thus all the fast sinusoidal transforms are KL transforms of nonstationary Markov processes; whereas, the (slow) transform of (2) is the KL transform of the stationary Markov process of (7).

V. Asymptotic Equivalence

The transforms of the sinusoidal family can be shown to be asymptotically equivalent. To establish this equivalence we first consider some definitions following Gray [20] and Pearl [21].

1) Let A_N be a real $N \times N$ matrix. Then

$$\eta_w(A_N) = \eta_w \triangleq \left(\frac{1}{N}\sum_{i=1}^{N}\sum_{j=1}^{N} a_{i,j}^2\right)^{1/2} \quad (44)$$

is called the *weak (or Hilbert-Schmidt) norm* of A_N. If $\{\lambda_k\}$ are the eigenvalues of $A_N^T A_N$, then

$$\eta_s(A_N) = \max_k \{\lambda_k\} \quad (45)$$

is called the *strong norm* of A_N.

2) Two sequences of matrices $\{A_N\}$ and $\{B_N\}$, $N = 1, 2, \cdots$ are considered asymptotically equivalent, written as $A_N \sim B_N$, if

$$\lim_{N\to\infty} \eta_w(A_N - B_N) = 0. \quad (46)$$

3) The matrix A_N is weakly or strongly bounded if $\eta_w(A_N) < \infty$ or $\eta_s(A_N) < \infty$, respectively. Based on these definitions we can state the following useful theorem [20].

A. Theorem 1

Let $\{A_N\}$ and $\{B_N\}$ be strongly bounded, real, symmetric matrix sequences with

$$\lambda_{\min} \leq \eta_s(A_N), \quad \eta_s(B_N) \leq \lambda_{\max}. \quad (47)$$

Then, if $\{A_N\}$ and $\{B_N\}$ are asymptotically equivalent, and for an arbitrary function $F(x)$, continuous on $[\lambda_{\min}, \lambda_{\max}]$

$$\lim_{N\to\infty} \frac{1}{N}\sum_{k=1}^{N} F(\alpha_{N,k}) = \lim_{N\to\infty} \frac{1}{N}\sum_{k=1}^{N} F(\beta_{N,k}) \quad (48)$$

where $\{\alpha_{N,k}\}$ and $\{\beta_{N,k}\}$ are the eigenvalues of A_N and B_N, respectively. This relation is valid if either of the two limits exists.

This theorem also implies, for any $m < \infty$

$$\lim_{N\to\infty} \frac{1}{N} \sum_{k=m+1}^{N} F(\alpha_{N,k}) = \lim_{N\to\infty} \frac{1}{N} \sum_{k=m+1}^{N} F(\beta_{N,k}). \quad (49)$$

Thus, if $F(x) = x$, (49) states that the basis restriction errors in data compression of two classes of random processes whose covariances are A_N and B_N are asymptotically equivalent. For $F(x) = \log_2 (x/\theta)$, $(0 < \theta < x, \forall x > 0)$ the same conclusion can be drawn from (48) for the rate-distortion functions of these two random processes.

4) Let $\{A_N\}$ and $\{B_N\}$ be a sequence of real strongly bounded convariance matrices with corresponding KL transform unitary matrices $\{\Phi_N\}$ and $\{\psi_N\}$, respectively. Define

$$\hat{\Lambda}_B = \text{diag } \{\Phi_N B_N \Phi_N^{*T}\}$$
$$\hat{B}_N = \Phi_N^{*T} \hat{\Lambda}_B \Phi_N.$$

The matrix $\hat{B}_N$ is called the projection of B_N on the sequence $\{A_N\}$. Note that both $\hat{B}_N$ and A_N are diagonalized by Φ_N.

5) The transform sequence $\{\Phi_N\}$ is said to asymptotically cover the sequence $\{A_N\}$ if A_N and $\hat{B}_N$ are asymptotically equivalent, i.e., $A_N \sim \hat{B}_N$.

6) The transform sequences $\{\Phi_N\}$ and $\{\psi_N\}$ are said to be asymptotically equivalent if $A_N \sim \hat{B}_N$ and $B_N \sim \hat{A}_N$.

We are now ready to prove the following asymptotic equivalence results.

a) The J matrices of the sinusoidal family are asymptotically equivalent. This is easily seen by noting that

$$\eta_W^2(\Delta) = \frac{\alpha^2}{N} \{(k_1' - k_1)^2 + (k_2' - k_2)^2 + (k_3' - k_3)^2 + (k_4' - k_4)^2\} \quad (50)$$

where

$$\Delta \triangleq J - J' \triangleq J(k_1, k_2, k_3, k_4) - J(k_1', k_2', k_3', k_4').$$

Clearly, $\eta_W(\Delta) \to 0$ as $N \to \infty$.

b) All the sinusoidal transforms are asymptotically equivalent. Let J_N and J_N' be two J matrices with Φ_N and ψ_N the corresponding transforms.

Define

$$\hat{\Lambda}_N' = \text{diag } \{\Phi_N J_N' \Phi_N^{*T}\}.$$

Since

$$J_N' = J_N - \Delta$$

we obtain

$$\Lambda_N' = \Phi_N J_N \Phi_N^{*T} - \text{diag } [\Phi_N \Delta \Phi_N^{*T}].$$

This gives

$$\lambda_N'(k) = \lambda_N(k) + \alpha(k_1 - k_1') \,|\phi_N(k,1)|^2 + \alpha(k_2 - k_2') \cdot |\phi_N(k,N)|^2 + \alpha(k_3' - k_3)\, \phi_N(k,1)\, \phi_N^*(k,N) + \alpha(k_4' - k_4)\, \phi_N^*(k,1)\, \phi_N(k,N). \quad (51)$$

From this, one readily obtains

$$\lim_{N\to\infty} \frac{1}{N} \sum_{k=1}^{N} \lambda_N'(k) = \lim_{N\to\infty} \frac{1}{N} \sum_{k=1}^{N} \lambda_N(k)$$

i.e., $\Lambda_N \sim \hat{\Lambda}_N'$, which implies $J_N \sim \hat{J}_N'$. Similarly, we can show $J_N' \sim \hat{J}_N$. From definition 6), this implies $\{\Phi_N\}$ and $\{\psi_N\}$ are asymptotically equivalent.

c) The Markov realizations of the sinusoidal family are asymptotically equivalent.

In the steady state, i.e., for $N = \infty$, the coefficients r_k satisfy

$$(1 - \alpha r) r - \alpha = 0.$$

Letting $\alpha \triangleq \rho/(1 + \rho^2)$, we get

$$r = \rho \text{ or } 1/\rho. \quad (52)$$

Since $|\alpha| < \frac{1}{2}$ for a valid covariance matrix J, $r = \rho$, $|\rho| < 1$ is the only admissible solution. Indeed, if $|r_N| < 1$, then from (37) $|r_k| < 1$, $\forall k$, so that r_k converges to ρ. Also, in the steady state

$$E[s_k^2] = p_k = \frac{\beta^2 \rho}{\alpha} = (1 - \rho^2). \quad (53)$$

Hence, (35) becomes the homogeneous Markov process

$$x_k = \rho x_{k-1} + s_k \quad (54)$$

which will be stationary as $k \to \infty$ for arbitrary initial conditions.

For the case of the DFT, $(k_3 = k_4 = -1)$. Letting $\alpha = \rho/(1 + \rho^2)$, we can obtain a circular factorization of $J(0, 0, -1, -1)$ to give a stationary representation

$$x_k = \rho x_{k-1} + s_k, \quad 2 \leqslant k \leqslant N \quad (55)$$

$$x_1 = \rho x_N + s_1 \quad (56)$$

$$E[s_k s_l] = \beta^2 \delta_{k,l}.$$

As $N \to \infty$, for any arbitrary initial value x_1, this will be equivalent to (54) as $k \to \infty$.

d) The sinusoidal transforms are asymptotically equivalent to the KL transform of any finite-order, stationary Markov process.

The spectral density function of a pth-order, stationary Markov process is given by

$$S(\omega) = \frac{2}{\left(1 - 2 \sum_{k=1}^{p} a_k \cos \omega k\right)}. \quad (57)$$

We will assume $S(\omega)$ is strictly positive. From (23) to (25), the $N \times N$ covariance matrix of a Markov sequence of length N satisfies

$$R_N^{-1} = f(J) + \tilde{H}_b \quad (58)$$

where $\tilde{H}_b$ is a sparse matrix with at most $4p^2$ nonzero terms. Defining

$$R_N^0 = [f(J)]^{-1}$$

it is easy to show that $R_N^{-1} \sim (R_N^0)^{-1}$. The eigenvalues of R_N^0 are given by

$$\lambda_N^0(k) = \beta^2 \Big/ \left(1 - 2 \sum_{l=1}^{p} a_l \cos kl\theta_N\right) \tag{59}$$

where θ_N depends on N and the J matrix.

Defining

$$\hat{D}_N = \text{diag}\,[\Phi R_N^{-1} \Phi^{*T}], \qquad \hat{H}_N = \Phi^{*T} \hat{D}_N \Phi$$

we obtain

$$\hat{D}_N = (\Lambda_N^0)^{-1} + \text{diag}\,(\Phi \tilde{H}_b \Phi^{*T}). \tag{60}$$

Using (25), one obtains $\hat{D}_N \sim (\Lambda_N^0)^{-1}$ or $\hat{H}_N \sim f(J)$. Similarly, we can show that $\hat{f}(J) \sim R_N^{-1}$. This means Φ is asymptotically equivalent to the KL transform of a pth-order Markov process.

V. Conclusions

In conclusion, a family of sinusoidal transforms has been introduced. All the transform family members are generated by a class of J matrices that is a slight variation of the Jacobi matrices. Equivalently, all the sinusoidal transforms are solutions of a second-order eigenvalue difference equation, subject to different boundary conditions. Each sinusoidal transform is the KL transform of a first-order Markov process. The fast sinusoidal transforms are the KL transforms of nonhomogeneous (and hence nonstationary) random processes, whereas, for the stationary Markov process, the corresponding KLT is not fast. Applications of the transform family discussed here are in the modeling of random processes such that their KL transforms are fast transforms and are also in fast solutions of sparse Toeplitz or nearly Toeplitz systems of equations. Both of these applications could be useful in many signal-processing problems, such as data compression and digital filtering of signals. Although all the sinusoidal transforms are asymptotically equivalent, their performance differences for signals with finite support can be significant for certain ranges of signal parameters.

References

[1] J. W. Cooley and J. W. Tukey, "An algorithm for machine calculation of complex Fourier series," *Math. Comput.*, vol. 19, pp. 297–301, 1965.

[2] E. O. Brigham, *The Fast Fourier Transform.* Englewood Cliffs, NJ: Prentice-Hall, 1974.

[3] N. Ahmed and K. R. Rao, *Orthogonal Transforms for Digital Signal Processing.* New York: Springer, 1975.

[4] W. K. Pratt, *Image Processing.* Reading, MA: Addison-Wesley, 1977.

[5] W. D. Ray and R. M. Driver, "Further decomposition of the Karhunen-Loeve series representation of a stationary random process," *IEEE Trans. Inform. Theory*, vol. IT-16, pp. 663–668, Nov. 1970.

[6] N. Ahmed, T. Natarajan, and K. R. Rao, "Discrete cosine transform," *IEEE Trans. Comput.*, vol. C-23, pp. 90–93, Jan. 1974.

[7] M. Hamidi and J. Pearl, "Comparison of the cosine and Fourier transforms of Markov-1 signals," *IEEE Trans. Acoust., Speech, Signal Processing*, vol. ASSP-24, pp. 428–429, Oct. 1976.

[8] Y. Yemini and J. Pearl, "Asymptotic properties of discrete unitary transforms," School of Eng. and Appl. Sci., Univ. of Calif., Los Angeles, UCLA-ENG-Rep. 7566, Nov. 1975.

[9] A. K. Jain, "Computer program for fast Karhunen Loeve transform algorithm," Dep. Elec. Eng., State Univ. of New York, Buffalo, Final Rep. NASA Contract NAS8-31434, Feb. 1976.

[10] —, "A fast Karhunen Loeve transform for a class of stochastic processes," *IEEE Trans. Commun.*, vol. COM-24, pp. 1023–1029, Sept. 1976.

[11] A. K. Jain, S. H. Wang, and Y. Z. Liao, "Fast Karhunen Loeve transform data compression studies," in *Proc. National Telecommunications Conf.*, Dallas, TX, Nov. 1976.

[12] A. K. Jain, "Some new techniques in image processing," in *Image Science Mathematics*, C. O. Wilde and E. Barrett, Eds. North Hollywood: Western Periodicals, Nov. 1976, pp. 201–223.

[13] A. Z. Meiri, "The pinned Karhunen-Loeve transform of a two-dimensional Gauss Markov field," in *Proc. 1976 SPIE Conf. Image Processing*, San Diego, CA, Aug. 1976.

[14] R. Bellman, *Introduction to Matrix Analysis.* New York: McGraw-Hill, 1970.

[15] W. C. Chen and C. H. Smith, "Adaptive coding of color images using cosine transform," in *Proc. Int. Communications Conf.*, June 1976, pp. 47-7–47-13.

[16] A. K. Jain, "An operator factorization method for restoration of blurred images," *IEEE Trans. Comput.*, vol. C-25, pp. 1061–1071, Nov. 1977.

[17] —, "Fast inversion of banded Toeplitz matrices by circular decompositions," *IEEE Trans. Acoust., Speech, Signal Processing*, vol. ASSP-26, pp. 121–127, 1978.

[18] L. R. Rabiner *et al.*, "The chirp Z transform algorithm," *IEEE Trans. Audio Electroacoust.*, vol. AU-17, pp. 86–92, 1969.

[19] H. J. Whitehouse, R. W. Means, and J. M. Speiser, "Signal processing using transversal filter technology," Naval Undersea Center, San Diego, CA, Tech. Rep., 1975.

[20] R. M. Gray, "Toeplitz and circulant matrices: A review," Stanford Univ., Stanford, CA, Tech. Rep. SU-SEL-71-032, June 1971; also —, "On the asymptotic eigenvalue distribution of Toeplitz matrices," *IEEE Trans. Inform. Theory*, vol. IT-18, pp. 725–30, Nov. 1972.

[21] J. Pearl, "Asymptotic equivaluence of spectral representations," *IEEE Trans. Acoust. Speech, Signal Processing*, vol. ASSP-23, pp. 547–551, Dec. 1975.

[22] B. J. Fino and R. Algazi, "A unified treatment of fast unitary transforms," *SIAM J. Comput.*, vol. 6, Dec. 1977.

[23] A. K. Jain, *Multidimensional Techniques in Digital Image Processing*, to be published.

11

Reprinted from *Comput. Electr. Eng.* **7**:45–55 (1980)

ON THE COMPUTATION AND THE EFFECTIVENESS OF DISCRETE SINE TRANSFORM†

P. YIP‡ and K. R. RAO
Department of Electrical Engineering, University of Texas at Arlington, Arlington, TX 76019, U.S.A.

(*Received* 25 *June* 1979; *received for publication* 11 *January* 1980)

Abstract—The effectiveness of the discrete sine transform (DST) in terms of residual correlation as developed by Hamidi and Pearl[1] is investigated. Based on this criterion, the DST is compared with other discrete transforms such as cosine (DCT)[2] and Fourier (DFT)[3]. The odd-even property of the DST is established. Based on this, the DST is computed through the Walsh–Hadamard transform (WHT). This development parallels that of Jones *et al.*[4] for other transforms such as slant (ST), DFT, DCT, etc.

INTRODUCTION

The discrete sine transform (DST) was first introduced by Jain[5], as the optimal fast transform for a first order Markov process, where the boundary conditions are known. Subsequently, Yip and Rao[6] have developed a sparse-matrix factorization for the DST leading to a computational algorithm that is faster than that for the discrete cosine transform (DCT)[2, 7–13]. Hamidi and Pearl[1] have evaluated the effectiveness of DCT and discrete Fourier transform (DFT)[3] in decorrelating a signal based on Markov-1 process. This evaluation is now extended to the DST. Hein and Ahmed[14] have shown that the DCT can be computed through the Walsh-Hadamard transform (WHT) and have suggested a hardware implementation for processing TV signals in real time. Similar techniques have been developed by Tadokoro and Higuchi[15] for computation of DFT through WHT and by Ohira *et al.*[16] for computation of slant transform (ST) through WHT. This concept is generalized by Jones *et al.*[4] for any transform that has even/odd structure. Specifically implementation of DST through WHT is now shown.

DISCRETE SINE TRANSFORM

Jain[5] defined the basis set for DST as

$$\phi_{ij} = \sqrt{\left(\frac{2}{M+1}\right)} \sin\left(\frac{(ij\pi)}{(M+1)}\right) \qquad i, j = 1, 2, \ldots, M. \tag{1}$$

The first 15 basis functions for the DST are shown in [17].

Kekre and Solanki, [18] on the other hand, defined the DST as

$$\begin{aligned}\phi_{ij} &= \sqrt{\left(\frac{2}{M}\right)} \sin\left(\frac{\pi(i+1)(2j+1)}{2M}\right) \qquad && i = 0, 1, \ldots, M-2 \\ &= (-1)^j (M)^{-1/2} && i = M-1\end{aligned} \tag{2}$$

for $j = 0, 1, \ldots, M-1$.

A. DST of Jain[5], $(\mathrm{DST})_\mathrm{I}$. The DST of a data sequence $x(i)$, $i = 1, 2, \ldots, M$ and its inverse can be respectively defined as

$$X(j) = \sqrt{\left(\frac{2}{M+1}\right)} \sum_{i=1}^{M} x(i) \sin\left(\frac{ij\pi}{M+1}\right), \qquad j = 1, 2, \ldots, M \tag{3}$$

†This research is supported in part by NSERC Grant No. A3635. A paper based on part of this research was presented at the 22nd Midwest Symposium on Circuits and Systems, Philadelphia, PA, 17–19 June (1979).

‡Associate Professor of Applied Mathematics on leave of absence, Applied Math Department, McMaster University, Hamilton, Ontario, Canada, L8S-4K1.

and

$$x(i) = \sqrt{\left(\frac{2}{M+1}\right)} \sum_{j=1}^{M} X(j) \sin\left(\frac{ij\pi}{M+1}\right), \qquad i = 1, 2, \ldots, M$$

where $X(j)$, $j = 1, 2, \ldots, M$ is the transform sequence.

The basis functions in (1) are orthonormal, i.e.

$$\sum_{k=1}^{M} \sin\left[\frac{ki\pi}{(M+1)}\right] \sin\left[\frac{kj\pi}{(M+1)}\right] = \left(\frac{M+1}{2}\right) \delta_{ji} \tag{4}$$

where δ_{ji} is the Kronecker delta. Jain[5] has shown that the N-point DST can be implemented via a $2(N+1)$-point FFT.

B. DST of Kekre and Solanki[18], $(\text{DST})_2$. Corresponding to $(\text{DST})_1$, $(\text{DST})_2$ and its inverse can be expressed as

$$X(j) = \sqrt{\left(\frac{2}{M}\right)} \sum_{i=1}^{M-1} x(i) \sin\left[\frac{\pi(i+1)(2j+1)}{2M}\right] \qquad j = 0, 1, \ldots, M-2$$

$$X(M-1) = \frac{1}{\sqrt{(M)}} \sum_{i=0}^{M-1} (-1)^i x(i) \tag{5}$$

and

$$x(i) = \sqrt{\left(\frac{2}{M}\right)} \left[\sum_{j=0}^{M-2} X(j) \sin\left[\frac{\pi(i+1)(2j+1)}{2M}\right] + \frac{(-1)^i}{\sqrt{(2)}} X(M-1)\right], \qquad i = 0, 1, \ldots, M-1 \tag{6}$$

since

$$\sum_{k=0}^{M-1} \sin\left(\frac{\pi(i+1)(2k+1)}{2M}\right) \sin\left(\frac{\pi(j+1)(2k+1)}{2M}\right) = \left(\frac{M}{2}\right) \delta_{ij}, \qquad i \neq M-1$$
$$= M\delta_{ij}, \qquad i = M-1.$$

Yip and Rao[6] have shown that by recursive decomposition of the DST matrix into sparse matrix factors with real elements a fast algorithm similar to that of Chen *et al.*[2] for the DCT can be developed. Both the number of additions and number of multiplications are much less for the DST compared to those for DCT. Wagh and Ganesh[19] have also developed a fast DST algorithm based on multidimensional convolutions and group theory. Computationally the algorithm based on sparse-matrix factorization[6] is simpler compared to this technique (Table 1).

EFFECTIVENESS OF DST

Hamidi and Pearl[1] have defined the degree of decorrelation for an orthogonal transform and have developed expressions for the correlation left undone for the DCT and DFT when the signal statistic is governed by a first order Markov process. Similar development is now carried out for the DST to test how effective it is in decorrelating a Markov-1 signal. As is well known, Karhunen–Loêve transform (KLT)[7, 17, 20, 21] diagonalizes a covariance matrix, Σ Hence there is no residual correlation in the KLT domain. Following the development in [1], the Hilbert-Schmidt norm (weak Norm) of $T - T_U$ is defined as

$$|T - T_U|^2 = \frac{1}{M} \sum_{m,n=0}^{M-1} |(T - T_U)_{nm}|^2 \tag{7}$$

where T = Toeplitz correlation matrix, U = orthogonal transform, $T' = UTU^{-1}$, $T'_U =$

Table 1. Comparison of # of multiplications required for fast implementation of DST

Length M	Yip and Rao[6]	Wagh and Ganesh[20]
3	2	1
4		6
5		4
7	8	5
8		12
9		11
13		16
15	30	18
20		36
21		40
29		47
30		80
31	54	58
33		95
63	130	179
65		168
99		255
127	310	543
251		586
255	730	1636

diag $(UTU^{-1}) = \text{diag}(T'_{11}, T'_{22}, \ldots, T'_{MM})$ and $T_U = U^{-1}T'_U U$. The norm in (7) can be reduced to

$$|T - T_U|^2 = |T|^2 - \frac{1}{M}\sum_{m=0}^{M-1} |(UTU^{-1})_{mm}|^2 \tag{8}$$

which implies that the higher the norm of the diagonal vector of the transformed matrix the lower is $|T - T_U|$ and the better is the transform. Also defining $|T - I|^2$ as the amount of cross correlation in the original signal, (I is the unit matrix) the fractional correlation left "undone" by U can be represented as

$$(|T - T_U|^2)/(|T - I|^2). \tag{9}$$

For a Markov-1 signal, the covariance matrix and its weak norm respectively are

$$T = [\rho^{|j-k|}], \qquad j, k = 0, 1, \ldots, M-1$$

and

$$|T|^2 = \frac{1+\rho^2}{1-\rho^2} - \frac{2\rho^2(1-\rho^{2M})}{M(1-\rho^2)} \tag{10}$$

where $|\rho| < 1$ is the covariance coefficient between adjacent samples. Also[1]

$$|(T - I)|^2 = \frac{2\rho^2}{M(1-\rho^2)^2}[M - 1 - M\rho^2 + \rho^{2M}]. \tag{11}$$

Residual correlation in Markov-1 *signal for* (DST)$_1$

We shall derive here the expression

$$|T - T_{s_1}|^2 = |T|^2 - \frac{1}{M}\sum_{m=0}^{M-1} |(S_1 T S_1^{-1})_{mm}|^2 \tag{12}$$

where S_1 stands for $(\mathrm{DST})_1$ matrix.

$$(S_1TS_1^{-1})_{mm}=\frac{1}{2}\left\{\frac{(1+\beta\rho)}{(1-\beta\rho)}+\frac{(1+\beta^{-1}\rho)}{(1-\beta^{-1}\rho)}-\frac{[1+(-1)^m\rho^{M+1}]}{(M+1)}\cdot\left\{\frac{1}{(1-\beta\rho)}-\frac{1}{(1+\beta\rho)}\right\}^2\right. \tag{13}$$

where $\beta=e^{i(m+1)\pi/(M+1)}$, $i=\sqrt{(-1)}$, and $\beta^{M+1}=(-1)^{m+1}$.

It is not difficult to show that $(S_1TS_1^{-1})_{mm}$ is real and therefore

$$|(S_1TS_1^{-1})_{mm}|^2=\{(S_1TS_1^{-1})_{mm}\}^2=(i)_{mm}+(ii)_{mm}+(iii)_{mm} \tag{14}$$

where

$$(i)_{mm}=\frac{1}{4}\left\{\frac{(1+\beta\rho)}{(1-\beta\rho)}+\frac{(1+\beta^{-1}\rho)}{(1-\beta^{-1}\rho)}\right\}^2 \tag{15a}$$

$$(ii)_{mm}=\frac{[1+(-1)^m\rho^{M+1}]^2}{(M+1)^2}\left\{\frac{1}{(1-\beta\rho)}-\frac{1}{(1+\beta\rho)}\right\}^4 \tag{15b}$$

and

$$(iii)_{mm}=-\frac{[1+(-1)^m\rho^{M+1}]}{(M+1)}\left\{\frac{(1+\beta\rho)}{(1-\beta\rho)}+\frac{(1+\beta^{-1}\rho)}{(1-\beta^{-1}\rho)}\right\}\cdot\left\{\frac{1}{(1-\beta\rho)}-\frac{1}{(1+\beta\rho)}\right\}. \tag{15c}$$

We can perform the sum over m on each of these terms to obtain

$$\sum_{n=0}^{M-1}(i)_{mm}=(M+1)\frac{(1+\rho^2)}{(1-\rho^2)}-\frac{(1+\rho^2)^2}{(1-\rho^2)^2}-\frac{4\rho^2}{(1-\rho^2)^2} \tag{16a}$$

$$\sum_{m=0}^{M-1}(ii)_{mm}=\frac{2[1+\rho^{2(M+1)}]}{(M+1)}\left\{1-\frac{(1-3\rho^2)}{(1-\rho^2)^3}\right\} \tag{16b}$$

and

$$\sum_{m=0}^{M-1}(iii)_{mm}=\frac{4\rho^2}{(1-\rho^2)^2}. \tag{16c}$$

Combining these results we obtain for $M=2^k-1$, $k\geqslant 2$,

$$\sum_{m=0}^{M-1}|(S_1TS_1^{-1})_{mm}|^2=(M+1)\frac{(1+\rho^2)}{(1-\rho^2)}-\frac{(1+\rho^2)^2}{(1-\rho^2)^2}+\frac{2[1+\rho^{2(M+1)}]}{(M+1)}\left\{1-\frac{(1-3\rho^2)}{(1-\rho^2)^3}\right\}. \tag{17}$$

Substitution of (10) and (17) in (12) leads to

$$|T-T_{s_1}|^2=\frac{2\rho^2(1+\rho^{2M})}{M(1+\rho^2)^2}-\frac{2[1+\rho^{2(M+1)}]}{M(M+1)}\left[1-\frac{1-3\rho^2}{(1-\rho^2)^3}\right]. \tag{18}$$

Substitution of (11) and (18) in (9) yields the fractional residual correlation for $(\mathrm{DST})_1$. This is compared with that of DCT in Fig. 1. Inspection of (18) leads to the following comments

$$\text{(i)}\ \lim_{M\to\infty}|T-T_{s_1}|=0 \tag{19}$$

i.e. $(\mathrm{DST})_1$ is asymptotically equivalent to the Karhunen–Loêve transform (KLT)[20–22] of Markov-1 signals.

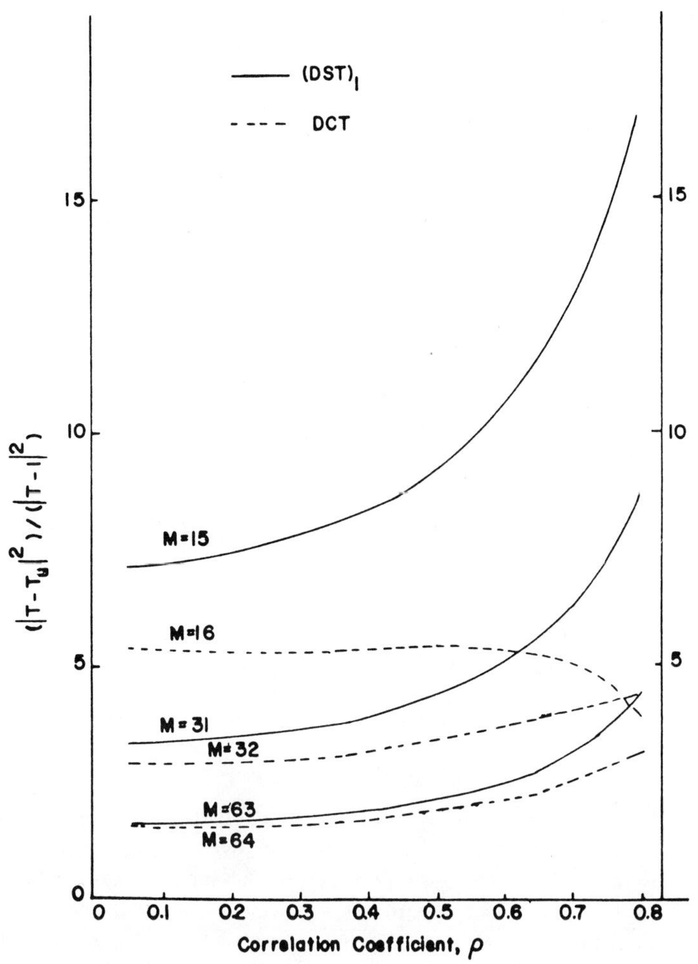

Fig. 1. Fractional correlation left undone by DCT and DST for Markov-1 signal.

(ii) For large M, and $|\rho|| < 1$

$$\|T - T_{s_I}| \approx \sqrt{(2)}\frac{\rho(1+\rho^2)^{1/2}}{(1-\rho^2)}0(M^{-1/2}). \tag{20}$$

This behavior is similar to those of DCT and DFT[1].

(iii) The differences in residual correlation when compared with DCT and (DST)$_I$: Strictly speaking an expression for the difference in residual correlation cannot be derived, because of the different restrictions on the sample size—DST here requires $M = 2^k - 1$, whereas DCT and DFT require $M = 2^k$. But for the large M, we may replace $(M + 1)$ by M to obtain

(a)

$$|T - T_F|^2 - |T - T_{s_I}|^2 \approx \frac{-2\rho^2}{M^2(1-\rho^2)} \tag{21}$$

(b)

$$|T - T_c|^2 - |T - T_{s_I}|^2 \approx \frac{-2\rho^2(3+4\rho+\rho^4)}{M^2(1-\rho^2)^3}. \tag{22}$$

For $|\rho| < 1$, both (21) and (22) are always negative but decrease as M^{-2}. The implication here is that, based on this definition of performance, the DST will not measure up to that of DFT or that of DCT.

However, for small ρ, the performance of (DST)$_I$ is not drastically different from those of DFT and DCT.

Residual correlation in Markov-1 signal for (DST)$_2$

Denoting the (DST)$_2$ matrix as S_2 we get

$$(S_2TS_2^{-1})_{mm} = \frac{(1+\rho)}{(1-\rho)} - \frac{1}{M(1-\rho^2)}[2\rho(1-\rho^M)], \qquad m = M-1 \tag{23a}$$

$$(S_2TS_2^{-1})_{mm} = \frac{1}{2}\left\{\frac{(1+\beta\rho)}{(1-\beta\rho)} + \frac{(1+\beta^{-1}\rho)}{(1-\beta^{-1}\rho)}\right\} - \frac{\rho}{M}[1+(-1)^m\rho^M]\cdot\left\{\frac{\beta^{-1/2}}{(1-\beta^{-1}\rho)} - \frac{\beta^{1/2}}{(1-\beta\rho)}\right\}^2,$$
$$m \neq M-1 \tag{23b}$$

where

$$\beta = e^{(m+1)i\pi/M}, \qquad i = \sqrt{(-1)} \text{ and } \beta^M = (-1)^{m+1}.$$

Based on these expressions we get, for $M = 2^k$, $k \geq 2$

$$\sum_{m=0}^{M-1} |(S_2TS_2^{-1})_{mm}|^2 = M\frac{(1+\rho^2)}{(1-\rho^2)} + \frac{2\rho(2\rho^3+\rho^2+\rho+4)}{(1-\rho^2)^2} + \frac{2\rho}{M}\left\{\frac{\rho(3\rho^2-1)(1+\rho^{2M})}{(1-\rho^2)^3} - \frac{2(1-\rho^M)}{(1-\rho^2)}\right.$$
$$\left. - \frac{2(1-\rho)(1-\rho^M)}{(1+\rho)^3}\right\} + \frac{4\rho^2}{M^2}(1-\rho^M)^2\left[\frac{1}{(1-\rho^2)^2} - \frac{2}{(1+\rho)^4}\right]. \tag{24}$$

We note that the comments indicated regarding (DST)$_1$ are also valid here.

COMPUTATION OF DST

Jones *et al.*[4] have shown that any even/odd transform can be computed by sparse-matrix multiplication of any other even/odd transform. An even/odd transform is one which has one-half even and one-half odd transform vectors (rows). Several transforms such as DCT, WHT, ST, DST and DFT belong to this category. As WHT can be implemented with subtractions/additions only, the hardware realization of DCT and other transforms via WHT has been investigated and DCT–WHT processor is being built at NASA Ames Research Center[4]. As the DST has the even/odd structure it can also be computed via the WHT. This is illustrated with few examples and the computational requirements are tabulated.

Even/odd structure of (DST)$_1$. From (1), when $k = (M-j+1)$ for $j = 1, 2, \ldots, M/2$

$$\phi_{ik} = \sqrt{\left(\frac{2}{M+1}\right)} \sin\left[\frac{i(M-j+1)\pi}{M+1}\right] = (-1)^{i+1}\phi_{ij} \qquad i = 1, 2, \ldots, M. \tag{25}$$

This means that the odd numbered rows i.e. $i = 1, 3, 5, \ldots, M-1$ are even vectors and the rest are odd vectors. Hence the conversion matrix relating (DST)$_1$ and any other even/odd transform has block diagonal structure with both blocks of size $((M/2)\times(M/2))$. To investigate whether further structure can be extracted, we consider the rearranged transform matrix as

$$S_1 = \begin{bmatrix} E_1 & \tilde{E}_1 \\ D_1 & -\tilde{D}_1 \end{bmatrix} \tag{26}$$

where E_1 and D_1 are matrices of size $((M/2)\times(M/2))$ made up of the first $N/2$ columns of even and odd rows respectively.

$\tilde{E}_1 = E_1\tilde{I}$ and $\tilde{D}_1 = D_1\tilde{I}$ where $\tilde{I}$ is the opposite diagonal identity matrix. If either E_1 or D_1 has even/odd structure, further partitioning, that will lead to smaller sub-blocks in the sparseness of the conversion matrix is possible. Now

$$(E_1)_{ij} = \sqrt{\left(\frac{2}{M+1}\right)} \sin[(2i-1)j\pi/(M+1)] \qquad i, j = 1, \ldots, M/2 \tag{27}$$

and

$$(D_1)_{ij} = \sqrt{\left(\frac{2}{M+1}\right)} \sin [2ij\pi/(M+1)], \qquad i, j = 1, \ldots, M/2.$$

If we examine the mirror images of $(E_1)_{ij}$ and $(D_1)_{ij}$ we find

$$(E_1)_{i,((M/2)-j+1)} = \sqrt{\left(\frac{2}{M+1}\right)} \sin \left(\left[(2i-1)\left(\frac{M}{2}+1\right) - (2i-1)j\right]\pi/(M+1)\right) \tag{28}$$

and

$$(D_1)_{i,((M/2)-j+1))} = \sqrt{\left(\frac{2}{M+1}\right)} \sin \left(\left[2i\left(\frac{M}{2}+1\right) - 2ij\right]\pi/(M+1)\right).$$

The even/odd structure is not contained either in E_1 or D_1 and therefore the $(M/2) \times (M/2)$ block diagonal structure for the conversion matrix is the best, one can get for $(\mathrm{DST})_1$.

The conversion matrix for $M = 4$ and 8 can be expressed as

$$[C_1(2)] = \left[\begin{array}{cc|cc} 1.9466 & -0.4595 & & \\ 0.4595 & 1.9466 & \multicolumn{2}{c}{\bigcirc} \\ \hline & & 1.9466 & 0.4595 \\ \multicolumn{2}{c|}{\bigcirc} & -0.4595 & 1.9466 \end{array}\right] \tag{29}$$

$$[C_1(3)] = \mathrm{diag}\left[\begin{bmatrix} 2.673 & -0.817 & -0.172 & -0.0396 \\ 0.817 & 2.449 & -0.817 & 0.817 \\ 0.396 & 0.817 & 2.673 & -1.172 \\ 1.172 & -0.817 & 0.396 & 2.673 \end{bmatrix} \begin{bmatrix} 2.673 & 0.396 & -0.817 & 0.172 \\ -0.172 & 2.673 & 0.817 & 0.396 \\ 0.817 & -0.817 & 2.449 & 0.817 \\ -0.396 & -0.172 & -0.817 & 2.673 \end{bmatrix}\right]$$

where

$$\begin{aligned} [\hat{S}_1(2)] &= [C_1(2)][\hat{H}_w(2)] \\ [\hat{S}_1(3)] &= [C_1(3)][\hat{H}_w(3)]. \end{aligned} \tag{30}$$

$[H_w(m)]$ is the $(2^m \times 2^m)(\mathrm{WHT})_w$ matrix whose rows are rearranged in bit-reversed order and $M = 2^m$. $[\hat{S}_1(m)]$ is the $(2^m \times 2^m)(\mathrm{DST})_1$ matrix with rows rearranged such that the first half represents even vectors and the second half represents odd vectors. As the conversion matrix $[C_1(m)]$ is not highly sparse, $(\mathrm{DST})_1$ via $(\mathrm{WHT})_w$ requires significant number of multiplications (Table 2). There seems to be no apparent gain in computation of $(\mathrm{DST})_1$ via $(\mathrm{WHT})_w$.

The Even/Odd Structure for $(\mathrm{DST})_2$. We again examine ϕ_{ik} (see (2)) where $k = (M - j - 1)$, $j = 0, \ldots, (M/2) - 1$,

$$\phi_{ij} = \sqrt{\left(\frac{2}{M}\right)} \sin [(i+1)(2M - 2j - 2 + 1)\pi/2M] = (-1)^i \, (\phi_{ij}. \tag{31}$$

Table 2. Number of multiplications required for $(\mathrm{DST})_1$

Length of sequence M	8	16	32	64
Direct (sparse matrix factorization)[6]	8	24	68	184
$(\mathrm{DST})_1$ via $(\mathrm{WHT})_w$	32	128	512	2098

Thus, the even indiced rows are even and the odd-indiced rows are odd. If we now look at the elements of the lower sub-block of the rearranged transform matrix as before, i.e.

$$\hat{S}_2 = \begin{bmatrix} E_2 & \tilde{E} \\ D_2 & -\tilde{D}_2 \end{bmatrix}$$

$$(D_2)_{ij} = \sqrt{\left(\frac{2}{M}\right)} \sin\left[\frac{(i+1)(2j+1)\pi}{M}\right] \qquad i, j = 0, 1, \ldots, \frac{M}{2} - 1 \tag{33}$$

and

$$(D_2)_{i,((M/2)-j-1)} = (-1)^i (D)_{ij} \qquad i, j = 0, 1, \ldots, \frac{M}{2} - 1.$$

Therefore, it is evident that there are smaller sub-blocks in the $((M/2)\times(M/2))$ lower block of the conversion matrix C_2. This matrix has the structure

$$\text{diag}\,[(E_1)_{M/2}, (E_2)_{M/4}, (E_3)_{M/8} \ldots, (E_m)_1],$$

which is very similar to that for the DCT.

RATE DISTORTION

Another criterion for evaluating the orthogonal transforms is the rate distortion. The rate distortion function yields the minimum information rate in bits per transform component needed for coding such that the average distortion is less than or equal to a chosen value D for any specified source probability distribution. It has been shown [23] that for Gaussian distribution and for mean square error D as a fidelity criterion, the rate distortion $R(D)$ can be expressed as

$$R(D) = \frac{1}{2N} \sum_{j=1}^{N} \max\left[0, \log_e\left(\frac{\lambda_j}{\theta}\right)\right] \tag{34}$$

where λ_1 is the ith-eigenvalue for the KLT or ith diagonal element of the covariance matrix in the transform domain, $\tilde{\Sigma}_x$, for any other transform and θ is any parameter satisfying the relation

$$D(\theta) = \frac{1}{N} \sum_{j=1}^{N} \min\,[\theta, \lambda_j]. \tag{35}$$

The rate distortion is a measure of the decorrelation of the transform components as the distortion can be spread uniformly in the transform domain thus minimizing the rate required for transmitting the information. The rate vs distortion for a Markov-1 process is shown in Figs. 2–5 for DST, DCT and IT (identity transform). For $N > 16$ or for $\rho < 0.7$, the DST performance approaches that of DCT.

CONCLUSIONS

Althought DCT is superior to all the other transforms (except the KLT) including the DST, based on the residual correlation, for low correlation coefficient ($\rho \leqslant 0.6$) and for large sequence length ($N \geqslant 32$), DST compares favorably well with the DCT. This comparison is also valid on the basis of the rate distortion. As the DST has the even/odd structure, it can be computed via $(\text{WHT})_w$ and other transforms having the same structure. However computation of $(\text{DST})_1$ via $(\text{WHT})_w$ results in significant increase in the number of multiplications compared to its implementation based on the sparse matrix factorization. In terms of the computational complexity, DST requires significantly less number of multiplications compared to DCT and the number of additions for the DST are slightly less than that of DCT [6]. The fast algorithm for the DST enables an efficient implementation of the Karhunen–Loêve transform for a stationary Gauss–Markov signal with known boundary conditions [5].

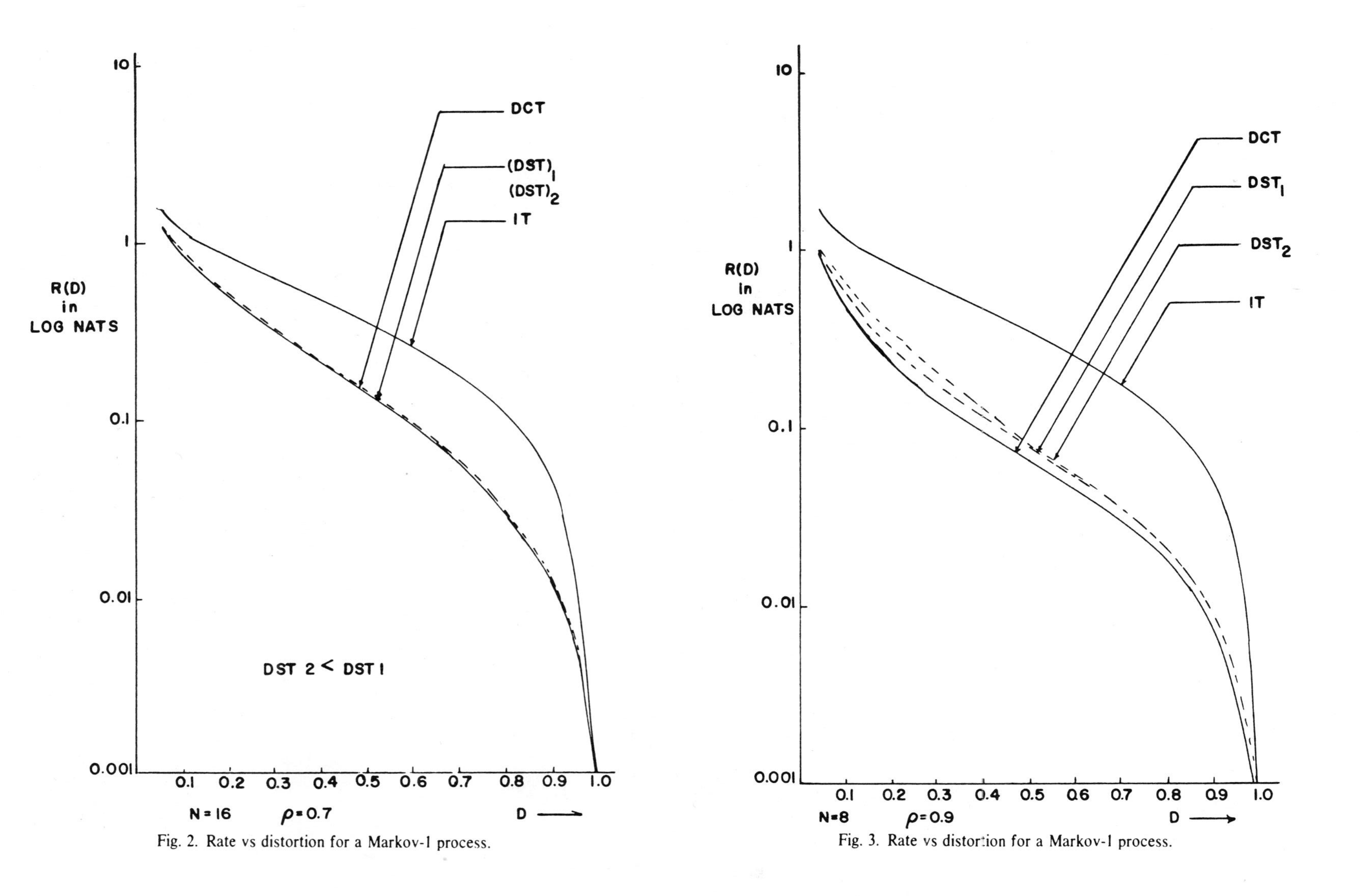

Fig. 2. Rate vs distortion for a Markov-1 process.

Fig. 3. Rate vs distortion for a Markov-1 process.

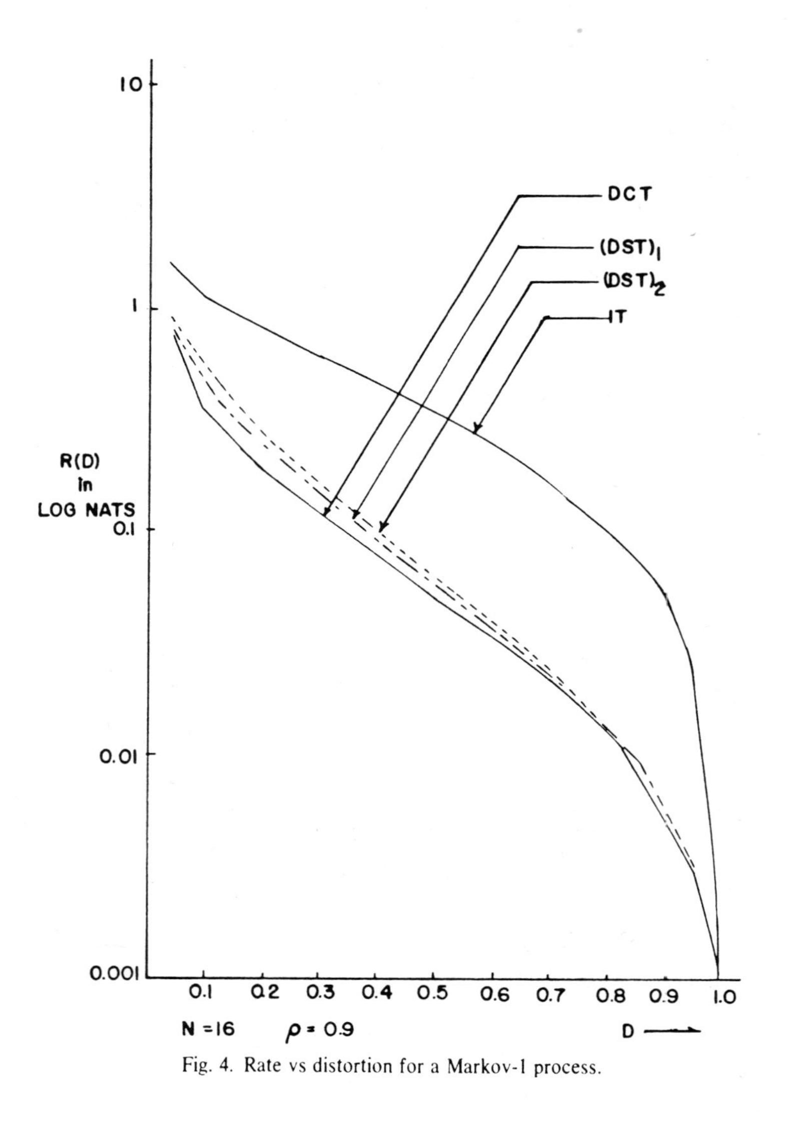

Fig. 4. Rate vs distortion for a Markov-1 process.

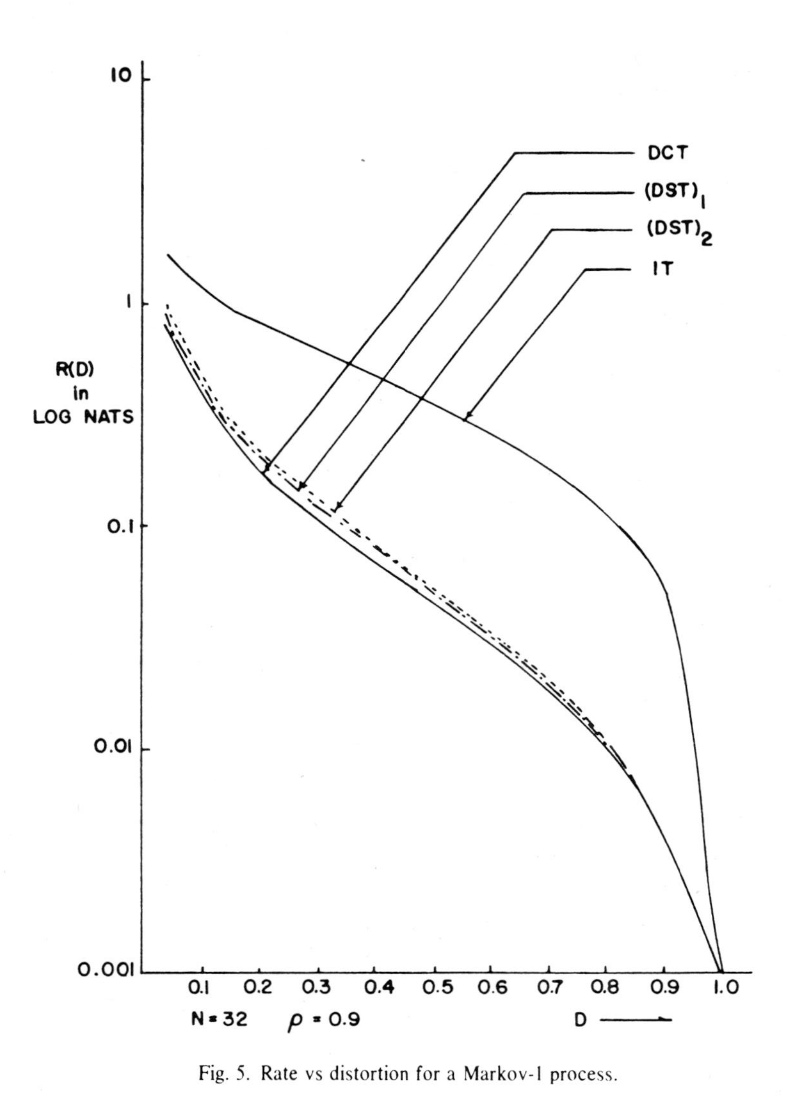

Fig. 5. Rate vs distortion for a Markov-1 process.

REFERENCES

1. M. Hamidi and J. Pearl, Comparison of the cosine and Fourier transforms of Markov-1 signals. *IEEE Trans. Acoust., Speech, Signal Proc.* **ASSP-24**, 428–429 (1976).
2. W. H. Chen, C. H. Smith and S. C. Fralick, A fast computational algorithm for the discrete cosine transform. *IEEE Trans. Commun.* **COM-25**, 1004–1009 (1977).
3. E. O. Brigham, *The Fast Fourier Transform.* Prentice-Hall, Englewood Cliffs, New Jersey (1974).
4. H. W. Jones, D. N. Hein and S. C. Knauer, The Karhunen–Loêve, discrete cosine and related transforms obtained via the Hadamard transform. *Int. Telemetering Conf.*, Los Angeles 14–16 November (1978).
5. A. K. Jain, A Fast Karhunen–Loêve transform for a class of stochastic processes. *IEEE Trans. Commun.* **COM-24**, 1023–1029 (1976).
6. P. Yip and K. R. Rao, Sparse-matrix factorization of discrete sine transform. *Proc. 12th Annual Asilomar Conf. on Circuits, Sys. and Comput.* **608**, 549–555. Pacific Grove, California, November (1978).
7. N. Ahmed and K. R. Rao, *Orthogonal Transforms for Digital Signal Processing.* Springer-Verlag, New York (1975).
8. N. Ahmed, T. Natarajan and K. R. Rao, Discrete cosine transform. *IEEE Trans. Comput.* **C-23**, 90–93 (1974).
9. W. Chen and C. H. Smith, Adaptive coding of color images using cosine transform. *Proc. ICC*, pp. 47-7–47-13, Philadelphia, 14–16 June (1976).
10. M. S. Corrington, Implementation of fast cosine transforms using real arithmetic. *NAECON*, Dayton, Ohio, 16–18 May (1978).
11. M. J. Narasimha and A. M. Peterson, On the computation of the discrete cosine transform. *IEEE Trans. Comm.* **COM-26**, 934–946 (1978).
12. G. G. Murray, Microprocessor systems for TV imaginary compression. *Proc. of SPIE Conf. on Appl. of Digital Image Proc.* **119**, 121–129, San Diago, 22–26 August (1977).
13. R. A. Belt, R. V. Keele and G. G. Murray, Digital TV microprocessor system, *National Telecommun. Conf., Proc.* pp. 10:6-1–16:6-6, Los Angeles, 3–5 December (1977).
14. D. Hein and N. Ahmed, On a real-time Walsh–Hadamard/cosine transform image processing. *IEEE Trans. Electromag. Compat.* **EMC-20**, 453–457 (1978).
15. Y. Tadokoro and T. Higuchi, Discrete Fourier transform computation via the Walsh transform. *IEEE Trans. Acoust., Speech, Signal Proc.* **ASSP-26**, 236–240 (1978).
16. T. Ohira, M. Hayakawa and K. Matsumoto, Orthogonal transform coding system for NTSC color television signals. *IEEE Trans. Commun.* **COM-26**, 1454–1463 (1978).
17. W. K. Pratt, *Digital Image Processing.* Wiley, New York (1978).
18. H. B. Kekre and J. K. Solanki, Comparative performance of various trigonometric unitary transforms for transform image coding. *Int. J. Electronics* **44**, 305–315 (1978).
19. M. Wagh and H. Ganesh, Fast Karhunen–Loêve transform via convolution. To be published.
20. H. P. Kramer and M. V. Mathews, *A Linear Coding for Transmitting a Set of Correlated Signals.*
21. H. C. Andrews, *Computer Techniques in Image Processing.* Academic Press, New York (1970).
22. W. K. Pratt, Generalized Wiener filtering computation techniques. *IEEE Trans. Comput.* **C-21**, 636–641 (1972).
23. J. Pearl, H. C. Andrews and W. K. Pratt, Performance measures for transform data coding. *IEEE Trans. Commun. Tech.* **COM-20**, 411–415 (1972).
24. H. C. Andrews, Multidimensional rotations in feature selection. *IEEE Trans. Comput.* **C-20**, 1045–1051 (1971).
25. H. F. Harmuth, *Sequency Theory.* Academic Press, New York (1977).
26. H. F. Silverman, An introduction to programming the Winograd fourier transform algorithm (WFTA). *IEEE Trans. Acoust., Speech, Signal Proc.* **ASSP-25**, 125–165 (1977).
27. J. A. Roese, W. R. Pratt and G. S. Robinson, Interframe cosine transform image coding. *IEEE Trans. Commun.* **COM-25**, 1329–1338 (1977).
28. A. Habibi and G. S. Robinson, A survey of digital picture coding. *Computer* **7**, 22–23 (1974).
29. H. Whitehouse, *et al.*, A digital real time intraframe video bandwidth compression system. *Proc. of SPIE Conf. on Appl. of Digital Image Proc.* **119**, 64–78, San Diego, 22–26 August (1977).
30. L. W. Martinson, A 10 MHz image bandwidth compression model. *IEEE Conf. on Pattern Recognition and Image Proc.*, pp. 132–136, 31 May–2 June, Chicago (1978).
31. A. K. Jain, A fast Karhunen–Loêve transform for digital restoration of images degraded by white and colored noise. *IEEE Trans. Comput.* **C-26**, 560–571 (1977).

12

A FAST COMPUTATIONAL ALGORITHM FOR THE DISCRETE SINE TRANSFORM

P. Yip and K. R. Rao

***Abstract*—A sparse-matrix factorization is developed for the discrete sine transform (DST). This factorization has a recursive structure and leads directly to an efficient algorithm for implementing the DST, a feature most**

Paper approved by the Editor for Communication Theory of the IEEE Communications Society for publication after presentation at the 12th Annual Asilomar Conference on Circuits, Systems, and Computers, Pacific Grove, CA, November 1978. Manuscript received April 10, 1979; revised September 11, 1979.

P. Yip is with the Department of Applied Mathematics, McMaster University, Hamilton, Ont., Canada.

K. R. Rao is with the Department of Electrical Engineering, University of Texas, Arlington, TX 76019.

desirable and very similar to that of the DCT. This algorithm requires fewer arithmetic operations compared to that for the discrete cosine transform (DCT).

INTRODUCTION

Jain [1] has shown that for a first-order Markov sequence, with given boundary conditions, the Karhunen-Loêve transform (KLT), which is known to be statistically optimal, reduces to the discrete sine transform (DST). It is this fact that makes the DST a rather unique transform. The object of this paper is to develop an efficient algorithm for fast implementation of DST based on recursive sparse matrix factorization of the DST matrix. This development is similar to that of Chen *et al.* [2] for the DCT and involves real arithmetic only. The DST fast algorithm can lead to its hardware realization, as in the case of the DCT for applications in digital image processing [3]-[9].

DISCRETE SINE TRANSFORM

Jain [1] has defined the orthonormal basis set for DST as

$$\phi_{ij} = \sqrt{\frac{2}{N+1}} \sin \frac{(ij\pi)}{(N+1)} \tag{1}$$

where $i, j, = 1, 2, \cdots, N$ for N sample points. The first 15 basis functions for the DST are shown in [8]. The DST and its inverse can be defined as

$$X(j) = \sum_{i=1}^{N} x(i)\phi_{ij} \quad \text{and} \quad x(i) = \sum_{j=1}^{N} X(j)\phi_{ji} \tag{2}$$

$$j = 1, 2, \cdots, N \qquad i = 1, 2, \cdots, N.$$

Let $N = 2^m - 1$ and let $[S_N]$ denote the DST matrix so that

$$[S_N] = \left(\frac{2}{N+1}\right)^{1/2} [P_N][A_N] \tag{3}$$

where $[P_N]$ is the row-permutation matrix defined by bit-reversal of $(i-1)$, i being the row index $i = 1, 2, \cdots, N$ and the element of $[S_N]$ in the ith row and jth column is ϕ_{ij} described in (1). Also, $[P_N][P_N] = I_N$, the identity matrix. $[A_N]$ then contains all the structures that are necessary for the factorization. The first step is a recursive relation:

$$[A_N] = \left[\begin{array}{c|c} A_{(N+1)/2} & \bigcirc \\ \hline \bigcirc & A_{(N-1)/2} \end{array}\right] [B_N] \tag{4}$$

where $[B_N]$ and $[B_N]^{-1}$ have the structure as illustrated here for $N = 5$:

$$[B_5] = \begin{bmatrix} 1 & & & & 1 \\ & 1 & & 1 & \\ & & 1 & & \\ & 1 & & -1 & \\ 1 & & & & -1 \end{bmatrix}, \quad [B_5]^{-1} = \frac{1}{2}\begin{bmatrix} 1 & & & & 1 \\ & 1 & & 1 & \\ & & 2 & & \\ & 1 & & -1 & \\ 1 & & & & -1 \end{bmatrix} \tag{5}$$

The recursive structure lies in $[A_{(N-1)/2}]$. Using $S(i/k)$ to denote $\sin(i\pi/k)$, the recursion algorithm is developed starting with $[A_3]$:

$$[A_3] = \left[\begin{array}{c|c} [A_2] & 0 \\ \hline 0 & 1 \end{array}\right] [B_3], \quad [A_2] = \left[\begin{array}{c|c} S(\frac{1}{4}) & S(\frac{1}{2}) \\ \hline S(\frac{1}{4}) & -S(\frac{1}{2}) \end{array}\right]$$

$$= \frac{1}{\sqrt{2}} \begin{bmatrix} 1 & \sqrt{2} \\ 1 & -\sqrt{2} \end{bmatrix}$$

and

$$[A_7] = \left[\begin{array}{c|c} [A_4] & 0 \\ \hline 0 & [A_3] \end{array}\right] [B_7].$$

The matrix

$$[A_{(N+1)/2}] = [P_{(N+1)/2}] \left[\sin\left(\frac{(2i+1)(j+1)\pi}{N+1}\right) \right],$$

$$i, j = 0, \cdots, \frac{N-1}{2}. \tag{6}$$

It can be factorized into

$$[A_{(N+1)/2}] = [B_{(N+1)/2}][D_{(N+1)/2}][C_{(N+1)/2}]^T \tag{7}$$

where $[B_{(N+1)/2}]^{-1}$ is the same butterfly matrix as $[B_5]^{-1}$ except that there is not a center element of 2 since $[(N+1)/2] = 2^{m-1}$. $[C_{(N+1)/2}]$ is a column permutation matrix grouping all odd columns to the left and all even columns to the right, i.e.,

$$[C_{(N+1)/2}] = \begin{bmatrix} 1 & 0 & . & . & . & . & . \\ 0 & 0 & 1 & . & . & . & 0 \\ 0 & . & . & . & . & 1 & 0 \\ 0 & 1 & 0 & . & . & . & . \\ . & . & . & . & . & 0 & 1 \end{bmatrix}.$$

Matrix $[D_{(N+1)/2}]$ is block diagonal and is given by

$$[D_{(N+1)/2}] = \left[\begin{array}{c|c} [F_{(N+1)/4}] & 0 \\ \hline 0 & [G_{(N+1)/4}] \end{array}\right] \tag{8}$$

Here, another recursive relation occurs in that $[G_{(N+1)/4}]$ is related to $[A_{(N+1)/2}]$ by simply row operations, i.e., $[G_{(N+1)/4}] = [E_{(N+1)/4}]\,[A_{(N+1)/4}]$ where

$$[E_{(N+1)/4}]$$

$$= -\delta_{\langle (i+1)/2-j \rangle}, 0 \qquad \text{for odd } i \text{ and } j > (N+1)/8$$

$$= \delta_{\langle (i/2+j) \rangle}, 1 \qquad \text{for even } i \text{ and } j \leqslant (N+1)/8$$

$$= 0 \qquad \text{otherwise.}$$

Here, $\langle\rangle$ is evaluated modulo $(N+1)/8$. As an example, for $N = 15$,

$$[G_4] = \begin{bmatrix} 0 & 0 & -1 & 0 \\ 0 & 1 & 0 & 0 \\ 0 & 0 & 0 & -1 \\ 1 & 0 & 0 & 0 \end{bmatrix} [A_4].$$

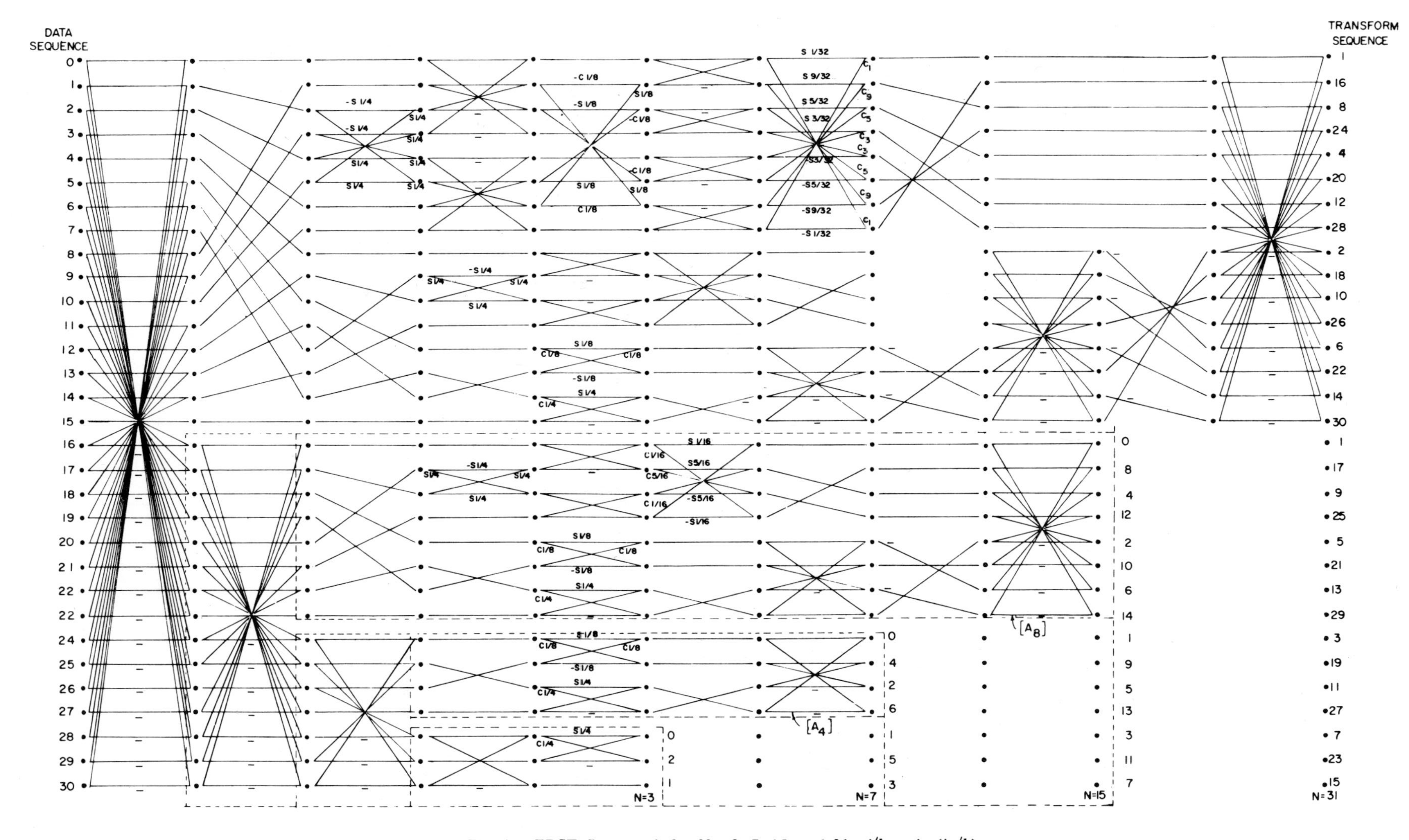

Fig. 1. FDST flow graph for $N = 3, 7, 15$, and 31, $sj/k \equiv \sin(j\pi/k)$, $cj/k \equiv \cos(j\pi/k)$.

The factorization of $[F_{(N+1)/2}]$ is similar to that of Chen *et al.* [2], i.e.,

$$[F_{(N+1)/4}] = [R_{(N+1)/4}]^T [Ml]_{(N+1)/4} \cdot [M2]_{(N+1)/4} \cdots [ML]_{(N+1)/4}. \quad (9)$$

$[R_{(N+1)/4}]$ is a row-permutation matrix which groups the odd rows in increasing order in the upper half and the even rows in reversed order in the lower half. The remaining factors can be classified into four types. The number of $[M]$ factors is given by $2 \log_2 ((N+1)/2) - 3$ for $N > 7$.

Type 1: $[M1]_{(N+1)/4}$, the first matrix factor in (9) after $[R]$, has nonzero elements only along the two diagonals. Elements on the main diagonal are given by $\sin (a_j\pi/N + 1)$ where a_j is the bit-reversal of $((N+1)/4) + j - 1$ for $j = 1, 2, 3, \cdots, (N+1)/8$ from the upper left to the center and are antisymmetric about the center for the remaining elements. Elements on the opposite diagonal are given by $\cos (a_j\pi/N + 1)$ from the upper right to the center and symmetric about the center for the remainder.

Type 2: $[ML]_{(N+1)/4}$, the last matrix in (9), is given by concatenating $[I_{(N+1)/16}]$, $-\sin(\pi/4)[I_{(N+1)/16}]$, $\sin(\pi/4)[I_{(N+1)/16}]$ and $[I_{(N+1)/16}]$ along the main diagonal from the upper left to the lower right. The opposite diagonal is obtained by concentenating $[0_{(N+1)/16}]$, $\sin(\pi/4)[\bar{I}_{(N+1)/16}]$, $\sin(\pi/4)[\bar{I}_{(N+1)/16}]$, and $[0_{(N+1)/16}]$ from the upper right to the lower left. Here $[0]$ represents the null matrix and $[\bar{I}]$ is the opposite diagonal unit matrix.

Type 3: $[MQ]_{(N+1)/16}$, the remaining odd numbered matrix factors are obtained by concatenating from the upper left to the center the matrix sequence $[I_{(N+1)/4i}]$, $-\cos(k_j\pi/i)[I_{(N+1)/4i}]$, $-\sin(k_j\pi/i)[I_{(N+1)/4i}]$, and $[I_{(N+1)/4i}]$ where $i = ((N+1)/2) \times [2^{(q-1)/2}]^{-1}$, $j = 1, 2, \cdots, i/8$ and k_j is the bit-reversal of $(i/4 + j - 1)$, the number of bits in the representation being the minimum required to represent q. The remainder of the main diagonal is antisymmetric about the center. Similarly, the opposite diagonal is obtained by concatenating from the upper right to the center the matrix sequence

$$[0_{(N/2i)}],\ \sin\left(\frac{k_j\pi}{i}\right)[I_{(N/2i)}],$$

$$-\cos\left(\frac{k_j\pi}{i}\right)[I_{(N/2i)}], \qquad \text{and } [0_{(N/2i)}].$$

The remainder of the opposite diagonal is symmetric about the center.

Type 4: $[MP]$, the even numbered matrix factors are binary matrices formed by alternating B_i and B_i^* from the upper left to the lower right of the main diagonal where $i = 2^{P/2}$ and

$$B_i = \begin{bmatrix} 1 & \cdot & \cdot & & \cdot & \cdot & \cdot & 1 \\ 0 & 1 & \cdot & & & \cdot & 1 & 0 \\ & & 1 & 1 & & & & \\ & & 1 & -1 & & & & \\ & \cdot^{\cdot^{\cdot}} & & & \ddots & & & \\ 1 & & & & & & & 1 \end{bmatrix} \qquad \text{and } B_i^*$$

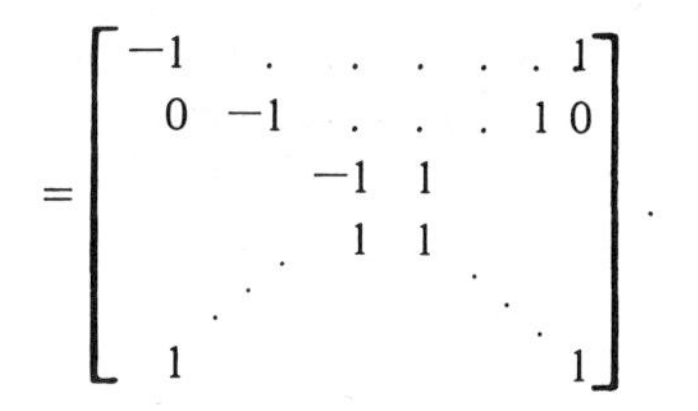

TABLE I
COMPARISON OF NUMBER OF ADDITIONS AND MULTIPLICATIONS REQUIRED FOR DST AND DCT

N	# of Additions		# of Multiplications	
	DCT	DST	DCT	DST
4 (3)*	8	4	6	2
8 (7)	26	22	16	8
16 (15)	74	62	44	30
32 (31)	194	166	116	54
64 (63)	482	422	292	130
128 (127)	1154	1030	708	310
256 (255)	2690	2438	1668	730

* Number in brackets is the number of sample points for DST

Fig. 1 is a signal flow graph for DST when $N = 31$. The main recursive features of the factorization are evident. Table I compares the arithmetic computations required for DST and DCT.

CONCLUSIONS

A sparse-matrix factorization for the DST is developed. The fast computational algorithm that results is superior to the fast DCT developed by Chen *et al.* [2]. The scheme presented is not the only possible one, nor is there any indication that this is optimal. One may proceed along the lines of Corrington [10], using real arithmetic. Hopefully, the fast algorithms for implementating the DST developed here can lead to actual hardware realization for potential applications in signal and image processing.

REFERENCES

[1] A. K. Jain, "A fast Karhunen-Loève transform for a class of random processes," *IEEE Trans. Commun.*, vol. COM-24, pp. 1023–1029, 1976.
[2] W. H. Chen, C. H. Smith, and S. C. Fralick, "A fast computational algorithm for the discrete cosine transform," *IEEE Trans. Commun.*, vol. COM-25, pp. 1004–1009, Sept. 1977.
[3] J. A. Roese, W. K. Pratt, and G. S. Robinson, "Interframe cosine transform image coding," *IEEE Trans. Commun.*, vol. COM-25, pp. 1329–1338, Nov. 1977.
[4] A. Habibi and G. S. Robinson, "A survey of digital picture coding" *Computer*, vol. 7, pp. 22–23, May 1974.
[5] H. Whitehouse *et al.*, "A digital real time intraframe video bandwidth compression system," in *Proc. SPIE Conf. Appl. Digital Image Processing* (San Diego, CA), vol. 119, Aug. 22–26, 1977, pp. 64–78.
[6] G. G. Murray, "Microprocessor system for TV imagery compression," in *Proc. SPIE Conf. Appl. Digital Image Processing* (San Diego, CA), Aug. 22–26, 1977, pp. 121–129.
[7] L. W. Martinson, "A 10 MHz image bandwidth compression model," in *Proc. IEEE Conf. Pattern Recognition and Image Processing* (Chicago, IL), May 31–June 2, 1978, pp. 132–136.
[8] W. K. Pratt, *Digital Image Processing*. New York: Wiley, 1978.
[9] P. Camana, "Video-bandwidth compression: A study in tradeoffs," *IEEE Spectrum*, vol. 16, pp. 24–29, June 1979.
[10] M. S. Corrington, "Implementation of fast cosine transforms using real arithmetic," presented at NAECON, Dayton, OH, May 16–18, 1978.

A FAST ALGORITHM FOR THE DISCRETE SINE TRANSFORM IMPLEMENTED BY THE FAST COSINE TRANSFORM

Z. D. Wang

***Abstract*–A method of composing the discrete sine transform from the discrete cosine transform is demonstrated. As a result of this method. the fast sine transform is accomplished by the existing implementation of the fast cosine transform.**

INTRODUCTION

The discrete sine transform (DST) was first introduced by Jain [1] and subsequently redefined by Kekre and Solanki [2]. The basis set defined by Jain is [1]

$$\phi_{ij} = \sqrt{\frac{2}{M+1}} \sin\left(\frac{ij\pi}{M+1}\right) \qquad i, j = 1, 2, \cdots, M. \tag{1}$$

This will be referred to as DSTI. On the other hand, Kekre and Solanki [2] defined the DST as

$$\begin{aligned} \phi_{ij} &= \sqrt{\frac{2}{M}} \sin\left[\frac{(i+1)(2j+1)\pi}{2M}\right] && i = 0, 1, \cdots, M-2 \\ & && j = 0, 1, \cdots, M-1 \\ &= (-1)^j (M)^{-1/2} && i = M-1 \\ & && j = 0, 1, \cdots, M-1. \end{aligned} \tag{2}$$

This will be referred to as DSTII.

Jain [1] has shown that for a first-order Markov sequence with certain boundary conditions, the Karhunen–Loeve transform (KLT), which is known to be optimal statistically with respect to energy concentration, reduces to the DST. Yip and Rao [5] have proven that for large sequence length ($N \geqslant 32$) and low correlation coefficient ($\rho < 0.6$) the DST performs even better than the DCT, which is widely accepted as the best substitute for the KLT, on the basis of the residual correlation and of the rate distortion. Therefore, the DST is sometimes used as a substitute for the KLT.

Yip and Rao have developed a fast algorithm for DSTI [3] which is similar to the fast algorithm for the discrete cosine

Manuscript received September 16, 1981; revised March 22, 1982.
The author is with the Kunming Physics Research Institute, China, on leave at the Department of Electrical Engineering, University of Arizona, Tucson, AZ 85721.

transform (FDCT) developed by Chen *et al.* [4] which involves real arithmetic only. The Yip and Rao algorithm for DSTI is suitable for size $2^m - 1$, while the algorithm for DSTII shown in this correspondence is suitable for size 2^m, which is adopted as the size of most pictures to be processed, because it fits the commonly used radix-2 FFT program.

CONSTRUCTION OF DSTII BY DCT

Let $x(n)$, $n = 0, 1, 2, \cdots, N - 1$, be a sequence of N data values. The discrete cosine transform of it is defined as [6]

$$C(m) = c_m \sqrt{\frac{2}{N}} \sum_{n=0}^{N-1} x(n) \cos \frac{(2n+1)m\pi}{2N}, \quad m = 0, 1, \cdots, N-1 \tag{3}$$

where

$$c_m = \frac{\sqrt{2}}{2} \quad \text{for } m = 0$$
$$c_m = 1 \quad \text{for } m \neq 0. \tag{4}$$

Substituting

$$m = N - k \quad k = 1, 2, \cdots, N \tag{5}$$

into (3), we have

$$C(N-k) = c_{N-k} \sqrt{\frac{2}{N}} \sum_{n=0}^{N-1} x(n) \cos \frac{(2n+1)(N-k)\pi}{2N}, \quad k = 1, 2, \cdots, N. \tag{6}$$

Since

$$\cos \frac{(2n+1)(N-k)\pi}{2N} = \cos (2n+1)\frac{\pi}{2} \cos \frac{(2n+1)k\pi}{2N} + \sin (2n+1)\frac{\pi}{2} \sin \frac{(2n+1)k\pi}{2N} = (-1)^n \sin \frac{(2n+1)k\pi}{2N}, \tag{7}$$

therefore

$$C(N-k) = c_{N-k} \sqrt{\frac{2}{N}} \sum_{n=0}^{N-1} (-1)^n x(n) \sin \frac{(2n+1)k\pi}{2N}, \quad k = 1, 2, \cdots, N. \tag{8}$$

Let

$$\bar{x}(n) = (-1)^n x(n) \quad n = 0, 1, \cdots, N-1 \tag{9}$$

and $\bar{C}(m)$ be the discrete cosine transform of sequence $\bar{x}(n)$. Then

$$\bar{C}(N-k) = c_{N-k} \sqrt{\frac{2}{N}} \sum_{n=0}^{N-1} (-1)^n \bar{x}(n) \sin \frac{(2n+1)k\pi}{2N}$$
$$= c_{N-k} \sqrt{\frac{2}{N}} \sum_{n=0}^{N-1} x(n) \sin \frac{(2n+1)k\pi}{2N}$$
$$= S(k), \quad k = 1, 2, \cdots, N \tag{10}$$

where $S(k)$ is the discrete sine transform (DSTII) of the sequence $x(n)$. Therefore, the procedure for obtaining the sine transform of the sequence $x(n)$ is composed of three steps.

1) Change the signs of all odd numbered data to the opposite sign to form a new sequence $\bar{x}(n)$. (Notice that the sequence number is counted from zero.)

2) Compute the discrete cosine transform on the sequence $\bar{x}(n)$. The FDCT developed by Chen *et al.* [4] is a good implementation of this step.

3) By reversing the sequence order of data which were produced by step 2), the discrete sine transform of the sequence $x(n)$ is obtained.

This procedure may be represented in the form of matrix multiplication. Let $[S_N]$ and $[C_N]$ be the discrete sine and cosine transforms of order N, respectively. Then

$$[S_N] = [\bar{I}_N][C_N][D_N], \tag{11}$$

where $[\bar{I}_N]$ is the opposite diagonal identity matrix, $[D_N]$ is the odd sign-changing matrix defined as

$$[D_N] = \begin{bmatrix} 1 & & & & & & 0 \\ & -1 & & & & & \\ & & 1 & & & & \\ & & & -1 & & & \\ & & & & \ddots & & \\ & & & & & 1 & \\ 0 & & & & & & -1 \end{bmatrix}. \tag{12}$$

Adding two simple steps on the fast cosine transform, the fast sine transform is accomplished. The signal flowgraph for the fast discrete sine transform is just a duplication of Chen's graph [4] for the fast discrete cosine transform and will not be repeated here.

ACKNOWLEDGMENT

The author wishes to thank Prof. B. R. Hunt of the University of Arizona for his encouragement.

REFERENCES

[1] A. K. Jain, "A fast Karhunen-Loeve transform for a class of stochastic processes," *IEEE Trans. Commun.*, vol. COM-24, pp. 1023–1029, 1976.

[2] H. B. Kekre and J. K. Solanki, "Comparative performance of various trigonometric unitary transforms for transform image coding," *Int. J. Electron.*, vol. 44, pp. 305–315, 1978.

[3] P. Yip and K. R. Rao, "Sparse-matrix factorization of discrete sine transform," in *Proc. 12th Annu. Asilomar Conf. Circuit, Syst., Comput.*, vol. 608, pp. 549–555, Pacific Grove, CA, Nov. 1978.

[4] W. S. Chen, C. H. Smith, and S. C. Fralick, "A fast computational algorithm for the discrete cosine transform," *IEEE Trans. Commun.*, vol. COM-25, pp. 1004–1009, 1977.

[5] P. Yip and K. R. Rao, "On the computation and the effectiveness of discrete sine transform," *Comput. Electron.*, vol. 7, pp. 45–55, 1980.

[6] N. Ahmed, T. Natarajan, and K. R. Rao, "Discrete cosine transform," *IEEE Trans. Comput.*, vol. C23, pp. 90–93, 1974.

Editor's Comments on Paper 14

14 KEKRE, SAHASRABUDHE, and GOYAL
Image Bandwidth Compression Using Legendre Transforms

DISCRETE LEGENDRE TRANSFORM

Kekre, Sahasrabudhe, and Goyal (Paper 14) developed the discrete Legendre transform (DLT) and have applied the same for image bandwidth compression. By appropriately modifying the sampled values of Legendre polynomials, they have generated orthonormal Legendre sequences possessing a recursive property. DLT of any size, including integer power of two, can be developed. Also, as the DLT is an EOT it can be implemented via the WHT through a conversion matrix that has a sparse block-diagonal structure. The fast algorithm (Paper 14) developed for an N-point DLT reduces the number of multiplications from N^2 to $N(N + 1)/2$. As the comparison of DLT with the WHT is based on the mean square error (mse), it is difficult to grade the utility and effectiveness of the DLT in image processing. Further research on evaluating the DLT-processed reconstructed images based on subjective criteria is warranted. Also, other quantitative parameters such as the variance distribution, rate distorsion (Paper 38), and residual correlation (Paper 4) need to be computed for the DLT.

14

Reprinted from *Inst. Electron. Telecommun. Eng. J.* **28:**51-55 (1982)

IMAGE BANDWIDTH COMPRESSION USING LEGENDRE TRANSFORMS

H.B. KEKRE* (*Fellow*), S.C. SAHASRABUDHE† AND N.C. GOYAL* (*Member*)

The television pictures can be approximated by polynomials, so an orthogonal transform having polynomials as the basis vectors can be used for best representation of such images. The possibilities of Legendre Polynomial transform for data compression are explored in the present paper. This paper also presents a computationally fast algorithm for obtaining Legendre transforms of sequences.

Indexing Terms : Bandwidth Compression, Legendre Transforms, Orthogonal Transforms, Threshold Sampling

IN THE RECENT years, the transmission of broad-band TV signals by pulse code modulation (PCM) is becoming popular because the original signal generation can easily be done without decrease of signal-to-noise (S/N) ratio. But the disadvantage is that it calls for the larger bandwidth (Enomoto and Shibata 1971). Therefore, the main design considerations of digital image processing system, in general, have been the representation of images with acceptable fidelity and with appropriate coding to reduce the bit rate as much as possible. This can be achieved either by coding the picture during quantization and/or by applying data compression techniques after quantization. Applications of data compression are primarily in transmission and storage of information. Several techniques have been proposed for reduction of the redundancy in data and thereby reducing the bandwidth requirement. Most of them are based on the statistical properties of the images. Depending on the statistics of the data, various encoding schemes such as predictive quantization, adaptive-predictive quantization, delta modulation etc. have been presented (Huang *et al.* 1972, Davisson *et al.* 1976, Lei *et al.* 1977, Delp 1979, Zschunke 1977, Hung 1979). These schemes result in the reduction of bits per picture element, as such they are called as Entropy Reduction Transformation (ERT). The performance of such techniques is sensitive to changes in the statistics of the data and most of them are implemented on line.

An alternative to ERT techniques is transform coding. Here, a long sequence of data is divided into blocks of N samples and each block is treated as the vector. This vector is linearly transformed to produce another vector whose components are mutually uncorrelated. By retaining the few large energy components and discarding the rest, the data compression can be achieved. Although Karhunen-Loeve Transform (KLT) is the optimum one, it does not have a fast algorithm associated with it. The other unitary transforms such as trigonometric; including Fourier and Cosine, Walsh, Hadamard, Slant, Haar, etc., are commonly used because they have fast algorithms associated with them (Haralick 1972, Habibi *et al.* 1974, Jain 1974, Praft 1974, Shore 1973, Jain 1976, Kekre & Solanki 1978, Ahmed & Rao 1978, Jain 1981).

If one examines the scanning lines of a typical monochrome image, it is found that some of the lines are of nearly constant grey level over a considerable range, while a large number of lines are the lines that increase or decrease in brightness either linearly or higher power of variations (Enomoto and Shibata 1971). Thus the variation of brightness of TV samples may be approximated by some

Manuscript received 1980 October 30; revised 1981 October 26.
*Computer Centre, Indian Institute of Technology, Bombay 400 076.
†Department of Electrical Engineering, Indian Institute of Technology, Bombay 400 076.

polynomials. The polynomial approximation is used for the simplicity that they can be evaluated, differentiated and integrated easily in a finite number of steps using just the basic arithmetic operations of addition, subtraction and multiplication.

Thus a transform possessing polynomials as the basis vectors can efficiently represent such image lines. Shibata and Enomoto (1971) have introduced orthogonal transformation using slant basis vectors. The slant vector is a discrete sawtooth waveform decreasing in uniform steps over its length but they do not represent higher-order polynomials

The Legendre polynomials because of their orthogonal property can be used for representation of TV signals. In this paper, the applicability of the Legendre transform for data compression is being explored.

LEGENDRE POLYNOMIALS

The Legendre polynomials are the orthogonalization of the set of the time functions, $\{x^n\}$, $n = 0, 1, \ldots, N$; in the interval $[0 - T]$ they are defined by the recursive relation given by eqn. (1) (Sansone *et al.* 1959).

$$(n+1)\,P_{n+1}(x) = 2\,(2n-1)\left(\frac{x}{T} - \frac{1}{2}\right)P_n(x) - nP_{n-1}(x) \qquad n = 0, 1, \ldots, N-1 \tag{1}$$

where, $P_0(x) = 1$.
with the condition that

$$\int_0^T P_n(x)\,P_m(x)\,dx = \frac{T}{2n+1}, \quad n = m$$
$$= 0, \quad n \neq m. \tag{2}$$

These functions are given by Kekre (1968) and Lee (1960).

Discrete Legendre Transform

For the use and implementation of Legendre polynomials on digital computers, one needs to discretize these polynomials. However, when these polynomials are discretized by giving the discrete values to x, they lose their orthogonality property. Thus they need orthogonalization once more. This procedure will lead to the wastage of computational efforts. For a given length, N, of a sequence, the Legendre sequences can be generated by the following recursive relation

$$P_0(J) = 1,$$
$$P_n(J) = \left[\frac{N-1}{2} - J\right]P_{n-1}(J) - \frac{(n-1)^2\,[N^2-(n-1)^2]}{4\,[4(n-1)^2 - 1]}\,P_{n-2}(J) \tag{3}$$

where, $J = 0, 1, 2, 3, \ldots, N-1$
$n = 1, 2, 3, \ldots, N-1,$

with orthogonality condition that

$$< P_n(J) \cdot P_m(J) > = \mu_n\,\delta_{mn} \tag{4}$$

where, δ_{mn} is the Kronecker delta defined by

$$\delta_{mn} = 1, \quad m = n$$
$$= 0, \quad m \neq n. \tag{5}$$

The energy of the corresponding function is given by

$$\mu_n = \frac{N}{2n+1}(\sigma_n)^2 \tag{6}$$

where, σ_n can be obtained once again by the recursive relation

$$\sigma = \sigma_{n-1}\,\frac{n}{2\,(2n-1)}\,(N^2 - n^2)^{\frac{1}{2}} \tag{7}$$

for

$\sigma_0 = 1, \quad n = 1, 2, 3, \ldots, N-1.$

By dividing each relation by corresponding energy co-efficient μ_n, it is possible to get a recursion formula for generating orthonormal Legendre sequences. It is given by

$$P_0(J) = \sqrt{N},$$
$$n\sqrt{\left(\frac{N^2 - n^2}{2n+1}\right)}\,P_n(J) = 2\sqrt{(2n-1)}\left[\frac{N-1}{2} - J\right]P_{n-1}(J) - (n-1)\sqrt{\frac{N^2-(n-1)^2}{(2n-3)}}\,P_{n-2}(J) \tag{8}$$

where, $n = 1, 2, 3, 4, \ldots, N-1$
$J = 0, 1, 2, 3, \ldots, N-1$

with the condition that

$$< P_n(J) \,.\, P_m(J) > = \delta_{mn}. \tag{9}$$

The first eight orthogonal discrete Legendre function as given by eqn. (8) are shown in Fig. 1.

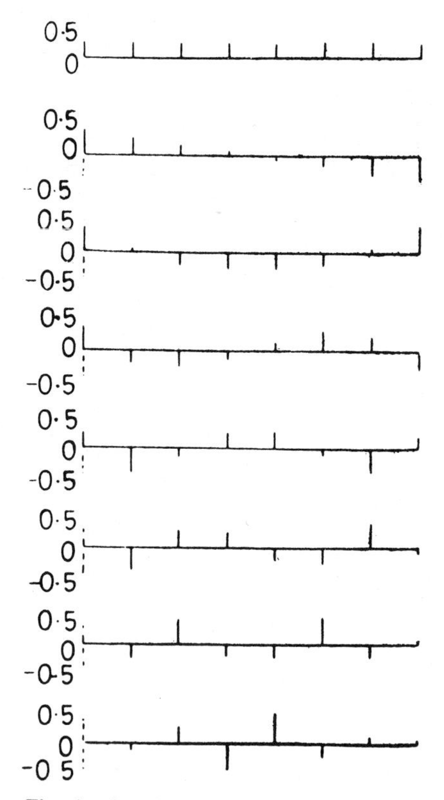

Fig. 1. *8 × 8 Discrete Legendre functions*

FAST ALGORITHM FOR LEGENDRE TRANSFORMS

The set of orthogonal Legendre sequences $\{P_n\}$, $n = 0, 1, 2, \ldots, N-1$ is complete for the sequence of length N.

Thus every finite energy sequence $s(n)$ of length N can be expanded into a series of the form

$$s(n) = \sum_{i=0}^{N-1} \frac{S_i}{\mu_i} P_i(n) \quad n = 0, 1, 2, \ldots, N-1 \tag{10}$$

where,

$$\mu_i = \sum_{n=0}^{N-1} P_i^2(n)$$

and the spectral (transformed) co-efficients S_i's are given by

$$S_i = \sum_{n=0}^{N-1} s(n)\, P_i(n) \tag{11}$$

which can be written in the matrix form as

$$\mathbf{S} = [\mathbf{P}]\, \mathbf{s}. \tag{12}$$

One can restore the original samples by taking inversion of equation (12)

$$\mathbf{s} = [\mathbf{P}]^{-1}\, \mathbf{S}. \tag{13}$$

If $[\mathbf{P}]$ represents an orthonormal transform matrix then $[\mathbf{P}]^{-1} = [\mathbf{P}]^t$ therefore the eqn. (13) can be modified as

$$\mathbf{s} = [\mathbf{P}]^t \mathbf{S}. \tag{14}$$

However, when **P** is not a normalized matrix, the spectral co-efficients are to be divided by the energy of each function as in eqn. (10) hence

$$\mathbf{s} = [\mathbf{P}]^t [\Lambda]\, \mathbf{S} \tag{15}$$

where, $[\Lambda]$ is a diagonal matrix whose diagonal elements are given by

$$\lambda_{ii} = \frac{1}{\mu_i}, \; i = 0, 1, \ldots N-1.$$

The transform matrix $[\mathbf{P}]$ is case of N-points sequence is an $N \times N$ matrix. In case of $N = 2$, it is same as the Hadamard matrix and in case of $N = 4$, it is same as Slant matrix and for other values of N, they are different. Eqn. (16) and eqn. (17) show a 4 point and 8 point Legendre matrices respectively:

$$[P_4] = \begin{bmatrix} 1 & 1 & 1 & 1 \\ 1 & \frac{1}{3} & -\frac{1}{3} & 1 \\ 1 & -1 & -1 & 1 \\ 1 & -3 & 3 & -1 \end{bmatrix} \tag{16}$$

$$[P_8] = \begin{bmatrix} 1 & 1 & 1 & 1 & 1 & 1 & 1 & 1 \\ 1 & \frac{5}{7} & \frac{3}{7} & \frac{1}{7} & -\frac{1}{7} & -\frac{3}{7} & -\frac{5}{7} & -1 \\ 1 & \frac{1}{7} & -\frac{3}{7} & -\frac{5}{7} & -\frac{5}{7} & -\frac{3}{7} & \frac{1}{7} & 1 \\ 1 & -\frac{5}{7} & -1 & -\frac{3}{7} & \frac{3}{7} & 1 & \frac{5}{7} & -1 \\ 1 & -\frac{13}{7} & -\frac{3}{7} & \frac{9}{7} & \frac{9}{7} & -\frac{3}{7} & -\frac{13}{7} & 1 \\ 1 & -\frac{23}{7} & \frac{17}{7} & \frac{15}{7} & -\frac{15}{7} & -\frac{17}{7} & \frac{23}{7} & -1 \\ 1 & -5 & 9 & -5 & -5 & 9 & -5 & 1 \\ 1 & -7 & 21 & -35 & 35 & -21 & 7 & -1 \end{bmatrix} \tag{17}$$

The above matrices show the even and odd symmetry, which can be exploited to generate faster algorithm for Legendre transform, by reordering the columns in such a way that the mth column on either side of the central line forms a pair. The flow diagram for four-point Legendre transform is shown in Fig. 2*a*.

This requires total of 8 additions and 4 multiplications in comparison to 16 additions and 16 multiplications. Similarly exploiting the even and odd symmetry, the flow diagram for an 8-point Legendre transform is shown in Fig. 2*b*.

This algorithm reduces the number of multiplications to $N(N+1)/2$ from N^2.

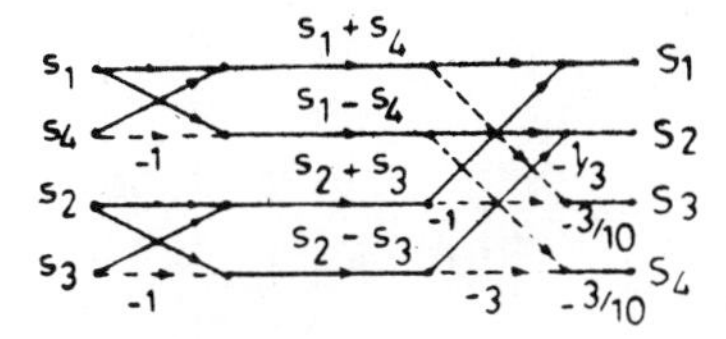

FIG 2a FLOW DIAGRAM FOR OBTAINING LEGENDRE TRANSFORM OF A 4 POINT SEQUENCE

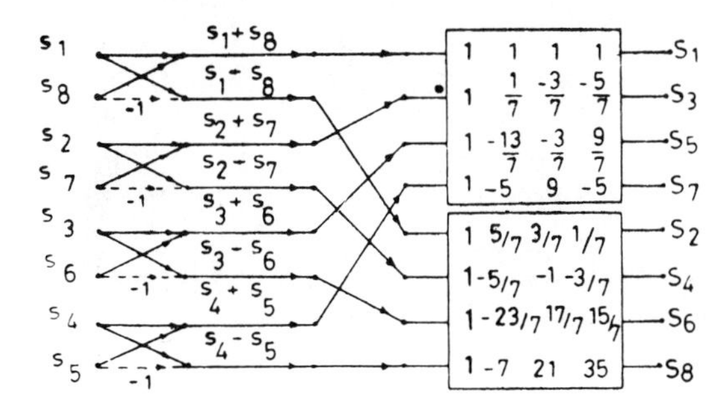

Fig. 2*b*. *Flow diagram for obtaining Legendre transform of a 8-point sequence.*

BANDWIDTH COMPRESSION USING LEGENDRE TRANSFORM

The Legendre transform coding can be used for bandwidth reduction or data compression in the manner similar to other orthogonal transforms (Ahmed and Rao 1975). The key for securing data compression is the signal representation which concerns the representation of a given class (or classes) of signals in an efficient manner. Let **s** be an

N-point vector representing the N-sampled values of the discrete signal. For efficient representation, the orthogonal transform of s may be obtained which results in

$$\mathbf{S} = [\mathbf{P}]\mathbf{s}, \tag{18}$$

where S and [P] denote the transform vector and transform matrix respectively. The objective is to select a subset of M components of S, where M is substantially less than N. The remaining $(N—M)$ components can then be discarded without introducing the objectionable error, when the signal is reconstructed using the retained M components of S. Smaller the value of M for the specified value of mean-square error, more efficient is the transform for data compression.

Threshold Sampling

One of the methods of discarding $(N—M)$ co-efficients of the transformed vector S is the threshold sampling, where M spectral co-efficients having largest magnitude are retained and the remaining $(N—M)$ co-efficients are given a preselected values b_i. The mean-square error in estimating the original vector is given by

$$e = \sum_{i=M+1}^{N} E[(S_i - b_i)^2]. \tag{19}$$

The optimum transform, for which the value of e is minimum for a class of input sequences is known as Karhunen-Loeve Transform (KLT). However, if b_i's are set equal to zero then the error estimate comes out to be

$$e = \sum_{i=M+1}^{N} E[(S_i)^2]. \tag{20}$$

The percentage mean-square error in the space domain is given by

$$MSE\,(\%) = \frac{\sum_{i=0}^{N-1} \sum_{j=0}^{N-1} [s(i,j) - \hat{s}(i,j)]^2}{\sum_{i=0}^{N-1} \sum_{j=0}^{N-1} [s(i,j)]^2} \times 100 \tag{21}$$

where, $s(i,j)$ is the element of the original image.

$\hat{s}(i,j)$ is the element of the estimated image obtained from M spectral co-efficients.

RESULTS AND DISCUSSION

A small image, sampled and digitized in an array of size 64 × 128 pixels, is used for investigating the bandwidth compression techniques using Legendre Transform. The original image is shown in Fig. 3*a*. The image is the photograph of the computer line printer output obtained by character over-printing for 16 grey levels. Data compression ratio 2 : 1 is used by applying the Legendre transform and the image is retrieved back from the retained co-efficients. Figure 3*b* shows such image. The processing is done in the block of 8 × 8 pixels using 8-point Legendre transform. The results are compared with the well-known fast Walsh Hadamard Transform (WHT) for the same compression ratio. Figure 3*c* shows such image.

The percentage mean-square error for 2 : 1 data compression for these transforms is given below for comparison

Transform	*Error*
WHT	1.971 %
Legendre Transform	0.627 %

Thus the discrete Legendre transforms have the following advantages over other unitary transforms.

(*i*) Since the variation of the transform rows matches the polynomial variation of the television picture, they are found more suitable for image data compression.

(*ii*) The size of transform matrix N need not to be an integer power of two unlike WHT, Slant, Haar etc. It can be a square matrix of any size. Utilizing the even and odd symmetry, there is always a possibility to find a fast algorithm for Legendre transforms of any value of N.

Fig. 3*a*. *Original Image, Picture of the Computer Line Printer Printout.*

Fig. 3*b*. *Legendre Transforms* (*MSE*= 0.625%). **Fig.** 3*c*. *WHT* (*MSE* = 1.91%).

Reconstructed Images. Processing is done in the blocks of 8×8 pixels. Threshold sampling is used. Compression Rato 2 : 1.

REFERENCES

1. **Ahmed** (N) & **Rao** (K R). *Orthogonal Transforms for Digital Signal Processing*. 1975. Springer — Verlag, Berlin, P 200-218.
2. **Davisson** (L D) & **Robert** (M G) Eds. *Data Compression*. 1976. Dowder Hatchinson and Ross, Inc. Stroundsburg, Pennsylvania, P 10-25.
3. **Delp** (E J) & **Mitchell** (O R). Image Compression Using Block Truncation Coding. *IEEE Trans.* **COM-27**, 1979; 1335-1342.
4. **Enomoto** (H) & **Shibata** (K). Orthogonal Transform Coding System for Television Signals. *IEEE. Trans.* **EMC-13**, 1; 1971; 11-17.
5. **Habibi** (A) & **Robinson** (G S). A Survey of Digital Picture Coding *IEEE Computer*. 7, 5; 1974; 22-34.
6. **Haralick** (R M) & **Shanmugam** (K). Comparative Study of a Discrete Linear Basis for Image Data Compression. *IEEE Trans.* **SMC-4**, 1974; 16-27.
7. **Huang** (T S) & **Tretiak** (O J) Eds. *Picture Bandwidth Compression*. 1972. Gordon and Breach Science Pub., New York, P 555-573.
8. **Hung** (S H Y). A Generalization of DPCM for Digital Image Compression. *IEEE Trans.* **PAMI-1**, 1979; 100-108.
9. **Jain** (A K). A Fast Karhunen — Loeve Transform for Finite Discrete Images. *Proc. Nat. Electronics Conf.* **29**, 1974; 323-328.
10. **Jain** (A K). A Fast Karhunen — Loeve Transform for a Class of Random Process. *IEEE Trans.* **COM-24**, 1976; 1023-1029.
11. **Jain** (A K). Image Data Compression : A Review. *Proc. IEEE*. **69**, 1981; 349-389.
12. **Kekre** (H B). Evaluation of Impulsive Response of Linear System by Using Legendre Polynomials. *Inter J. Control.* 7, 1; 1968; 81-83.
13. **Kekre** (H B) & **Solanki** (J K). Image Model and Spectral Extrapolation in Transform Image Coding. *Int. J. Electronics.* **45**, 4; 1978; 465-474.
14. **Kekre** (H B) & **Solanki** (J K). Comparative Performance of Various Trigonometric Unitary Transforms for Transform Image Coding. *Int. J. Electronics.* **45**, 3, 1978; 305-315.
15. **Lee** (Y W). *Statistical Theory of Communications.* J Wiley and Sons, Inc., New York. 1960, P 476-480.
16. **Lei** (T R), **Scheinberg** (N) & **Schilling** (D L). Adaptive Delta Modulation System for Video Encoding. *IEEE Trans.* **COM-23**, 1977; 1302-1314.
17. **Pratt** (W K). Transform Domain Signal Processing Techniques. *Proc. Nat. Electronis Conf.* **29**, 1974; 317-322.
18. **Sansone** (G) & **Dimand** (A H). *Orthogonal Functions.* 1959. Interscience Publishers, Inc., New York, P 169.
19. **Shore** (J E). On the Application of Haar Functions. *IEEE Trans.* **COM-21**, 1973; 209-216.
20. **Zschunke** (W). DPCM Picture Coding with Adaptive Prediction. *IEEE Trans.* **COM-25**, 1977; 1295-1302.

Editor's Comments on Papers 15 Through 20

WALSH-HADAMARD TRANSFORM

Walsh-Hadamard transform (WHT) is based on Walsh functions that form a complete orthogonal set. They can be generated easily from Rademacher functions. WHT matrices can also be generated recursively. The elements of these matrices are binary in nature (± 1) (Henderson, 1964; Paper 15). Also, sparse matrix factorization or matrix partitioning of the WHT matrices leads to the fast algorithms. These algorithms require only $N \log_2 N$ additions/subtractions for implementing an N-point WHT. Also, as the computation involves real arithmetic and as the fast algorithm has in-place structure (this implies reduced memory requirements), implementation of WHT on a computer or in hardware is quite simple.

By rearranging the Walsh functions, they can be grouped as sequency or Walsh ordering, dyadic or Paley ordering, natural or Hadamard ordering, and cal-sal ordering. The transform matrices resulting from this rearrangement are named accordingly. These transformations are outlined in several papers. (Bhagavan and Polge, 1973; Cheng and Liu 1977; Lackey and Meltzer, 1971; Rao et al., 1978). Also relations between the WHT and HT have been developed (Paper 16).

WHT power spectrum, which is invariant to circular shift of a sequence, can be developed. These compacted spectra represent groups of sequencies and have been utilized in pattern recognition (Nemcek and Lin 1974). Also, a compacted phase spectrum with the same sequency composition has been developed (Paper 15). Fast algorithms for composition of these spectra, even without computing the WHT, also exist (Paper 15). The energy-packing efficiency (EPE) (ratio of energy in the first few transform cofficients to the total energy of a sequence) of WHT has been evaluated by Kitajima (Paper 17). This concept was later extended to the generalized discrete transforms by Yip and Rao (1978). The WHT power spectra is invariant to a number of shifts other than circular (Paper 18). Arazi (Paper 18) also has developed various interesting properties of WHT when a sequence undergoes certain circular shifts and when the rows and columns of a 2D-data undergo certain interchanges. Needless to mention, the WHT is separable and can be easily extended to multidimensions. Another unique property of the WHT is that the 2D-WHT of a matrix is equivalent to the 1D-WHT of a vector resulting from a lexicographic ordering of the elements of the matrix (Paper 19). The concept of invariance of the WHT power spectra to a circular shift has been extended further by Hama and Yamashita (Paper 20). They extend this invariance to circular shifting, enlarging, reducing, rotating by integer multiples of 90 degrees, and some other transformations of input patterns. Thus WHT can be applied to feature extraction and pattern recognition.

Because of its simplicity WHT has been used in speech (Zelinski and Noll, 1977, 1979; Paper 37) and image coding (Pratt, Kane, and Andrews, 1969). WHT processors have been built and utilized for bandwidth compression of images both for transmission (Paper 40) and for storage (Peters and Kanters, 1983). Also, a speaker-dependent LSI speech recognition chip based on the WHT has been built (Matsushita Electronics Industry Co., Ltd., 1983).

REFERENCES

Bhagavan, B. K., and R. J. Polge, 1973, Sequencing the Hadamard Transform, *IEEE Audio Electroacoust. Trans.* **AU-21:**472–473.

Cheng, D. K., and J. J. Liu, 1977, An Algorithm for Sequency Ordering of Hadamard Functions, *IEEE Comput. Trans.* **C-26:**308–309.

Henderson, K. W., 1964, Some Notes on the Walsh Functions, *IEEE Electron. Comput. Trans.* **EC-13:**50–52.

Lackey, R. B., and D. Meltzer, 1971, A Simplified Definition of Walsh Functions, *IEEE Comput. Trans.* **C-20:**211–213.

Manz, J. W., 1972, A Sequency Ordered Fast Walsh Transform, *IEEE Audio Electroacoust. Trans.* **AU-20:**204–205.

Nemcek, W. F., and W. C. Lin, 1974, Experimental Investigation of Automatic Signature Verification, *IEEE Syst., Man, Cybern. Trans.* **SMC-4:**121–126.

Pratt, W. K., J. Kane, and H. C. Andrews, 1969, Hadamard Transform Image Coding, *IEEE Proc.* **57:**58–68.

Peters, J. H., and J. T. Kanters, 1983, *Hadamard Transform of Composite Video for Consumer Recording,* Picture Coding Symposium Proceedings, University of California, Davis, Calif., pp. 32–33.

Rao, K. R. et al. 1978, Cal-Sal Walsh-Hadamard Transform, *IEEE Acoust., Speech, Signal Process. Trans.* **ASSP-26:**605–607.

Shanks, J. L., 1969, Computation of the Fast Walsh-Fourier Transform, *IEEE Comput. Trans.* **C-18:**457–459.

Yip, P., and K. R. Rao, 1978, Energy Packing Efficiency for the Generalized Discrete Transform, *IEEE Commun. Trans.* **COM-26:**1257–1262.

Zelinski, R., and P. Noll, 1977, Adaptive Transform Coding of Speech Signals, *IEEE Acoust., Speech, Signal Process. Trans.* **ASSP-25:**299–309.

Zelinski, R., and P. Noll, 1979, Approaches to Adaptive Transform Speech Coding at Low-bit Rates, *IEEE Acoust., Speech, Signal Process. Trans.* **ASSP-27:**89–95.

BIFORE or Hadamard Transform

NASIR AHMED, Member, IEEE
Department of Electrical Engineering
Kansas State University
Manhattan, Kans.

K. R. RAO, Member, IEEE
Department of Electrical Engineering
University of Texas at Arlington
Arlington, Tex.

A. L. ABDUSSATTAR, Student Member, IEEE
Department of Electrical Engineering
Kansas State University
Manhattan, Kans.

Abstract

BIFORE or Hadamard transform[1] is defined and several of its properties are developed. BIFORE power and phase spectra are developed and their frequency–sequency composition is explored. Using matrix partitioning, fast algorithms for efficient computation of BIFORE coefficients and power and phase spectra are developed. Multidimensional BIFORE transform is defined and a physical interpretation of its power spectrum is presented. Advantages and as well the limitations of the BIFORE transform in its application in information processing are cited.

Introduction

The notion of binary Fourier representation (BIFORE) was introduced by Ohnsorg [1]. BIFORE resembles the Fourier harmonic analysis in both geometrical and analytical characteristics. While the Fourier bases are sinusoids with harmonic frequencies, the BIFORE bases are Walsh functions. Since the Walsh functions are square waves, they take only two values, namely, +1 or −1. Thus, in the case of finite systems, these square waves are represented by binary n-tuples. The simplicity of square waves or binary n-tuples relative to sinusoids or sampled sinusoids permits relatively easy information processing in several applications which include signal representation and classification [2], image coding [3], spectral analysis of digital systems [4], speech processing [5], and sequency analysis and synthesis of voice signals [6]. Since only real number operations are involved, BIFORE transform (BT) can save additional computer time. Other advantages are possibly data compression, tolerance to channel errors, and reduced bandwidth transmission [3].

Walsh Functions

Walsh functions [7] are best introduced by referring to the following Fourier sinusoids.

$$\begin{aligned} U_0(t) &= 1 \\ U_{m,1}(t) &= \cos (2\pi mt) \\ U_{m,2}(t) &= \sin (2\pi mt), \qquad m = 1, 2, 3, \cdots . \end{aligned} \tag{1}$$

The first five of the Fourier sinusoids are shown in Fig. 1 over the interval (0, 1). The corresponding Walsh functions $\{\psi_m(t)\}$ are shown in Fig. 2. Superimposition of these figures leads to the fact that:

1) $\psi_0(t) = U_0(t)$;
2) the zero crossings of the remaining pairs of pictured functions are identical.

This seems to indicate that the Walsh functions are an infinitely clipped version of the sinusoids. In general this is not the case, as the sign changes are not equidistant (Fig. 3). The total number of zero crossings of Walsh functions and corresponding sinusoids, however, are the same. Walsh functions can be generated recursively, are orthogonal, and form a closed set [7]–[9]. Signal analysis and synthesis can be carried out by expanding functions in Walsh series [6]. An expository article on applications of Walsh function has appeared recently [11].

Hadamard Matrices

Sampling of the Walsh functions shown in Fig. 3 at 2^3 equally spaced sample points $t_1, t_2, \cdots, t_8$ in (0, 1) results in an 8×8 array of +1's and −1's. As each sample represents a constant section of the Walsh function, the information content is preserved. The rows of any array obtained by this method can be rearranged to form a particular matrix of

[1] Also called Walsh–Hadamard transform.

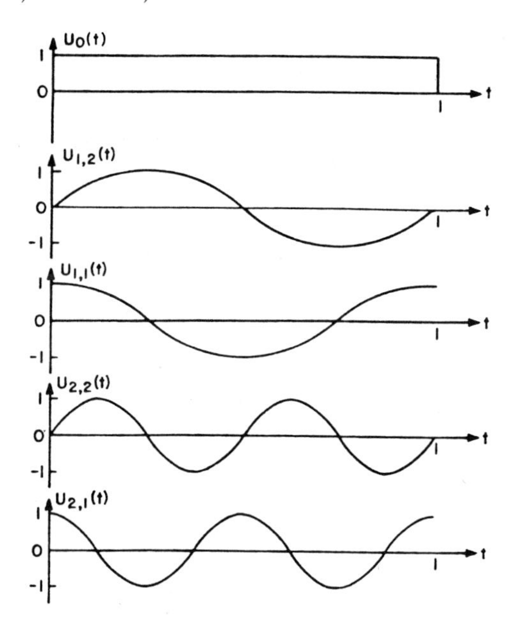

Fig. 1. Fourier harmonics.

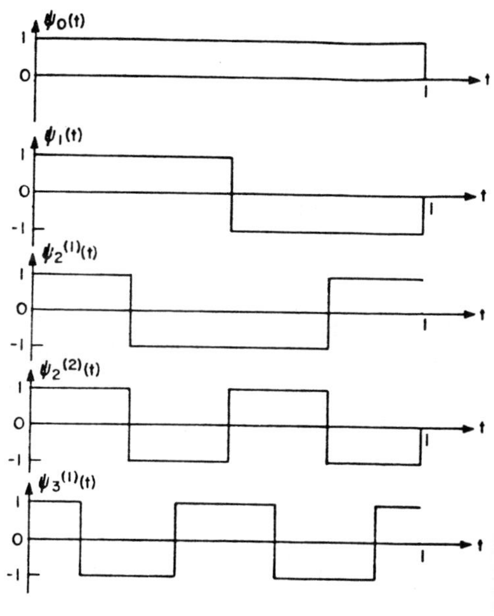

Fig. 2. Walsh functions.

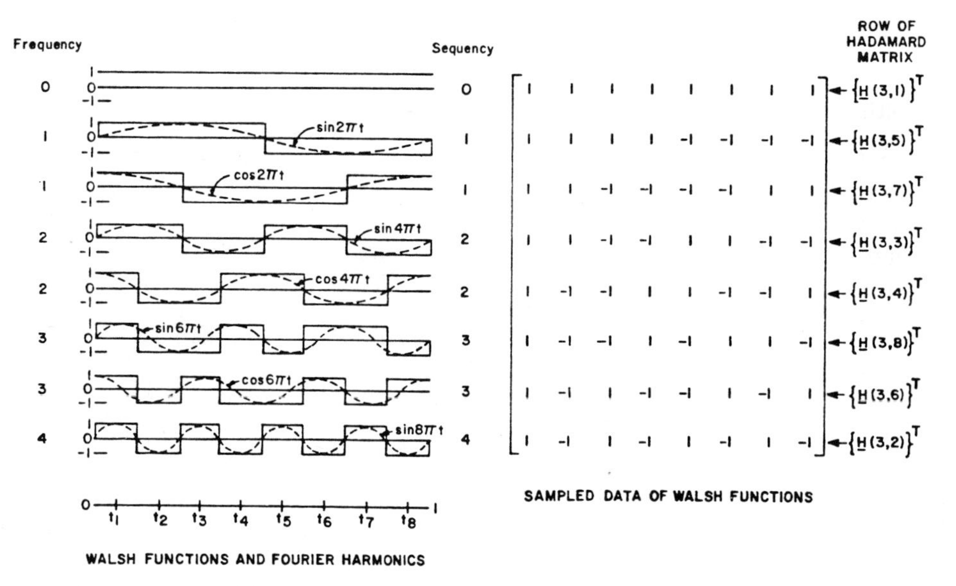

Fig. 3. Fourier harmonics, Walsh functions, and Hadamard matrices for $n=8$.

the type studied by Hadamard [12]. One such rearrangement is

$$[H(3)] = \left[\begin{array}{cccc|cccc} 1 & 1 & 1 & 1 & 1 & 1 & 1 & 1 \\ 1 & -1 & 1 & -1 & 1 & -1 & 1 & -1 \\ 1 & 1 & -1 & -1 & 1 & 1 & -1 & -1 \\ 1 & -1 & -1 & 1 & 1 & -1 & -1 & 1 \\ \hline 1 & 1 & 1 & 1 & -1 & -1 & -1 & -1 \\ 1 & -1 & 1 & -1 & -1 & 1 & -1 & 1 \\ 1 & 1 & -1 & -1 & -1 & -1 & 1 & 1 \\ 1 & -1 & -1 & 1 & -1 & 1 & 1 & -1 \end{array}\right] \quad (2)$$

where $[H(3)]$ is a $2^3 \times 2^3$ Hadamard matrix.

Thus, the periodic sampling of Walsh functions yields, after rearrangment of the rows, Hadamard matrices which can be recursively generated as follows:

$$[H(0)] = [1]$$

$$[H(k+1)] = \left[\begin{array}{c|c} [H(k)] & [H(k)] \\ \hline [H(k)] & -[H(k)] \end{array}\right]$$

$$k = 0, 1, 2, \cdots, g; \; g = \log_2 n. \quad (3)$$

If $\{H(g, l)\}$, $l=1, 2, \cdots, n$, $g=\log_2 n$, denotes the lth column of $[H(g)]$, then its rows $\{H(g, l)\}^T$ serve as a basis set in the finite dimensional vector space of dimension n.

These matrices possess transform properties since $|H(g)|$ is both symmetric and orthogonal, i.e.,

$$[H(g)]^T[H(g)] = n[I(g)]$$

where $[I(g)]$ is the $n \times n$ identity matrix.

BIFORE or Hadamard Transform

BIFORE or Hadamard transform (BT or HT) of an n-periodic sequence $\{x(k)\} = \{x(0), x(1), \cdots, x(n-1)\}$ is defined as

$$\{B_x(g)\} = \frac{1}{n}[H(g)]\{x(g)\} \tag{4}$$

where $\{x(g)\}^T = \{x(0)x(1) \cdots x(n-1)\}$ is the vector representation of the sequence $\{x(k)\}$ and $\{B_x(g)\}^T = \{B_x(0)B_x(1) \cdots B_x(n-1)\}$, the $B_x(\ )$ being the BT coefficients.

The signal can be recovered uniquely from the inverse BIFORE transform (IBT), i.e.,

$$\{x(g)\} = [H(g)]\{B_x(g)\}. \tag{5}$$

Fast BIFORE Transform (FBT)

Using matrix factoring [13] or matrix partitioning [14], algorithms for fast and efficient computation of BT can be developed. Similar techniques are applicable to discrete Fourier and other orthogonal transforms [13]–[17]. As an example, for $n=8$, (4) can be expressed in matrix form as follows:

$$\begin{Bmatrix} B_x(0) \\ B_x(1) \\ B_x(2) \\ B_x(3) \\ \hline B_x(4) \\ B_x(5) \\ B_x(6) \\ B_x(7) \end{Bmatrix} = \frac{1}{8} \left[\begin{array}{cc|cc} \begin{pmatrix}1 & 1\\1 & -1\end{pmatrix} & 1\begin{pmatrix}1 & 1\\1 & -1\end{pmatrix} & \begin{pmatrix}1 & 1\\1 & -1\end{pmatrix} & 1\begin{pmatrix}1 & 1\\1 & -1\end{pmatrix} \\ \begin{pmatrix}1 & 1\\1 & -1\end{pmatrix} & -1\begin{pmatrix}1 & 1\\1 & -1\end{pmatrix} & \begin{pmatrix}1 & 1\\1 & -1\end{pmatrix} & -1\begin{pmatrix}1 & 1\\1 & -1\end{pmatrix} \\ \hline \begin{pmatrix}1 & 1\\1 & -1\end{pmatrix} & 1\begin{pmatrix}1 & 1\\1 & -1\end{pmatrix} & -1\begin{pmatrix}1 & 1\\1 & -1\end{pmatrix} & -1\begin{pmatrix}1 & 1\\1 & -1\end{pmatrix} \\ \begin{pmatrix}1 & 1\\1 & -1\end{pmatrix} & -1\begin{pmatrix}1 & 1\\1 & -1\end{pmatrix} & -1\begin{pmatrix}1 & 1\\1 & -1\end{pmatrix} & 1\begin{pmatrix}1 & 1\\1 & -1\end{pmatrix} \end{array} \right] \begin{Bmatrix} x(0) \\ x(1) \\ x(2) \\ x(3) \\ \hline x(4) \\ x(5) \\ x(6) \\ x(7) \end{Bmatrix} \tag{6}$$

The structure of (6) suggests repeated applications of matrix partitioning, and the related sequence of computations is shown in the signal flow graph in Fig. 4. Apart from the $\frac{1}{8}$ multiplier, the total number of arithmetic operations (real additions and subtractions) required for computing the BT is $8 \times 3 = 24$.

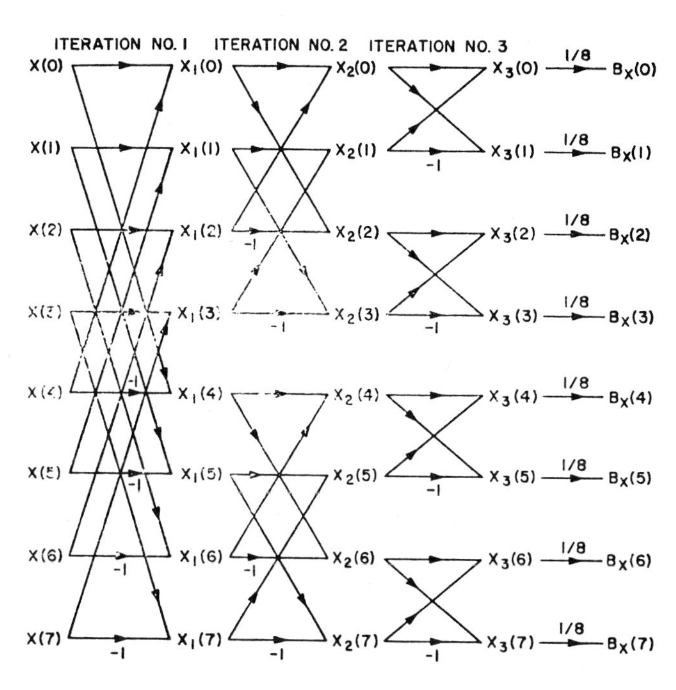

Fig. 4. Signal flow graph illustrating the computation of BT coefficients for $n=8$.

Generalizations

The generalizations pertinent to the FBT are straightforward. The overall structure of the signal flow graph for any n is similar to that of Fig. 4. For the fast algorithm the following observations can be made.

1) The total number of iterations is given by $g = \log_2 n$. Thus, if r is an iteration index, then $r = 1, 2, \cdots, g$.

2) The rth iteration results in 2^{r-1} groups with $n/2^{r-1}$ members in each group. Half the members in each group are associated with an addition operation while the remaining half are associated with a subtraction operation.

3) The total number of arithmetic operations to compute all the transform coefficients is approximately $n \log_2 n$, compared to n^2 as implied by (4).

4) The algorithm can also be used to compute the IBT by merely omitting the constant multiplier $1/n$.

BIFORE Power Spectrum [18]

To entertain the idea of a BIFORE power spectrum, Parseval's theorem is developed. The transpose of (4) yields

$$\{B_x(g)\}^T = \frac{1}{n}\{x(g)\}^T[H(g)]^T. \tag{7}$$

From (4) and (7) it follows that

$$\{B_x(g)\}^T\{B_x(g)\} = \frac{1}{n^2}\{x(g)\}^T[H(g)]^T[H(g)]\{x(g)\},$$

$$\{B_x(g)\}^T\{B_x(g)\} = \frac{1}{n}\{x(g)\}^T\{x(g)\}, \tag{8}$$

or

$$\frac{1}{n}\sum_{h=0}^{n-1} x^2(h) = \sum_{h=0}^{n-1} B_x^2(h). \tag{9}$$

BT of $\{x^{(1)}(3)\}$ is

$$\{B_x^{(1)}(3)\} = \tfrac{1}{8}[H(3)][M(3)]\{x(3)\}. \tag{12}$$

That is,

$$\{B_x^{(1)}(3)\} = [A(3)]\{B_x(3)\} \tag{13}$$

where

$$[A(3)] = \tfrac{1}{8}[H(3)][M(3)][H(3)]. \tag{14}$$

Evaluation of the shift matrix $[A(3)]$ in (14) results in

$$[A(3)] = \frac{1}{2}\left[\begin{array}{cc|cc|cccc} 2 & 0 & 0 & 0 & 0 & 0 & 0 & 0 \\ 0 & -2 & 0 & 0 & 0 & 0 & 0 & 0 \\ \hline 0 & 0 & 0 & -2 & 0 & 0 & 0 & 0 \\ 0 & 0 & 2 & 0 & 0 & 0 & 0 & 0 \\ \hline 0 & 0 & 0 & 0 & 1 & -1 & -1 & -1 \\ 0 & 0 & 0 & 0 & 1 & -1 & 1 & 1 \\ 0 & 0 & 0 & 0 & 1 & 1 & 1 & -1 \\ 0 & 0 & 0 & 0 & -1 & -1 & 1 & -1 \end{array}\right]. \tag{15}$$

Thus, if the sequence $\{x(k)\}$ represents the sampled values of a current or voltage signal $x(t)$ and a 1 Ω pure resistive load is assumed, then the left-hand side of (9) represents the average power dissipated. The summation on the right-hand side of (9) implies that the signal power is conserved in the transform domain. The set $\{B_x^2(k)\}$, however, does not represent the individual spectral points as it is not invariant to the shift of the sampled data.

Development of the Spectrum

The development of the power spectrum is best illustrated by considering the case when $n=8$. Let $\{x^{(l)}(k)\}$ denote $\{x(k)\}$ shifted to the left by l positions. That is,

$$\{x^{(l)}(3)\}^T = \{x(l)x(l+1)\cdots x(l-2)x(l-1)\}, \quad l = 1, 2, \cdots, 7$$

which with $l=1$ yields

$$\{x^{(1)}(3)\} = [M(3)]\{x(3)\} \tag{10}$$

where

$$[M(3)] = \begin{bmatrix} 0 & 1 & 0 & 0 & 0 & 0 & 0 & 0 \\ 0 & 0 & 1 & 0 & 0 & 0 & 0 & 0 \\ 0 & 0 & 0 & 1 & 0 & 0 & 0 & 0 \\ 0 & 0 & 0 & 0 & 1 & 0 & 0 & 0 \\ 0 & 0 & 0 & 0 & 0 & 1 & 0 & 0 \\ 0 & 0 & 0 & 0 & 0 & 0 & 1 & 0 \\ 0 & 0 & 0 & 0 & 0 & 0 & 0 & 1 \\ 1 & 0 & 0 & 0 & 0 & 0 & 0 & 0 \end{bmatrix}. \tag{11}$$

Repetitive application of (14) yields

$$\{B_x^{(l)}(3)\} = [A(3)]^l\{B_x(3)\}, \qquad l = 1, 2, \cdots, 7. \tag{16}$$

$[A(3)]$ is made up of square matrices of increasing order along the diagonal. From the "block diagonal" structure of $[A(3)]$ and (16), the following set of equations is obtained:

$$B_x^{(l)}(0) = B_x(0)$$

$$B_x^{(l)}(1) = (-1)^l B_x(1)$$

$$\begin{bmatrix} B_x^{(l)}(2) \\ B_x^{(l)}(3) \end{bmatrix} = [D(1)]^l \begin{bmatrix} B_x(2) \\ B_x(3) \end{bmatrix}$$

and

$$\begin{bmatrix} B_x^{(l)}(4) \\ B_x^{(l)}(5) \\ B_x^{(l)}(6) \\ B_x^{(l)}(7) \end{bmatrix} = \frac{1}{2^l}[D(2)]^l \begin{bmatrix} B_x(4) \\ B_x(5) \\ B_x(6) \\ B_x(7) \end{bmatrix} \tag{17}$$

where

$$[D(1)] = \begin{bmatrix} 0 & -1 \\ 1 & 0 \end{bmatrix}$$

and

$$[D(2)] = \frac{1}{2}\begin{bmatrix} 1 & -1 & -1 & -1 \\ 1 & -1 & 1 & 1 \\ 1 & 1 & 1 & -1 \\ -1 & -1 & 1 & -1 \end{bmatrix} \tag{18}$$

are orthogonal. That is, $[D(1)]^T[D(1)]=[I(1)]$ and $[D(2)]^T[D(2)]=[I(2)]$.

Equation (17) implies that

$$(B_x^{(l)}(k))^2 = B_x^2(k), \qquad k = 0, 1;$$

$$\sum_{k=2}^{3} (B_x^{(l)}(k))^2 = \sum_{k=2}^{3} B_x^2(k) \tag{19}$$

and

$$\sum_{k=4}^{7} (B_x^{(l)}(k))^2 = \sum_{k=4}^{7} B_x^2(k), \qquad l = 1, 2, \cdots, 7.$$

The set of invariants defined in (19) represents the power spectrum. In general, denote this spectrum as follows:

$$P_0 = B_x^2(0)$$

$$P_s = \sum_{k=2^{s-1}}^{2^s-1} B_x^2(k), \qquad s = 1, 2, 3, \cdots, g. \tag{20}$$

Fast Algorithm for Power Spectrum

By suitably modifying the FBT approach, the power spectrum can be computed without having to actually compute all the coefficients $B_x(k)$. The modification is best illustrated for the case when $n=8$.

From the signal flow graph in Fig. 4, it follows that

$$B_x(0) = x_3(0)/8$$

$$B_x(1) = x_3(1)/8$$

$$\begin{bmatrix} B_x(2) \\ B_x(3) \end{bmatrix} = \tfrac{1}{8}[H(1)] \begin{bmatrix} x_2(2) \\ x_2(3) \end{bmatrix}$$

and

$$\begin{bmatrix} B_x(4) \\ B_x(5) \\ B_x(6) \\ B_x(7) \end{bmatrix} = \tfrac{1}{8}[H(2)] \begin{bmatrix} x_1(4) \\ x_1(5) \\ x_1(6) \\ x_1(7) \end{bmatrix}. \tag{21}$$

As $[H(g)]$ is orthogonal, one can obtain from (21)

$$\sum_{k=2}^{3} B_x^2(k) = \frac{2}{8^2} \sum_{m=2}^{3} x_2^2(m)$$

and

$$\sum_{k=4}^{7} B_x^2(k) = \frac{4}{8^2} \sum_{m=4}^{7} x_1^2(m). \tag{22}$$

The power spectrum then is

$$P_0 = \frac{1}{8^2} x_3^2(0)$$

$$P_1 = \frac{1}{8^2} x_3^2(1)$$

$$P_2 = \frac{2}{8^2} \sum_{m=2}^{3} x_2^2(m)$$

and

$$P_3 = \frac{2^2}{8^2} \sum_{m=4}^{7} x_1^2(m). \tag{23}$$

The generalization of the above FBT modification to compute the power spectrum is straightforward as indicated below.

$$P_0 = x_g^2(0)/n^2$$

$$P_s = \frac{2^{s-1}}{n^2} \sum_{m=2^{s-1}}^{2^s-1} x_{g+1-s}^2(m), \qquad s = 1, 2, \cdots, g. \tag{24}$$

The signal flow graph corresponding to (24) is shown in Fig. 5.

Physical Interpretation of the BIFORE Power Spectrum

BIFORE power spectrum has dual significance. The spectrum points P_s represent the average powers in a set of $(g+1)$ sequences. Each spectral point also represents the power content of a group of frequencies rather than that of a single frequency. The n-periodic sequence $\{x(k)\}$ can be decomposed into an $n/2$-periodic sequence $\{F_1(k)\}$ and an $n/2$-antiperiodic[2] sequence $\{G_1(k)\}$ as follows:

$$\{x(k)\} = \{F_1(k)\} + \{G_1(k)\}$$

where

$$\{F_1(k)\} = \frac{1}{2}\left\{x(k) + x\left(k + \frac{n}{2}\right)\right\}$$

and

$$\{G_1(k)\} = \frac{1}{2}\left\{x(k) - x\left(k + \frac{n}{2}\right)\right\}. \tag{25}$$

$\{F_1(k)\}$ can be further decomposed into an $n/4$-periodic sequence $\{F_2(k)\}$ and an $n/4$-antiperiodic sequence $\{G_2(k)\}$, i.e., $F_1(k)=\{F_2(k)\}+\{G_2(k)\}$ where

$$\{F_2(k)\} = \frac{1}{2}\left\{F_1(k) + F_1\left(k + \frac{n}{4}\right)\right\}$$

and

$$\{G_2(k)\} = \frac{1}{2}\left\{F_1(k) - F_1\left(k + \frac{n}{4}\right)\right\}. \tag{26}$$

Continuing this process, it follows that the given n-periodic sequence $\{x(k)\}$ can be decomposed into the following $(g+1)$ subsequences:

$$\{x(k)\} = \{F_g(k)\} + \{G_g(k)\} + \{G_{g-1}(k)\} + \cdots + \{G_1(k)\} \tag{27}$$

where $\{F_g(k)\}$ is a 1-periodic sequence and $\{G_{g-r}(k)\}$ is a 2^r-antiperiodic sequence, $r=0, 1, \cdots, (g-1)$.

This process of decomposition is illustrated in Fig. 6 for $N=8$. Thus the sequence $\{x(k)\}$ can be decomposed to obtain

$$\{x(k)\} = \{F_3(k)\} + \{G_3(k)\} + \{G_2(k)\} + \{G_1(k)\} \tag{28}$$

where $\{F_3(k)\}$ is 1-periodic, while $\{G_3(k)\}$, $\{G_2(k)\}$, and

[2] A sequence $\{x(k)\}$ is said to be M-antiperiodic if $x(m) = -x(M+m)$.

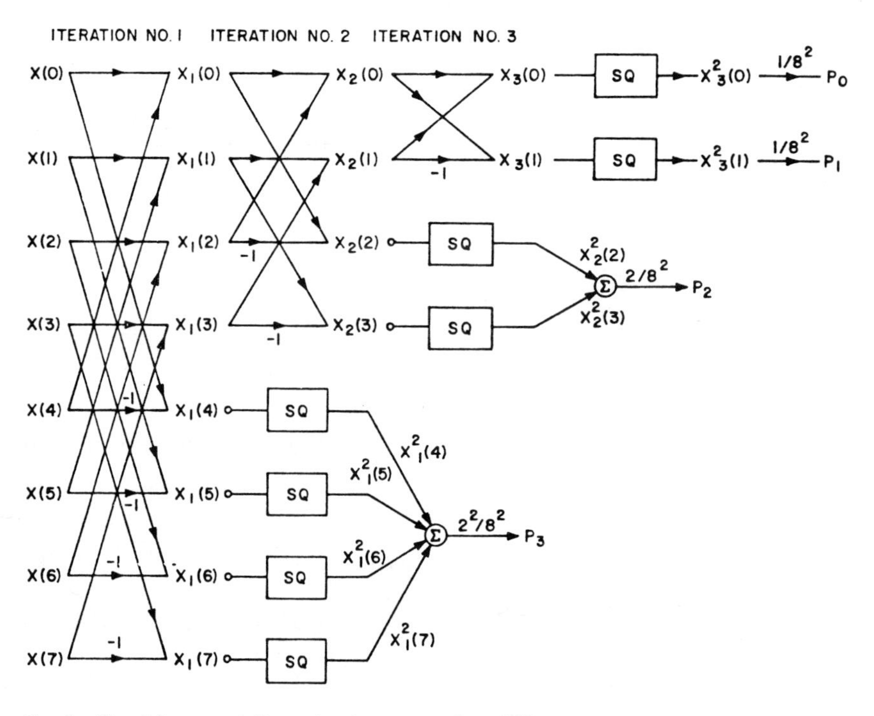

Fig. 5. Signal flow graph illustrating the computation of BT power spectrum.

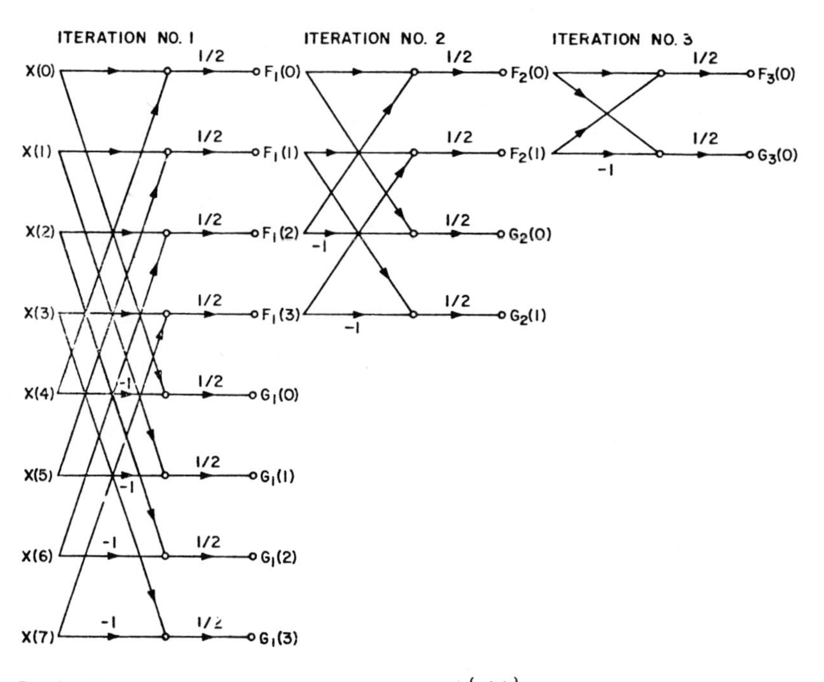

Fig. 6. Flow graph illustrating the decomposition of $\{x(k)\}$ into subsequences.

$\{G_1(k)\}$ are respectively 1-, 2-, and 4-antiperiodic. Therefore, these sequences can be denoted by vectors as follows:

$$
\begin{aligned}
\{F_3(3)\}^T &= \{b_x(0) \quad b_x(0) \quad b_x(0) \quad b_x(0) \quad b_x(0) \quad b_x(0) \quad b_x(0) \quad b_x(0)\} \\
\{G_3(3)\}^T &= \{b_x(1) \; -b_x(1) \quad b_x(1) \; -b_x(1) \quad b_x(1) \; -b_x(1) \quad b_x(1) \; -b_x(1)\} \\
\{G_2(3)\}^T &= \{b_x(2) \quad b_x(3) \; -b_x(2) \; -b_x(3) \quad b_x(2) \quad b_x(3) \; -b_x(2) \; -b_x(3)\} \\
\{G_1(3)\}^T &= \{b_x(4) \quad b_x(5) \quad b_x(6) \quad b_x(7) \; -b_x(4) \; -b_x(5) \; -b_x(6) \; -b_x(7)\}
\end{aligned}
\qquad (29
$$

where the coefficients $b_x(k)$ will be expressed in terms of the BIFORE coefficients $B_x(k)$ at a later stage.

Inspection of the vectors in (29) shows that they are all mutually orthogonal and hence it follows that:

$$\|\{x(3)\}\|^2 = \|\{F_3(3)\}\|^2 + \|\{G_3(3)\}\|^2 + \|\{G_2(3)\}\|^2 + \|\{G_1(3)\}\|^2 \quad (30)$$

where $\|\cdot\|$ represents the norm.

$$\|\{F_3(3)\}\|^2 = 8b_x^2(0), \quad \|\{G_3(3)\}\|^2 = 8b_x^2(1)$$

$$\|\{G_2(3)\}\|^2 = 4\sum_{k=2}^{3} b_x^2(k)$$

and

$$\|\{G_1(3)\}\|^2 = 2\sum_{k=4}^{7} b_x^2(k). \quad (31)$$

The average power of $\{x(k)\}$ is

$$P_{av} = \frac{1}{8}\sum_{m=0}^{7} x^2(m) = \frac{1}{8}\|\{x(3)\}\|^2.$$

From (30) and (31), there results

$$P_{av} = b_x^2(0) + b_x^2(1) + \frac{1}{2}\sum_{k=2}^{3} b_x^2(k) + \frac{1}{4}\sum_{k=4}^{7} b_x^2(k). \quad (32)$$

In terms of the decomposed sequences of $\{x(k)\}$, its BT can be expressed as

$$\begin{bmatrix} B_x(0)\\ B_x(1)\\ B_x(2)\\ B_x(3)\\ B_x(4)\\ B_x(5)\\ B_x(6)\\ B_x(7) \end{bmatrix} = \frac{1}{8}\left[\begin{array}{cc|cc|cc|cc} 1 & 1 & 1 & 1 & 1 & 1 & 1 & 1\\ \hline 1 & -1 & 1 & -1 & 1 & -1 & 1 & -1\\ \hline \multicolumn{2}{c|}{[H(1)]} & \multicolumn{2}{c|}{-[H(1)]} & \multicolumn{2}{c|}{[H(1)]} & \multicolumn{2}{c}{-[H(1)]}\\ \hline \multicolumn{4}{c|}{[H(2)]} & \multicolumn{4}{c}{[H(2)]} \end{array}\right] \left[\begin{bmatrix} b_x(0)\\ b_x(0)\\ b_x(0)\\ b_x(0)\\ b_x(0)\\ b_x(0)\\ b_x(0)\\ b_x(0) \end{bmatrix} + \begin{bmatrix} b_x(1)\\ -b_x(1)\\ b_x(1)\\ -b_x(1)\\ b_x(1)\\ -b_x(1)\\ b_x(1)\\ -b_x(1) \end{bmatrix} + \begin{bmatrix} b_x(2)\\ b_x(3)\\ -b_x(2)\\ -b_x(3)\\ b_x(2)\\ b_x(3)\\ -b_x(2)\\ -b_x(3) \end{bmatrix} + \begin{bmatrix} b_x(4)\\ b_x(5)\\ b_x(6)\\ b_x(7)\\ -b_x(4)\\ -b_x(5)\\ -b_x(6)\\ -b_x(6) \end{bmatrix}\right]. \quad (33)$$

From (33) it follows that

$$B_x(0) = b_x(0), \quad B_x(1) = b_x(1),$$

$$\begin{bmatrix} B_x(2)\\ B_x(3) \end{bmatrix} = \tfrac{1}{2}[H(1)]\begin{bmatrix} b_x(2)\\ b_x(3) \end{bmatrix}$$

$$\begin{bmatrix} B_x(4)\\ B_x(5)\\ B_x(6)\\ B_x(7) \end{bmatrix} = \tfrac{1}{4}[H(2)]\begin{bmatrix} b_x(4)\\ b_x(5)\\ b_x(6)\\ b_x(7) \end{bmatrix} \quad (34)$$

Since the matrices $[H(1)]$ and $[H(2)]$ are orthogonal, it follows that

TABLE I

Frequency–Sequency Composition of BIFORE Coefficients

BIFORE Component	Frequency	Sequency
$B_x(0)$	0	0
$B_x(1)$	4	4
$B_x(2)$	2	2
$B_x(3)$	2	2
$B_x(4)$	1	1
$B_x(5)$	3	3
$B_x(6)$	1	1
$B_x(7)$	3	3

$$\sum_{k=2}^{3} b_x^2(k) = 2\sum_{k=2}^{3} B_x^2(k)$$

and

$$\sum_{k=4}^{7} b_x^2(k) = 4\sum_{k=4}^{7} B_x^2(k).$$

Thus

$$P_{av} = B_x^2(0) + B_x^2(1) + \sum_{k=2}^{3} B_x^2(k) + \sum_{k=4}^{7} B_x^2(k) = P_0 + P_1 + P_2 + P_3. \quad (35)$$

P_0 represents the average power in the 1-periodic sequence $\{F_3(k)\}$, while P_1, P_2, and P_3 represent the average powers in $\{G_3(k)\}$, $\{G_2(k)\}$, and $\{G_1(k)\}$ which are respectively 1-, 2-, and 4-antiperiodic. In general, P_r $(r = 1, 2, \cdots, g)$ represents the average power in $\{G_{g+1-r}(k)\}$.

An additional interpretation of the power spectrum is found in Table I which gives the frequency–sequency[3] structure of BIFORE transform derived from Fig. 3.

Denoting the frequency content of the spectral point P_s by $F(P_s)$, from Table I it follows that

$$F(P_0) = 0; \quad F(P_3) = 1, 3; \quad F(P_2) = 2; \quad F(P_1) = 4.$$

[3] "Sequency" is defined as one-half the average number of zero-crossings per unit time [11].

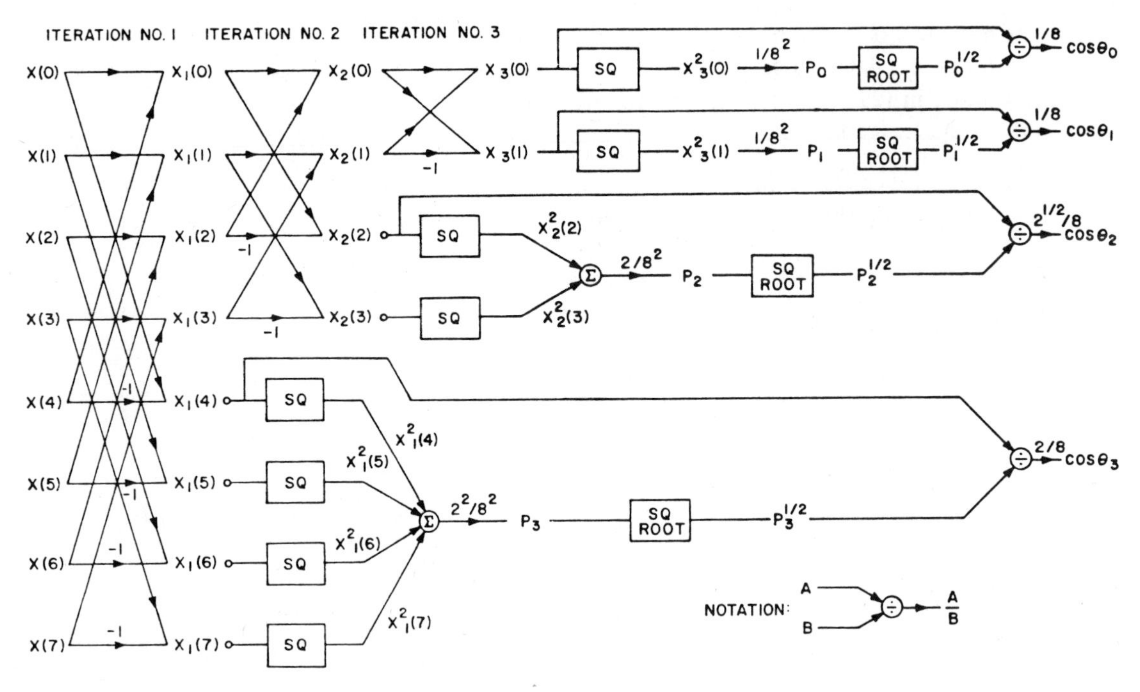

Fig. 7. Signal flow graph for computation of BIFORE power and phase spectra.

In general, then

$$F(P_0) = 0; \quad F(P_1) = n/2; \quad \text{and} \quad F(P_s) = 2^{g-s}(2k+1),$$
$$s = 2, 3, \cdots, g, \quad k = 0, 1, \cdots, (2^{s-2} - 1). \quad (36)$$

Clearly, each spectral point represents the power content of a group of frequencies rather than that of a single frequency as in the case of the Fourier transform. This frequency grouping, however, is not arbitrary. Each group contains a fundamental and the set of all odd harmonics relative to that fundamental. This corresponds to half-wave symmetry structure. Furthermore, there is a one-to-one correspondence between the $F(P_s)$ and the sequences $\{G_{g-r}(k)\}$.

BIFORE Phase Spectrum [19]

Analogous to the power spectrum, a phase spectrum of BT can be developed. Using the concepts of average power and phase angle, the phase spectrum is defined in multidimensional space in terms of a reference vector.

Average Power and Phase Angle

The average power P_{av} of and the angle θ between two n-dimensional vectors $\{x_1(g)\}$ and $\{x_2(g)\}$ are defined as

$$P_{av} = \frac{1}{n} \langle \{x_1(g)\}, \{x_2(g)\} \rangle$$
$$= \frac{1}{n} \|\{x_1(g)\}\| \, \|\{x_2(g)\}\| \cos\theta$$
$$\cos\theta = \frac{\langle \{x_1(g)\}, \{x_2(g)\} \rangle}{\|\{x_1(g)\}\| \, \|\{x_2(g)\}\|} \quad (37)$$

where the notation $\langle \cdot \rangle$ denotes the inner product.

Following the decomposition of $\{x(k)\}$ as in (27), the BIFORE phase spectrum can be developed [19] to yield

$$\cos\theta_0 = \frac{B_x(0)}{P_0^{1/2}}$$
$$\cos\theta_s = \frac{\sum_{k=2^{s-1}}^{2^s-1} B_x(k)}{2^{(s-1)/2} P_s^{1/2}}, \quad s = 1, 2, \cdots, g. \quad (38)$$

For $n=8$, (38) reduces to

$$\cos\theta_0 = \frac{B_x(0)}{P_0^{1/2}},$$
$$\cos\theta_1 = \frac{B_x(1)}{P_1^{1/2}},$$
$$\cos\theta_2 = \frac{\sum_{k=2}^{3} B_x(k)}{(2P_2)^{1/2}}$$

and

$$\cos\theta_3 = \frac{\sum_{k=4}^{7} B_x(k)}{(4P_3)^{1/2}}.$$

This phase spectrum together with the power spectrum can be evaluated rapidly using the signal flow graph shown in Fig. 7. The phase spectrum is invariant to multiplication of the data sequence $\{x(k)\}$ by a real number but changes as the sequence is shifted. In fact it can be shown that the spectrum point $\cos\theta_s$ is 2^{s-1}-antiperiodic. Clearly, these properties are analogous to those of the DFT phase spectrum. However, the concept of phase is defined for groups of frequencies whose composition is the same as that of the power spectrum. Because of this frequency grouping and consequent data compression, the original sequence $\{x(k)\}$ cannot be recovered from the phase and power spectra as summarized in (20) and (38). This is in contrast to the DFT where the phase and power spectra are defined for individual frequencies and where the time signal can be reconstructed through them. Another phase spectrum based on sequency is developed by Boeswetter [6].

Multidimensional BT [20]

Analogous to DFT, BT can be generalized to any number of dimensions. The r-dimensional BT is defined as

$$B_f(u_1, u_2, \cdots, u_r) = \prod_{i=1}^{r} \sum_{x_i=0}^{N_i-1} f(x_1, x_2, \cdots, x_r)(-1)^{\langle x,u \rangle} \quad (39)$$

where

$B_f(u_1, u_2, \cdots, u_r)$ is the transform coefficient.

$f(x_1, x_2, \cdots, x_r)$ is an input data point.

$$u_1, x_i = 0, 1, 2, \cdots, N_i - 1, \qquad k_i = \log_2 N_i.$$

$$\langle x, u \rangle = \sum_{j=1}^{r} \langle x_j, u_j \rangle.$$

$$\langle x_i, u_i \rangle = \sum_{m=0}^{k_i-1} x_i(m) u_i(m), \qquad i = 1, 2, \cdots, r.$$

The terms $u_i(m)$ and $x_i(m)$ are the binary representations of u_i and x_i, respectively, i.e.,

$$[u_i]_{\text{decimal}} = [u_i(k_i - 1), u_i(k_i - 2), \cdots, u_i(1), u_i(0)]_{\text{binary}}.$$

$f(x_1, x_2, \cdots, x_r)$ can be recovered uniquely from the inverse BT, i.e.,

$$f(x_1, x_2, \cdots, x_r) = \frac{1}{N} \prod_{i=1}^{r} \sum_{u_i=0}^{N_i-1} B_f(u_1, u_2, \cdots, u_r)(-1)^{\langle u,x \rangle} \quad (40)$$

where

$$N = \prod_{i=1}^{r} N_i.$$

Power Spectra

An extension of the power spectra of the 1-dimensional BT to the multidimensional case leads to the following:

$$P(z_1, z_2, \cdots, z_r) = \prod_{i=1}^{r} \sum_{u_i=2^{z_i-1}}^{2^{z_i}-1} B_f^2(u_1, u_2, \cdots, u_r) \quad (41)$$

where $z_i = 0, 1, 2, \cdots, k_i$. The total number of spectral points are $\prod_{i=1}^{r} (1+k_i)$. The frequency composition of the power spectra consists of all possible combinations of the groups of frequencies based on the odd-harmonic structure (half-wave symmetry) in each dimension. For example, the frequency grouping in the ith dimension is

$$0$$

$$1, 3, 5, \cdots, \left(\frac{N_i}{2} - 1\right)$$

$$2, 6, 10, \cdots, \left(\frac{N_i}{2} - 2\right)$$

$$\cdots$$

$$\frac{N_i}{2}.$$

As an illustration, the frequency content for $N_1 = 8$, $N_2 = 16$, and $N_3 = 32$ consists of all possible combinations of the following groups.

N_1	N_2	N_3
0	0	0
1, 3	1, 3, 5, 7	1, 3, 5, 7, 9, 11, 13, 15
2	2, 6	2, 6, 10, 14
4	4	4, 12
	8	8
		16

The total number of spectral points is $\prod_{i=1}^{3} (1+k_i) = 120$. The power spectra defined in (41) is invariant to cyclic shift of the sampled data in any or all dimensions.

Properties

Other properties of the multidimensional BT can easily be derived. Some of these are listed as follows.

Parseval's Theorem:

$$\frac{1}{N} \prod_{i=1}^{r} \sum_{x_i=0}^{N_i-1} f^2(x_1, x_2, \cdots, x_r) = \prod_{i=1}^{r} \sum_{u_i=0}^{N_i-1} B_f^2(u_1, u_2, \cdots, u_r). \quad (42)$$

Convolution: If

$$v(m_1, m_2, \cdots, m_r) = \frac{1}{N} \prod_{i=1}^{r} \sum_{x_i=0}^{N_i-1} f(x_1, x_2, \cdots, x_r) \cdot h(m_1 - x_1, m_2 - x_2, \cdots, m_r - x_r) \quad (43)$$

where $m_i = 0, 1, 2, \cdots, N_i - 1$, then

$$\prod_{i=1}^{r} \sum_{u_i=2^{z_i-1}}^{2^{z_i}-1} B_v(u_1 u_2, \cdots, u_r) = \prod_{i=1}^{r} \sum_{u_i=2^{z_i-1}}^{2^{z_i}-1} B_f(u_1, u_2, \cdots, u_r) B_h(u_1, u_2, \cdots, u_r). \quad (44)$$

Relationships similar to (44) are valid for cross correlation and autocorrelation.

Conclusions

The BIFORE transform and several of its properties are developed. These properties have been summarized and compared with those of DFT [21]. Fast algorithms for evaluation of BT, power and phase spectra are also developed. FBT can save computer time and result in reduced storage space as it requires about $n \log_2 n$ arithmetic operations to evaluate all the BT coefficients. Although the analysis in this paper has been restricted to real-valued sequences $\{x(k)\}$, it can be easily extended to complex-valued sequences.

The BT has already found applications in several areas of information processing [11]. Greater utilization and exploitation of the BT, however, requires recognition of the ease and efficiency offered by the fast algorithms and development of methods for signal recovery from the spectra. This can lead to evolution of special purpose digital hardware tailored to specific application areas.

References

[1] F. R. Ohnsorg, "Binary Fourier representation," presented at the Spectrum Analysis Techniques Symp., Honeywell Res. Cen., Hopkins, Minn., Sept. 20–21, 1966.

[2] J. E. Whelchel, Jr., and D. F. Guinn, "The fast Fourier–Hadamard transform and its use in signal representation and classification," in *Aerospace Electronics Conf., EASCON Rec.*, Sept. 1968, pp. 561–573.

[3] W. K. Pratt, J. Kane, and H. C. Andrews, "Hadamard transform image coding," *Proc. IEEE*, vol. 57, Jan. 1969, pp. 58–68.

[4] N. Ahmed and K. R. Rao, "Spectral analysis of linear digital systems using BIFORE," *Electron. Lett.*, vol. 6, Jan. 22, 1970, pp. 43–44.

[5] G. S. Robinson and S. J. Campanella, "Digital sequency decomposition of voice signals," presented of the Walsh Function Symp., Naval Res. Lab., Washington, D. C., Mar. 31–Apr. 3, 1970.

[6] C. Boeswetter, "Analog sequency analysis and synthesis of voice signals," presented at the Walsh Function Symp.. Naval Res. Lab., Washington, D. C., Mar. 31–Apr. 3, 1970.

[7] J. L. Walsh, "A closed set of normal orthogonal functions," *Amer. J. Math.*, vol. 55, 1923, pp. 5–24.

[8] R. E. A. C. Paley, "On orthogonal matrices," *J. Math. Phys.*, vol. 12, 1933, pp. 311–320.

[9] N. J. Fine, "On the Walsh functions," *Trans. Amer. Math. Soc.*, vol. 65, 1949, pp. 372–414.

[10] K. W. Henderson, "Some notes on the Walsh function," *IEEE Trans. Electron. Comput.* (Corresp.), vol. EC-13, Feb. 1964, pp. 50–52.

[11] H. F. Harmuth, "Applications of Walsh functions in communications," *IEEE Spectrum*, vol. 6, Nov. 1969, pp. 82–91.

[12] J. Hadamard, "Resolution d'une question relative aux determinants," *Bull. Sci. Math.*, ser. 2, vol. 17, 1893, pp. 240–246.

[13] H. C. Andrews and K. L. Caspari, "A generalized technique for spectral analysis," *IEEE Trans. Comput.*, vol. C-19, Jan. 1970, pp. 16–25.

[14] N. Ahmed and S. M. Cheng, "On matrix partitioning and a class of algorithms," *IEEE Trans. Educ.*, vol. E-13, Aug. 1970, pp. 103–105.

[15] E. O. Brigham and R. E. Morrow, "The fast Fourier transform," *IEEE Spectrum*, vol. 4, Dec. 1967, pp. 63–70.

[16] J. A. Glassman, "A generalization of the fast Fourier transform," *IEEE Trans. Comput.*, vol. C-19, Feb. 1970, pp. 105–116.

[17] F. Theilheimer, "A matrix version of the fast Fourier transform," *IEEE Trans. Audio Electroacoust.*, vol. AU-17, June 1969, pp. 158–161.

[18] N. Ahmed and K. R. Rao, "Convolution and correlation using binary Fourier representation," in *Proc. 1st Annu. Houston Conf. Circuits, Systems and Computers*, May 1969, pp. 182–191.

[19] N. Ahmed, K. R. Rao, and P. S. Fisher, "BIFORE Phase spectrum," presented at the 13th Midwest Symp. Circuit Theory, Univ. Minnesota, Minneapolis, Minn., May 7–8, 1970.

[20] N. Ahmed, R. M. Bates, and K. R. Rao, "Multidimensional BIFORE transform," *Electron. Lett.*, vol. 6, Apr. 16, 1970, pp. 237–238.

[21] N. Ahmed and K. R. Rao, "Discrete Fourier and Hadamard transforms," *Electron. Lett.*, Apr. 2, 1970, pp. 221–224.

Relations Between Haar and Walsh/Hadamard Transforms

B. J. Fino

Abstract—Relations between the Haar and Walsh/Hadamard (W/H) transforms, which are proved, show that for some applications the Haar transform performs as well as, and faster than, the W/H transform. These relations yield a family of orthogonal transforms including the Haar and W/H transforms with a common fast algorithm.

Introduction

The Haar and Walsh/Hadamard (W/H) functions are formally defined by Alexits[1] and Fig. 1 shows the eight Haar and eight W/H functions of order 8. The W/H functions are ordered by their sequencies (number of sign changes). The square orthogonal matrices $[H_{2^n}]$ and $[W_{2^n}]$ of order 2^n, the rows of which are the 2^n Haar and 2^n W/H functions normalized by $1/\sqrt{2^n}$, are called the Haar and W/H matrices.[2]

In the following section, we wish to show that simple relations exist between Haar and W/H submatrices, which are obtained by partitioning the matrices $[H_{2^n}]$ and $[W_{2^n}]$, respectively, into $(n+1)$ rectangular matrices $[MH_{2^n}{}^k]$ and $[MW_{2^n}{}^k]$ of dimension $(2^n \times 2^{k-1})$. $[MH_{2^n}{}^0]$ is the first row H_0, $[MH2_n{}^k]$ is formed from the Haar functions of rank r, $2^{k-1} \leq r < 2^k$, for $k=1, \cdots, n$, and similarly for the W/H submatrices $[MW_{2^n}{}^k]$. The submatrices of order 8 are shown in Fig. 1. We define the matrix operators $\mathcal{H}$ and $\mathcal{W}$ which express simply the recursive generations of the Haar and W/H submatrices. Then, using some properties of these operators, we prove by induction on n simple relations between the Haar and W/H submatrices.

Basic Relations

Alexits[1] has suggested that a matrix relation exists between the submatrices $[MH_{2^n}{}^k]$ and $[MW_{2^n}{}^k]$. We now prove that this relation is

$$[MW_{2^n}{}^k] = [S_{2^{k-1}}] \cdot [W_{2^{k-1}}] \cdot [MH_{2^n}{}^k], \qquad k = 1, \cdots, n \tag{1}$$

in which $[W_{2^{k-1}}]$ is the ordered W/H matrix of order 2^{k-1}, and $[S_{2^{k-1}}]$ is the following permutation matrix of order 2^{k-1}:

$$[S_{2^{k-1}}] = \begin{bmatrix} 0 & & & 1 \\ & & \cdot & \\ & 1 & & \\ 1 & & & 0 \end{bmatrix}.$$

If (1) holds, as $[W_{2^{k-1}}]$ and $[S_{2^{k-1}}]$ are symmetric and orthogonal, we have the converse relation:

$$[MH_{2^n}{}^k] = [W_{2^{k-1}}] \cdot [S_{2^{k-1}}] \cdot [MW_{2^n}{}^k]. \tag{2}$$

Recursive Definitions of the Haar and W/H Functions

We first define the matrix operators $\mathcal{H}$, $\mathcal{W}$, and $\mathcal{W}'$ which, applied to a $(m \times p)$ matrix $[M]$, generate the following $(2m \times 2p)$ matrices.

Operator $\mathcal{H}$:

$$[\mathcal{H}([M])] = [M] \otimes \begin{bmatrix} 1 & 0 \\ 0 & 1 \end{bmatrix} = \left[\begin{array}{c|c} [M] & [0] \\ \hline [0] & [M] \end{array}\right]$$

where $\otimes$ denotes the Kronecker product.

Operators $\mathcal{W}$ and $\mathcal{W}'$: each row $(a_1, a_2, \cdots, a_m)$ of $[M]$ is replaced by two rows

$$\begin{vmatrix} \mathcal{L} \\ \mathcal{L}' \end{vmatrix} \text{ for } \mathcal{W} \quad \text{and} \quad \begin{vmatrix} \mathcal{L}' \\ \mathcal{L} \end{vmatrix} \text{ for } \mathcal{W}',$$

in which

$$\mathcal{L} = (a_1, a_2, \cdots, a_m, \operatorname{sgn}(a_1 \times a_m) \times (a_1, a_2, \cdots, a_m))$$

and

$$\mathcal{L}' = (a_1, a_2, \cdots, a_m, \operatorname{sgn}(-a_1 \times a_m) \times (a_1, a_2, \cdots, a_m)).$$

Using these operators it can be shown that:

$$[MH_{2^n}{}^k] = [\mathcal{H}([MH_{2^{n-1}}{}^{k-1}])], \qquad k = 2, \cdots, n \tag{3}$$

$$[MW_{2^n}{}^k] = 1/\sqrt{2}[\mathcal{W}([MW_{2^{n-1}}{}^{k-1}])], \qquad k = 1, \cdots, n \tag{4}$$

and globally for the W/H matrices:

$$[W_{2^n}] = 1/\sqrt{2}[\mathcal{W}([W_{2^{n-1}}])]. \tag{5}$$

These relations are alternative recursive definitions, given the matrices of order 2, for the Haar and W/H matrices and, from them, for the Haar and W/H functions. They are similar to the difference equations introduced by Harmuth.[3]

The following properties of the operators $\mathcal{H}$, $\mathcal{W}$, and $\mathcal{W}'$ will be used:

$$[S_{2m}] \cdot [\mathcal{W}([M])] = [\mathcal{W}'([S_m] \cdot [M])] \tag{6}$$

and

$$[S_{2m}] \cdot [\mathcal{W}'([M])] = [\mathcal{W}([S_m] \cdot [M])]. \tag{7}$$

Let $[M] = \{m_{ij}\}$ be an $(m \times p)$ matrix, $[N]$ a $(p \times q)$ matrix, and $[P] = (M]. [N] = \{p_{ij}\}$ of dimension $(m \times q)$. We say that the product matrix $[P]$ is sign invariant (contravariant) if for all i $(p_{i1} \times p_{iq})$ has same (opposite) sign as $(m_{i1} \times m_{ip})$. Then, if $[P]$ is sign invariant,

$$[\mathcal{W}([M])] \cdot [\mathcal{H}([N])] = [\mathcal{W}([P])]. \tag{8}$$

If $[P]$ is sign contravariant,

$$[\mathcal{W}([M])] \cdot [\mathcal{H}([N])] = [\mathcal{W}'([P])]. \tag{9}$$

---

[1] G. Alexits, *Convergence Problems of Orthogonal Series*. New York: Pergamon, 1961, pp. 46-62.
[2] H. C. Andrews, *Computer Techniques in Image Processing*. New York: Academic Press, 1970, pp. 73-90.
[3] H. F. Harmuth, *Transmission of Information by Orthogonal Functions*. New York: Springer, 1970, pp. 19-48.

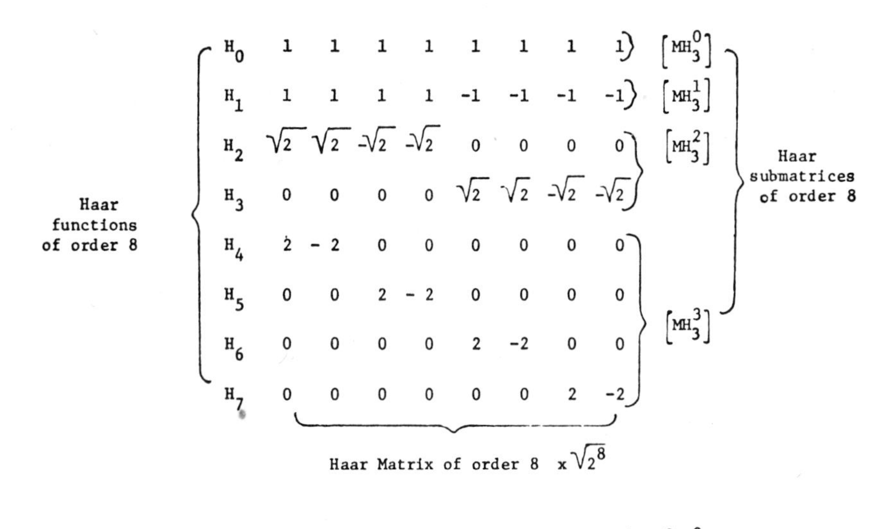

W/H functions of order 8

W_0	1	1	1	1	1	1	1	1	$[MW_3^0]$
W_1	1	1	1	1	-1	-1	-1	-1	$[MW_3^1]$
W_2	1	1	-1	-1	-1	-1	1	1	
W_3	1	1	-1	-1	1	1	-1	-1	$[MW_3^2]$
W_4	1	-1	-1	1	1	-1	-1	1	
W_5	1	-1	-1	1	-1	1	1	-1	
W_6	1	-1	1	-1	-1	1	-1	1	
W_7	1	-1	1	-1	1	-1	1	-1	$[MW_3^3]$

W/H submatrices of order 8

W/H Matrix of order 8 $\times\sqrt{2^8}$

Fig. 1. Haar and W/H functions, matrices, and submatrices.

Proof of (1)

Assuming that (1) is true at the order $(n-1)$, we have:

$$[MW_{2^{n-1}}{}^{k}] = [S_{2^{k-1}}]\cdot[W_{2^{k-1}}]\cdot[MH_{2^{n-1}}{}^{k}], \quad k=1,\cdots,(n-1). \quad (10)$$

Applying $\mathcal{W}$ to both sides of (10), and using (4) for the left hand side and (7) for the other side we obtain:

$$\sqrt{2}[MW_{2^n}{}^{k+1}] = [S_{2^k}]\cdot[\mathcal{W}'([W_{2^{k-1}}]\cdot[MH_{2^{n-1}}{}^{k}])].$$

We can verify that the product of a matrix by a Haar submatrix $[MH_{2^n}{}^{k}]$ is sign contravariant; so, using (9), (3), and (5), we obtain (1) for $k=2,\cdots,n$. As (1) is obvious for $k=1$, this completes the proof of (1), and consequently (2).

Application to Transform Elements

The matrix relations (1) and (2) imply relations between the transform vectors $V_H(V_{H_0}, V_{H_1},\cdots, V_{H_{2^n-1}})$ and $V_W(V_{W_0}, V_{W_1},\cdots, V_{W_{2^n-1}})$ of a vector V. If we right-multiply (1) by V, we obtain:

$$\begin{pmatrix} V_{W_{2^{k-1}}} \\ \vdots \\ V_{W_{2^k-1}} \end{pmatrix} = [S_{2^{k-1}}]\cdot[W_{2^{k-1}}]\cdot\begin{pmatrix} V_{H_{2^{k-1}}} \\ \vdots \\ V_{H_{2^k-1}} \end{pmatrix}$$

$$k = 1,\cdots,n. \quad (11)$$

The converse relation is similarly obtained from (2).

Let a "zone" be the set of coefficients of the transform vectors which appear in these relations. We see that a zone in a transform vector determines the corresponding zone in the other transform vector. This property shows that, if we approximate V by the same subset of zones of the transform vectors V_H and V_W or, in particular, if we truncate these vectors at the end of a zone, we obtain the same approximate vector after inverse transforms.

We see that corresponding zones in (11) are related by an orthogonal transform. It follows by Parseval's theorem that the energies in corresponding zones of the transform vectors are identical.

The computations of the Haar and W/H transforms of a vector of order 2^n require, respectively, $2(2^n-1)$ and $n2^n$ elementary operations with a fast algorithm.[2] Thus when relations (1) and (2) can be used, the Haar transform performs as well as and faster than, the W/H transform (about 4 times faster for $2^n=256$). These results have been applied to image processing.[4]

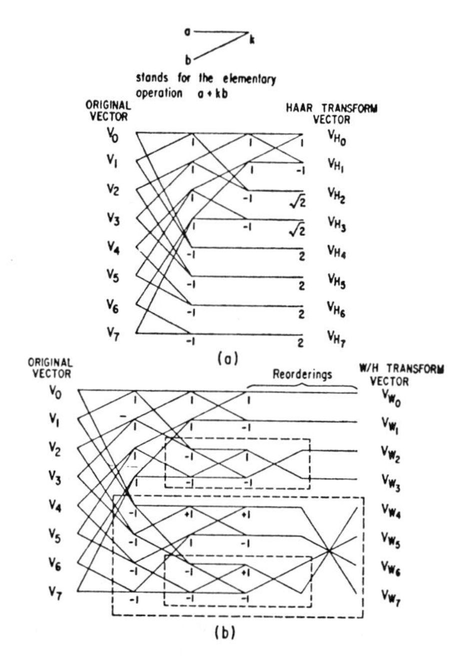

Fig. 2. (a) Haar transform fast algorithm. (b) Haar and W/H transforms and fast algorithm.

Application to Fast Algorithms

The diagram of a simple fast algorithm for the Haar transform of order 8 is shown in Fig. 2(a). By recursive use of relation (11), we derive from this diagram the diagram of a fast algorithm for the W/H transform of order 8 given in Fig. 2(b). The W/H transforms of lower orders are shown inside dotted lines and are followed by the reorderings of the matrices $[S]$.

Note that in the decomposition of a W/H transform into W/H transforms of lower orders, replacing some W/H transforms by corresponding Haar transforms yields a family of orthogonal transforms. Obviously the fast algorithm of Fig. 2(b) computes all the transform elements of all the transforms of this family including the Haar and W/H transforms. For each transform of this family the transform elements appear at appropriate nodes of Fig. 2(b).

Acknowledgment

The author wishes to thank Prof. V. R. Algazi for his advice and encouragement.

[4] B. J. Fino, "Etude expérimentale du codage d'images par transformations de Haar et de Hadamard complexe," *Ann. Télécommun.*, May–June 1972.

17

Energy Packing Efficiency of the Hadamard Transform

HIDEO KITAJIMA

***Abstract*–This concise paper presents a theoretical evaluation of energy packing of the Hadamard transform which is often used in signal processing. It is shown that energy contained in the lowest $1/2^j$ (j: positive integer) of a signal's sequency spectrum can be explicitly evaluated in terms of covariances of the signal.**

INTRODUCTION

Image processing often requires the utilization of discrete unitary transforms such as Karhunen–Loeve, Fourier, Hadamard, etc. The Hadamard transform is known for its fairly good performance and computational simplicity. Suggestive experimental results of the use of the transform will be found in [1] and [2].

In [3] it is compared with other transforms by using the minimum bit rate as a performance measure. Crude justification for the use of a unitary transform is that we can pack most signal energy into low-frequency components of a generalized spectrum. High-frequency components, on the other hand, have so little energy that they can be neclected in most circumstances. The identity transform, which is obviously an orthogonal transform, does not have this property, and therefore is not practically used.

The objective of this concise paper is to show that the fraction of energy contained in the lowest $1/2^j$ of the sequency [4] spectrum can be simply evaluated.

THE HADAMARD TRANSFORM OF THE DATA VECTOR

Let $\{x_0,x_1, \cdots, x_{N-1}\}$ be a finite sequence of random variables with zero mean and covariances given by

$$E\{x_k \cdot x_s\} = c_{ks}, \qquad 0 \leqslant k,s \leqslant N-1 \tag{1}$$

where $E\{\cdot\}$ denotes the expectation. Let $(y_0,y_1, \cdots, y_{N-1})$ be the $N = 2^n$ point Hadamard transform of $(x_0,x_1, \cdots, x_{N-1})$, i.e.,

$$y_p = \sum_{k=0}^{N-1} a_{pk}x_k, \qquad p = 0, 1, 2, \cdots, N-1 \tag{2}$$

where

$$a_{pk} = \frac{1}{\sqrt{N}} (-1)^{q(p,k)} \tag{3}$$

$$q(p,k) = \sum_{i=0}^{n-1} g_i(p)k_i \text{ (modulo two additions)}$$

$$g_0(p) = p_{n-1}$$

$$g_1(p) = p_{n-1} + p_{n-2}$$

$$g_2(p) = p_{n-2} + p_{n-3}$$

$$\vdots$$

$$g_{n-1}(p) = p_1 + p_0.$$

The terms p_i and k_i are the ith bit in the binary representations of p and k, respectively, i.e.,

$$p = (p_{n-1}, p_{n-2}, \cdots, p_1, p_0)_{\text{binary}}$$

$$k = (k_{n-1}, k_{n-2}, \cdots, k_1, k_0)_{\text{binary}}.$$

It is known that the transform (2) is in sequency order.

Now define

$$d_p = E\{y_p{}^2\}, \qquad p = 0, 1, 2, \cdots, N-1. \tag{4}$$

We shall define energy packing efficiency $\eta(m)$ of a unitary transform as the ratio of energy contained in the first m terms of the transform to the total energy. In coding applications, if the m terms are retained, then the normalized mean-square truncation error is given by $1 - \eta(m)$. Let us assume that the process is stationary in the wide sense. Then c_{ks} defined by (1) assumes the form

$$c_{ks} = c(\,|\,k - s\,|\,) = c(\,|\,t\,|\,)$$

where $t = s - k$ ranging from $-(N-1)$ to $(N-1)$. For the transform under consideration, $\eta(m)$ can be written as

$$\eta(m) = \frac{\sum_{p=0}^{m-1} E\{y_p{}^2\}}{\sum_{p=0}^{N-1} E\{y_p{}^2\}} = \frac{\sum_{p=0}^{m-1} d_p}{Nc(0)} \tag{5}$$

which follows from (4), (1), and the above assumption.

We maintain the following.

Theorem: For the Hadamard transform, $\eta(m = N/2^j)$ is given by

$$\eta(N/2^j) = \frac{1}{2^j}\left\{1 + 2c(0)^{-1} \sum_{t=1}^{2^j-1} c(t)\left(1 - \frac{t}{2^j}\right)\right\},$$

$$N = 2^n,\ j = 1, 2, \cdots, n. \tag{6}$$

Proof: Substituting (2) into (4), we obtain

$$d_p = \sum_{s=0}^{N-1} \sum_{k=0}^{N-1} a_{pk} a_{ps} c_{ks}. \tag{7}$$

Setting $c_{ks} = c(|\,t\,|)$, $t = s - k$, we can rewrite (7) as

$$d_p = c(0) + 2 \sum_{t=1}^{N-1} c(t) \sum_{k=0}^{N-t-1} a_{pk} a_{pk+t}. \tag{8}$$

It is worth noting that modulo two additions in the exponent in (3) can be freely interchanged with usual additions. Hence, the product in the inside summation in (8) can be written as

$$a_{pk} a_{pk+t} = \frac{1}{N} (-1)^{p_0(t_{n-1} + w_{n-2})} \cdot (-1)^{\sum_{i=0}^{n-1} p_i(t_{n-i} + t_{n-i-1} + w_{n-i-1} + w_{n-i-2})} \tag{9}$$

where $t = (t_{n-1}, t_{n-2}, \cdots, t_1, t_0)_{\text{binary}}$ and w_r is the carry generated at the rth bit in the addition $(k + t)$.

We now sum up the first 2^m, $0 \leqslant m \leqslant n-1$ terms of d_p's. From (8)

$$S(2^m) \triangleq \sum_{p=0}^{2^m-1} d_p = 2^m c(0) + 2 \sum_{t=1}^{N-1} c(t) \sum_{k=0}^{N-t-1} \sum_{p=0}^{2^m-1} a_{pk} a_{pk+t}. \tag{10}$$

Noting that the innermost summation can be performed bitwise, we have from (9)

$$\sum_{p=0}^{2^m-1} a_{pk} a_{pk+t} = \frac{1}{N}\{1 + (-1)^{t_{n-1} + w_{n-2}}\} \cdot \{1 + (-1)^{t_{n-1} + t_{n-2} + w_{n-2} + w_{n-3}}\} \cdots \{1 + (-1)^{t_{n-m+1} + t_{n-m} + w_{n-m} + w_{n-m-1}}\}$$

$$= \begin{cases} \dfrac{2^m}{N}, & \text{for } t \text{ and } k \text{ such that } t \leqslant 2^{n-m} - 1 \text{ and } w_{n-m-1} = 0 \\ 0, & \text{elsewhere.} \end{cases} \tag{11}$$

For the above summation to be different from zero, obviously each factor must be nonzero. This is satisfied if and only if $t_{n-1} = t_{n-2} = \cdots = t_{n-m} = 0$ and $w_{n-2} = w_{n-3} = \cdots = w_{n-m-1} = 0$. The latter requirement is equivalent to $w_{n-m-1} = 0$. Fussy details are omitted. Substituting (11) into (10), we have

$$S(2^m) = c(0)2^m + \frac{2^{m+1}}{N} \sum_{t=1}^{2^{n-m}-1} c(t)K(t) \tag{12}$$

where

$$K(t) = \sum_{k=0}^{N-1-t} 1\,|_{w_{n-m-1}=0}, \qquad \text{for given } t.$$

In other words, $K(t)$ is the number of k's such that $w_{n-m-1} = 0$, $0 \leqslant k \leqslant N-1-t$ for $1 \leqslant t \leqslant 2^{n-m} - 1$. It is not difficult to show that

$$K(t) = 2^n - 2^m t.$$

Hence, from (12)

$$S(2^m) = c(0)2^m + \frac{2^{m+1}}{N} \sum_{t=1}^{2^{n-m}-1} c(t)(2^n - 2^m t).$$

Define $j = n - m$, $1 \leqslant j \leqslant n$. Then

$$S(2^{n-j}) = S(N/2^j) = \frac{c(0)N}{2^j} + \frac{2N}{2^j} \sum_{t=1}^{2^j-1} c(t) \left(1 - \frac{t}{2^j}\right). \tag{13}$$

From (5)

$$\eta(N/2^j) = \frac{S(N/2^j)}{Nc(0)} = \frac{1}{2^j} \cdot \left\{1 + 2c(0)^{-1} \sum_{t=1}^{2^j-1} c(t) \left(1 - \frac{t}{2^j}\right)\right\},$$

which proves the theorem.

Finally, we shall illustrate the theorem with an example.

Example: If the process is modeled by

$$c(t) = c_0 \rho^t, \qquad t \geqslant 0,$$

we obtain from (6)

$$\eta(N/2) = \frac{1 + \rho}{2}$$

$$\eta(N/4) = \frac{1}{4}\left(1 + \frac{3}{2}\rho + \rho^2 + \frac{1}{2}\rho^3\right).$$

CONCLUSION

We have introduced energy packing efficiency $\eta(m)$ as a possible performance measure for unitary transforms and have shown that, for the Hadamard transform, $\eta(m = N/2^j)$ can be simply evaluated. We remark that it is independent of the transform size N. Although the theorem deals with limited cases, i.e., only for $m = N/2^j$, it serves to give lower and upper bounds for a more general case.

REFERENCES

[1] H. J. Landau and D. Slepian, "Some computer experiments in picture processing for band width reduction," *Bell Syst. Tech. J.*, vol. 50, pp. 1525-1546, May-June 1971.

[2] A. Wintz, "Transform picture coding," *Proc. IEEE*, vol. 60, pp. 809-820, July 1972.

[3] J. Pearl, H. Andrews, and W. K. Pratt, "Performance measures for transform data coding," *IEEE Trans. Commun.*, vol. COM-20, pp. 411-415, June 1972.

[4] H. F. Harmuth, "A generalized concept of frequency and some applications," *IEEE Trans. Inform. Theory*, vol. IT-14, pp. 375-382, May 1968.

18

ASSP-23:580–583 (1975)

Hadamard Transforms of Some Specially Shuffled Signals

BENJAMIN ARAZI

Abstract–**Changes in the Hadamard transform of a vector are investigated under two conditions: 1) shifting the elements of the vector cyclically for 2^l places and 2) interchanging any two elements of the vector. The two-dimensional case is also investigated including shifting and interchanging of rows or columns.**

Manuscript received June 17, 1974; revised December 23, 1974.
The author is with the Signal Processing Division, National Electrical Engineering Research Institute, Council for Scientific and Industrial Research, Pretoria, South Africa.

I. Introduction

A natural Hadamard matrix H_n of order $2^n \times 2^n$ is obtained by n successive Kronecker products of the matrix $\begin{bmatrix} 1 & 1 \\ 1 & -1 \end{bmatrix}$. The Hadamard transform of a vector X of dimension 2^n is obtained by the operation $X \cdot H_n$. (The Kronecker product is denoted by $\otimes$.)

The connection between the Hadamard transforms of a vector before and after cyclic shifting of its elements, has already been investigated [1]. It was shown that if X is a vector of real numbers of dimension 2^n and $X^{(p)}$ is obtained from X by shifting the elements of X cyclically to the right by p places, then $X \cdot H_n = X^{(p)} \cdot H_n \cdot A(P)$. It has been shown that $A(P)$ has a block diagonal orthonormal structure, and the order of the diagonal matrices is a power of 2, where the power increases along the diagonal. It is of interest to check what features of $A(P)$ can be deduced when only the value of p is known. In other words: knowing p, what can be known about the connection between the Hadamard transforms of X and $X^{(p)}$? This question will be answered partially in Section II. Changes in the two-dimensional Hadamard transform of a matrix under certain shiftings of its rows or columns are investigated also.

Section III deals with certain shufflings of the elements of a one- and two-dimensional discrete signal, which shufflings leave part of the Hadamard coefficients of the signal unchanged.

Reference [1] also shows that if B is the Hadamard transform of a vector A of dimension 2^n and P is a vector of dimension $n+1$ such that

$$P_s = \sum_{k=2^{s-1}}^{2^s-1} B_k^2 \qquad 1 \leqslant s \leqslant n$$

$$P_0 = B_0^2,$$

then P remains unchanged under cyclic shiftings of A. Since the dimension of P is only $n+1$, it is apparent that there are other shiftings of the elements of A under which P is invariant. Such shiftings are demonstrated in Section IV.

II. Changes in the Hadamard Transform of a Discrete Signal Under Certain Cyclic Shiftings of its Elements

Notations

1) $A_n^{(m,l)}$ denotes a matrix of order $2^n \times 2^n$ obtained from a matrix A_n of the same order in the following way. The rows of A_n are divided into 2^m groups, $0 \leqslant m \leqslant n$, such that each group contains 2^{n-m} rows of A_n and in each group the first 2^l rows are placed after the last row of the group.

2) $X^{(m,l)}$ denotes a vector of dimension 2^n obtained from a vector X of the same dimension in the following way. The elements of X are divided into 2^m groups, $0 \leqslant m \leqslant n$ and in each group the elements are shifted cyclically to the right by 2^l places.

Illustration

1) For the case where A is a Hadamard matrix the following illustration demonstrates how $H_3^{(1,0)}$ is obtained from H_3.

$$\begin{bmatrix} + & + & + & + & + & + & + & + \\ + & - & + & - & + & - & + & - \\ + & + & - & - & + & + & - & - \\ + & - & - & + & + & - & - & + \\ + & + & + & + & - & - & - & - \\ + & - & + & - & - & + & - & + \\ + & + & - & - & - & - & + & + \\ + & - & - & + & - & + & + & - \end{bmatrix}$$

2) If $X = (1,2,3,4 \mid 5,6,7,8 \mid 9,10,11,12 \mid 13,14,15,16)$, then $X^{(2,0)} = (4,1,2,3 \mid 8,5,6,7 \mid 12,9,10,11 \mid 16,13,14,15)$.

For reasons of convenience $X^{(0,l)}$ and $H_n^{(0,l)}$ are denoted by $X^{(l)}$ and $H_n^{(l)}$, respectively. Under this notation, $H_n^{(0)}$ means a Hadamard matrix of order $2^n \times 2^n$ whose first row was transferred to the end. Let us denote by A, B, C, D the first four columns of H_n, respectively. For every $n \geqslant 2$ $A, B, C,$ and D each consist of a group of four numbers which are repeated 2^{n-2} times; the groups are: $++++$, $+-+-$, $++--$ and $+--+$, respectively. Taking into account the forms of these groups, the following conclusion can be stated.

Conclusion 1: Let A, B, C, and D be the first four columns of H_n, respectively. Then the first four columns of $H_n^{(0)}$ are, $A, -B, D, -C$, respectively.

Illustration

$$H_2 = \begin{bmatrix} + & + & + & + \\ + & - & + & - \\ + & + & - & - \\ + & - & - & + \end{bmatrix} \qquad H_2^{(0)} = \begin{bmatrix} + & - & + & - \\ + & + & - & - \\ + & - & - & + \\ + & + & + & + \end{bmatrix}$$

$$\begin{matrix} A & B & C & D \end{matrix} \qquad\qquad \begin{matrix} A & -B & D & -C \end{matrix}$$

It can be shown that $H_n^{(m,l)} = H_m \otimes H_{n-m-l}^{(0)} \otimes H_l$. Therefore, using Conclusion 1, the following can be stated.

Conclusion 2: Let $H_{ij}^{(m,l)}$ be the element i, j of $H_n^{(m,l)}$ where $0 \leqslant m \leqslant n$ and $0 \leqslant l \leqslant m-2$. Then

$$H_{ij}^{(m,l)} = \begin{cases} H_{ij} & \text{for } p \cdot 2^{n-m} \leqslant j \leqslant p \cdot 2^{n-m} + 2^l - 1 \\ -H_{ij} & \text{for } p \cdot 2^{n-m} + 2^l \leqslant j \leqslant p \cdot 2^{n-m} + 2 \cdot 2^l - 1 \\ H_{i,j+2^l} & \text{for } p \cdot 2^{n-m} + 2 \cdot 2^l \leqslant j \leqslant p \cdot 2^{n-m} + 3 \cdot 2^l - 1 \\ -H_{i,j-2^l} & \text{for } p \cdot 2^{n-m} + 3 \cdot 2^l \leqslant j \leqslant p \cdot 2^{n-m} + 4 \cdot 2^l - 1 \end{cases}$$

$$0 \leqslant i \leqslant 2^n - 1 \qquad 0 \leqslant p \leqslant 2^m - 1.$$

Illustration

$$H_3 = \begin{bmatrix} + & + & + & + & + & + & + & + \\ + & - & + & - & + & - & + & - \\ + & + & - & - & + & + & - & - \\ + & - & - & + & + & - & - & + \\ + & + & + & + & - & - & - & - \\ + & - & + & - & - & + & - & + \\ + & + & - & - & - & - & + & + \\ + & - & - & + & - & + & + & - \end{bmatrix}$$

$$\begin{matrix} a & b & c & d & e & f & g & h \end{matrix}$$

$$H_3^{(1,0)} \begin{bmatrix} + & - & + & - & + & - & + & - \\ + & + & - & - & + & + & - & - \\ + & - & - & + & + & - & - & + \\ + & + & + & + & + & + & + & + \\ + & - & + & - & - & + & - & + \\ + & + & - & - & - & - & + & + \\ + & - & - & + & - & + & + & - \\ + & + & + & + & - & - & - & - \end{bmatrix}$$

$$\begin{matrix} a & -b & d & -c & e & -f & h & -g \end{matrix}$$

Let X be a real vector of dimension 2^n. Since $X \cdot H_n^{(m,l)} = X^{(m,l)} H_n$, the following can be concluded from Conclusion 2.

Conclusion 3: If $X \cdot H_n = Y$, and $X^{(m,l)} \cdot H_n = Z$, then

$$Z_i = \begin{cases} Y_i & \text{for } p \cdot 2^{n-m} \leqslant i \leqslant p \cdot 2^{n-m} + 2^l - 1 \\ -Y_i & \text{for } p \cdot 2^{n-m} + 2^l \leqslant i \leqslant p \cdot 2^{n-m} + 2 \cdot 2^l - 1 \\ Y_{i+2^l} & \text{for } p \cdot 2^{n-m} + 2 \cdot 2^l \leqslant i \leqslant p \cdot 2^{n-m} + 3 \cdot 2^l - 1 \\ -Y_{i-2^l} & \text{for } p \cdot 2^{n-m} + 3 \cdot 2^l \leqslant i \leqslant p \cdot 2^{n-m} + 4 \cdot 2^l - 1 \end{cases}$$

$$0 \leqslant i \leqslant 2^n - 1 \qquad 0 \leqslant p \leqslant 2^m - 1.$$

For $m = 0$, Conclusion 3 deals with the case where the signal vector is shifted cyclically to the right 2^l places. (As mentioned previously, the shifted vector is denoted by $X^{(l)}$.)

Illustration

$$X = (4, 6, 9, 8, 7, 4, 3, 2, 9, 7, 3, 9, 3, 6, 7, 9)$$

$$X^{(2)} = (3, 6, 7, 9, 4, 6, 9, 8, 7, 4, 3, 2, 9, 7, 3, 9)$$

$$X \cdot H_4 = (\overbrace{96, -6, -4, 6,}^{a} \overbrace{14, -4, -2, 4,}^{b} \overbrace{-10, 12, 2, -8,}^{c} \overbrace{8, -6, -24, -14}^{d})$$

$$X^{(2)} \cdot H_4 = (\overbrace{96, -6, -4, 6,}^{a} \overbrace{-14, 4, 2, -4,}^{-b} \overbrace{8, -6, -24, -14,}^{d} \overbrace{10, -12, -2, 8}^{-c}).$$

It should be mentioned here that the connection between the Hadamard transforms of a vector before and after *left* cyclic shifting of its elements can be derived from Conclusion 3 using the fact that X is obtained from $X^{(m,l)}$ by left shifting of the elements of $X^{(m,l)}$.

It was shown [2] that the two-dimensional Hadamard transform of a matrix can be obtained by writing the rows of the matrix successively to form a vector and then taking the one-dimensional transform of the vector. Using this property, Conclusion 3 can be applied to the two-dimensional case where z_i and y_i represent rows or columns rather than elements.

III. Changes in the Hadamard Transform of a Discrete Signal Under Certain Interchanges of its Elements

Remark: In this section the notation H_n denotes a natural as well as an ordered Hadamard matrix (a matrix whose rows are Walsh functions arranged in an increasing sequency order). All properties shown here imply both kinds of matrices.

Let H^i denote the ith row of H_n. It is known that $H^i \cdot H^j = H^{i \oplus j}$, where $\cdot$ means a point-by-point multiplication and $\oplus$ means that the elements of the binary representations of i and j are added modulo 2. For example, the third and fifth rows result in the sixth row (since $(011) \oplus (101) = (110)$).

The property mentioned above means that H^i and H^j have the same element in places where $H^{i \oplus j}$ has +1 elements. Therefore, the following can be concluded.

Conclusion 4: Interchanging the ith and jth rows of a Hadamard matrix leaves half of the columns unchanged. The unchanged columns are those whose indices are the indices of +1 elements in the kth row of the matrix, where $k = i \oplus j$.

Conclusion 5: Let X' be a vector of dimension 2^n obtained from a vector X of the same dimension by interchanging the ith and jth elements of X.

Let $i \oplus j = k$ and let $\{p\} = \{l \mid H_{kl} = +1\}$ where H_{kl} is entry k, l of the Hadamard matrix of order $2^n \times 2^n$. Let Y and Y' be the Hadamard transforms of X and X', respectively. Then $Y_q = Y'_q$ for $q \in \{p\}$.

Remark: Since Conclusion 5 deals with the interchange of two elements, one can perform several interchanges at a time and still have half of the Hadamard coefficients unchanged as long as the sum modulo 2 of the indices of each interchanged pair gives the same result.

Illustration

$$X = \begin{bmatrix} 7 & 5 & 2 & 1 & 2 & 4 & 5 & 6 \\ 3 & 2 & 0 & 1 & 2 & 4 & 5 & 8 \\ 7 & 5 & 4 & 2 & 0 & 1 & 2 & 4 \\ 5 & 6 & 8 & 7 & 5 & 3 & 2 & 6 \\ 7 & 5 & 4 & 8 & 9 & 8 & 5 & 6 \\ 3 & 2 & 0 & 1 & 2 & 4 & 5 & 7 \\ 8 & 9 & 6 & 5 & 7 & 8 & 5 & 3 \\ 2 & 1 & 4 & 5 & 7 & 8 & 5 & 3 \end{bmatrix} \begin{matrix} A \\ B \\ C \\ D \\ E \\ F \\ G \\ H \end{matrix} \qquad Y = \begin{bmatrix} 286 & -10 & 16 & 12 & -16 & 20 & 22 & 6 \\ 34 & 10 & 32 & -4 & 36 & 0 & 14 & 6 \\ -20 & -12 & -6 & 14 & -46 & 6 & 36 & 12 \\ 36 & 4 & -2 & 10 & 2 & -10 & -40 & 0 \\ -38 & -2 & -20 & 4 & 28 & 20 & 42 & -18 \\ -54 & 6 & -8 & -8 & -12 & 12 & 22 & 10 \\ 0 & 4 & 14 & -18 & -26 & -2 & 0 & 12 \\ 12 & 0 & 6 & -10 & 18 & 2 & 16 & -12 \end{bmatrix} \begin{matrix} S \\ T \\ U \\ V \\ W \\ X \\ Y \\ Z \end{matrix}$$

$$X^{(1,0)} = \begin{bmatrix} 3 & 2 & 0 & 1 & 2 & 4 & 5 & 8 \\ 7 & 5 & 4 & 2 & 0 & 1 & 2 & 4 \\ 5 & 6 & 8 & 7 & 5 & 3 & 2 & 6 \\ 7 & 5 & 2 & 1 & 2 & 4 & 5 & 6 \\ 3 & 2 & 0 & 1 & 2 & 4 & 5 & 7 \\ 8 & 9 & 6 & 5 & 7 & 8 & 5 & 3 \\ 2 & 1 & 4 & 5 & 7 & 8 & 5 & 3 \\ 7 & 5 & 4 & 8 & 9 & 8 & 5 & 6 \end{bmatrix} \begin{matrix} B \\ C \\ D \\ A \\ F \\ G \\ H \\ E \end{matrix} \qquad Z = \begin{bmatrix} 286 & -10 & 16 & 12 & -16 & 20 & 22 & 6 \\ -34 & -10 & -32 & 4 & -36 & 0 & -14 & -6 \\ -36 & -4 & 2 & -10 & -2 & 10 & 40 & 0 \\ -20 & -12 & -6 & 14 & -46 & 6 & 36 & 12 \\ -38 & -2 & -20 & 4 & 28 & 20 & 42 & -18 \\ 54 & -6 & 8 & 8 & 12 & -12 & -22 & -10 \\ -12 & 0 & -6 & 10 & -18 & -2 & -16 & 12 \\ 0 & 4 & 14 & -18 & -26 & -2 & 0 & 12 \end{bmatrix} \begin{matrix} S \\ -T \\ -V \\ U \\ W \\ -X \\ -Z \\ Y. \end{matrix}$$

Illustration

$X = (1, 4, 7, 3, 2, 5, 8, 6).$

The elements with the following indices are interchanged.

$7 \longleftrightarrow 4, 5 \longleftrightarrow 6, 3 \longleftrightarrow 0 \quad (\text{since } 7 \oplus 4 = 5 \oplus 6 = 3 \oplus 0)$

$X' = (3, 4, 7, 1, 6, 8, 5, 2).$

For an ordered Hadamard transform

$$Y = (\overbrace{36, -6}^{A}, 2, -12, \overbrace{-12, -2}^{B}, 2, 0)$$

$$Y' = (\overbrace{36, -6}^{A}, -8, 6, \overbrace{-12, -2}^{B}, -16, 6).$$

Y and Y' have equal elements in places where row 3 of an ordered Hadamard matrix has +1 elements. Conclusion 5 can be extended for the two-dimensional case.

Conclusion 6: Interchanging the ith and jth columns (rows) of a matrix leaves half of the columns (rows) of its two-dimensional Hadamard transform unchanged. The unchanged columns (rows) are those whose indices are the indices of +1 elements in the kth row of the Hadamard matrix of the same order as the signal matrix, where $k = i \oplus j$.

Remark: As in the one-dimensional case, one can perform several interchanges of columns (rows) at a time, and still have half of the columns (rows) of the transform matrix unchanged, as long as the sum modulo 2 of the indices of each interchanged pair gives the same result.

IV. Certain Grouping of the Squares of Hadamard Transform Coefficients

It was described in the Introduction that if B is the natural Hadamard transform of A, where A is of dimension 2^n, then a vector P of dimension $n + 1$ is invariant under cyclic shiftings of A, where P is obtained by grouping the squares of the elements of B in the following way.

$$P_s = \sum_{k=2^{s-1}}^{2^s-1} B_k^2 \qquad 1 \leqslant s \leqslant n$$

$$P_0 = B_0^2.$$

Since the dimension of P is $n + 1$, there should be many other shiftings of the elements of A, under which P is invariant. Some of these shiftings are demonstrated here.

Any dyadic shifting of the elements of A results in only some sign changes in the elements of B. It follows that P is invariant under the 2^n possible dyadic shiftings of A.

If A_k denotes the kth element of the input vector, it follows from the orthogonality of H_n that

$$\sum_{k=0}^{2^n-1} B_k^2 = 2^n \sum_{k=0}^{2^n-1} A_k^2.$$

Since

$$\sum_{s=0}^{n} P_s = \sum_{k=0}^{2^n-1} B_k^2$$

it follows that the sum of the elements of P is invariant under any shuffling of the elements of the input vector. In view of Conclusion 5 it follows that if A_k and $A_{k+2^{n-1}}$ are interchanged, then B_k remains unchanged for $0 \leqslant k \leqslant 2^{n-1} - 1$. ($i \oplus (i + 2^{n-1}) = 2^{n-1}$. The row with index 2^{n-1} of a natural Hadamard matrix has +1 elements in its first 2^{n-1} places.) It follows that under such an interchange the first n elements of P remain unchanged. But since the sum of all the elements is invariant, it follows that the last element also remains unchanged. Therefore, the following can be stated.

Conclusion 7: The vector P defined above is invariant under interchange of those elements of A whose indices are k and $k + 2^{n-1}$, for any k such that $0 \leqslant k \leqslant 2^{n-1} - 1$.

The vector P is invariant under any number of interchanges described in Conclusion 7. If $N = 2^{n-1}$, the number of possibilities for such interchanges is

$$\sum_{i=0}^{N} \binom{N}{i} = 2^N.$$

Thus Conclusion 7 suggests $2^{2^{n-1}}$ shiftings beside cyclic shiftings, under which the vector P is invariant. To be more accurate, two of these shiftings ($i = 0$ and $i = N$) are included in the cyclic shiftings.

References

[1] N. Ahmed, K. R. Rao, and A. L. Abdussattar, "BIFORE or Hadamard transform," *IEEE Trans. Audio Electroacoust.*, vol. AU-19, pp. 225–234, Sept. 1971.

[2] B. Arazi, "Two-dimensional digital processing of one-dimensional signal," *IEEE Trans. Acoust., Speech, Signal Processing*, vol. ASSP-22, pp. 81–86, Apr. 1974.

Two-Dimensional Digital Processing of One-Dimensional Signal

BENJAMIN ARAZI

I. INTRODUCTION

IT HAS been shown by several authors (see [1] for example) that with the Hadamard transform it is possible to process a one-dimensional signal vector by changing it into a two-dimensional signal matrix. This is done by dividing the input signal into groups to form the rows of a square matrix. A two-dimensional Hadamard transform is performed on this matrix. If the rows of the resulting matrix are written successively in the order of the rows, the resulting vector is the transform of the original one. Some applications of this result are also given in [1]. Another practical application is now mentioned.

The importance of two-dimensional processing of a one-dimensional signal arises in those cases where the input vector has length 2^{2n} while the working memory area is limited to 2^n. In those cases the described property of the transform makes it possible to minimize the number of transfers from storage to working area. In order to make clearer what follows, we remind the reader of the following.

A two-dimensional Hadamard transform of a signal matrix involves the following operations: the one-dimensional Hadamard transform is performed on each row of the matrix, and again performed on each column of the resulting matrix [2]. This is possible because a Hadamard matrix is a symmetric matrix [3] (for this reason this paper deals mainly with symmetric matrices). This implies that while processing a one-dimensional signal in the form of a matrix, the one-dimensional signal is stored and divided into 2^n groups, each consisting of 2^n elements. The groups are transferred to the working area one at a time, and their transform is transferred back and stored in place. For performing the whole two-dimensional transform, 2^n groups are transferred to and from the working area twice.

Let us compare this to the case where the input vector is transformed without two-dimensional processing. Using the usual fast algorithm, one finds that there are n steps of decimation from the beginning of the process up to the stage where one has 2^n groups, each group consisting of 2^n elements. (From this stage on, a fast algorithm is performed on each group separately, so the procedure is identical in both cases.) If the working area has a length of only 2^n, this means that at each stage of decimation 2^n groups are transferred from storage area to memory, and back. Including the last stage of computation, one has altogether $n + 1$ transfers compared with the two transfers in the first case. (During each transfer 2^n groups are transferred from storage area to working area and back.)

On the other hand, in those cases where the working area is not limited, it is much more convenient to process a two-dimensional signal as one long vector.

It is of importance to find the necessary and sufficient properties of any transform so that its one-dimensional and two-dimensional processing should be equivalent.

In those cases where the transformation matrix is of order $p^{2k} \times p^{2k}$ (p- any positive integer) it is possible to reduce the order k times. This reduction means changing the input vector into a two-dimensional signal matrix, processing each row of this matrix as a two-dimensional signal and so on, up to the stage where one has to process vectors of length p. Since in many cases it is more convenient to process vectors of length p rather than vectors of length p^{2k}, it is of importance to check what properties the matrix must have in order that the described procedure will be possible. These properties will be found.

Remarks

1) Performing two-dimensional instead of one-dimensional processing does not involve fewer multiplications or less storage area. Let us assume we have a fast algorithm of dimension $N^2 = 2^{2k}$, k any integer > 0. In the one-dimensional case the number of multiplications is $2N^2 \log_2 N^2$. In the two-dimensional case the fast algorithm is performed N times, then N times again, and the matrix is of order N. Thus, in this case the number of multiplications is $2N2N \log_2 N$. But, $2N2N \log_2 N = 2N^2 \log_2 N^2$. This means that the number of multiplications is the same in both dimensions. If one has no fast algorithm it can be checked that one also arrives at the same number of multiplications in both cases.

2) A practical case is that where the transformation matrix $[M]$ has the property $[M][M] = k[I]$, where k is any integer > 0 and $[I]$ is the identity matrix.

This property means that the same matrix is used for coding and decoding, which is very convenient. The Hadamard matrix possesses this property, the Fourier

Manuscript received August 14, 1973; revised November 6, 1973.
The author is with the Signal Processing Division, National Electrical Engineering Research Institute, Council for Scientific and Industrial Research, Pretoria, South Africa.

matrix not. It may be noted here that the general concept of a spectrum [5] is defined only when the transformation matrix is orthogonal, and this of course includes the case where $[M][M] = k[I]$. Therefore, if a vector is multiplied by $[M]$ the resulting vector is the spectrum of the input vector.

II. GENERAL THEORY

Definition

A matrix $[X]$ of order $N \times N$ is the matrix representation of a vector $\boldsymbol{X}$ of dimension N^2 iff:

$$X_{ij} = X_{(i-1)N+j}.$$

(Two indices refer to the matrix, one index to the vector.) This definition means that $\boldsymbol{X}$ is divided into N ordered groups, each consisting of N elements, and these groups form the rows of the matrix.

It was pointed out in the introduction that the situation $\boldsymbol{X}[R] = \boldsymbol{Y}$ and $[S][X][S] = [Y]$ has a practical meaning only if $[S]$ is symmetric. Therefore this general theory will concern this case particularly. ($\boldsymbol{X}[R] = \boldsymbol{Y}$ means one-dimensional processing of the vector $\boldsymbol{X}$. $[S][X][S] = [Y]$ means two-dimensional processing of the matrix $[X]$. The situation $\boldsymbol{X}[R] = \boldsymbol{Y}$ and $[S][X][S] = [Y]$ means that the one-dimensional and two-dimensional processing are equivalent.)

Since this paper is concerned largely with the Kronecker product of matrices the reader is referred to two articles that show the connection between Kronecker product and fast techniques for spectral analyses [4], [5].

The following definition is a special case of the general definition of the Kronecker product.

Definition

Let $[Q]$ be a matrix of order $N \times N$, then $[Q] \otimes [Q] = [M]$, iff:

$$M_{(i-1)N+j,(k-1)N+l} = Q_{jl} \cdot Q_{ik}.$$

$i,j,k,l = 1,2,\cdots N$. $[M]$ of order $N^2 \times N^2$, where $\otimes$ denotes the Kronecker product.

Example

In the case where $N = 2$

$$Q = \begin{bmatrix} a & b \\ c & d \end{bmatrix}$$

$$M = \begin{bmatrix} a\begin{bmatrix} a & b \\ c & d \end{bmatrix} & b\begin{bmatrix} a & b \\ c & d \end{bmatrix} \\ c\begin{bmatrix} a & b \\ c & d \end{bmatrix} & d\begin{bmatrix} a & b \\ c & d \end{bmatrix} \end{bmatrix}.$$

In the general case:

$$M = \begin{bmatrix} Q_{11}\cdot[Q] & \cdots & Q_{1N}\cdot[Q] \\ Q_{N1}\cdot[Q] & \cdots & Q_{NN}\cdot[Q] \end{bmatrix}$$

since

$$Q_{ik} \cdot Q_{jl} = Q_{jl} \cdot Q_{ik}$$

one obtains from the definition:

$$M_{(i-1)N+j,(k-1)N+l} = M_{(j-1)N+i,(l-1)N+k}.$$

In the case where M is of order 4×4 this means:

$$M_{12} = M_{13}, M_{23} = M_{32}, N_{21} = M_{31}, M_{24} = M_{34}, M_{22} = M_{33}, M_{42} = M_{43}.$$

From the definition one also obtains the following:

$$M_{ij} \cdot M_{i,(N^2+1-j)} = M_{i,(j+k)} \cdot M_{i,(N^2+1-j-k)}$$

$$k = 0,1,2\cdots N^2 - j$$

$$M_{ij} \cdot M_{(N^2+1-i),j} = M_{(i+k),j} \cdot M_{(N^2+1-i-k),j}$$

$$k = 0,1,2\cdots N^2 - i$$

$$i = 1,2\cdots N^2, \qquad j = 1,2,\cdots N^2.$$

In the above example these formulas mean:

$$M_{12} \cdot M_{13} = M_{11} \cdot M_{14} \quad M_{11} \cdot M_{41} = M_{21} \cdot M_{31}$$

$$M_{22} \cdot M_{23} = M_{21} \cdot M_{24} \quad M_{12} \cdot M_{42} = M_{22} \cdot M_{32}$$

$$M_{32} \cdot M_{33} = M_{31} \cdot M_{34} \quad M_{13} \cdot M_{43} = M_{23} \cdot M_{33}$$

$$M_{42} \cdot M_{43} = M_{41} \cdot M_{44} \quad M_{14} \cdot M_{44} = M_{24} \cdot M_{34}.$$

From the definition it follows that for Walsh–Hadamard matrices

$$[H_{2n}] = [H_n] \otimes [H_n].$$

($[H_n]$ is the Hadamard matrix of order $2^n \times 2^n$.)

This is understood immediately if one inspects the definition of the Walsh–Hadamard transform.

$$[H_1] = \begin{bmatrix} 1 & 1 \\ 1 & -1 \end{bmatrix} \quad [H_{n+1}] = \begin{bmatrix} [H_n] & [H_n] \\ [H_n] & -[H_n] \end{bmatrix}$$

Theorem I

Let $[X],[Y]$ be the matrix representations of $\boldsymbol{X},\boldsymbol{Y}$ respectively, and let $[S]$, $[R]$ be two matrices of order $N \times N$, $N^2 \times N^2$ respectively. $[X],[Y]$ are of order $N \times N$. $\boldsymbol{X},\boldsymbol{Y}$ are of dimension N^2.

1a) If $[S]$ is a symmetric matrix, then: $\mathbf{X}[R] = \mathbf{Y}$, and $[S][X][S] = [Y]$ for any $\boldsymbol{X}$, iff: $[R] = [S] \otimes [S]$.

b) If $\boldsymbol{X}\ [R] = \boldsymbol{Y}$ and $[S][X][S] = [Y]$ for any $\boldsymbol{X}$ then: $[S][S] = \pm K[I]$ iff: $[R][R] = K^2[I]$ where $[I]$ is the identity matrix and $[K]$ is any integer > 0.

2) If $\boldsymbol{X}\ [R] = \boldsymbol{Y}$ and $[S][X][S] = [Y]$ for any $\boldsymbol{X}$ then: $[S]$ is symmetric iff: $[R]$ is symmetric (unless all the diagonal elements of $[S]$ are 0). (The case $[R][R] = K^2[I]$ is treated in detail since it has practical importance as was mentioned in the introduction.)

The proof of this theorem is given in Appendix A.

III. THE FOURIER TRANSFORM CASE

It is not possible to have $X[R] = Y$ and $[S][X][S] = [Y]$ for any X where $[S]$ and $[R]$ are Fourier matrices [6]. This is due to the fact that although $[S]$ is a symmetric matrix, the condition $[R] = [S] \otimes [S]$ cannot be satisfied in the case of the Fourier matrices. This condition must, however, be met in view of Theorem I. As a matter of fact, it is not possible to find any general matrix $[S]$ such that $[R] = [S] \otimes [S]$ since $[R]$ lacks the property:

$$R_{(i-1)N+j,(k-1)N+l} = R_{(j-1)N+i,(l-1)N+k}$$

which is a basic property of the Kronecker product as was mentioned before.

If, however, the input vector is very long and it is of interest to process it in parts, the question arises as to whether it is possible to have any two matrices $[Q]$, $[P]$ of order $N \times N$, so that $X[R] = Y$ and $[Q][X][P] = [Y]$ while $[R]$ is still a Fourier matrix.

The answer is given by Theorem II which excludes the following conditions.

Condition 1.

$$R_{(k-1)N+l,(i-1)N+j} = 0 \qquad \text{for } l = j. \quad i,j,k,l = 1,2,\cdots N.$$

Condition 2.

$$R_{(k-1)N+l,(i-1)N+j} = 0 \qquad \text{for } i = k. \quad i,j,k,l = 1,2,\cdots N.$$

Condition 3. (1) and (2) together.

Condition 1 means that if $[R]$ is divided into $N \times N$ submatrices, each one of order $N \times N$, then each submatrix has zeros along the mean diagonal. Condition 2 means that if $[R]$ is divided as above, then all the submatrices whose main diagonals are on the main diagonal of $[R]$, are identically zero.

Theorem II

Let $[X]$,$[Y]$ be the matrix representations of X,Y respectively. $[X]$,$[Y]$ are of order $N \times N$, and X,Y of dimension N^2. Let $[R]$ be a symmetric matrix of order $N^2 \times N^2$, and let $[P]$,$[Q]$ be any two matrices of order $N \times N$.
If

$$X[R] = Y$$

and

$$[P][X][Q] = [Y]$$

for any vector X, then $[P]$ and $[Q]$ are symmetric matrices and

$$R_{(k-1)N+l,(i-1)N+j} = R_{(i-1)N+l,(k-1)N+j}.$$

The proof is given in Appendix B.

Conclusion

Under the conditions of Theorem II, $[R] = [P] \otimes [Q]$.

Example

If $[R]$ is of order 4×4, it follows from this theorem that $R_{14} = R_{32}$. But $[R]$ is symmetric, therefore it follows that all the elements in the diagonal from R_{14} to R_{41} are equal.

Conclusion

$[R]$ cannot be a Fourier matrix. (Since a Fourier matrix is symmetric, it must have the property

$$R_{(k-1)N+l,(i-1)N+j} = R_{(i-1)N+l,(k-1)N+j}.)$$

One may now reverse the question. Given the two-dimensional Fourier transform $[S][X][S] = [Y]$, where $[S]$ is a Fourier matrix of order $N \times N$, is it possible to find a matrix $[R]$ of order $N^2 \times N^2$ such that $X[R] = Y$?

According to Theorem I the answer is YES. Since $[S]$ is symmetric, $[R] = [S] \otimes [S]$ (which implies that $[R]$ is not a Fourier matrix). The operation $X[R] = Y$ can be performed by a fast algorithm since $[R]$ is obtained by the Kronecker product of two Fourier matrices which can be reduced further to the Kronecker product of matrices [5].

Example

In order to process a matrix $[X]$ of order 2×2

$$[S] = \begin{bmatrix} 1 & 1 \\ 1 & A \end{bmatrix}$$

$$[R] = \begin{bmatrix} 1 & 1 & 1 & 1 \\ 1 & A & 1 & A \\ 1 & 1 & A & A \\ 1 & A & A & A^2 \end{bmatrix} \qquad A = \exp(2\Pi j/2)\,(= -1))$$

while the Fourier matrix of order 4×4 is:

$$[F] = \begin{bmatrix} 1 & 1 & 1 & 1 \\ 1 & W & W^2 & W^3 \\ 1 & W^2 & W^4 & W^6 \\ 1 & W^3 & W^6 & W^9 \end{bmatrix} \qquad W = \exp(2\Pi j/4)\,(= j).$$

Both operations $X[R] = Y$ and $X[F] = Z$ can be performed by a fast algorithm as shown by Fig. 1.

It is interesting to examine more closely the difference between the two operations. The operation $[S][X][S]$ means: take the Fourier transform of each row of $[X]$ and then take the Fourier transform of each column of the resulting matrix. The operation $X[R]$ means the following: Take a vector of dimension N whose elements are:

$$\sum_{k=1}^{N} S_{ik}X_{k1}, \quad \sum_{k=1}^{N} S_{ik}X_{k2},\cdots, \quad \sum_{k=1}^{N} S_{ik}X_{kN}.$$

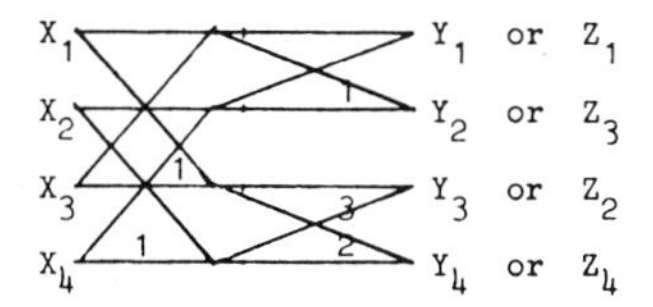

Multiplier Key

Code	X [R] = Y	X [F] = Z
1	A	W^2
2	A	W^3
3	1	W

Fig. 1. Algorithm flow chart.

The fast Fourier transform (FFT) of this vector is the ith row of $[Y]$.

In words: In order to get the ith row of $[Y]$ multiply the rows of $[X]$ by the ith row of $[S]$ in such a sequence that the jth row of $[X]$ is multiplied by S_{ij}. Then the elements of each column of the resulting matrix are added, and yield a vector of dimension N, whose FFT is the ith row of $[Y]$.

Example

$$[X] = \begin{bmatrix} 1 & 2 \\ 3 & 4 \end{bmatrix} \quad [S] = \begin{bmatrix} 1 & 1 \\ 1 & W \end{bmatrix} \quad W = \exp(2\Pi j/2).$$

The normal way:

$$\begin{bmatrix} 1 & 1 \\ 1 & W \end{bmatrix} \begin{bmatrix} 1 & 2 \\ 3 & 4 \end{bmatrix} \begin{bmatrix} 1 & 1 \\ 1 & W \end{bmatrix} = \begin{bmatrix} 10 & 4 + 6W \\ 3 + 7W & 1 + 5W + 4W^2 \end{bmatrix}.$$

The second way: For example to obtain only the second row of $[Y]$

$$1 \cdot (1,2) + W \cdot (3,4) = (1 + 3W, 2 + 4W).$$

$$(1 + 3W, 2 + 4W) \begin{bmatrix} 1 & 1 \\ 1 & W \end{bmatrix} = (3 + 7W, 1 + 5W + 4W^2).$$

IV. THE CASE WHERE THE TRANSFORMATION MATRIX IS OF ORDER $2^{2k} \times 2^{2k}$

It was said before that performing the operation $[S][X][S]$ has an advantage over the operation $X[R]$. $[X]$ was obtained from X by a process which has already been described. It is possible to continue this process by performing on the rows of $[X]$ the same operation as was performed on X, and so on. If $[R]$ is of order $a^{2k} \times a^{2k}$ and it has certain properties, it is possible to divide the vector X into a^{2k-1} groups, each one of length a, and the calculation of $X[R]$ will involve only processing groups of order a, with a minimum number of transfers from storage area to working area. It was shown that the two possibilities of processing involve the same number of multiplications. Taking the practical case where $[R]$ is symmetric, then in accordance with Theorem I the described k-stage reduction is possible iff $[R]$ is obtained from a matrix $[P]$ (of order $a \times a$) by k successive Kronecker products. Successive Kronecker products here mean that at each state the matrix obtained is multiplied by itself and not by the first basic matrix.

The problem is: given a symmetric matrix $[R]$ of order $a^{2k} \times a^{2k}$ how can one determine if it can be obtained from a matrix $[P]$ of order $a \times a$ by means of k successive Kronecker products?

For practical reasons the discussion will from now on deal only with the case $a = 2$ and $[R][R] = K[I]$, which is also the most interesting case. However, one can generalize this discussion also for the case where a is any positive integer.

A direct conclusion from Theorem I is as follows:

1) If $[Q] \otimes [Q] = [R]$ then $[Q]$ is symmetric iff $[R]$ is symmetric.

2) If $[Q] \otimes [Q] = [R]$ and $[Q]$ or $[R]$ are symmetric then $[Q][Q] = \pm K[I]$ iff $[R][R] = K^2[I]$.

From this it follows: if $[R]$ is a symmetric matrix of order $2^{2k} \times 2^{2k}$ and $[R][R] = K[I]$ and $[R]$ is obtained from a matrix $[P]$ of order 2×2 by means of k successive Kronecker products, then all the intermediate matrices including $[P]$ are symmetric and are their own inverse.

If $[P]$ is a symmetric own inverse matrix it must be of the form

$$[P] = \begin{bmatrix} a & b \\ b & -a \end{bmatrix}.$$

From the definition of the Kronecker product it follows that all the intermediate matrices including $[R]$ have the following properties.

1) The signs of the elements of these matrices are the signs of the elements of the Hadamard matrix of the same order.

2) Except for the sign, the elements of each row in these matrices equal the elements in any other row, and they are the elements of the polynomial expansion of $(a + b)^{2l}$ where l is the number of successive Kronecker products of $[P]$ until one arrives at a certain intermediate matrix.

3) Except for the sign, these matrices are diadic translated matrices [7]. One way of defining a diadic translated matrix is as follows:

a) The matrix is of order $2^n \times 2^n$, $n = 1,2,\cdots$.

b) The elements of each row in this matrix equal the elements of any other row.

c) If the matrix is divided into 2^k submatrices, $k = 0,1,\cdots 2(n-1)$, then all the submatrices are symmetric and the elements in each of the two diagonals of these submatrices equal one another.

The following is an example of a matrix which possesses properties 1–3.

$$\begin{bmatrix} 4 & 6 & 6 & 9 \\ 6 & -4 & 9 & -6 \\ 6 & 9 & -4 & -6 \\ 9 & -6 & -6 & 4 \end{bmatrix}.$$

V. SUMMARY

Necessary and sufficient conditions are found under which the one-dimensional and two-dimensional processing of any general transform should be equivalent. It is shown that the Hadamard transform possesses these properties, while the Fourier transform does not.

The implications of the possibility to process a one-dimensional signal as a two-dimensional signal and vica versa are discussed in detail. Since the Fourier transform does not possess the required properties, it was checked whether it is possible to find general transforms such that one-dimensional processing of the data matrix will be equivalent to two-dimensional Fourier processing and two-dimensional processing of a vector will be equivalent to one-dimensional Fourier processing. Firstly, it was proved that it is impossible to process the one-dimensional Fourier transform as a two-dimensional transform of another kind. However, it was shown that it is possible to process a two-dimensional Fourier transform as a one-dimensional transform of another kind. It is shown that this operation can be performed by a fast algorithm.

Finally, the properties which a transform must have in order that a vector of dimension p^{2k} can be processed in groups of p^{2^n} elements each, ($n = 0$ to k,) are derived.

APPENDIX A

Proof of Theorem I

1a) $\Rightarrow$:

$$\boldsymbol{X}[R] = \boldsymbol{Y} \Rightarrow Y_p = \sum_{q=1}^{N} X_q R_{qp} \qquad p = 1,2,\cdots N^2$$

$$[S][X][S] = [Y] \Rightarrow Y_{ij} = \sum_{k=1}^{N} \sum_{l=1}^{N} X_{kl} S_{ik} S_{lj}$$

$$i,j = 1,2,\cdots N.$$

Since $[X],[Y]$ are the matrix representation of $\boldsymbol{X},\boldsymbol{Y}$, $\Rightarrow$:

$$p = (i-1)N + j, \quad q = (k-1)N + l.$$

The coefficients of the elements of $\boldsymbol{X}$ are equal in both representations: $\Rightarrow$

$$R_{(k-1)N+l,(i-1)N+j} = S_{ik} S_{lj}.$$

$[S]$ is symmetric $\Rightarrow$

$$S_{ik} = S_{ki} \Rightarrow R_{(k-1)N+l,(i-1)N+j} = S_{ki} \cdot S_{lj}$$

which means $[R] = [S] \otimes [S]$.

$\Leftarrow$: $[R] = [S] \otimes [S] \Rightarrow R_{(k-1)N+l,(i-1)N+j}$

$$= S_{ki} \cdot S_{lj} = S_{ik} \cdot S_{lj} \text{ ([S] symmetric)}.$$

Let us assume:

$\boldsymbol{X}[R] = \boldsymbol{Y}$ and $[S][X][S] = [Z]$

$$\Rightarrow Y_p = \sum_{q=1}^{N^2} X_q R_{qp} \qquad p = 1,2,\cdots N^2$$

$$Z_{ij} = \sum_{k=1}^{N} \sum_{l=1}^{N} X_{kl} S_{ik} S_{lj} \qquad i,j = 1,2,\cdots N.$$

If $[X]$ is written in the form of a vector one has:

$$Z_{ij} = \sum_{k=1}^{N} \sum_{l=1}^{N} X_{(k-1)N+l} \cdot R_{(k-1)N+l,(i-1)N+j}$$

$$= \sum_{q=1}^{N^2} X_q R_{qp} \qquad \begin{matrix} q = (k-1)N + l \\ p = (i-1)N + j \end{matrix}$$

$\Rightarrow [Z]$ is the matrix representation of $\boldsymbol{Y}$. Q.E.D.

1b) $\Rightarrow$:

$$\boldsymbol{X}[R] = \boldsymbol{Y} \Rightarrow \boldsymbol{X}[R][R] = \boldsymbol{Y}[R]$$

if

$$[R][R] = K^2[I]$$

then

$$\boldsymbol{Y}[R] = K^2 \boldsymbol{X}.$$

Since $\boldsymbol{X}[R] = \boldsymbol{Y}$ and $[S][X][S] = [Y]$ for any $\boldsymbol{X}$ holds also for $\boldsymbol{Y}$,

$$\Rightarrow \boldsymbol{Y}[R] = \boldsymbol{Z} \quad \text{and} \quad [S][Y][S] = [Z]$$

$$\Rightarrow [S][Y][S] = K^2[X].$$

$$[Y] = [S][X][S]$$

$$\Rightarrow [S][S][X][S][S] = K^2[X] \Rightarrow [S][S] = \pm K[I].$$

$\Leftarrow$:

$$[S][X][S] = [Y] \Rightarrow [S][S][X][S][S] = [S][Y][S]$$

if

$$[S][S] = \pm K[I]$$

then

$$[S][Y][S] = K^2[X].$$

Since $\boldsymbol{X}[R] = \boldsymbol{Y}$ and $[S][X][S] = [Y]$ for any $\boldsymbol{X}$ it holds for $\boldsymbol{Y}$

$$\Rightarrow \boldsymbol{Y}[R] = \boldsymbol{Z} \quad \text{and} \quad [S][Y][S] = [Z]$$

$$\Rightarrow \boldsymbol{Y}[R] = K^2\boldsymbol{X} \Rightarrow \boldsymbol{Y}[R][R] = K^2\boldsymbol{X}[R]$$

$$\boldsymbol{X}[R] = \boldsymbol{Y} \Rightarrow \boldsymbol{Y}[R][R] = K^2\boldsymbol{Y} \Rightarrow [R][R] = K^2[I].$$

Q.E.D.

2) $\Rightarrow$:

$$[S] \text{ symmetric} \Rightarrow [R] = [S] \otimes [S]$$

[see part (a)]

$$\Rightarrow [R] - \text{symmetric}.$$

$\Leftarrow$:

$$\boldsymbol{X}[R] = \boldsymbol{Y}$$

and

$$[S][X][S] = [Y] \Rightarrow R_{(k-1)N+l,(i-1)N+j} = S_{ik} \cdot S_{lj}$$

[see proof of part (a)].

Since $[R]$ is symmetric $\Rightarrow$

$$R_{(k-1)N+l,(i-1)N+j} = R_{(i-1)N+j,(k-1)N+l} \Rightarrow S_{ik} \cdot S_{lj} = S_{ki} \cdot S_{jl}.$$

Since this equation holds for any i,j,k,l, it holds also for $i = k$.

$$\Rightarrow S_{ii} \cdot S_{lj} = S_{ii} \cdot S_{jl} \Rightarrow S_{lj} = S_{jl}$$

(unless $S_{ii} = 0$ for all $i.\ i = 1, \cdots N$). Q.E.D.

APPENDIX B

Proof of Theorem II

$$\boldsymbol{X}[R] = \boldsymbol{Y} \Rightarrow Y_p = \sum_{q=1}^{N^2} X_q R_{qp} \qquad p = 1,2, \cdots N^2$$

$$[P][X][Q] = [Y] \Rightarrow Y_{ij} = \sum_{k=1}^{N} \sum_{l=1}^{N} X_{kl} P_{ik} Q_{lj}$$

$$i,j = 1,2, \cdots N.$$

Since $[X],[Y]$ are the matrix representation of $\boldsymbol{X},\boldsymbol{Y}$ respectively it follows that:

$$p = (i-1)N + j, \quad q = (k-1)N + l.$$

The coefficients of $\boldsymbol{X}$ are equal in both representations: $\Rightarrow$

$$R_{(k-1)N+l,(i-1)N+j} = P_{ik} \cdot Q_{lj}$$

$$R_{(i-1)N+j,(k-1)N+l} = P_{ki} \cdot Q_{jl}.$$

Since $[R]$ is symmetric

$$P_{ik} \cdot Q_{lj} = P_{ki} \cdot Q_{jl} \cdot \qquad i,j,k,l, = 1,2, \cdots N.$$

This equation holds also for $k = i, l = j \Rightarrow$

$$Q_{lj} = Q_{jl}$$

$$P_{ik} = P_{ki}.$$

Unless:

1) $Q_{kk} = 0 \qquad k = 1,2, \cdots N$

2) $P_{kk} = 0 \qquad k = 1,2, \cdots N$

3) Q_{kk} and $P_{kk} = 0$.

These cases were excluded above.
Thus, $[P],[Q]$ are symmetric $\Rightarrow$

$$R_{(k-1)N+l,(i-1)N+j} = R_{(i-1)N+l,(k-1)N+j}. \qquad \text{Q.E.D.}$$

ACKNOWLEDGMENT

The author wishes to thank J.H.J. Filter of the National Electrical Engineering Research Institute for his useful suggestions and remarks.

REFERENCES

[1] G. G. Murray, "Modified transforms in imagery analysis," in *Proc. Symp. Applications Walsh Functions*, 1972, pp. 235–239.
[2] W. K. Pratt, J. Kane, and H. C. Andrews, "Hadamard transform image coding," *Proc. IEEE*, vol. 57, pp. 58–68, Jan. 1969.
[3] T. S. Huang, W. F. Schreiber, and O. J. Tretiak, "Image processing," *Proc. IEEE*, vol. 59, pp. 1586–1609, Nov. 1971.
[4] H. C. Andrews and J. Kane, "Kronecker matrices, computer implementation and generalized spectra," *J. Ass. Comput. Mach.*, Apr. 1970.
[5] H. C. Andrews and K. L. Caspari, "A generalized technique for spectral analysis," *IEEE Trans. Comput.*, vol. C-19, pp. 16–25, Jan. 1970.
[6] I. J. Good, "The interaction algorithm and practical Fourier analysis," *J. Roy. Stat. Soc.*, vol. 20, no. 2, pp. 361–372, 1958.
[7] F. Pichler, "Walsh functions and optimal linear systems," in *Proc. Symp. Applications Walsh Functions*, 1970, pp. 17–22.

Walsh–Hadamard Power Spectra Invariant to Certain Transform Groups

HIROMITSU HAMA, MEMBER, IEEE, AND KAZUMI YAMASHITA, SENIOR MEMBER, IEEE

Abstract—**Changes in Walsh–Hadamard power spectrum of an input pattern are investigated under several transformations: 1) shifting the elements of the input pattern cyclically, 2) enlarging and reducing the input pattern, 3) rotating the input pattern by multiples of 90°, and so on. Then the Walsh–Hadamard power spectrum is developed to be unchangeable by all the transformations through an introduced composing process. It may be considered as one of geometrical features. Every interesting geometrical property is generally invariant under some transformation groups. The composing process is available for obtaining functions having such group-invariant properties. The main idea is to make a linear combination of group-equivalent functions. First a G_1-invariant power spectrum and next a permutation group on the G_1-invariant power spectrum caused by G_2 operating on the input pattern is found, thus arriving at a power spectrum being constant under both G_1 and G_2 through the composing process, where G_1 and G_2 are some transformation groups. Continuing this process, a power spectrum can be derived from the Walsh–Hadamard power spectrum that is invariant under a more general group of transformations.**

INTRODUCTION

WALSH FUNCTIONS were introduced by Walsh in 1923 [1]. They can be generated recursively, are orthonormal, and form a closed set [1]–[5]. They have been often used in several applications because of the simplicity of square waves. One of the applications is feature extraction for geometrical patterns. Every interesting geometrical property is invariant to any element of some transformation group. The theory of computational geometry was first developed by Minsky and Papert [6], and next extended by Uesaka [7]. As a method of realizing transformation group-invariant functions, Minsky and Papert [6] have already presented a method of equating the coefficients of partial functions in the same equivalence class.

In this paper a composing process is introduced that produces functions which are invariant to some transformation groups. Such functions may be considered as geometrical features. According to the process, Walsh–Hadamard power spectrum is developed to be unchangeable by some transformation groups. Walsh–Hadamard transform $(\mathrm{WHT})_h$ [8], [9] has the advantage of computational simplicity when compared with Fourier transform. Let $B_X(k)$ be the kth $(\mathrm{WHT})_h$ coefficient. Then the set $\{B_X^2(k)\}$ is not invariant to cyclic shifts, nor is it invariant to enlargements, reductions, rotations by multiples of 90°, and so on.

The $(\mathrm{WHT})_h$ for image processing has been discussed by Andrews and others [10]. The axis-symmetry-histograms developed by Alexandridis and Klinger [11] are invariant to cyclic shifts and rotations by multiples of 90°, but they require normalization of an input pattern through the Fourier transform. The $(\mathrm{WHT})_h$ power spectrum developed by Ahmed, Rao, and Abdussattar [8], [9] is invariant to cyclic shifts, which is obtained directly through the $(\mathrm{WHT})_h$. The fast algorithm for the power spectrum was also presented by them. It is shown by Arazi [12] that the power spectrum is invariant under many other shufflings besides cyclic shiftings. If a family of functions is invariant to a group G and is able to distinguish G-nonequivalent patterns, then we call it a G-invariant complete system. This is not an exact definition, which will be given later, but a rough idea. The power spectrum is a cyclic shift-invariant system, but not a cyclic shift-invariant complete system. According to the process presented here, a G-invariant complete system seems easy to make mathematically, but in general, difficult to make practically because of the huge number of functions required. Power spectra which we chiefly adopt may be regarded as subsets of a G-invariant complete system. This paper describes how to develop the $(\mathrm{WHT})_h$ power spectrum to be invariant to all of cyclic shifts, enlargements, reductions, rotations by multiples of 90°, and some other transformations.

WALSH–HADAMARD TRANSFORMATION

A set of standards for notation and definition in the area of nonsinusoidal complete orthogonal sets of functions is proposed by Ahmed *et al.* [9], [13]. There are three types of orderings for the Walsh functions. One of them is used here, that is, natural or Hadamard ordering. Hadamard matrices can be recursively generated as follows:

$$H(0) = [1]$$

$$H(n) = \begin{pmatrix} H(n-1) & H(n-1) \\ H(n-1) & -H(n-1) \end{pmatrix} = H^n(1) \tag{1}$$

where $H^n(1)$ is the n successive Kronecker products of $H(1)$. Let $h_k^{(n)}$ denote the kth row vector of $H(n)$, $k = 0, 1, \cdots, 2^n - 1$. The Hadamard matrices satisfy the following:

$$H(n) = H(n-m) \otimes H(m)$$

$$(U \otimes V) \cdot (H(n-m) \otimes H(m)) = (U \cdot H(n-m)) \otimes (V \cdot H(m)) \tag{2}$$

where $0 \le m \le n$; U and V are a 2^{n-m} vector and a 2^m vector, respectively; and the notation $\otimes$ denotes the Kronecker

Manuscript received November 21, 1977; revised December 11, 1978.
The authors are with the Department of Electrical Engineering, Faculty of Engineering, Osaka City University, Sumiyoshiku, Osaka, Japan 558.

product. These can be proved by taking the definitions of the Hadamard matrices and the Kronecker product into consideration.

Let $\{X(k)\}$ denote an N-periodic sequence $X(k)$, $k = 0, 1, \cdots, N-1$, of real numbers,[1] and $\{X(k)\}$ be represented by means of an N-vector $X(n)$:

$$X(n) = [X(0), X(1), \cdots, X(N-1)] \tag{3}$$

where $N = 2^n$. Walsh–Hadamard transform $(\mathrm{WHT})_h$ of an input pattern $X(n)$ is defined as

$$\boldsymbol{B}_X(n) = (1/N)\boldsymbol{X}(n) \cdot \boldsymbol{H}(n) \tag{4}$$

where $\boldsymbol{B}_X(n) = [B_X(0), B_X(1), \cdots, B_X(N-1)]$ and $B_X(k)$ is the kth $(\mathrm{WHT})_h$ coefficient. The inverse transform is defined as

$$\boldsymbol{X}(n) = \boldsymbol{B}_X(n) \cdot \boldsymbol{H}(n). \tag{5}$$

From (4) it follows that

$$\boldsymbol{B}_X(n) \cdot \boldsymbol{B}_X(n)^T = (1/N)\boldsymbol{X}(n) \cdot \boldsymbol{X}(n)^T \tag{6}$$

where T denotes the transposed vector. The right side of (6) represents the average power of the input pattern. Although Fourier power spectrum is invariant to cyclic shifts, the set $\{B_X^2(k)\}$ is not. A composing process is proposed in preparation for the development of the $(\mathrm{WHT})_h$ power spectrum.

Composing Process

The process to obtain a $G_1 \otimes G_2$-invariant power spectrum in several stages is shown by Fig. 1. First, an input pattern is transformed through the $(\mathrm{WHT})_h$, and the $(\mathrm{WHT})_h$ coefficients are squared. Second, the results are combined according to a certain function to obtain a G_1-invariant power spectrum. Lastly, we find a permutation group on the G_1-invariant power spectrum caused by G_2 operating on the input pattern, and then we combined the G_1-invariant power spectrum to be G_2-invariant in the same way and arrive at a $G_1 \otimes G_2$-invariant power spectrum. G_1 and G_2 are some transformation groups, and the product $G_1 \otimes G_2$ is defined by

$$G_1 \otimes G_2 = \{g \mid g = g_1 g_2, \quad g_1 \text{ and } g_2 \text{ are in } G_1 \cup G_2\}. \tag{7}$$

Several mathematical terms and two theorems are introduced in this section, using basic definitions of the theory of groups. Let G be a finite transformation group on an input pattern X. A member g in G operating on an input pattern X yields another pattern $X' = g \circ X$. We will say that two patterns X and X' are *G-equivalent* (and we write $X \overset{G}{\equiv} X'$), if there is a member g in G for which $X' = g \circ X$. When a group is chosen, a classification of patterns into equivalence classes is automatically set up. Given a function f and a member g, $g \circ f$ will be defined as the function that has the value $f(g \circ X)$ for every X, that is, $g \circ f(X) = f(g \circ X)$. We will say that two functions f and f' are *G-equivalent* (and we write $f \overset{G}{\equiv} f'$), if there is a member g in G such that $f'(X) = f(g \circ X)$ for every X. It is said that a function f is *G-invariant*, if f is invariant to any member of G,

[1] Most of the results developed here can be extended to complex number sequences.

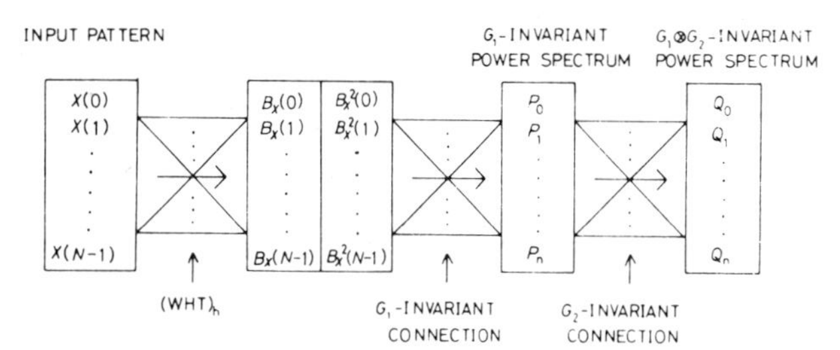

Fig. 1. Process to obtain $G_1 \otimes G_2$-invariant power spectrum.

that is, $f = g \circ f$ for any g in G. In other words, the value of a G-invariant function f depends only upon the G-equivalence classes of input patterns. Assume that $f(X)$ is expressed by a linear combination of $\phi_i(X)$ $(i = 1, 2, \cdots, M)$ as follows:

$$f(X) = \sum_{i=1}^{M} c_i \phi_i(X) \tag{8}$$

for any X (abbreviated as $f = \sum_{i=1}^{M} c_i \phi_i$). Then (8) is called a *linear expression* of f by $\Phi = \{\phi_i\}$ $(i = 1, 2, \cdots, M)$. We will say Φ is *closed under* G if for every ϕ_i $(i = 1, 2, \cdots, M)$ in Φ and every g in G the function $g \circ f$ is also in Φ. If one can choose $\{c_i\}$ $(i = 1, 2, \cdots, M)$ so that $\sum_{i=1}^{M} c_i \phi_i(X) = 0$ and $(c_1, c_2, \cdots, c_M) \neq (0, 0, \cdots, 0)$, then it is said that Φ is *linearly dependent*, otherwise *linearly independent*. When there is at least one linear expression of every function by Φ, Φ goes by the name of a *complete system*. For instance, $\Phi = \{X^{i_0}(0), X^{i_1}(1), \cdots, X^{i_{N-1}}(N-1)\}$ $(i_0 = 0, 1, \cdots, i_1 = 0, 1, \cdots, \cdots, i_{N-1} = 0, 1, \cdots)$ is a well-known linearly independent complete system in N-dimensional Hilbert space.

Now we can state a group-invariance theorem which is available for making G-invariant connections. The theorem was first introduced by Minsky and Papert [6]. Equating the coefficients of G-equivalent partial functions was *sufficient* because there was tolerance with respect to the expression of a function owing to use of a linear threshold function. But in this paper it is *necessary* and *sufficient*, because there is no tolerance with regard to the expression on account of use of a linear expression. The theorem is described in the following. But the proof is omitted, since it can be obtained in the same manner as that in [6].

Theorem 1 (Group-Invariance Theorem): Suppose that $\Phi = \{\phi_i\}$ $(i = 1, 2, \cdots, M)$ is closed under G and is a linearly independent finite set and that f is a G-invariant function whose linear expression is (8). Then the coefficients c_i $(i = 1, 2, \cdots, M)$ depend only on G-equivalence classes of Φ; that is,

$$\text{if } \phi_i \overset{G}{\equiv} \phi_j, \quad \text{then } c_i = c_j. \tag{9}$$

The reverse is also true; that is, if (9) is satisfied, then f is G-invariant. The facts are the same with an infinite set Φ if Φ is an orthogonal system and is closed under G. We should notice that in the case of an infinite set Φ it is not sufficient for Φ to be only linearly independent and closed under G and it is necessary to be orthogonal.

Corollary 1: For any function ϕ, f written as follows is G-invariant:

$$f = \sum_{g \in G} g \circ \phi. \tag{10}$$

Corollary 2: f expressed as follows is also G-invariant:

$$f = \sum_{i=1}^{L} c_i f_i \tag{11}$$

where the f_i $(i = 1, 2, \cdots, L)$ are G-invariant and the c_i $(i = 1, 2, \cdots, L)$ are arbitrary real numbers. We have a more general expression than (11):

$$f = F(f_1, f_2, \cdots, f_L) \tag{12}$$

where F is an arbitrary functional.

A *G-invariant complete system* Φ_G is defined by the following conditions: i) *Φ_G is a family of G-invariant functions* and ii) *every G-invariant function has at least one linear expression by Φ_G*. The next theorem shows how to compose a G-invariant linearly independent complete system. The proof is also omitted, which can be seen in [14].

Theorem 2 (G-Invariant Linearly Independent Complete System): Let Φ be a linearly independent complete system and be closed under G; then Φ_G expressed by the following is a G-invariant linearly independent complete system:

$$\Phi_G = \left\{ f \mid f = \sum_{g \in G} g \circ \phi^*, \phi^* \in \Phi \right\} \tag{13}$$

where ϕ^* is a representative of a G-equivalence class of Φ.

Summation $\sum$ in *Corollaries 1 and 2* and *Theorem 2* is replaceable by some other operations. For example, group-invariant complete systems on an n-cube $\{0, 1\}^n$ were obtained by replacing $\sum$ with $\wedge$ and $\vee$ [15], [16], where $\wedge$ and $\vee$ are notations of conventional Boolean functions.

Let functions $F_l(X_1, X_2, \cdots, X_L)$ $(l = 1, 2, \cdots, L)$ be defined by

$$F_l(X_1, X_2, \cdots, X_L) = \begin{cases} \sum_{i=1}^{L} X_i^2/A, & \text{if } A \neq 0,\ l = 1 \\ 2 \sum_{i=1}^{L-l+1} X_i X_{i+l-1}/A, & \text{if } A \neq 0,\ l = 2, 3, \cdots, L \\ 0, & \text{if } A = 0,\ l = 1, 2, \cdots, L \end{cases} \tag{14}$$

where $A = \sum_{i=1}^{L} X_i$, then we have

$$\sum_{l=1}^{L} F_l(X_1, X_2, \cdots, X_L) = \sum_{i=1}^{L} X_i$$

$$F_l(X_1, X_2, \cdots, X_L) \geq 0 \ (l = 1, 2, \cdots, L), \quad \text{if } X_i \geq 0 \ (i = 1, 2, \cdots, L). \tag{15}$$

Hence $\{F_l(X_1, X_2, \cdots, X_L)\}$ $(l = 1, 2, \cdots, L)$ may be regarded as a developed power spectrum, if $\{X_i\}$ $(i = 1, 2, \cdots, L)$ is a power spectrum.

Cyclic Shift (G_1)

A cyclic shift-invariant power spectrum is illustrated by considering the case when $n = 3$. Using *Corollary 1* the second degree terms are combined to make cyclic shift-invariant functions:

$$\phi_l = \sum_{k=0}^{7} X(k) X(k \oplus l) \tag{16}$$

where $l = 0, 1, \cdots, 4$, and $\oplus$ is used for modulo eight addition. The ϕ_l $(l = 0, 1, \cdots, 4)$ can be expressed by the set $\{B_X(k)\}$ $(k = 0, 1, \cdots, N-1)$:

$$\phi_0 = 8 \sum_{k=0}^{7} B_X^2(k)$$

$$\phi_1 = 4\{2B_X^2(0) - 2B_X^2(1) + B_X^2(4) - B_X^2(5) + B_X^2(6) - B_X^2(7) - 2B_X(4)B_X(7) + 2B_X(5)B_X(6)\}$$

$$\phi_2 = 8\{B_X^2(0) + B_X^2(1) - B_X^2(2) - B_X^2(3)\}$$

$$\phi_3 = 4\{2B_X^2(0) - 2B_X^2(1) - B_X^2(4) + B_X^2(5) - B_X^2(6) + B_X^2(7) + 2B_X(4)B_X(7) - 2B_X(5)B_X(6)\}$$

$$\phi_4 = 4\{B_X^2(0) + B_X^2(1) + B_X^2(2) + B_X^2(3) - B_X^2(4) - B_X^2(5) - B_X^2(6) - B_X^2(7)\}. \tag{17}$$

Suppose a function f has a linear expression by the second degree terms of the $X(k)$ $(k = 0, 1, \cdots, 7)$. From *Theorem 1*, f has a linear expression by $\{\phi_l\}$, $(l = 0, 1, \cdots, 4)$, if it is cyclic shift invariant. Inversely f cannot be cyclic shift invariant, if it cannot have a linear expression by $\{\phi_l\}$ $(l = 0, 1, \cdots, 4)$.

The adequate linear combination of the ϕ_l $(l = 0, 1, \cdots, 4)$ leads to a cyclic shift-invariant power spectrum. Adding ϕ_1 to ϕ_3 removes the cross terms $B_X(4)B_X(7)$ and $B_X(5)B_X(6)$:

$$\phi_1 + \phi_3 = 16\{B_X^2(0) - B_X^2(1)\}. \tag{18}$$

Then $\{B_X^2(k)\}$ is grouped as

$$\{B_X^2(0)\} \cup \{B_X^2(1)\} \cup \{B_X^2(2), B_X^2(3)\} \cup \{B_X^2(4), B_X^2(5), B_X^2(6), B_X^2(7)\}.$$

The $B_X^2(k)$ in the same group have the same coefficients in the expressions of ϕ_0, ϕ_2, ϕ_4, and $\phi_1 + \phi_3$. Thus we obtain

$$P_0 = (1/8^2)(\phi_0 + 2\phi_1 + 2\phi_2 + 2\phi_3 + 2\phi_4) = B_X^2(0)$$

$$P_1 = (1/8^2)(\phi_0 - 2\phi_1 + 2\phi_2 - 2\phi_3 + 2\phi_4) = B_X^2(1)$$

$$P_2 = (2/8^2)(\phi_0 - 2\phi_2 + 2\phi_4) = B_X^2(2) + B_X^2(3)$$

$$P_3 = (4/8^2)(\phi_0 - 2\phi_4) = B_X^2(4) + B_X^2(5) + B_X^2(6) + B_X^2(7). \tag{19}$$

The remainder may be expressed by

$$P_4 = (8/8^2)(\phi_1 - \phi_3) = B_X^2(4) - B_X^2(5) + B_X^2(6) - B_X^2(7) - 2B_X(4)B_X(7) + 2B_X(5)B_X(6). \tag{20}$$

$\{P_i\}$ $(i = 0, 1, 2, 3, \neq 4)$ turns out to be the very same power spectrum as that developed by Ahmed, *et al.* [8], [9].

From the first-degree terms we obtain only one cyclic shift-invariant function which is linearly independent:

$$P_5 = (1/8) \sum_{k=0}^{7} X(k) = B_X(0). \tag{21}$$

As pointed out in [12], $\{P_i\}$ $(i = 0, 1, 2, 3)$ is not a cyclic shift-invariant complete system; that is, there are other shiftings under which $\{P_i\}$ $(i = 0, 1, 2, 3)$ is invariant. This is the same even if $\{P_i\}$ $(i = 0, 1, \cdots, 5)$ is used instead of $\{P_i\}$ $(i = 0, 1, 2, 3)$. In order to obtain a complete system we must usually use the higher degree terms besides the first- and the second-degree terms.

The generalization of the power spectrum mentioned above is straightforward as seen in [8] and [9]:

$$P_0 = B_X^2(0)$$

$$P_s = \sum_{i=2^{s-1}}^{2^s-1} B_X^2(i), \qquad s = 1, 2, \cdots, n. \tag{22}$$

The generalization in the two-dimensional case is as follows:

$$P_{00} = B_X^2(0, 0), \qquad P_{0t} = \sum_{j=2^{t-1}}^{2^t-1} B_X^2(0, j)$$

$$P_{s0} = \sum_{i=2^{s-1}}^{2^s-1} B_X^2(i, 0), \qquad P_{st} = \sum_{i=2^{s-1}}^{2^s-1} \sum_{j=2^{t-1}}^{2^t-1} B_X^2(i, j). \tag{23}$$

where $s = 1, 2, \cdots, n$ and $t = 1, 2, \cdots, m$. The developed $(\mathrm{WHT})_h$ power spectra are expressed as

$$\boldsymbol{P}(n) = [P_0, P_1, \cdots, P_n]$$

$$\boldsymbol{P}(n, m) = \begin{pmatrix} P_{00}, P_{01}, \cdots, P_{0m} \\ \vdots \qquad \vdots \\ P_{n0}, P_{n1}, \cdots, P_{nm} \end{pmatrix}. \tag{24}$$

When we want to take notice of the input pattern, we write as follows:

$$\boldsymbol{P}_X(n) = [P_0(X), P_1(X), \cdots, P_n(X)]$$

$$\boldsymbol{P}_X(n, m) = \begin{pmatrix} P_{00}(X), P_{01}(X), \cdots, P_{0m}(X) \\ \vdots \qquad \vdots \\ P_{n0}(X), P_{n1}(X), \cdots, P_{nm}(X) \end{pmatrix} \tag{25}$$

This can be extended to any number of dimensions [8], [9].

Enlargement and Reduction (G_2)

The changing aspects of the $(\mathrm{WHT})_h$ coefficients under enlargements and reductions of an input pattern are investigated. An input pattern X is called a 2^i-time enlargeable pattern, if

$$X = [\underbrace{U, U, \cdots, U}_{2^i}]$$

and $U = [X(0), X(1), \cdots, X(2^{n-i} - 1)]$. Similarly an input pattern Y is called a $1/2^i$-time reducible pattern, if $Y = [Y(0)\mathbf{1}, Y(2^i)\mathbf{1}, \cdots, Y(2^n - 2^i)\mathbf{1}]$ and

$$\mathbf{1} = [\underbrace{1, 1, \cdots, 1}_{2^i}].$$

The 2^i-time enlarged pattern of X and the $1/2^i$-time reduced pattern of Y are defined as $[X(0)\mathbf{1}, X(1)\mathbf{1}, \cdots, X(2^{n-i} - 1)\mathbf{1}] = U \otimes \mathbf{1}$ and

$$[\underbrace{U', U', \cdots, U'}_{2^i}] = \mathbf{1} \otimes U',$$

respectively, where

$$\mathbf{1} = [\underbrace{1, 1, \cdots, 1}_{2^i}]$$

and $U' = [Y(0), Y(2^i), \cdots, Y(2^n - 2^i)]$. For example, the 2-time enlarged pattern of $X = [01230123]$ is $[00112233]$, the 2^2-time enlarged pattern of $X = [01010101]$ is $[00001111]$. The 1/2-time reduced pattern of $Y = [00112233]$ is $[01230123]$ and so on.

From (1) the kth row vector $\boldsymbol{h}_k^{(n)}$ of $\boldsymbol{H}(n)$ is expressed as

$$\boldsymbol{h}_k^{(n)} = \begin{cases} [\boldsymbol{h}_k^{(n-1)}, \ \boldsymbol{h}_k^{(n-1)}], & k = 0, 1, \cdots, 2^{n-1} - 1 \\ [\boldsymbol{h}_k^{(n-1)}, \ -\boldsymbol{h}_k^{(n-1)}], & k = 2^{n-1}, 2^{n-1} + 1, \cdots, 2^n - 1. \end{cases} \tag{26}$$

The upper half row vectors of $\boldsymbol{H}(n)$ are 2-time enlargeable patterns. The fact of (26) induces recursively the following expression:

$$\boldsymbol{h}_k^{(n)} = \begin{cases} [\boldsymbol{h}_k^{(0)}, \cdots, \boldsymbol{h}_k^{(0)}] = [1, 1, \cdots, 1], \\ \qquad k = 0, \ 2^n\text{-time enlargeable}, \\ [\boldsymbol{h}_k^{(1)}, \cdots, \boldsymbol{h}_k^{(1)}] = [1, -1, \cdots, 1, -1], \\ \qquad k = 1, \ 2^{n-1}\text{-time enlargeable}, \\ [\boldsymbol{h}_k^{(2)}, \cdots, \boldsymbol{h}_k^{(2)}], \quad k = 2, 3, \ 2^{n-2}\text{-time enlargeable}, \\ [\boldsymbol{h}_k^{(3)}, \cdots, \boldsymbol{h}_k^{(3)}], \\ \qquad k = 4, 5, 6, 7, \ 2^{n-3}\text{-time enlargeable}, \\ \quad \vdots \\ [\boldsymbol{h}_k^{(n-1)}, \boldsymbol{h}_k^{(n-1)}], \ k = 2^{n-2}, 2^{n-2} + 1, \\ \qquad \cdots, 2^{n-1} - 1, \ 2\text{-time enlargeable}, \\ [\boldsymbol{h}_k^{(n-1)}, -\boldsymbol{h}_k^{(n-1)}], \\ \qquad k = 2^{n-1}, 2^{n-1} + 1, \cdots, 2^n - 1. \end{cases} \tag{27}$$

On the other hand from (2) $\boldsymbol{H}(n)$ is also expressed as

$$\boldsymbol{H}(n) = \boldsymbol{H}(n-i) \otimes \boldsymbol{H}(i) = \boldsymbol{H}(i) \otimes \boldsymbol{H}(n-i). \tag{28}$$

Taking account of that

$$\boldsymbol{h}_0^{(i)} = [\underbrace{1, 1, \cdots, 1}_{2^i}],$$

we obtain that for a multiple k of 2^i $\boldsymbol{h}_k^{(n)}$ is a $1/2^i$-time reducible pattern. For an even number k the kth row vector $\boldsymbol{h}_k^{(n)}$ of $\boldsymbol{H}(n)$ is a 1/2-time reducible pattern. From (28) the 2^i-time enlarged pattern of $\boldsymbol{h}_l^{(n)}$ is $\boldsymbol{h}_{2^i l}^{(n)}$, where $l = 0, 1, \cdots, 2^{n-i} - 1$. The 2-time enlarged pattern of $\boldsymbol{h}_l^{(n)}$ is $\boldsymbol{h}_{2l}^{(n)}$. These facts are well illustrated by considering the case when $n = 3$:

4-time enlargement

$$\boldsymbol{H}(3) = \begin{pmatrix} 1 & 1 & 1 & 1 & 1 & 1 & 1 & 1 \\ 1 & -1 & 1 & -1 & 1 & -1 & 1 & -1 \\ 1 & 1 & -1 & -1 & 1 & 1 & -1 & -1 \\ 1 & -1 & -1 & 1 & 1 & -1 & -1 & 1 \\ 1 & 1 & 1 & 1 & -1 & -1 & -1 & -1 \\ 1 & -1 & 1 & -1 & -1 & 1 & -1 & 1 \\ 1 & 1 & -1 & -1 & -1 & -1 & 1 & 1 \\ 1 & -1 & -1 & 1 & -1 & 1 & 1 & -1 \end{pmatrix} \begin{matrix} \boldsymbol{h}_0^{(3)} \\ \boldsymbol{h}_1^{(3)} \\ \boldsymbol{h}_2^{(3)} \\ \boldsymbol{h}_3^{(3)} \\ \boldsymbol{h}_4^{(3)} \\ \boldsymbol{h}_5^{(3)} \\ \boldsymbol{h}_6^{(3)} \\ \boldsymbol{h}_7^{(3)} \end{matrix}$$

2-time enlargement

2-time enlargeable patterns

1/2-time reducible patterns (29)

A 2^i-time enlargeable pattern X and the 2^i-time enlarged pattern Y of it are expressed as

$$\begin{aligned}
X &= [\underbrace{U, U, \cdots, U}_{2^i}] = [\underbrace{1, 1, \cdots, 1}_{2^i}] \otimes U = \mathbf{1} \otimes U \\
Y &= [X(0)\mathbf{1}, X(1)\mathbf{1}, \cdots, X(2^{n-i}-1)\mathbf{1}] \\
&= [X(0), X(1), \cdots, X(2^{n-i}-1)] \otimes \mathbf{1} = U \otimes \mathbf{1} \\
U &= [X(0), X(1), \cdots, X(2^{n-i}-1)] \\
\mathbf{1} &= [\underbrace{1, 1, \cdots, 1}_{2^i}].
\end{aligned} \tag{30}$$

The $(\mathrm{WHT})_h$'s of X and Y are as follows:

$$\begin{aligned}
B_X(n) &= (1/N)X \cdot H(n) = (1/N)[U, U, \cdots, U] \\
&\quad \cdot (H(i) \otimes H(n-i)) \\
&= (1/N)[U, U, \cdots, U] \\
&\quad \cdot \begin{pmatrix} \begin{matrix} H(n-i), & H(n-i) \\ H(n-i), & -H(n-i) \\ & \cdots \end{matrix} & H(n-1) \\ H(n-1) & -H(n-1) \end{pmatrix} \\
&= (1/N)(\mathbf{1} \cdot H(i)) \otimes (U \cdot H(n-i)) \\
&= (1/2^{n-i})[\underbrace{1, 0, 0, \cdots, 0}_{2^i}] \otimes (U \cdot H(n-i)) \\
B_Y(n) &= (1/N)Y \cdot H(n) = (1/N)(U \otimes \mathbf{1})(H(n-i) \otimes H(i)) \\
&= (1/N)[X(0)\mathbf{1}, X(1)\mathbf{1}, \cdots, X(2^{n-i}-1)\mathbf{1}] \\
&\quad \cdot \begin{pmatrix} \begin{matrix} H(i), & H(i) \\ H(i), & -H(i) \\ & \cdots \end{matrix} & H(n-1) \\ H(n-1) & -H(n-1) \end{pmatrix} \\
&= (1/N)(U \cdot H(n-i)) \otimes (\mathbf{1} \cdot H(i)) \\
&= (1/2^{n-i})(U \cdot H(n-i)) \otimes [\underbrace{1, 0, 0, \cdots, 0}_{2^i}].
\end{aligned} \tag{31}$$

From (31) we conclude that

$$B_X(k) = \begin{cases} \text{value of the } k\text{th element of } (1/2^{n-i})U \cdot H(n-i), \\ \qquad 0 \le k \le 2^{n-i}-1 \\ 0, \; 2^{n-i} \le k \le 2^n - 1 \end{cases}$$

$$B_X(k) = \begin{cases} \text{value of the } k\text{th element of } (1/2^{n-i})U \cdot H(n-i), \\ k = l \cdot 2^i (l = 0, 1, \cdots, 2^{n-i}-1) \\ 0, k \quad \text{is not a multiple of } 2^i \end{cases} \tag{32}$$

and furthermore

$$B_X(k) = B_Y(2^i k), \qquad 0 \le k \le 2^{n-i}-1. \tag{33}$$

These aspects are well summarized on a $n \times 2^{n-1}$ matrix $C_X(n) = [\{C_X(i, j)\}]$ defined recursively by

$$C_X(1, j) = B_X(2^{n-1} + j - 1)$$

$$C_X(i, j) = \begin{cases} B_X(k/2), & \text{if } C_X(i-1, j) = B_X(k), \\ & \qquad k \text{ is an even number} \\ 0, \text{ otherwise} \end{cases} \tag{34}$$

where $i = 2, 3, \cdots, n$ and $j = 1, 2, \cdots, 2^{n-1}$. 2^i-time enlargement and $1/2^i$-time reduction of an input pattern cause upward shift by i rows and downward shift by i rows, respectively. $B_X(0)$ is invariant under any shuffling of the elements of the input pattern. In the case when $n = 3$, the matrix is as follows:

$$C_X(3) = \begin{pmatrix} B_X(4), & B_X(5), & B_X(6), & B_X(7) \\ B_X(2), & 0, & B_X(3), & 0 \\ B_X(1), & 0, & 0, & 0 \end{pmatrix} \quad \begin{matrix} \text{enlargement} \\ \uparrow \quad \downarrow \\ \text{reduction} \end{matrix} \tag{35}$$

Let Y be a 2-time enlarged pattern of a two-time enlargeable pattern X, and then

$$C_Y(3) = \begin{pmatrix} B_Y(4), & 0, & B_Y(6), & 0 \\ B_Y(2), & 0, & 0, & 0 \\ 0, & 0, & 0, & 0 \end{pmatrix} = \begin{pmatrix} B_X(2), & 0, & B_X(3), & 0 \\ B_X(1), & 0, & 0, & 0 \\ 0, & 0, & 0, & 0 \end{pmatrix}. \tag{36}$$

As seen in (36) 2-time enlargement of an input pattern causes upward shift of the elements of $C_X(3)$. Then an enlargement and reduction-invariant power spectrum $\{Q_i\}$ $(i = 0, 1, \cdots, 7)$ can be obtained by the method introduced in the section on the composing process:

$$\begin{aligned}
Q_0 &= B_X^2(0) & Q_3 &= F_1(B_X^2(3), B_X^2(6)) \\
Q_1 &= F_1(B_X^2(1), B_X^2(2), B_X^2(4) & Q_6 &= F_2(B_X^2(3), B_X^2(6)) \\
Q_2 &= F_2(B_X^2(1), B_X^2(2), B_X^2(4)) & Q_5 &= B_X^5(5) \\
Q_4 &= F_3(B_X^2(1), B_X^2(2), B_X^2(4)) & Q_7 &= B_X^2(7).
\end{aligned} \tag{37}$$

The average power of $\{X(k)\}$ is

$$P_{\mathrm{av}} = \sum_{i=0}^{7} Q_i = \sum_{k=0}^{7} B_X^2(k) = (1/8) \sum_{k=0}^{7} X^2(k). \tag{38}$$

There are many other functions which are enlargement and reduction invariant and have linear expressions by the second-degree terms of the $B_X(k)$ $(k = 0, 1, \cdots, 7)$. Taking the following into consideration in this case

$$\begin{aligned}
&P_0 = B_X^2(0), \; P_2 = B_X^2(2) + B_X^2(3) \\
&P_1 = B_X^2(1), \; P_3 = B_X^2(4) + B_X^2(5) + B_X^2(6) + B_X^2(7),
\end{aligned} \tag{39}$$

we have

$$\begin{aligned}
\boldsymbol{P}_X(3) &= [P_0(X), P_1(X), P_2(X), 0] \\
&= [P_0(Y), P_2(Y), P_3(Y), 0] \\
\boldsymbol{P}_Y(3) &= [P_0(Y), 0, P_2(Y), P_3(Y)] \\
&= [P_0(X), 0, P_1(X), P_2(X)] \\
\boldsymbol{P}(3) &= [P_0 \,\vdots\, P_1, P_2, P_3]
\end{aligned} \tag{40}$$

-------→ enlargement

←——— reduction.

This means that 2-time enlargement of an input pattern X causes shift of $\boldsymbol{P}_X(3)$ toward the right by one element except $P_0(X)$. $P_0(X)$ is always invariant to any permutation of an input pattern. Therefore we arrive at a power spectrum $\{Q_i\}$ ($i = 0, 1, 2, 3$) which is invariant to cyclic shifts, enlargements, and reductions:

$$Q_0 = P_0,\ Q_1 = F_1(P_1, P_2, P_3),\ Q_2 = F_2(P_1, P_2, P_3),$$
$$Q_3 = F_3(P_1, P_2, P_3)$$
$$P_{\mathrm{av}} = \sum_{i=0}^{3} Q_i = \sum_{i=0}^{3} P_i = \sum_{k=0}^{7} B_X^2(k) = (1/8) \sum_{k=0}^{7} X^2(k). \tag{41}$$

Generally the statements following (40) are also valid for any natural number n. This is known from (22) and (33) in the same way as used in the case when $n = 3$. Then a cyclic shift, enlargement, and reduction-invariant power spectrum can be obtained:

$$Q_0 = P_0,\ Q_l = F_l(P_1, P_2, \cdots, P_n), \qquad l = 1, 2, \cdots, n. \tag{42}$$

In the case when an input pattern X is a $2^n \times 2^m$ matrix, these results are developed as follows. Let G_2 denote the group of horizontal enlargements and reductions and vertical ones, and let G_{22}, the subgroup of G_2, denote any member of which is enlargement or reduction formed on the same scale horizontally and vertically. Examples of G_2-equivalent and G_{22}-equivalent patterns are shown in Fig. 2. The values in Fig. 2 are the P_{ij} multiplied by 64 for simplicity. Ones and negative ones are blacked and blanked, respectively. In the same manner, the variation of $\boldsymbol{P}(n, m)$ caused by G_{22} and G_2 are illustrated in Fig. 3 for the case when $n = m = 3$. We cite changes of $\boldsymbol{P}_X(3, 3)$ in Fig. 2 for more detailed explanation:

$$\boldsymbol{P}_{X_2}(3,3) = \left(\begin{array}{c|cccc}
P_{00}(X_1) & P_{02}(X_1) \leftarrow & P_{03}(X_1) \leftarrow & 0 \\
\hline
P_{20}(X_1) & P_{22}(X_1) & P_{23}(X_1) & 0 \\
\uparrow & & \nwarrow & \nwarrow \\
P_{30}(X_1) & P_{32}(X_1) & P_{33}(X_1) & 0 \\
\uparrow & & \nwarrow & \nwarrow \\
0 & 0 & 0 & 0
\end{array}\right). \tag{43}$$

$\{P_{ij}\}$ is classified into G_{22}-equivalence classes: $\{P_{00}\} \cup \{P_{01}, P_{02}, P_{03}\} \cup \{P_{10}, P_{20}, P_{30}\} \cup \{P_{11}, P_{22}, P_{33}\} \cup \{P_{12}, P_{23}\} \cup \{P_{21}, P_{32}\} \cup \{P_{13}\} \cup \{P_{31}\}$. The second-degree terms of P_{ij} in the same G_{22}-equivalence classes are also classified into G_{22}-equivalence classes, for example, $\{P_{01}, P_{02}, P_{03}\} \rightarrow \{P_{01}^2, P_{02}^2, P_{03}^2\} \cup \{P_{01} \cdot P_{02}, P_{02} \cdot P_{03}\} \cup \{P_{01} \cdot P_{03}\}$. From *Theorem 1* the Q_{0i} ($i = 1, 2, 3$) defined by the following expressions are G_{22}-invariant:

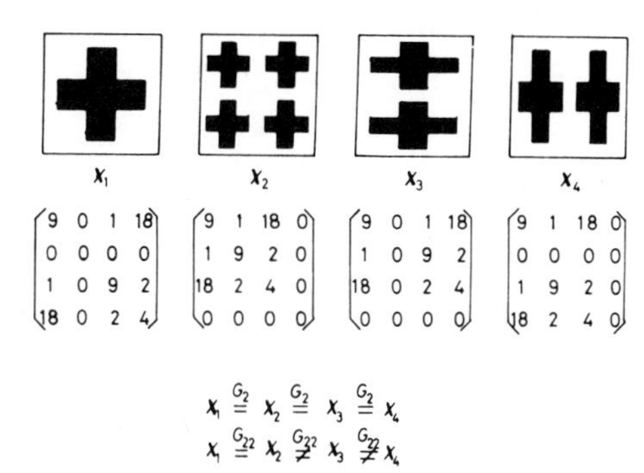

Fig. 2. G_2-equivalent patterns and G_{22}-equivalent patterns.

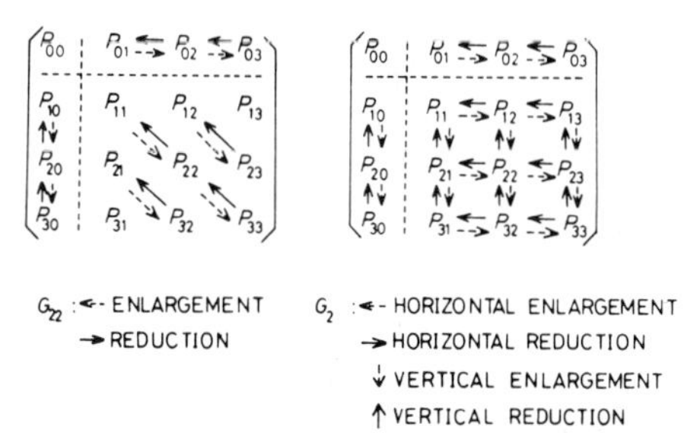

Fig. 3. Shifts of elements of $\boldsymbol{P}(3, 3)$ caused by G_{22} and G_2.

$$\begin{aligned}
Q_{01} &= (P_{01}^2 + P_{02}^2 + P_{03}^2)/(P_{01} + P_{02} + P_{03}) \\
Q_{02} &= 2(P_{01} \cdot P_{02} + P_{02} \cdot P_{03})/(P_{01} + P_{02} + P_{03}) \\
Q_{03} &= 2P_{01} \cdot P_{03}/(P_{01} + P_{02} + P_{03}).
\end{aligned} \tag{44}$$

Let $\Phi_{1\otimes 22} = \{Q_{ij}\}$ ($i = 0, 1, 2, 3, j = 0, 1, 2, 3$) be defined by

$$\begin{aligned}
&Q_{00} = P_{00},\ Q_{0j} = F_j(P_{01}, P_{02}, P_{03}), \\
&Q_{i0} = F_i(P_{10}, P_{20}, P_{30}) \\
&Q_{ij} = F_{j_1+1}(P_{i-j_1, j-j_1}, P_{i-j_1+1, j-j_1+1}, \\
&\qquad \cdots, P_{ij}, \cdots, P_{i+j_2, j+j_2})
\end{aligned} \tag{45}$$

where $1 \le i - j_1, 1 \le j - j_1, i + j_2 \le 3, j + j_2 \le 3, i = 1, 2, 3, j = 1, 2, 3$. $\Phi_{1\otimes 22}$ is a $G_1 \otimes G_{22}$-invariant power spectrum. For any numbers n and m we obtain

$$\begin{aligned}
&Q_{00} = P_{00},\ Q_{0j} = F_j(P_{01}, P_{02}, \cdots, P_{0m}), \\
&Q_{i0} = F_i(P_{10}, P_{20}, \cdots, P_{n0}) \\
&Q_{ij} = F_{j_1+1}(P_{i-j_1, j-j_1}, P_{i-j_1+1, j-j_1+1}, \\
&\qquad \cdots, P_{ij}, \cdots, P_{i+j_2, i+j_2})
\end{aligned} \tag{46}$$

where $1 \le i - j_1$, $1 \le j - j_1$, $i + j_2 \le n$, $j + j_2 \le m$, $i = 1, 2, \cdots, n$, $j = 1, 2, \cdots, m$. In the same way we obtain a

$G_1 \otimes G_2$-invariant power spectrum. Let $\Phi_{1\otimes 2} = \{Q_{ij}\}$ $(i = 0, 1, \cdots, 5, j = 0, 1, \cdots, 5)$ be defined by

$$Q_{00} = P_{00},\ Q_{0j} = F_j(P_{01}, P_{02}, P_{03}),$$
$$Q_{i0} = F_i(P_{10}, P_{20}, P_{30}),$$
$$Q_{11} = \sum_{k=1}^{3}\sum_{l=1}^{3} P_{kl}^2/A,\ Q_{12} = D_{0,1}/A,$$
$$Q_{13} = D_{0,2}/A,\ Q_{21} = D_{1,0}/A,$$
$$Q_{22} = D_{1,1}/A,\ Q_{23} = D_{1,2}/A,$$
$$Q_{24} = D_{1,-2}/A,\ Q_{31} = D_{2,0}/A,$$
$$Q_{32} = D_{2,1}/A,\ Q_{33} = D_{2,2}/A,$$
$$Q_{42} = D_{2,-1}/A,\ Q_{44} = D_{1,-1}/A,$$
$$Q_{55} = D_{2,-2}/A \tag{47}$$

where $i = 1, 2, 3, j = 1, 2, 3$, $A = \sum_{k=1}^{3}\sum_{l=1}^{3} P_{kl} \neq 0$, the Q_{ij} undefined in (47) are equal to zero, and the $D_{k,l}$ are given by

$$D_{k,l} = 2\sum_r\sum_s P_{r,s}\cdot P_{r+k,s+l}, \qquad 1 \le r, s, r+k, s+l \le 3. \tag{48}$$

The Q_{ij} $(i = 1, 2, \cdots, 5, j = 1, 2, \cdots, 5)$ are equal to zero, if $A = 0$. Then $\Phi_{1\otimes 2}$ is a $G_1 \otimes G_2$-invariant power spectrum. Let the ρ_{ij} $(i = 1, 2, \cdots, L_1, j = 1, 2, \cdots, L_2)$ be the G_2 equivalence classes of the second-degree terms of P_{ij} $(i = 1, 2, \cdots, n, j = 1, 2, \cdots, m)$, where some of them may be equal to the empty set and the ρ_{ij} are symmetric with the P_{ij} if $n = m$ and $L_1 = L_2$. Exceptionally, we define that $\rho_{11} = \{(1/2)P_{ij}^2\}$ $(i = 1, 2, \cdots, n, j = 1, 2, \cdots, m)$. For any numbers n and m we have

$$Q_{00} = P_{00}, \qquad Q_{0j} = F_j(P_{01}, P_{02}, \cdots, P_{0m}),$$
$$Q_{i0} = F_i(P_{10}, P_{20}, \cdots, P_{n0})$$
$$Q_{ij} = 2\sum_{\phi\in\rho_{ij}} \phi/A \tag{49}$$

where $i = 1, 2, \cdots, L_1, j = 1, 2, \cdots, L_2$,

$$A = \sum_{k=1}^{n}\sum_{l=1}^{m} P_{kl} \neq 0.$$

The Q_{ij} $(i = 1, 2, \cdots, L_1, j = 1, 2, \cdots, L_2)$ are equal to zero, if $A = 0$.

$$P_{av} = \sum_{Q\in\Phi_{1\otimes 22}} Q = \sum_{Q\in\Phi_{1\otimes 2}} Q = \sum_{i=0}^{n}\sum_{j=0}^{m} P_{ij}. \tag{50}$$

These aspects are shown in Fig. 4. The values in Fig. 4(b) and (c) are the Q_{ij} multiplied by 64. Each element $X(k, l)$ $(k = 0, 1, \cdots, 7, l = 0, 1, \cdots, 7)$ of X_i $(i = 1, 2, \cdots, 6)$ is located at the intersection of row k and column l. The ones and negative ones are blacked and blanked, respectively. The results in a two-dimensional case can be easily extended to any number of dimensions.

Rotation by Multiples of 90°, Symmetry Transformation, and Exchanging One for Negative One (G_3)

Let a horizontal symmetry transformation operator on an input pattern X be denoted by g_h, $G_h = \{g_h, g_h^2\}$, a vertical one be denoted by g_v, $G_v = \{g_v, g_v^2\}$, a diagonal one be denoted by g_d, $G_d = \{g_d, g_d^2\}$. Then g_h^2, g_v^2, and g_d^2 are identity operators. Symmetry transformation of an input pattern yields only sign change of some $(\mathrm{WHT})_h$ coefficients, and does not have any other changes because of the symmetry of the row vectors $\boldsymbol{h}_k^{(n)}$. Hence, squares of the coefficients are invariant to symmetry transformation. Therefore the set $\{B_X^2(i, j)\}$ is $G_h \otimes G_v$-invariant. Let a 90° rotation operator be denoted by g_r and $G_r = \{g_r, g_r^2, g_r^3, g_r^4\}$, then g_r^4 is an identity operator. We notice that g_d and g_r operate only on a square matrix $X(n, n)$.

It is seen in Fig. 5 that g_r is equivalent to $g_v g_d$ and $g_d g_h$, that is,

$$g_r \circ X = g_v g_d \circ X = g_d g_h \circ X. \tag{51}$$

Therefore we have

$$B_{X_1}(i, j) = B_{X_2}(i, j) = B_{X_3}(j, i)$$
$$= \begin{cases} B_X(j, i), & \text{if } \sum_{k=0}^{n-1} i_k \text{ is an even number} \\ -B_X(j, i), & \text{otherwise} \end{cases} \tag{52}$$

where $X_1 = g_r \circ X$, $X_2 = g_d \circ X_3$, $X_3 = g_h \circ X$,

$$i = \sum_{k=0}^{n-1} 2^k i_k,$$

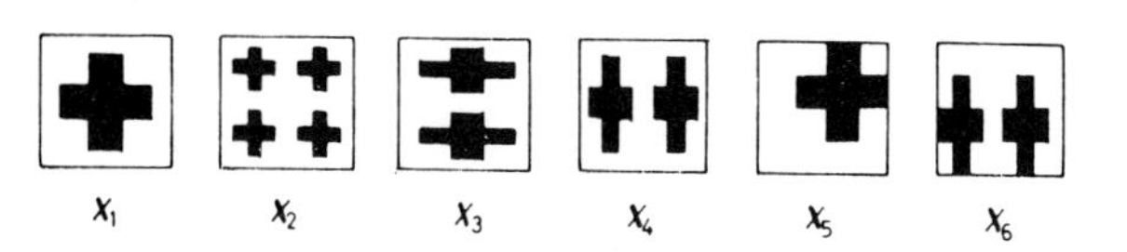

X_1 X_2 X_3 X_4 X_5 X_6

$$X_1 \overset{G_1\otimes G_{22}}{=\!=} X_2 \overset{G_1\otimes G_{22}}{=\!=} X_5 \overset{G_1\otimes G_{22}}{\neq} X_4 \overset{G_1\otimes G_{22}}{=\!=} X_6$$
$$X_1 \overset{G_1\otimes G_2}{=\!=} X_2 \overset{G_1\otimes G_2}{=\!=} X_3 \overset{G_1\otimes G_2}{=\!=} X_4 \overset{G_1\otimes G_2}{=\!=} X_5 \overset{G_1\otimes G_2}{=\!=} X_6$$

(a)

Q_{ij}	$i \backslash j$	0	1	2	3	$64\Sigma Q_{ij}$
X_1	0	9.00	17.11	1.89	0	64
X_2	1	17.11	7.46	2.00	0	
X_5	2	1.89	2.00	5.54	0	
	3	0	0	0	0	
X_4	0	9.00	17.11	1.89	0	64
X_6	1	17.11	2.00	0	0	
	2	1.89	7.46	0	0	
	3	0	2.00	5.54	0	

(b)

Q_{ij}	$i \backslash j$	0	1	2	3	4	5	$64\Sigma Q_{ij}$
X_1	0	9.00	17.11	1.89	0	–	–	64
X_2	1	17.11	6.18	3.06	0	–	–	
X_3	2	1.89	3.06	4.23	0	0	–	
X_4	3	0	0	0	0	–	–	
X_5	4	–	–	0	–	0.47	–	
X_6	5	–	–	–	–	–	0	

(c)

Fig. 4. $G_1 \otimes G_{22}$-invariant power spectrum $(\Phi_{1\otimes 22})$ and $G_1 \otimes G_2$-invariant power spectrum $(\Phi_{1\otimes 2})$. (a) Input patterns. (b) $G_1 \otimes G_{22}$-invariant power spectrum $(\Phi_{1\otimes 22})$. (c) $G_1 \otimes G_2$-invariant power spectrum $(\Phi_{1\otimes 2})$.

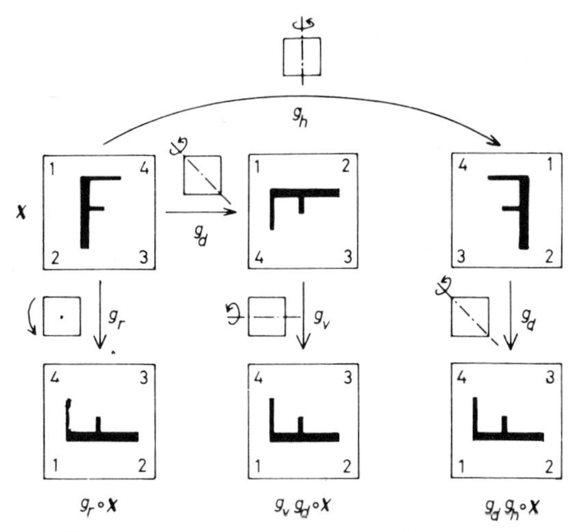

Fig. 5. Aspects of operations g_r, $g_v g_d$, and $g_d g_h$.

$i_k = 0$ or 1, $i = 0, 1, \cdots, 2^n - 1$, $j = 0, 1, \cdots, 2^n - 1$. Let $G_3 = G_h \otimes G_v \otimes G_d$, then it is known from (51) that G_r is included in G_3. From (46), (49), and (52) we obtain a $G_1 \otimes G_l \otimes G_3$-invariant power spectrum $\Phi_{1 \otimes l \otimes 3} = \{S_{ij}\}$ defined by

$$S_{ij} = \begin{cases} F_1(Q_{ij}, Q_{ji}), & \text{if } i > j \\ Q_{ij}, & \text{if } i = j \\ F_2(Q_{ij}, Q_{ji}), & \text{if } i < j \end{cases} \tag{53}$$

and

$$P_{av} = \sum_{S \in \Phi_{1 \otimes l \otimes 3}} S = \sum_{Q \in \Phi_{1 \otimes l}} Q = \sum_{i=0}^{n} \sum_{j=0}^{n} P_{ij} \tag{54}$$

where the Q_{ij} are in $\Phi_{1 \otimes l}$, $l = 2, 22$.

Geometric patterns are often drawn by using ones and negative ones which correspond to black and white points, respectively. Let g_e be an exchanging operator one for negative one and $G_4 = \{g_e, g_e^2\}$, then

$$g_e \circ X = -X, \qquad B_{X_1}(n, m) = -B_X(n, m) \tag{55}$$

where $X_1 = g_e \circ X$. From (55) it follows that the $B_X^2(i, j)$ are invariant to exchanging the sign of an input pattern.

The aspects of group-invariant power spectra described above are shown in Table I. The marks ○ and ◎ are used to show whether or not the families of some group-invariant functions have equal values with respect to input patterns $X_1, X_2, \cdots, X_{12}$ in the same column. To take G_1-invariant power spectrum $P_X(n, m)$, for example, $P_{X_1}(n, m) = P_{X_2}(n, m) = P_{X_4}(n, m) = P_{X_5}(n, m) = P_{X_6}(n, m)$ and $P_{X_1}(n, m) \neq P_{X_3}(n, m) \neq P_{X_7}(n, m)$ in general. It is observed from Table I that the $G_1 \otimes G_2 \otimes G_3$-invariant power spectrum $\Phi_{1 \otimes 2 \otimes 3}$ is a developed power spectrum to be invariant to cyclic shifts, enlargements, reductions, symmetry transformations, rotations by multiples of 90°, and exchanging the sign of an input pattern.

Enlargement and Reduction (G_2')

Enlargement and reduction have been already defined in the foregoing section. For instance, reduction of X_1 yields X_2

TABLE I
Aspects of Group-Invariant Power Spectra and other Group-Invariant Families

Transformation Groups / Input Patterns	Group-Invariant Power Spectra					Group-Invariant Families	
	G_1	$G_1 \otimes G_{22}$	$G_1 \otimes G_2$	$G_1 \otimes G_{22} \otimes G_3$	$G_1 \otimes G_2 \otimes G_3$	$G_1 \otimes G_{22}' \otimes G_3$	$G_1 \otimes G_2' \otimes G_3$
X_1	○	○	○	○	○	○	○
X_2	○	○	○	○	○	○	○
X_3	×	×	×	○	○	○	○
X_4	○	○	○	○	○	○	○
X_5	○	○	○	○	○	○	○
X_6	○	○	○	○	○	○	○
X_7	×	◎	◎	◎	◎	×	×
X_8	×	×	◎	×	◎	×	×
X_9	×	×	◎	×	◎	×	×
X_{10}	×	◎	◎	◎	◎	◎	◎
X_{11}	×	×	×	×	×	×	◎
X_{12}	×	×	×	×	×	◎	◎

in Fig. 6. Such a definition makes mathematical analysis easy, but does not make natural sense for man. Therefore the definition is newly introduced in this section as shown in Fig. 6. An input pattern X is called a 2^i-time enlargeable pattern, if $X = [U, \mathbf{0}]$, $U = [X(0), X(1), \cdots, X(2^{n-i} - 1)]$ and

$$\mathbf{0} = [\underbrace{0, 0, \cdots, 0}_{2^n - 2^{n-i}}].$$

For example, $X = [12000000]$ is a 2^2-time enlargeable pattern, but $X = [0120000]$ is not. A $1/2^i$-time reducible pattern is defined in the same manner as that of the foregoing section, that is, an input pattern Y is a $1/2^i$-time reducible pattern, if

$$Y = [Y(0)\mathbf{1}, Y(2^i)\mathbf{1}, \cdots, Y(2^n - 2^i)\mathbf{1}], \ \mathbf{1} = [\underbrace{1, 1, \cdots, 1}_{2^i}].$$

A 2^i-time enlarged pattern of X and $1/2^i$-time reduced pattern of Y are defined as $[X(0)\mathbf{1}, X(1)\mathbf{1}, \cdots, X(2^{n-i} - 1)\mathbf{1}]$ and $[U', \mathbf{0}]$, respectively, where

$$\mathbf{1} = [\underbrace{1, 1, \cdots, 1}_{2^i}], \ U' = [Y(0), Y(2^i), \cdots, Y(2^n - 2^i)]$$

$$\mathbf{0} = [\underbrace{0, 0, \cdots, 0}_{2^n - 2^{n-i}}].$$

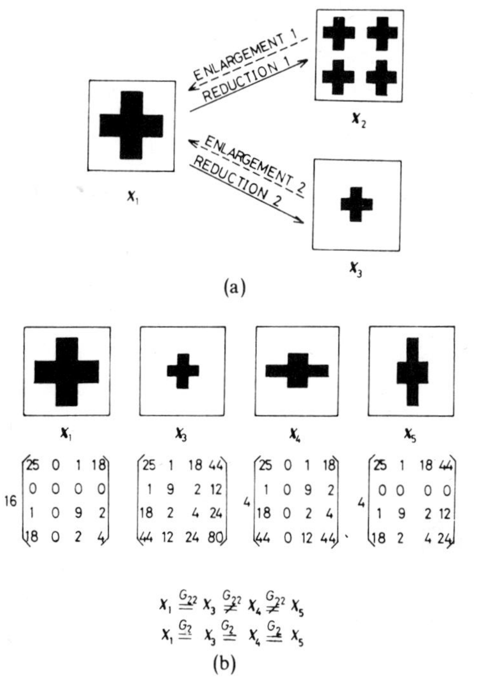

Fig. 6. Explanation of G'_2 and G'_{22}. (a) Two different definitions of enlargements and reductions. (b) G'_2-equivalent patterns and G'_{22}-equivalent patterns.

For instance, the 2-time enlarged pattern of $X = [12000000]$ is $[11220000]$, and the 2^2-time enlarged pattern of it is $[11112222]$. Reversely the 1/2-time reduced pattern of $Y = [11223300]$ is $[12300000]$ and so on. Under the definitions there is obviously no enlargement and reduction-invariant power spectrum, because the average power of an input pattern itself is changed by the transformations. Therefore we give up to obtain an enlargement and reduction-invariant power spectrum and try to get enlargement and reduction-invariant functions from the second-degree terms.

Let the 2^i-time enlarged pattern of 2^i-time enlargeable pattern X be Y, then Y is a $1/2^i$-time reducible pattern and $1/2^i$-time reduced pattern of Y is X. X and Y can be expressed as $X = [U, \mathbf{0}]$, $Y = U \otimes \mathbf{1}$, where

$$U = [X(0), X(1), \cdots, X(2^{n-i}-1)],\ \mathbf{0} = [\underbrace{0, 0, \cdots, 0}_{2^n - 2^{n-i}}]$$

$$\mathbf{1} = [\underbrace{1, 1, \cdots, 1}_{2^i}].$$

The $(\mathrm{WHT})_h$ of X is as follows:

$$\begin{aligned} B_X(n) &= (1/N)X \cdot H(n) \\ &= (1/N)[U, \mathbf{0}] \cdot (H(i) \otimes H(n-i)) \\ &= (1/N)[U, \mathbf{0}] \\ &\quad \cdot \left(\begin{array}{c|c} \begin{array}{cc} H(n-i), & H(n-i) \\ H(n-i), & -H(n-i) \end{array} \ \vdots & H(n-1) \\ \hline H(n-1) & -H(n-1) \end{array}\right) \\ &= (1/N)[U \cdot H(n-i), U \cdot H(n-i), \cdots, U \cdot H(n-i)] \\ &= (1/N)\mathbf{1} \otimes [U \cdot H(n-i)] \end{aligned} \tag{56}$$

where

$$\mathbf{1} = [\underbrace{1, 1, \cdots, 1}_{2^i}].$$

From (56) we have

$$B_X(k) = \begin{cases} \text{value of the } k\text{th element of } (1/N)U \cdot H(n-i), & 0 \le k \le 2^{n-i} - 1 \\ B_X(l), & 2^{n-i} \le k \le 2^n - 1 \end{cases} \tag{57}$$

where l is k modulo 2^{n-i}. Regarding Y, from (31) we have

$$B_Y(n) = (2^i/N)(U \cdot H(n-i)) \otimes (\underbrace{1, 0, \cdots, 0}_{2^i}). \tag{31'}$$

(56) and (31′) lead that

$$B_Y(k) = \begin{cases} 2^i B_X(l), & \text{if } k = 2^i l \\ 0, & \text{otherwise} \end{cases} \tag{58}$$

where $l = 0, 1, \cdots, 2^{n-i} - 1$.

These aspects are well-illustrated by utilizing the matrix $C_X(3)$ defined already. Let X be a 2-time enlargeable pattern, and let Y the 2-time enlarged pattern of it. Then we have

$$C_Y(3) = \begin{pmatrix} B_Y(4), & 0, & B_Y(6), & 0 \\ B_Y(2), & 0, & 0, & 0 \\ 0, & 0, & 0, & 0 \end{pmatrix} = \begin{pmatrix} 2B_X(2), & 0, & 2B_X(3), & 0 \\ 2B_X(1), & 0, & 0, & 0 \\ 0, & 0, & 0, & 0 \end{pmatrix}$$

$$B_Y(0) = 2B_X(0),\ B_X(0) = B_X(4),\ B_X(1) = B_X(5),$$

$$B_X(2) = B_X(6),\ B_X(3) = B_X(7),\ \sum_{k=0}^{7} B_Y^2(k) = 2 \sum_{k=0}^{7} B_X^2(k). \tag{59}$$

Let X be a 2^2-time enlargeable pattern, and let Z the 2^2-time enlarged pattern of it. Then we have

$$C_Z(3) = \begin{pmatrix} B_Z(4), & 0, & 0, & 0 \\ 0, & 0, & 0, & 0 \\ 0, & 0, & 0, & 0 \end{pmatrix} = \begin{pmatrix} 4B_X(1), & 0, & 0, & 0 \\ 0, & 0, & 0, & 0 \\ 0, & 0, & 0, & 0 \end{pmatrix}$$

$$B_Z(0) = 4B_X(0),\ B_X(0) = B_X(2) = B_X(4) = B_X(6),$$

$$B_X(1) = B_X(3) = B_X(5) = B_X(7),\ \sum_{k=0}^{7} B_Z^2(k) = 4 \sum_{k=0}^{7} B_X^2(k). \tag{60}$$

(59) and (60) are obtained from (57) and (58).

Furthermore, it follows that

$$P_s(X) = \sum_{j=0}^{s-1} P_j(X) \tag{61}$$

where X is a 2^i-time enlargeable pattern, and $n - i + 1 \le s \le n$. The following formulas are asymptotically led from (61):

$$P_{n-i+1}(X) = \sum_{j=0}^{n-i} P_j(X)$$

$$P_{n-i+2}(X) = \sum_{j=0}^{n-i+1} P_j(X) = 2P_{n-i+1}(X)$$

$$\vdots$$

$$P_n(X) = 2^i P_{n-i+1}(X). \tag{62}$$

From (58) we have

$$P_0(Y) = 2^{2i}P_0(X),\ P_s(Y) = \begin{cases} 2^{2i}P_{s-i}(X), & i+1 \le s \le n \\ 0, & 1 \le s \le i \end{cases}$$

$$\sum_{s=0}^{n} P_s(Y) = 2^i \sum_{s=0}^{n} P_s(X). \tag{63}$$

where Y is the 2^i-time enlarged pattern of X. In the case when $n = 3$ and $i = 1$, we obtain

$$\begin{aligned} P_X(3) &= [P_0(X), P_1(X), P_2(X), P_3(X)] \\ &= [P_0(X), P_1(X), P_2(X), P_0(X) + P_1(X) + P_2(X)] \\ P_Y(3) &= [P_0(Y), 0, P_2(Y), P_3(Y)] \\ &= [4P_0(X), 0, 4P_1(X), 4P_2(X)]. \end{aligned} \tag{64}$$

These aspects are illustrated in Fig. 6 in the two-dimensional case when $n = m = 3$. The values in Fig. 6 are the P_{ij} multiplied by 64^2 for simplicity. The ones and zeros are blacked and blanked, respectively, and we should notice that zeros are used in Fig. 6 instead of negative ones.

Let $P'(n) = [P'_0, P'_1, \cdots, P'_n]$ be defined by

$$P'_0 = P_0^2/A^2,\ P'_s = P_s\left(P_s - \sum_{j=0}^{s-1} P_j\right)\bigg/A^2 \tag{65}$$

where $s = 1, 2, \cdots, n$, and $A = \sum_{j=0}^{n} P_j$. Then the 2^i-time enlargement and $1/2^i$-time reduction of an input pattern cause shifts on $P'(n)$ by i elements toward the right and the left, respectively, except P'_0. P'_0 is invariant to the transformations. Therefore cyclic shift, enlargement, and reduction-invariant functions are obtained as

$$Q'_0 = P'_0,\ Q'_j = F_j(P'_1, P'_2, \cdots, P'_n), \qquad 1 \le j \le n. \tag{66}$$

We note that $\{Q'_j\}$ ($j = 0, 1, \cdots, n$) is not a power spectrum. If we treat an input pattern whose elements have only the values of positive one and negative one and redefine enlargement and reduction by changing zero for negative one, then we can get cyclic shift, enlargement, and reduction-invariant power spectrum. In the two-dimensional case $P'(n, m)$ is defined by

$$P'_{st} = P_{st}^2\left(P_{st} - \sum_{i<s} P_{it}\right)\left(P_{st} - \sum_{j<t} P_{sj}\right)\bigg/A^4 \tag{67}$$

where $s = 0, 1, \cdots, n$; $t = 0, 1, \cdots, m$; and

$$A = \sum_{i=0}^{n}\sum_{j=0}^{m} P_{ij}.$$

Let G'_2 and G'_{22} be defined in the similar method to those of the foregoing section. Then the $\{Q'_{ij}\}$, defined by replacing P_{ij} in (46) and (49) with the P'_{ij}, are $G_1 \otimes G'_{22}$-invariant and $G_1 \otimes G'_2$-invariant, respectively. $G_1 \otimes G'_{22} \otimes G_3$-invariant functions and $G_1 \otimes G'_2 \otimes G_3$-invariant functions can be also obtained in the same way as those of the foregoing section. The aspects of these group-invariant functions are shown in Table I.

CONCLUSION

A composing process of some transformation group-invariant functions and the application to the $(\mathrm{WHT})_h$ power spectrum have been presented. The main idea is to find a permutation group on a family of some functions caused by the transformation operating on an input pattern. Using the process, the $(\mathrm{WHT})_h$ power spectra are developed to be unchangeable by cyclic shifts, enlargements, reductions, rotations by multiples of 90°, and symmetry transformations. Using polar coordinates (r, θ) instead of orthogonal ones, we can define any rotations besides rotations by multiples of 90°. Then every rotation may be regarded as cyclic shifting toward θ-direction and enlargement and reduction as exponential shifting toward r-direction. With of this convenience the new problem arises, that is, how to define shiftings of the elements of an input pattern. We have the alternative of orthogonal or polar coordinates, complying with needs.

The above discussion has been restricted to real-valued input patterns and functions, but they can be extended to complex-valued inputs and functions [14], [15]. Although an input pattern is a vector (one-dimensional pattern) or a matrix (two-dimensional pattern) in this paper, the above analysis can be easily generalized to any number of dimensions of an input pattern, and furthermore, from the discrete input pattern to the continuous one.

Since the power spectra may be regarded as a proper subset of a group-invariant complete system, it cannot perfectly make distinctions between the group-nonequivalent patterns. For example, $P(n)$ is cyclic shift invariant and also sign exchanging invariant at the same time. But generally it seems almost impossible to make up a group-invariant complete system of hardware when the number of functions in the system is taken into account. Therefore it becomes very important to select appropriately a subset of the system. We mainly adopted power spectra, but it is also possible to adopt any other functions besides power spectra. As seen in the last section there is no group-invariant power spectrum in some cases. This depends on the transformation group. Although this paper is limited to the applications to the $(\mathrm{WHT})_h$ power spectra on a discrete input space, we hope the discussions will enhance the further research of group-invariant functions. The methods developed above are applicable to feature extraction for geometrical patterns. After extracting features, it is needed to recognize the patterns by using the features. Evidently people use more information, such as topology and context, rather than power spectra in recognizing an object. Under what transformation group do they regard that the objective pattern classes are invariant? This is not well-known yet. The group could help to design pattern recognition systems. This is left for future work.

ACKNOWLEDGMENT

The authors are grateful to Dr. Y. Kametaka for introducing them into topological analysis; to Prof. M. Kimura for sending papers; to Dr. M. Shimura, Prof. K. Tanaka, and the staff in the Electrical Communication Engineering Laboratory for their helpful suggestions during discussions.

REFERENCES

[1] J. L. Walsh, "A closed set of normal orthogonal functions," *Amer. J. Math.*, vol. 55, pp. 5–24, 1923.
[2] R. E. A. C. Paley, "On orthogonal matrices," *J. Math. Phys.*, vol. 12, pp. 311–320, 1933.
[3] N. J. Fine, "On the Walsh functions," *Trans. Amer. Math. Soc.*, vol. 65, pp. 372–414, 1949.
[4] H. F. Harmuth, *Transmission of information by orthogonal functions* New York: Springer-Verlag, 1972.
[5] M. Kimura, "Walsh functions and their applications," *JIEEJ. Proc.*, vol. 91, no. 4, pp. 619–622, Apr. 1971.
[6] M. Minsky and S. Papert, *Perceptrons—An introduction of computational geometry*. Cambridge, MA: MIT, 1969.
[7] Y. Uesaka, "Analogue perceptrons—On additive representation of functions," *Trans. of IECEJ.*, vol. 54-C, no. 7, pp. 586–593, July 1971.
[8] N. Ahmed, K. R. Rao, and A. L. Abdussatar, "BIFORE or Hadamard transform," *IEEE Trans. Audio Electroacoust.*, vol. AU-19, no. 3, pp. 225–234, Sept. 1971.
[9] N. Ahmed and K. R. Rao, *Orthogonal transforms for digital signal processing*. New York: Springer-Verlag, 1975.
[10] H. C. Andrews and B. R. Hunt, *Digital Image Restoration*. Englewood Cliffs, NJ: Prentice-Hall, 1977.
[11] N. A. Alexandridis and A. Klinger, "Walsh orthogonal functions in geometrical feature extraction," *Proc. 1971 Sym. on Applications of Walsh Functions*, Naval Research Lab., Washington, DC, pp. 18–25, Apr. 1971.
[12] B. Arazi, "Hadamard transforms of some specially shuffled signals," *IEEE Trans. Acoust., Speech, Signal Processing*, vol. ASSP-24, pp. 580–583, Dec. 1975.
[13] N. Ahmed, H. H. Schreiber, and P. V. Lopresti, "On notation and definition of terms related to a class of complete orthogonal functions," *IEEE Trans. Electromag. Compat.*, vol. EMC-15, no. 2, pp. 75–80, May 1973.
[14] K. Yamashita and H. Hama, "Walsh functions in feature extraction for geometric patterns (Part I)," *Papers of Technical Group on Pattern Recognition and Language, IECEJ.*, vol. PRL75-17, pp. 73–84, June 1975.
[15] H. Hama and K. Yamashita, "Walsh functions in feature extraction for geometric patterns (Part II)," *Papers of Technical Group on Pattern Recognition and Language, IECEJ.*, vol. PRL75-18, pp. 85–96, June 1975.
[16] H. Hama and K. Yamashita, "Families of partial functions for Perceptrons," *Trans. of IECEJ.*, vol. 57-A, no. 7, pp. 481–488, July 1974.

Editor's Comments on Paper 21

21 REIS, LYNCH, and BUTMAN
Adaptive Haar Transform Video Bandwidth Reduction System for RPV's

HAAR TRANSFORM

Similar to other discrete transforms, the Haar transform (HT) is orthogonal and separable. It has an efficient algorithm and has been applied to data compression of RPV images (Paper 21; Lynch and Reis, 1976*a*). Unlike other transforms, HT has both global and local (some elements are zeros) functions. These functions, therefore, represent both global and local properties respectively of sequence. It has, however, some irrational numbers, such as $(\sqrt{2})^n$, as multipliers. By adopting a rationalization process, these numbers are converted to integer powers of two. This modification results in what is called the rationalized HT (RHT) (Lynch and Reis, 1976*b*).

Because of the local nature of a majority of its basis functions, HT can code the edges much better than other transforms. An example of this is edge extraction as reported by Dixit (1980). Several relations between the HT and the WHT have been developed by Fino (see Paper 16). HT has also been applied to alphanumeric character recognition (see Paper 32).

REFERENCES

Dixit, V. V., 1980, Edge Extraction through Haar Transform, 14th Asilomar Conference on Circuits, Systems, and Computers, Pacific Grove, Calif., pp. 141–143.

Fino, B. J., 1972, Relations between Haar and Walsh/Hadamard Transforms, *IEEE Proc.* **60:**647–648.

Lynch, R. T., and J. J. Reis, 1976*a*, *Class of Transform Digital Processors for Compression of Multidimensional Data,* U.S. Patent 3981443.

Lynch, R. T., and J. J. Reis, 1976*b*, *Haar Transform Image Coding,* IEEE 1976 National Telecommunications Conference, Dallas, Tex., pp. 44.3-1–44.3-5.

Reprinted from *Photo-Opt. Instrum. Eng. Soc. Proc.* **87:**24-35 (1976)

ADAPTIVE HAAR TRANSFORM VIDEO BANDWIDTH REDUCTION SYSTEM FOR RPV's

J. J. Reis
R. T. Lynch
J. Butman*
Northrop Research and Technology Center
3401 West Broadway
Hawthorne, California 90250

Abstract

A practical video bandwidth compression system is described in detail. Video signals are Haar transformed in 2 dimensions and the resulting transform coefficients are adaptively filtered to achieve reduced transmission data rates to less than 1 bit/pel, while maintaining good picture quality.

Error detection and compensation is included into the transmission bit structure.

Introduction

The wideband video link of Remotely Piloted Vehicles (RPV's) is subject to both friendly and hostile interference. One way of minimizing the effects of jamming is to reduce the information bandwidth as much as possible before it is transmitted by a spread spectrum modem.

Northrop's approach for solving the RPV real time video bandwidth reduction problem is based upon a digital implementation of a 2-D fast rationalized Haar transform with a combination zonal and adaptive threshold compression algorithm. This concept is a natural outgrowth of Northrop's experience in designing and developing video processing systems based on the Haar transform. Figure 1 is a block diagram of the airborne portion of the proposed system. Input video, obtained from a TV camera, is preemphasized to reduce quantization noise before being digitized to 8 bits and transformed in two dimensions by using a fast rationalized version of the Haar transform. This transformation process maps a two-dimensional spatial picture of dimension 16 pels by 16 pels into a block of transform coefficients of dimension 16 x 16.

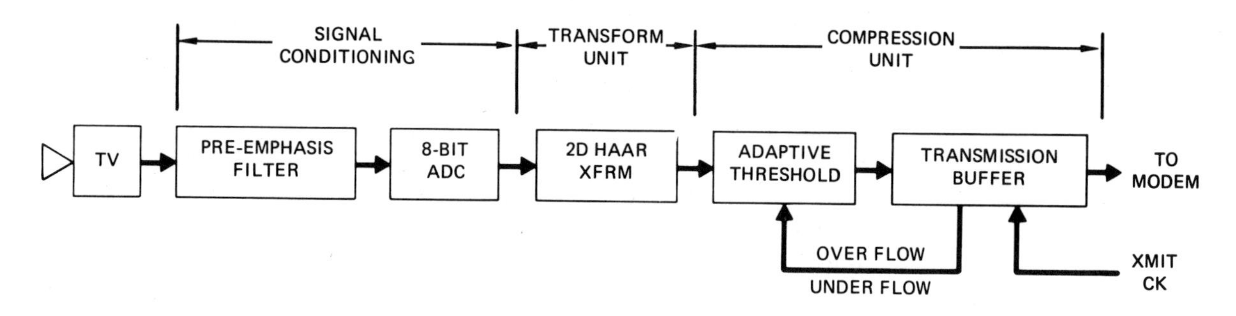

Figure 1. Block Diagram of Airborne Portion of the Northrop Real-Time Video Bandwidth Reduction System.

The purpose of the Haar transform is to transform a picture into a form where most of the picture energy is represented by a relatively small number of transform coefficients. These transform coefficients are then processed by a zonal filter and an adaptive threshold filter. The zonal filter always selects certain low order coefficients for transmission. The adaptive threshold filter selects from the remaining coefficients those coefficients whose amplitudes are greater than the threshold level. The number of coefficients to be selected by the adaptive threshold algorithm is determined by the transmission rate of the output modem and picture activity.

In addition to the spatial compression provided by the transformation and adaptive filtering, an additional 8:1 temporal compression is achieved by processing only 1/8 of each field as suggested by researchers at the Naval Undersea Center (NUC) in San Diego.[1,4] This 8:1 reduction is achieved by processing the picture in vertical stripes, 32 pels-wide by 240** lines long. Each successive field uses the next 32 pel wide stripe across the horizontal scan. A complete frame is sent after 8 fields. Because a frame store is not required, this method of obtaining temporal compression without introducing flicker in the reconstructed image permits a significant hardware reduction to be realized. Although this method produces temporal skewing of the reconstructed video data, under most operating conditions this artifacting should not be overly objectionable.

* Northrop Electronics Division, One Research Park, Palos Verdes Peninsula, California 90274
** 16 additional lines are taken up by the Vertical Interval

The airborne modem transmits the selected transform coefficients to the ground unit where they are inverse filtered and transformed and displayed on a TV monitor. Figure 2 shows the block diagram for the ground terminal of the real-time video bandwidth reduction system. Transform coefficients from the output of the receiver modem are accumulated by the input receiver buffer which arranges them into data blocks. The inverse compression unit takes these data blocks and expands them into transform coefficient blocks of dimension 16 x 16, filling in a zero for each coefficient discarded prior to transmission by the airborne compression unit. The 2-D inverse Haar transform re-maps these transform domain coefficients back into the spatial domain where they are transferred to the frame store.

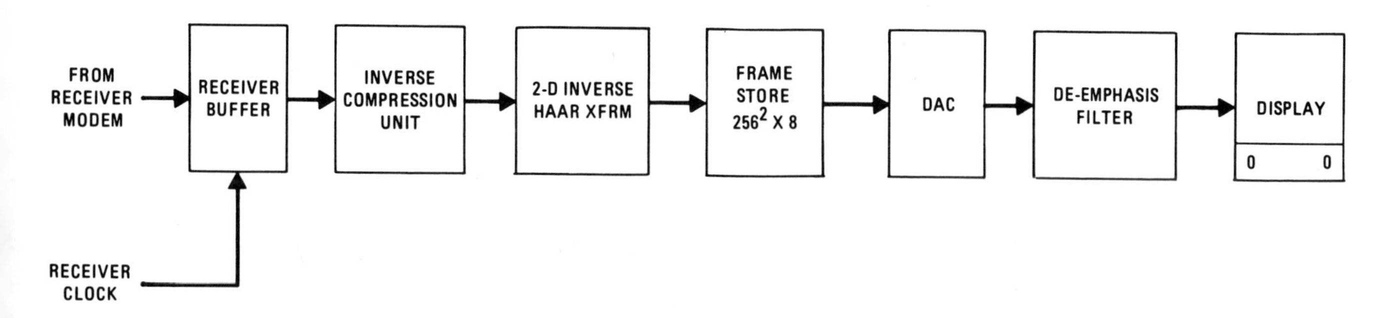

Figure 2. Block Diagram of Ground Terminal of the Northrop Real-Time Video Bandwidth Reduction System.

Airborne Subsystem

2-Dimensional Haar Transform Unit

The Northrop approach to developing a small, real time video bandwidth reduction system for RPV's utilizes a two-dimensional fast rationalized Haar transform to convert spatial domain video samples into transform domain coefficients. This transformation is given by equation 1 through 3.

$$B = HAH^T \tag{1}$$

where

B = Square Matrix of Transform Coefficients
A = Square Matrix of Input Spatial Video Samples
H = One-Dimensional Rationalized Haar Transform
T = Transpose

The one-dimensional Haar transform which is the basis for the 2-D transforms, has an extremely sparse matrix and a fast computational form as shown in Equation 2 and 3 .

$$H = H_4H_3H_2H_1 \tag{2}$$

where H, H_1, H_2, H_3, H_4 and P are defined by equation set 3.

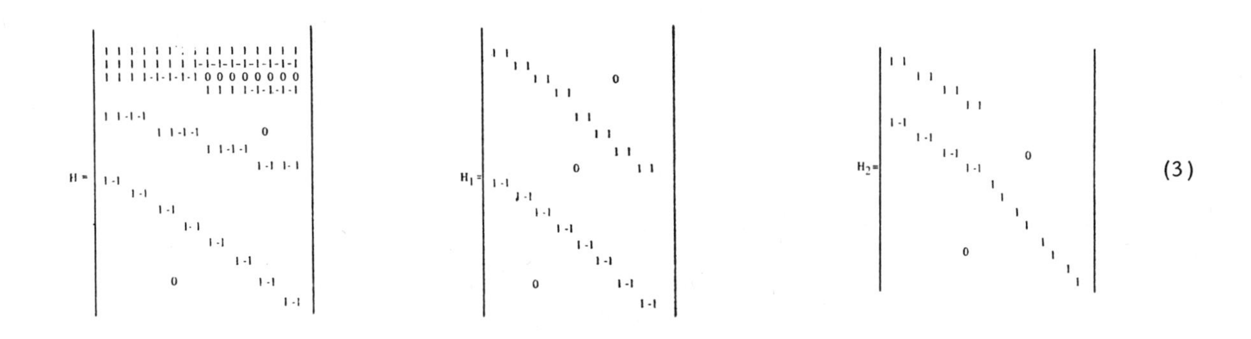

(3)

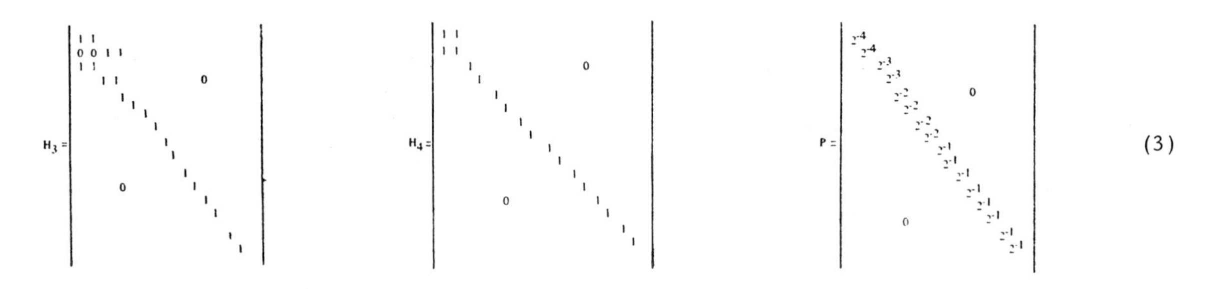

(3)

and the inverse transform is given by equation 4.

$$H^{-1} = H^{T}P = H_1^{T}H_2^{T}H_3^{T}H_4^{T}P \qquad (4)$$

Figure 3 shows the hardware required to mechanize the one-dimensional fast rationalized Haar transform in pipeline form.[2] It is presented here to illustrate how the fast algorithm actually works and to show that transform coefficients are available in real time for processing by the compression algorithm before the entire 16 point transform has been completely computed.

Operation of the pipeline transform unit is as follows. Input video data from the ADC enters the 1-D transform unit as shown in Figure 3. The first video sample is stored into latch L_1. The second video sample is presented to the B input of the first adder, Σ_1, whereupon the sum and difference of these first two samples are computed. The sum is temporarily stored in the second stage data latch, L_2, while the difference is outputted as transform coefficient #9 (see equation 3, definition of H_1). The 3rd video sample is stored into data latch, L_1, and the 4th input data sample is presented to the B input of the first adder, Σ_1, whereupon the sum and difference of these two samples are computed. The difference is outputted as transform coefficient #10 and the sum is fed to the B input of the second stage adder, Σ_2, whereupon the sum and difference of this term and the term in latch L_2, are computed. The (stage 2) difference is outputted as transform coefficient #5 and the sum is stored in the third stage latch, L_3. This process continues in a similar manner for each of the input video samples until all of the transform coefficients are computed.

The pipeline structure of Figure 3 is capable of operating at sample rates in excess of 15 MHz, which for word lengths of 8 bits, represents a data throughput rate of 120 Mbps. Northrop has constructed three real-time video processors using this pipeline architecture. The first real-time processor[3] incorporates an adaptive threshold compression algorithm. A second unit, built in the Spring of 1975, is an 8-bit arithmetic precision 1-D processor with an adaptive N-largest* compression system. This system will be demonstrated at this SPIE Technical Symposium. The third system, built in late 1975 and now undergoing debug, is a 2-D 8 bit Haar processor incorporating various signal processing algorithms.

The input data rate for one identified application of an RPV bandwidth reduction system,[1] is (8 bits/sample) (4.8 M samples per sec) = 38.4 Mbps. The average data rate is further reduced to 4.8 Mbps by 8:1 frame rate reduction, which is achieved by processing only 1/8 of each horizontal line per field in

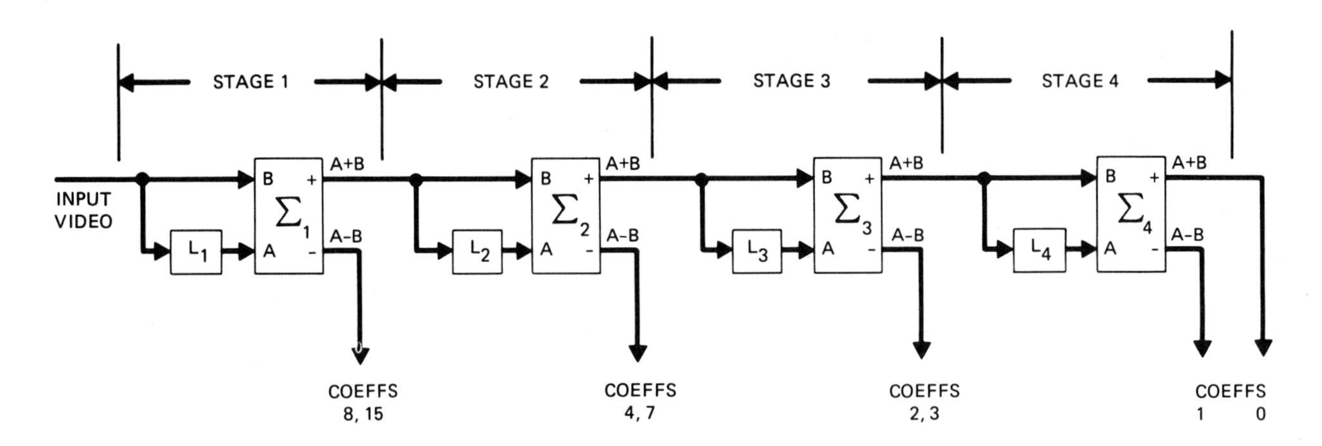

Figure 3. Block Diagram of Real-Time Fast Rationalized Haar Transform Unit (1-D).

* The largest "N" coefficients are selected for transmission where N, usually 1 or 2, is selected by front panel switches.

accordance with the concept developed by the researchers at NUC. The format used to obtain this frame rate reduction is shown in Figure 4. By reducing the data rate in this manner, the pipeline structure shown in Figure 3 can be telescoped into a more efficient architecture. This results in a considerable savings in hardware and power consumption. The new architecture is shown in Figure 5. This architecture is basically a highly specialized processor which consumes less than 5 watts of power and has been optimized to perform 2-D Haar transforms. Read-only memories (ROM's) control the operations of the adder/subtractor, data latches, shifters, and memory.

The 8:1 frame rate reduction requires 32 samples to be taken per horizontal line. Since Northrop uses a 16 point transform, two transforms per horizontal line must be computed as shown in Figure 4. The

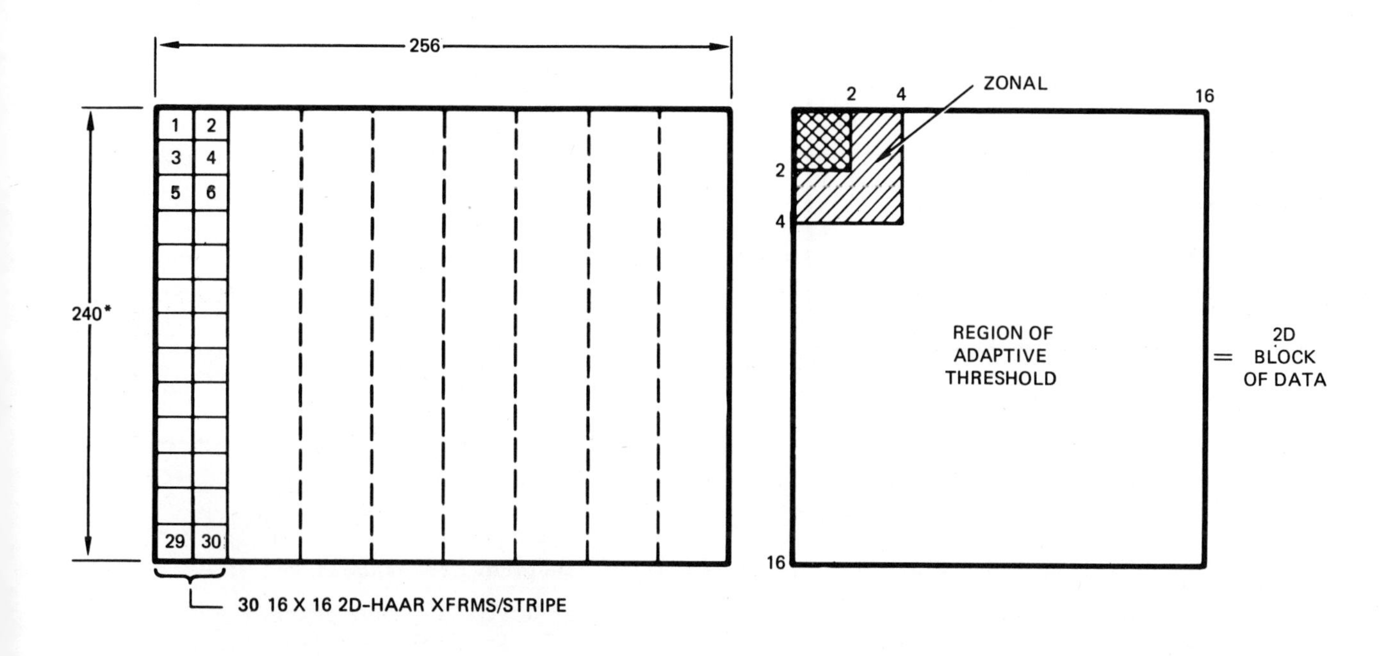

Figure 4. Scanning Format for Adaptive Compression.

processor of Figure 5 is programmed to take advantage of the fact that output transform coefficients are available in almost real time during the horizontal transformation process. This enables us to begin the computation of the vertical part of the 2-D transform before the horizontal transformation is completed. This feature of the hardware architecture allows us to reduce our storage requirements to 5 lines of 32 points each. Present methods of computing 2-D transformations of global transform (Walsh, Cosine, Fourier, etc.) require storage of 16 lines of data.

The details on how the vertical transforms are multiplexed between the horizontal transforms are relatively straightforward and will not be discussed here.

Hybrid Zonal/Adaptive Threshold Compression Unit

The compression unit is shown in Figure 6. The input video data from the Haar transform unit has been pre-whitened, digitized and transformed into coefficients in the Haar transform domain before entering the compression unit.

The compression unit consists of a cascade of operations. The transformed data is converted from 2's complement notation to magnitude and sign notation by an absolute value circuit. The converted data enters the threshold filter where only the largest coefficients are coded for transmission. The threshold level for a given stripe is the threshold value for the previous stripe plus a correction term which depends upon how full or empty the first-in-first-out (FIFO) buffer is. The correction term causes the total number of transmitted coefficients per stripe to be increased or decreased so that the FIFO will be (nearly) half full at the end of the next stripe. Provisions are provided to prevent buffer overflow or underflow by decreasing or increasing coefficients accordingly. The FIFO is large enough to store more than 1 1/4 full stripes at the highest data rate specified for the system (1.6 Mbps); therefore, buffer overflow will not be a problem.

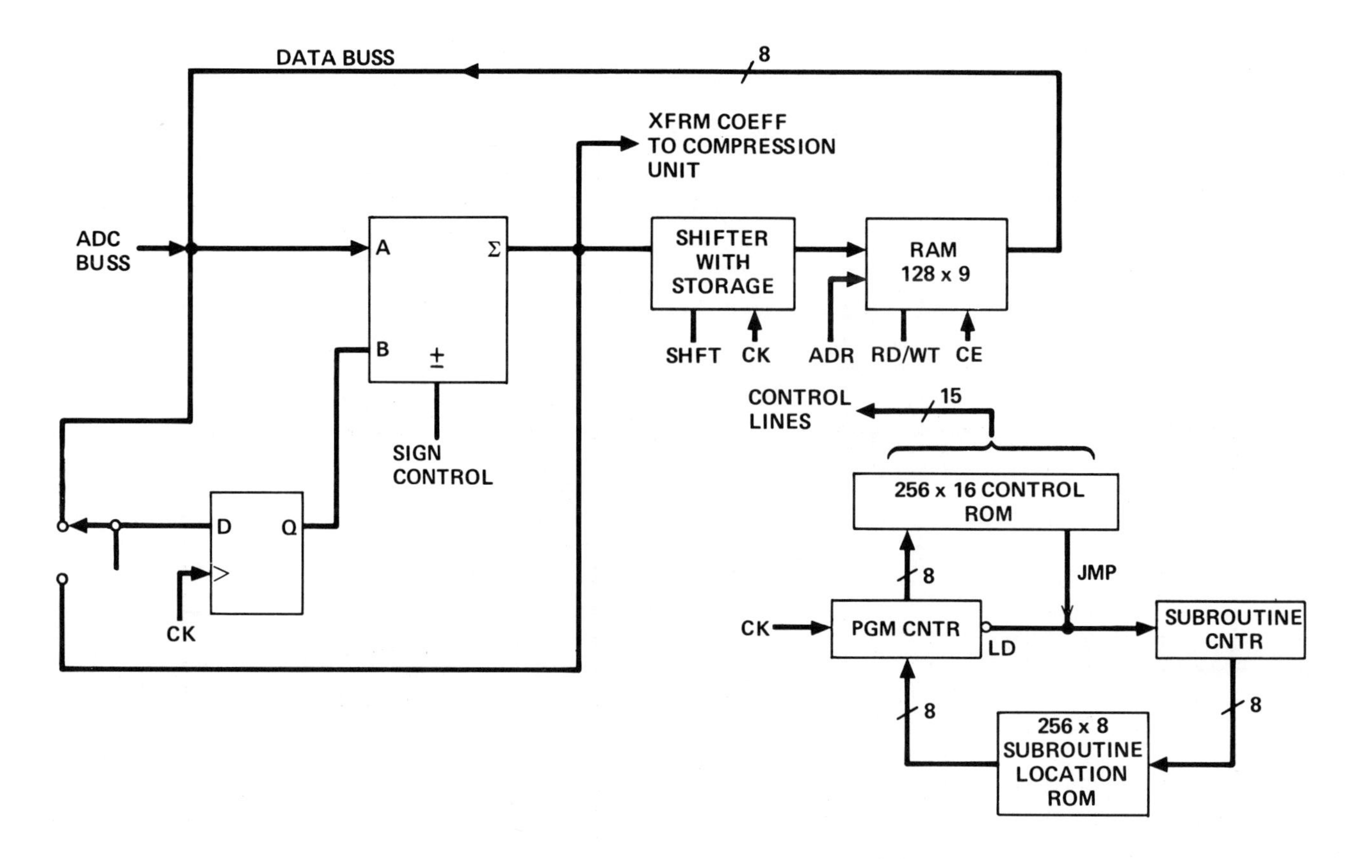

Figure 5. Block Diagram of 2-D Haar Transform Processor.

A fixed number of low frequency coefficients (zonal) in the upper left hand corner of the transform domain in Figure 4, are always sent and are not subject to thresholding. These zonal coefficients are almost always large and would almost always be selected for transmission by the adaptive threshold algorithm. It is therefore more efficient to encode these coefficients as a block rather than spend costly addressing bits to individually code them. The zonal coefficients are not companded. They are stored in order of arrival directly into the zonal burst buffer where they are accumulated for transmission as a continuous data block.

Transmission Format

The overall transmission format for the adaptive compression algorithms is shown in Figure 7. Two consecutive 16 point transforms are taken during each horizontal line of each stripe and 15 transforms are taken in the vertical direction for each stripe. There are 8 stripes per frame in accordance with the NUC temporal reduction concept. Figure 7 also shows that all the zonal coefficients for the entire stripe are sent as a fixed length contiguous block of data, followed by variable length data blocks composed of variable coefficients. Since most of the picture energy is contained in the low frequency zonal coefficients, we must make certain that channel errors do not cause them to be interpreted as variable coefficients.

To achieve this, the zonal coefficients are sent as a fixed block of ordered data to make them less sensitive to channel errors. The zonal coefficients are represented by a fixed number of bits and their relative position is always known with respect to frame and field sync. Channel errors can therefore only affect their amplitude, and not their location within the transform block. Because these low frequency terms are extremely important to the subjective quality of a video image, a parity bit is added to each zonal term. The format developed for error protection of the zonal coefficients is shown in Figure 8. The parity bit should be computed only over the 4 or 5 most significant bits, since errors in the lower bits are much less critical to the output picture quality. By including the parity bit, the majority of the transmission errors in the zonal coefficients can be detected. Although sufficient information is not available to correct detected errors in the zonal terms, errors can be compensated by replacement with the appropriate (possibly motion compensated) zonal coefficients from the previous frame. Error compensation for the zonal coefficient takes place in the ground terminal.

Non-zonal adaptive coefficients are transmitted in the format shown in Figure 9. Each companded coefficient is sandwiched between its 8 bit location address. This technique is used so the end-of-block sync word cannot be easily confused with the variable coefficient word. Note that each coefficient is absolutely addressed and each coefficient is represented by exactly 13 bits, including address, so by counting bits, address and amplitude information cannot be confused even when channel errors are present. The variable coefficients are transmitted in the order that the transform unit computes them so it is possible to

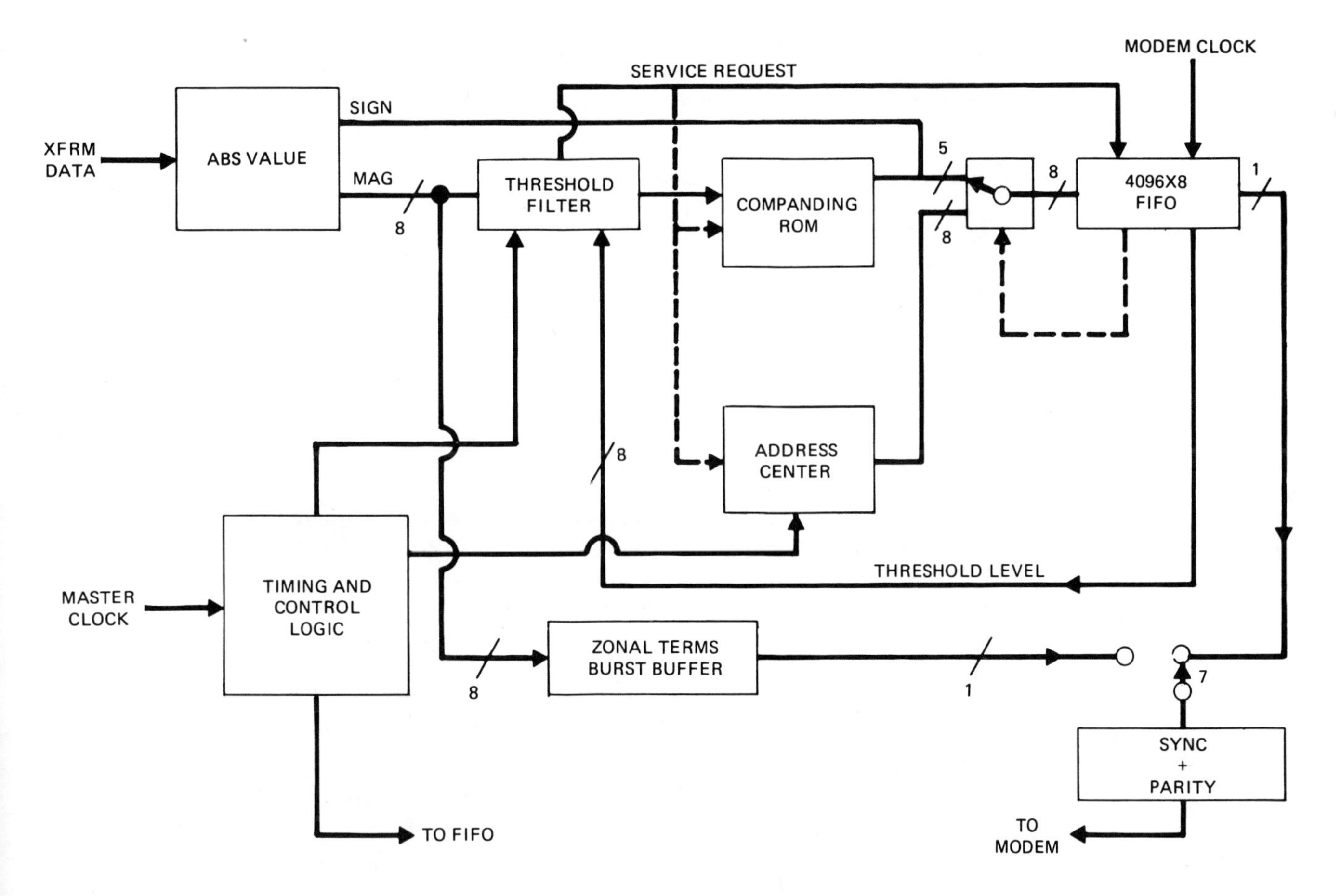

Figure 6. Block Diagram of Video Data Compression Unit.

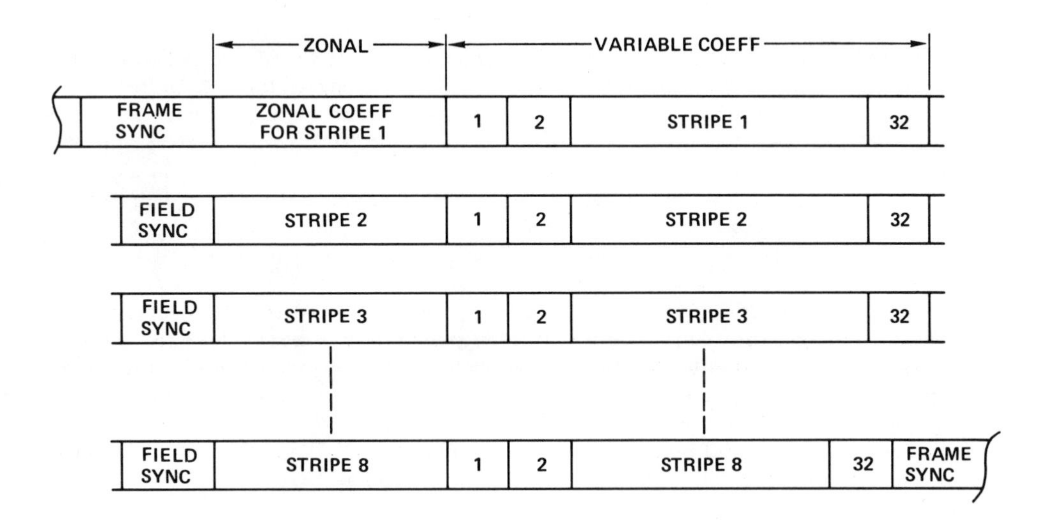

Figure 7. Transmission Format for Adaptive Compression Unit.

monitor the address of the coefficients being received on the ground to detect transmission errors in the address structure. Whenever errors are detected in the coefficients or address structure, these coefficients can be set to zero to minimize their effect on the reconstructed spatial picture. The presence of the onboard transmission buffer gives the Northrop system the opportunity to minimize the effects of burst errors on the received data by scrambling the order of the transmitted bits., i.e., spread out adjacent bits in time so that burst errors will not completely destroy several consecutive blocks of data.

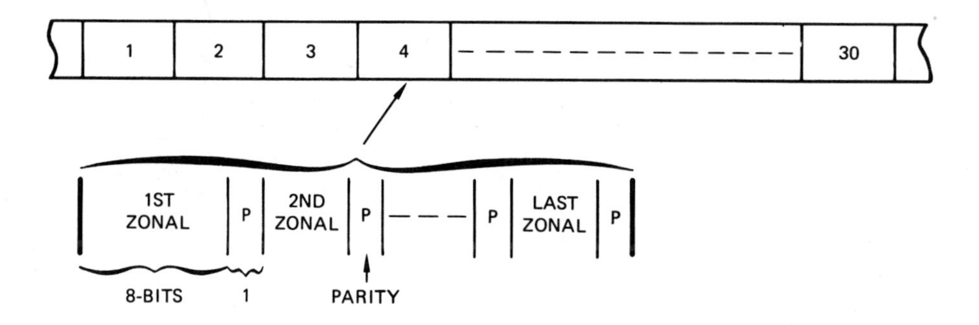

Figure 8. Format Detail for Zonal Coefficients.

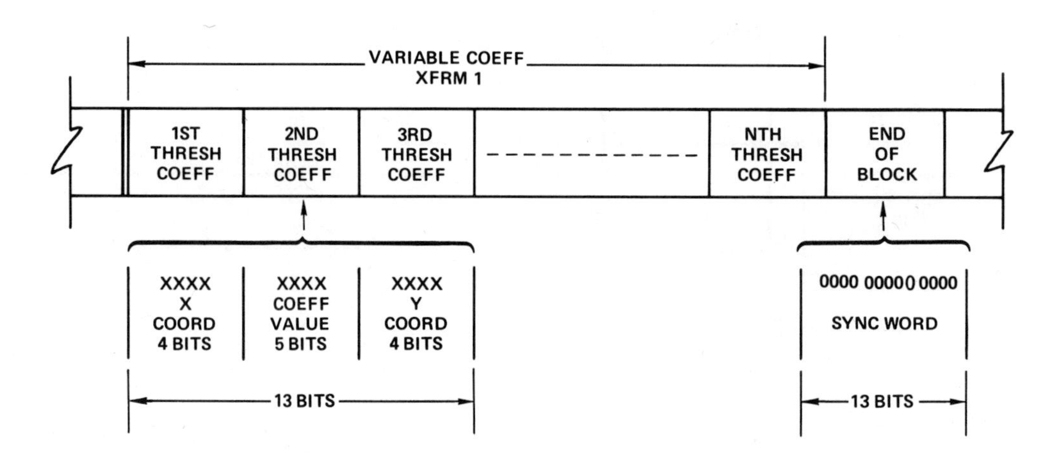

Figure 9. Format Details for Variable Coefficients.

Summary of Airborne Subsystem

The described 2-D Haar transform video bandwidth reduction system is designed to meet the size, weight and power constraints typical of real world RPV video communications links. This system also provides some inherent flexibility. For example, the frame rate can easily be reduced from the 7.5 frames/sec of the present design to 3.75 frames/sec by merely changing a few ROM's in the airborne unit so only one 16 point transform per line is computed instead of two per line. The frame rate can also be increased from 7.5 frames/sec to 12 frames/sec by taking three 16 point transforms per line instead of two per line. This modification of the transmission format can again be accomplished by changing a few ROM's in the airborne unit. In this manner it is possible to obtain almost any frame rate desired.

In designing the Northrop bandwidth reduction system, it was assumed that the input video has a continuous raster scan format. Though the basic Northrop video processor concept does not require that the input data have this format, the current design of the compression unit was influenced by the raster scan input. Northrop is aware that solid state sensors with snapshot capability are being developed and in fact has an IR&D program which has as its objectives the design and fabrication of visible and infrared CID and CCD sensor arrays with random access and non-destructive readout capability. The use of a snapshot mode sensor would eliminate the temporal skewing inherent in the stripe per field format.

Although the present design can operate with a raster scanned solid-state sensor array in a snapshot mode, when suitable random access sensors arrays become available, allowing greater flexibility in address techniques, the present bandwidth reduction system can be reconfigured to significantly reduce the size, weight and power consumption of the hardware.

The Northrop airborne subsystem provides significant flexibility to change data rates (compression ratios) on command from the ground terminal. In response to a (ground) command for a particular clock rate, the threshold in the compression unit will automatically increase or decrease as a function of how fast the transmission buffer empties. Thus, the adaptive compression scheme used in the airborne subsystem will respond to clock rates of almost any frequency by adaptively and automatically changing the threshold to achieve the required transmission rate.

Ground Subsystem

A detailed block diagram for the ground terminal is shown in Figure 10. The ground terminal consists of an input receiver data buffer, error compensation circuitry and an inverse compressor which feeds the inverse 2-D Haar transform unit. The 2-D inverse transform unit converts transform domain coefficients back into spatial domain video samples which are stored in the Frame Store unit to provide flicker-free display of the temporally compressed video. The output of the Frame Store is fed through a digital-to-analog converter (DAC) and through the de-emphasis filter to the TV display.

The following sections give a more detailed description of the operation of each of the major functional blocks in the ground subsystem.

Receiver Data Buffer, Error Compensation and Inverse Compression Unit

This unit serves to accumulate data from the receiver modem into 16 x 16 transform coefficient blocks suitable for inverse 2-D transformation. Provisions are included to error-compensate the zonal coefficients by replacing any coefficient having a detected parity error with the corresponding coefficient from the previous frame. This requires storing the zonal coefficients from the previous frame. The required storage is less than 1000 eight bit words for a 2 x 2 block of zonal coefficients per transform. Variable coefficients are inverse companded and clocked into the variable coefficient burst buffer according to the address specified by that particular coefficient. Any coefficient not addressed is set to zero. When a complete 16 x 16 coefficient block has been accumulated, the coefficients are transferred to the 2-D transform unit for further processing.

Although it is possible to provide error compensation for the variable coefficients, we find that from a viewpoint of subjective picture quality, it is more important to compensate for errors in the zonal coefficients than for errors in the variable coefficients. This is especially true for global/local transforms, such as the Haar. Though it would ultimately be desirable to also include error compensation for the variable coefficients, this compensation scheme is still under development at Northrop and will not be discussed at this time.

Inverse 2-D Transform Unit

The hardware used to mechanize the 2-D inverse Haar transform is similar to that used in the forward transform. The only major difference between the hardware implementation of the airborne and ground subsystems is that the arithmetic precision in the ground subsystem will be increased from 8 bits to 16 bits to provide sufficient dynamic range to prevent channel errors from causing integer overflows. This provision will also give the flexibility needed for incorporating image enhancement and contour extraction functions, if so desired. Power consumption will be greater in the ground subsystem since higher speed devices will be required to compensate for the increased propagation delays associated with the 16 bit arithmetic. The mathematical basis for the design of the inverse transform is described in Equation 3. In simple terms, the inverse transform is obtained by running the forward transform algorithm backwards.

Although 16 bit arithmetic will be used to compute the inverse transform, the resulting spatial video will only have a precision of 8 bits. In this design, the 8 bits directly above the LSB are the bits of interest (the LSB is an arbitrary scale factor of 2), and bits greater than these should only occur due to channel errors. Any reconstructed brightness value greater than the allowed 8 bit dynamic range is clipped to full scale brightness. Similarly, any negative reconstructed picture brightness value will be clipped to zero, since the original input video is defined for positive brightness only.

Frame Store Unit

The transmission format for the video signal requires that only 1/8 of each field be transmitted to achieve an 8:1 frame rate reduction. A frame store is required as part of the ground terminal to assemble 8 successive fields to produce one frame which is then displayed on the TV monitor. This type of transmission and storage concept provides a flicker-free image to the viewer. The display is updated with new 1/8 frame stripes for each successive field of the TV camera. A semiconductor memory for picture storage is used in the ground subsystem. This memory is packaged onto 4 PC boards, 7 inches x 7.3 inches. Each PC card is a memory plane, organized as shown in Figure 11. Each memory plane reads or writes four 8-bit words simultaneously as controlled by a simple 4:1 multiplexing scheme. Data can be accessed or loaded either sequentially or randomly and individual memory refresh is incorporated into each plane.

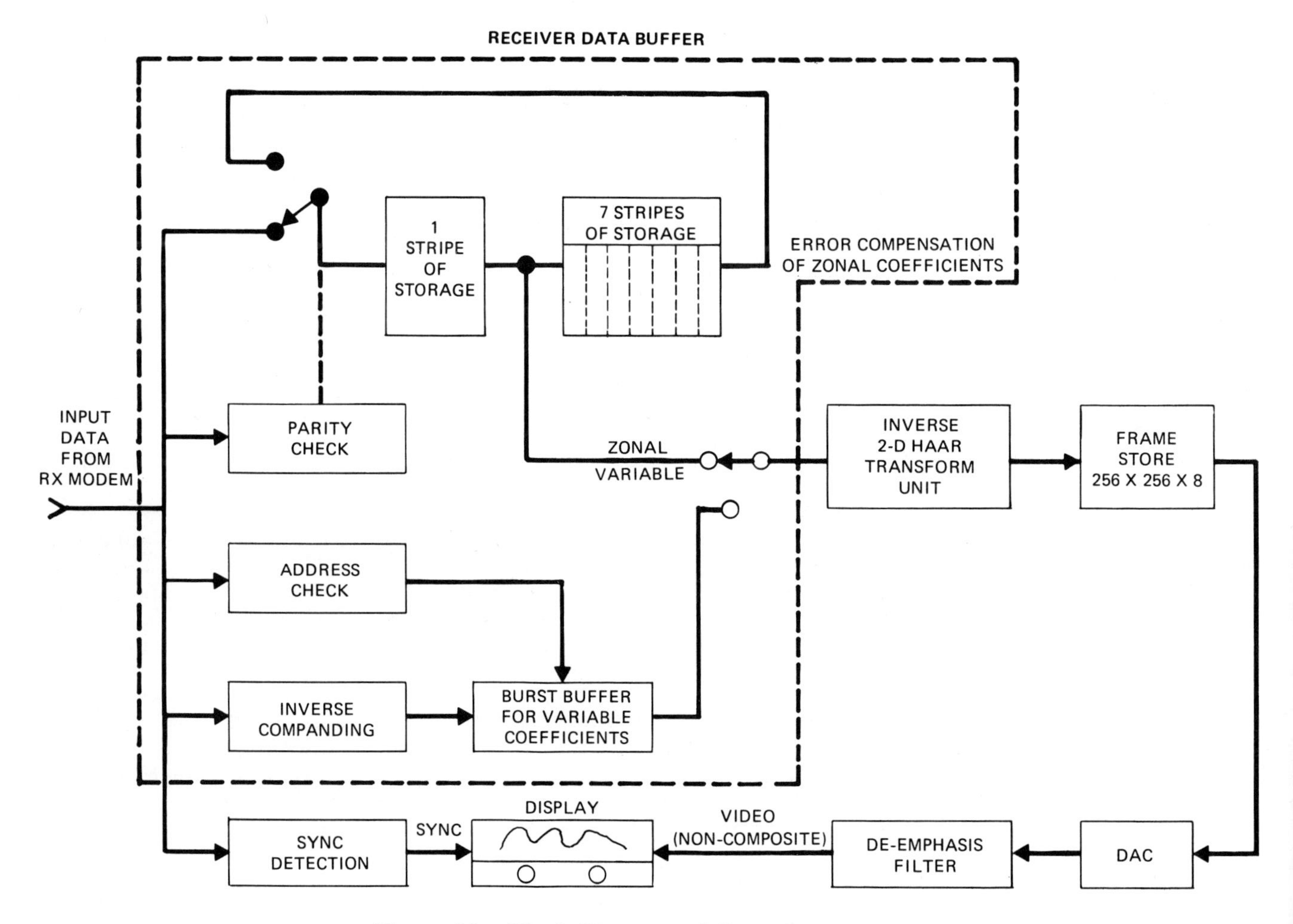

Figure 10. Block Diagram of Ground Terminal.

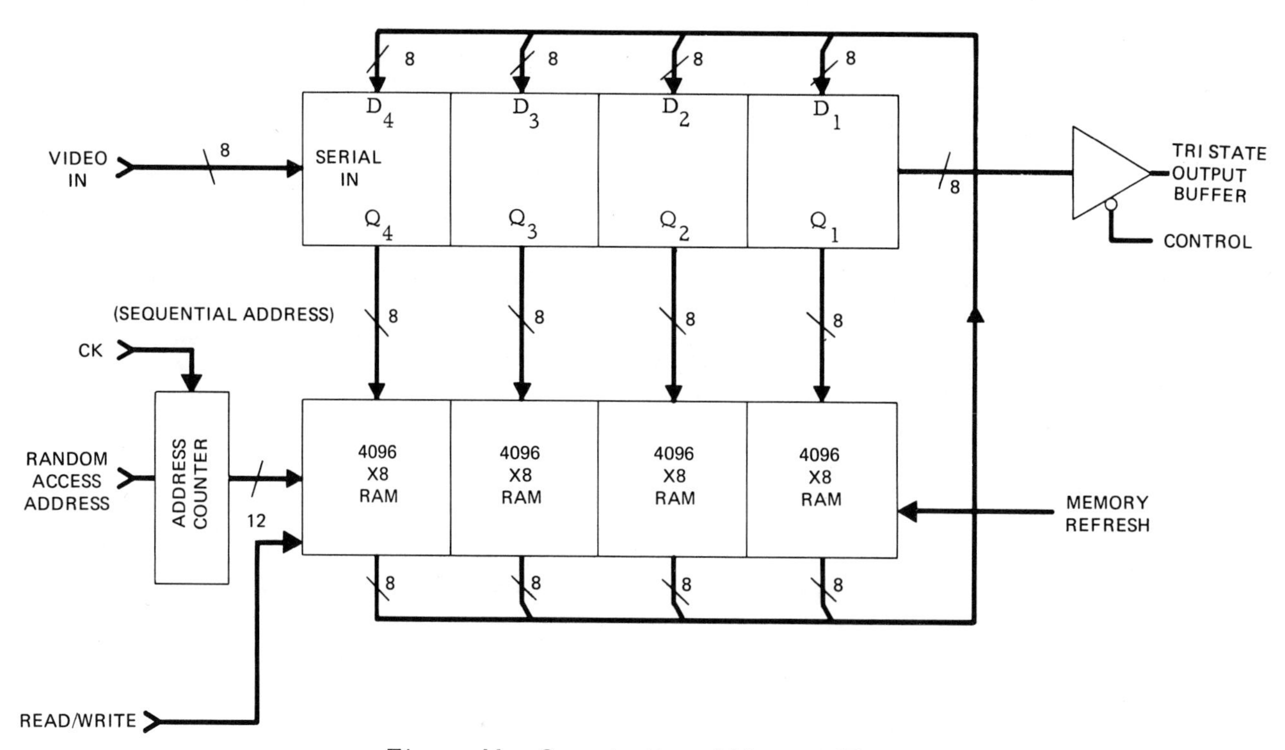

Figure 11. Organization of Memory Plane.

The output data from each memory plane drives a common tristate output buss so that the size of the memory can be easily expanded in increments of 16k, 8-bit words. The multiplexing scheme utilized in the design of the memory plane enables a frame store of size 256 pixel by 256 pixel by 8-bits/pixel to accept and deliver data at sample speeds in excess of 15 MHz.

DAC, De-emphasis, and Display

The DAC is a standard 10 bit 10 MHz commercial digital-to-analog converter. A 10 bit DAC is used because of its excellent linearity. This high quality DAC will provide an excellent quality 8-bit picture. The 2-LSB's are tied high to provide a slight black level which will help to produce a better quality image in the output display. The de-emphasis filter follows the DAC. Its function is to convert the frequency response of the reconstructed picture to the frequency response of the original picture. This filter will tend to remove some of the edge blockiness which tends to occur in highly compressed pictures (< 1 bit/pel). The bandwidth of the display should be greater than 10 MHz to ensure that the display does not degrade the high quality pictures obtained at the higher data rates.

Summary of Ground Hardware

The ground system contains a receiver data buffer unit, error compensation unit, inverse compression and Haar transform unit, frame store unit, DAC, de-emphasis circuitry and a high performance TV display. The basic differences between the airborne and ground subsystems are the error detection and compensation circuitry for the zonal terms which requires an additional storage of 1000 eight bit words and a frame store of 256^2 pels by 8 bits/pel to provide a flicker-free image on the TV display. The inverse Haar transform unit in the ground subsystem is implemented with 16 bit arithmetic to prevent integer overflows due to channel errors. A 10 bit 10 MHz DAC is used to provide the linearity required to obtain excellent quality 8 bit pictures.

Performance

The Northrop 2-D Haar video bandwidth reduction system has been designed to meet all of the performance, packaging and environmental specifications for some particular RPV applications.

As currently designed, the airborne subsystem will provide an 8 bit quality picture at the output of the 2-D Haar transform unit. The theoretical peak-to-peak signal to average RMS truncation error of the 2-D transform is 61.2 db and the average expected truncation error per transform coefficient is 0.19 quantization levels. This corresponds to an average roundoff error for the 2-D transform unit of 0.07% referenced to full scale signal, 2^8.

Figure 12 shows the mean squared error (MSE) performance of a computer simulation of the Haar Transform adaptive threshold compression system where MSE is defined by equation 5. Figure 13 shows some of the test pictures used in the simulation. It is Northrop's policy to always include the original picture whenever prints are made to minimize the effects of the developing process when a picture processing algorithm is being evaluated.

$$\mathrm{MSE} = \frac{\sum_{X=0}^{N-1} \sum_{Y=0}^{N-1} [f(X,Y) - \hat{f}(X,Y)]^2}{\sum_{X=0}^{N-1} \sum_{Y=0}^{N-1} [f(X,Y)]^2} \cdot 100\% \qquad (5)$$

where $f(X, Y)$ = original picture values

$\hat{f}(X, Y)$ = compressed picture values

N = number of horizontal and vertical samples.

Summary

A practical bandwidth compression system for RPV video links has been described. The system design is based on the fast 2-D Haar transform and an adaptive threshold compression technique to achieve a high level of system performance within the constraints imposed by current RPV requirements. A patent covering this system has been applied for.

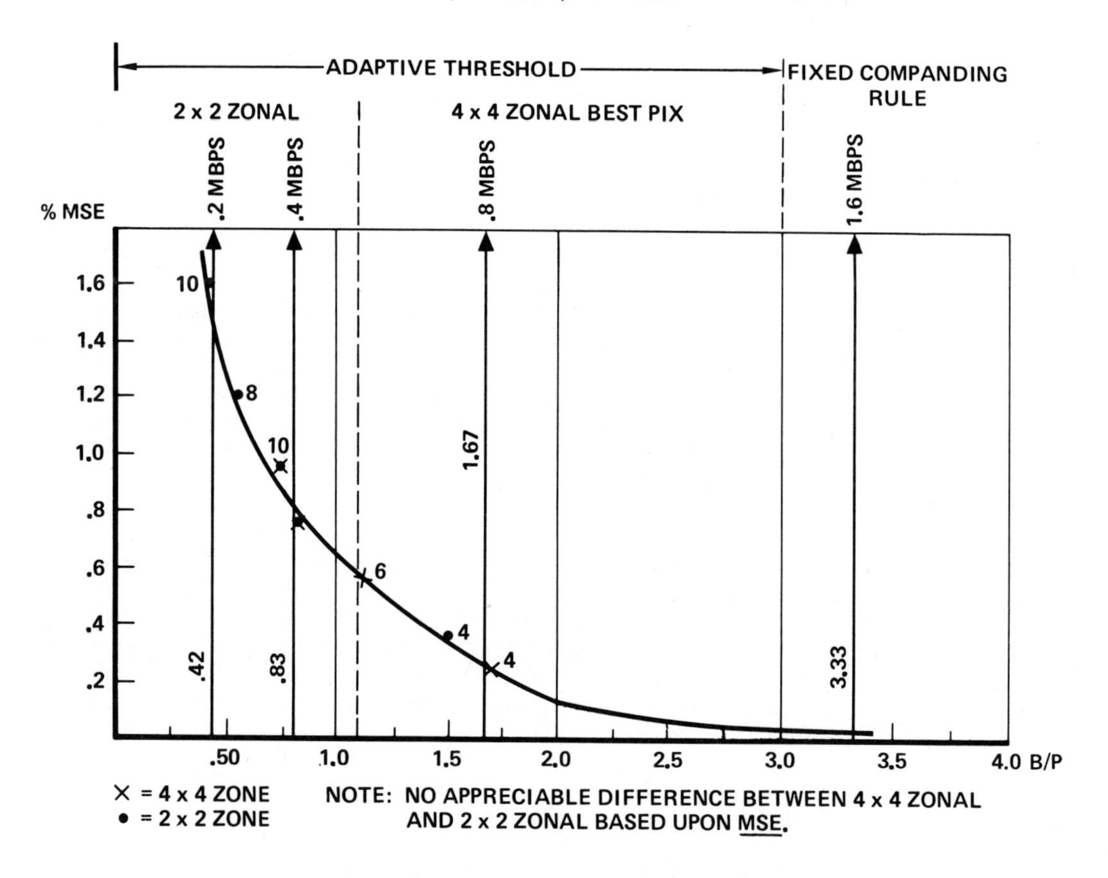

Figure 12. Mean Squared Error Performance of Adaptive Threshold Haar Compression System.

Acknowledgements

The authors wish to thank Mr. D. E. Miller of the Northrop Corporation for his assistance in preparing the demonstration of the 1-D real time Haar transform processor. We also wish to thank Mr. Donald J. Lynn and Mr. David Haas of the Space Image Processing Group at the Jet Propulsion Laboratory for their expert assistance in printing the pictures presented in this paper. The work reported in this paper was performed entirely under an on-going Northrop Corporation independent research and development (IR&D) program.

References

1. Naval Undersea Center (NUC) Solicitation No. N00123-76-R-0746, "Bandwidth Reduction System," February 1976.

2. J. J. Reis, R. T. Lynch, "Class of Transform Digital Processors for Compression of Multi-dimensional Data," Patent Application Serial No. 612, 090.

3. R. T. Lynch, J. J. Reis, "Transformation of Television Picture, "Applications of Walsh Functions and Sequency Theory, H. Schieber, F. Sandy, IEEE New York (74CHD861-5 EMC) pp 140, 146, 1974.

4. R. W. Means, J. M. Speiser, H. J. Whitehouse, "Image Transmission via Spread Spectrum Techniques," ARPA Quarterly Technical Report ARPA QR3, Oct. 1, 1973 - Jan. 1, 1974, p 2, AD 780805.

Compressed (1.5 B/P) Original (8 B/P)

Compressed (1.16 B/P) Original (8 B/P)

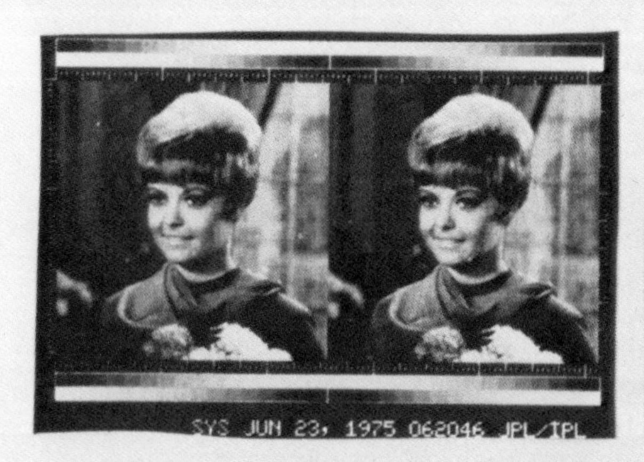

Compressed (1.7 B/P) Original (8 B/P)

Compressed (1.1 B/P) Original (8 B/P)

Figure 13. Adaptive Threshold Compression of 2-D Haar Transform Pictures.

Editor's Comments on Papers 22 and 23

22 PRATT, CHEN, and WELCH
Slant Transform Image Coding

23 WANG
New Algorithm for the Slant Transform

SLANT TRANSFORM

Slant transform (ST) was developed originallv by Enomoto and Shibata for $N = 8$ (1971). This was later extended by Pratt, Chen, and Welch (Paper 22) for any $N = 2^n$ using certain recursive properties. Similar to other transforms, the ST is separable and has a fast algorithm. A Cooley-Tukey type algorithm also has been developed for the ST (Ahmed and Chen, 1976). Other algorithms, including the progression from WHT to ST, also have been developed (Paper 23). As ST is an EOT, it can also be expressed via the WHT using a conversion matrix.

ST has a discrete sawtoothlike basis vector which can represent gradual brightness changes along an image line very efficiently. It has also other sawtooth line waveforms either increasing or decreasing over smaller ranges of its rows. In view of this, ST has been utilized in image coding (Paper 22). Also, a codec for digital transmission of NTSC TV signals based on WHT and ST has been built and tested (Ohira, Hayakawa, and Matsumoto, 1978; Paper 39).

REFERENCES

Ahmed, N., and M. C. Chen, 1976, A Cooley-Tukey Algorithm for the Slant Transform, *Int. J. Comput. Math.,* **B5:**331–338.

Enomoto, H., and K. Shibata, 1971, Orthogonal Transform Coding System for Television Signals, *IEEE Electromag. Compat. Trans.,* **EMC-13:**11–17.

Ohira, T., M. Hayakawa, and K. Matsumoto, 1978, Orthogonal Transform Coding System for NTSC Color Television Signals, *IEEE Commun. Trans.* **COM-26:**1454–1463.

22

Slant Transform Image Coding

WILLIAM K. PRATT, MEMBER, IEEE, WEN-HSIUNG CHEN, AND LLOYD R. WELCH

Abstract—**A new unitary transform called the slant transform, specifically designed for image coding, has been developed. The transformation possesses a discrete sawtoothlike basis vector which efficiently represents linear brightness variations along an image line. A fast computational algorithm has been found for the transformation.**

The slant transformation has been utilized in several transform image-coding systems for monochrome and color images. Computer simulation results indicate that good quality coding can be accomplished with about 1 to 2 bits/pixel for monochrome images and 2 to 3 bits/pixel for color images.

I. INTRODUCTION

DURING the past twenty years the applications of electronic imagery have grown enormously. This growth has placed severe demands on the capabilities of communication systems, since conventional television transmission requires exceptionally wide bandwidths. One means of bandwidth reduction that has shown particular promise is the transform image-coding process [1].

In 1968, the concept of coding and transmitting the two-dimensional Fourier transform of an image, computed by a fast computational algorithm, rather than the image itself, was introduced [2], [3]. This was followed shortly thereafter by the discovery that the Hadamard transform could be utilized in place of the Fourier transform with a considerable decrease in computational requirements [4], [5]. Investigations then began into the application of the Karhunen–Loeve [6], [7] and Haar [8] transforms for image coding. The Karhunen–Loeve transform provides minimum mean-square error coding performance but, unfortunately, does not possess a fast computational algorithm. On the other hand, the Haar transform has the attribute of an extremely efficient computational algorithm, but results in a larger coding error. None of the transforms mentioned above, however, has been expressly tailored to the characteristics of an image.

A major attribute of an image transform is that the transform compact the image energy to a few of the transform domain samples. A high degree of energy compaction will result if the basis vectors of the transform matrix "resemble" typical horizontal or vertical lines of an image. If the lines of a typical monochrome image are examined, it will be found that a large number of the lines are of constant grey level over a considerable length. The Fourier, Hadamard, and Haar transforms possess

Paper approved by the Associate Editor for Communication Theory of the IEEE Communications Society for publication after presentation at the 1973 Symposium on Applications of Walsh Functions. Manuscript received October 15, 1973; revised March 1, 1974. This work was supported in part by the Advanced Research Projects Agency of the Department of Defense and monitored by the Air Force Eastern Test Range under Contract F08606-72-C-0008.
W. K. Pratt and L. R. Welch are with the Electrical Engineering Department, Image Processing Institute, University of Southern California, Los Angeles, Calif. 90007.
W.-H. Chen was with the Electrical Engineering Department, Image Processing Institute, University of Southern California, Los Angeles, Calif. 90007. He is now with the Philco-Ford Corporation, Palo Alto, Calif.

a constant valued basis vector that provides an efficient representation for constant grey level image lines, while the Karhunen-Loeve transform has a nearly constant basis vector suitable for this representation. Another typical image line is one which increases or decreases in brightness over the length in a linear fashion. None of the transforms previously mentioned possess a basis vector that efficiently represents such image lines.

Shibata and Enomoto have introduced orthogonal transforms containing a "slant" basis vector for data of vector lengths of four and eight [9]. The slant vector is a discrete sawtooth waveform decreasing in uniform steps over its length, which is suitable for efficiently representing gradual brightness changes in an image line. Their work gives no indication of a construction for larger size data vectors, nor does it exhibit the use of a fast computational algorithm. In order to achieve a high degree of image-coding compression with transform coding techniques, it is necessary to perform the two-dimensional transform over block sizes 16×16 picture elements or greater. For large block sizes, computation is usually not feasible unless a fast algorithm is employed.

With this background, an investigation was undertaken to develop an image-coding slant-transform matrix possessing the following properties: 1) orthonormal set of basis vectors; 2) one constant basis vector; 3) one slant basis vector; 4) sequency property; 5) variable size transformation; 6) fast computational algorithm; and 7) high energy compaction. The following sections describe the construction of the slant-transformation matrix, present a fast computational algorithm, discuss its image-coding performance, and provide examples of its use for coding monochrome and color images.

Original images employed in the simulation of the coding systems studied are shown in Fig. 1. The monochrome images contain 256×256 pixels with each calibrated intensity value linearly quantized to 256 levels. The red, green, and blue tristimulus values of the color pictures have each been linearly quantized to 256 levels.

II. SLANT TRANSFORM

A slant-transform matrix possessing all of the above properties has been developed [10]. A description of its construction, fast algorithm, and application to image transformation follows.

A. Two-Dimensional Transform

Let $[F]$ be an $N \times N$ matrix of the picture element (pixel) brightness values of an image and let $[f_i]$ be an $N \times 1$ vector representing the ith column of $[F]$. The one-dimensional transform of the ith image line is then

$$[\mathfrak{f}_i] = [S][f_i] \tag{1}$$

where $[S]$ is the $N \times N$ unitary slant matrix. A two-dimensional slant transform is performed by sequential row and column transformations on $[F]$, yielding

$$[\mathfrak{F}] = [S][F][S]^T. \tag{2}$$

The inverse transformation to recover $[F]$ from the transform components $[\mathfrak{F}]$ is given by

$$[F] = [S]^T[\mathfrak{F}][S]. \tag{3}$$

It is also convenient to establish a series representation of the transform operation. The two-dimensional forward and inverse transforms in series form can be expressed as

$$\mathfrak{F}(u,v) = \sum_{j=1}^{N} \sum_{k=1}^{N} F(j,k) S(u,j) S(k,v) \tag{4}$$

and

$$F(j,k) = \sum_{u=1}^{N} \sum_{v=1}^{N} \mathfrak{F}(u,v) S(j,u) S(v,k). \tag{5}$$

Fig. 2 contains full size, two-dimensional slant-transform displays of the three monochrome original images of Fig. 1. The logarithm of the absolute value of each transform sample is displayed in Fig. 2(a)-(c) rather than the absolute value itself in order to reduce the dynamic range of the display. Fig. 2(d) shows a threshold display of Fig. 2(a) in which all samples whose magnitude is below a specified threshold are set to zero and all samples whose magnitude is above the threshold are set to a constant brightness. The typical energy distribution of the slant transform of an image is apparent from these pictures.

B. Slant-Transform Matrix

The slant transform is a member of a class of transforms whose matrices are orthogonal, have a constant function for the first row, and have a second row which is a linear (slant) function of the column index. The matrices are formed by an iterative construction that exhibits the matrices as products of sparse matrices, which in turn leads to a fast transform algorithm.

The slant-transform matrix of order two consisting of a constant and a slant basis vector is given by

$$S_2 = \frac{1}{2^{1/2}} \begin{bmatrix} 1 & 1 \\ 1 & -1 \end{bmatrix}. \tag{6}$$

The slant matrix of order four is obtained by the operation

$$S_4 = \frac{1}{2^{1/2}} \left[\begin{array}{cc|cc} 1 & 0 & 1 & 0 \\ a_4 & b_4 & -a_4 & b_4 \\ \hline 0 & 1 & 0 & -1 \\ -b_4 & a_4 & b_4 & a_4 \end{array} \right] \left[\begin{array}{c|c} S_2 & 0 \\ \hline 0 & S_2 \end{array} \right] \tag{7a}$$

or

$$S_4 = \frac{1}{2^{1/2}} \begin{bmatrix} 1 & 1 & 1 & 1 \\ a_4 + b_4 & a_4 - b_4 & -a_4 + b_4 & -a_4 - b_4 \\ 1 & -1 & -1 & 1 \\ a_4 - b_4 & -a_4 - b_4 & a_4 + b_4 & -a_4 + b_4 \end{bmatrix} \tag{7b}$$

(a) monochrome girl

(d) color girl

(b) monochrome couple

(e) color couple

(c) monochrome moonscape

Fig. 1. Original monochrome and color images.

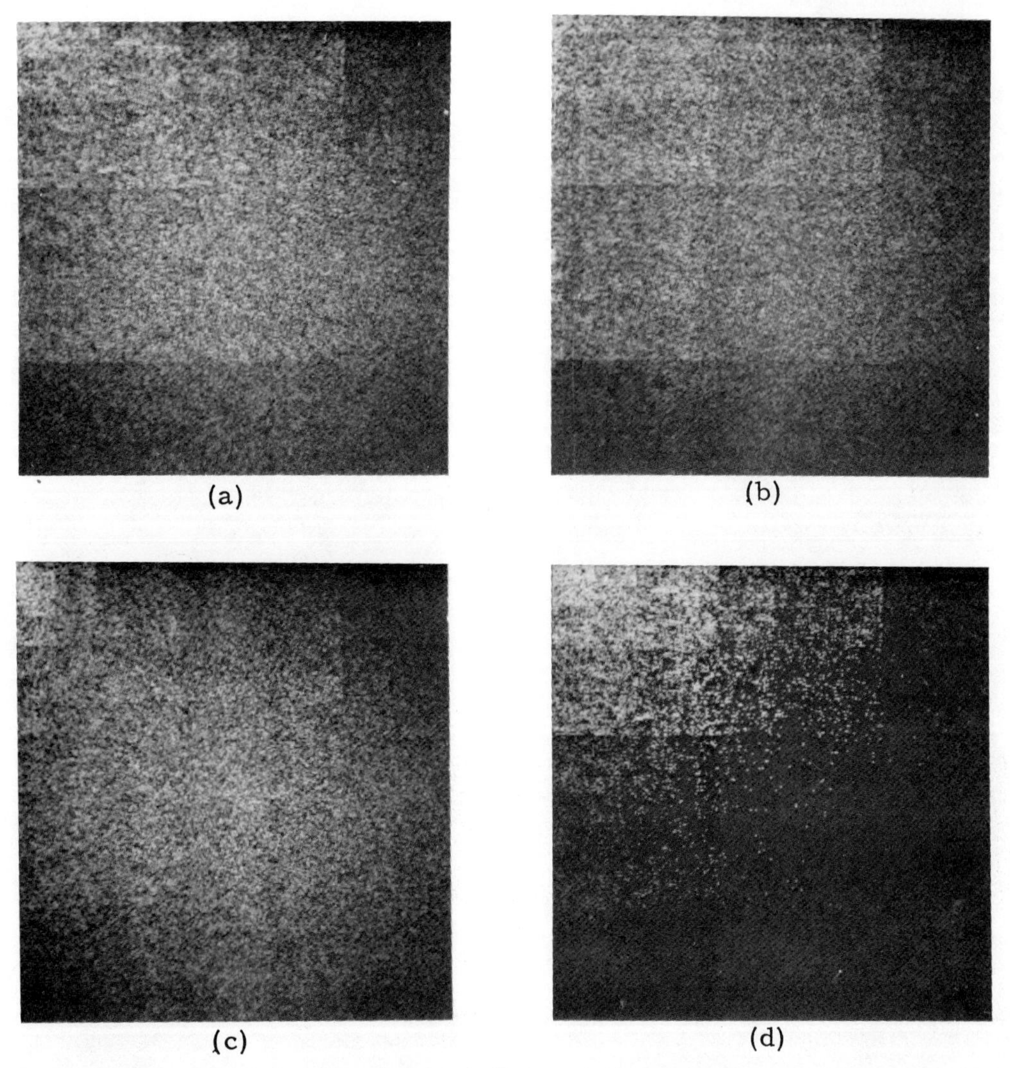

Fig. 2. Slant-transform domain logarithm of magnitude display. (a) Girl. (b) Couple. (c) Moonscape. (d) Transform threshold girl.

where a_4 and b_4 are scaling constants. The first two rows of the left factor of (7a) are uniquely determined by the orthogonality condition and the requirement that the first row of S_4 must be a positive constant and the second row must be linear with negative slope. The step sizes between adjacent elements of the slant vector of (7b) are $2b_4$, $2a_4 - 2b_4$, and $2b_4$. By setting these steps sizes equal, one finds that

$$a_4 = 2b_4.$$

The orthonormality condition $[S_4][S_4]^T = [I]$ leads to

$$b_4 = \frac{1}{5^{1/2}}.$$

The third and fourth rows of the left factor of (7a) form an orthonormal basis for vectors orthogonal to the first two rows and have the property that among all such bases, this basis requires the smallest number of nontrivial multiplications. In considering computational requirements, the scalar multiplier $1/\sqrt{2}$ is ignored; multiplication by ± 1 is considered trivial). When the values of a_4 and b_4 are substituted into (7b), the slant matrix of order 4 becomes

$$S_4 = \frac{1}{4^{1/2}} \begin{bmatrix} 1 & 1 & 1 & 1 \\ \frac{3}{5^{1/2}} & \frac{1}{5^{1/2}} & \frac{-1}{5^{1/2}} & \frac{-3}{5^{1/2}} \\ 1 & -1 & -1 & 1 \\ \frac{1}{5^{1/2}} & \frac{-3}{5^{1/2}} & \frac{3}{5^{1/2}} & \frac{-1}{5^{1/2}} \end{bmatrix}. \quad (8)$$

It is easily seen that the rows of S_4 form an orthonormal set. Furthermore, S_4 possesses the sequency property; each row has a distinct number of sign reversals and each integer from 0 to 3 is the number of sign reversals of some row.

An extension of the slant matrix to its next size of order 8 is given by

$$S_8 = \frac{1}{2^{1/2}} \left[\begin{array}{cccc:cccc} 1 & 0 & 0 & 0 & 1 & 0 & 0 & 0 \\ a_8 & b_8 & 0 & 0 & -a_8 & b_8 & 0 & 0 \\ 0 & 0 & 1 & 0 & 0 & 0 & 1 & 0 \\ 0 & 0 & 0 & 1 & 0 & 0 & 0 & 1 \\ \hdashline 0 & 1 & 0 & 0 & 0 & -1 & 0 & 0 \\ -b_8 & a_8 & 0 & 0 & b_8 & a_8 & 0 & 0 \\ 0 & 0 & 1 & 0 & 0 & 0 & -1 & 0 \\ 0 & 0 & 0 & 1 & 0 & 0 & 0 & -1 \end{array}\right] \left[\begin{array}{c:c} S_4 & 0 \\ \hdashline 0 & S_4 \end{array}\right]. \tag{9}$$

The first two rows of the left factor are determined by orthonormality and the requirement that the first row of S_8 be constant and the second row be linear with negative slope. Rows five and six of S_8 are also linear combinations of the first two rows of S_4. The particular combinations are chosen to minimize nontrivial multiplications while maintaining orthonormality. The remaining rows form a basis for the orthogonal complement of the space spanned by the four rows already treated. It is a basis which requires no nontrivial multiplications. It can be shown that S_8 has the sequency property.

Equation (9) can be generalized to give the slant matrix of order $N(n = 2^n, n = 3, \cdots)$ in terms of the slant matrix of order $N/2$ by the following recursive relation:

$$S_N = \frac{1}{2^{1/2}} \left[\begin{array}{cc:c:cc:c} 1 & 0 & & 1 & 0 & \\ a_N & b_N & 0 & -a_N & b_N & 0 \\ \hdashline \multicolumn{2}{c:}{0} & I_{(N/2)-2} & \multicolumn{2}{c:}{0} & I_{(N/2)-2} \\ \hdashline 0 & 1 & & 0 & -1 & \\ -b_N & a_N & 0 & b_N & a_N & 0 \\ \hdashline \multicolumn{2}{c:}{0} & I_{(N/2)-2} & \multicolumn{2}{c:}{0} & -I_{(N/2)-2} \end{array}\right] \left[\begin{array}{c:c} S_{N/2} & 0 \\ \hdashline 0 & S_{N/2} \end{array}\right]. \tag{10}$$

The matrix $I_{(N/2)-2}$ is the identity matrix of dimension $(N/2) - 2$ and the various partition blocks are determined by the same considerations as described above for S_8. The constants a_N, b_N may be computed from the recursive relation [7]:

$$a_2 = 1 \tag{11a}$$

$$b_N = [1 + 4(a_{N/2})^2]^{-1/2} \tag{11b}$$

$$a_N = 2b_N a_{N/2} \tag{11c}$$

or by the formulas

$$a_{2N} = \left(\frac{3N^2}{4N^2 - 1}\right)^{1/2}, \qquad b_{2N} = \left(\frac{N^2 - 1}{4N^2 - 1}\right)^{1/2}. \tag{11d}$$

Fig. 3 contains a plot of the slant-transform basis vectors of S_{16} represented as waveforms.

C. Slant-Transform Fast Computational Algorithm

The fast computational algorithm of the slant transform is based on the matrix factorization corresponding to (10). A column vector multiplication by S_N can be accomplished by multiplying by the left factor of (10). Letting A_N denote the number of additions and subtractions for S_N, and M_N the number of nontrivial multiplications in the above scheme, it can be shown that A_N and M_N satisfy the recursions $A_N = 2A_{N/2} + N + 4$ and $M_N = 2M_{N/2} + 8$. However, a more economical algorithm is obtained by a further factorization of (10) into

$$S_N = \frac{1}{2^{1/2}} \left[\begin{array}{cc|c|cc|c} 1 & 0 & & 1 & 0 & \\ & & 0 & & & 0 \\ 0 & b_N & & a_N & 0 & \\ \hline \multicolumn{2}{c|}{0} & I_{(N/2)-2} & \multicolumn{2}{c|}{0} & 0 \\ \hline 0 & 0 & & 0 & 1 & \\ & & 0 & & & 0 \\ 0 & a_N & & -b_N & 0 & \\ \hline \multicolumn{2}{c|}{0} & 0 & \multicolumn{2}{c|}{0} & I_{(N/2)-2} \end{array}\right] \left[\begin{array}{c|c} I_{N/2} & I_{N/2} \\ \hline I_{N/2} & -I_{N/2} \end{array}\right] \left[\begin{array}{c|c} S_{N/2} & 0 \\ \hline 0 & S_{N/2} \end{array}\right] \tag{12}$$

Multiplication by the middle factor involves only N additions or subtractions, while multiplication by the left factor involves only 2 additions and 4 multiplications. The recursions for the number of operations for S_N are now $A_N = 2A_{N/2} + N + 2$ and $M_N = 2M_{N/2} + 4$.

A special, more efficient factorization of S_4 has been discovered. With this factorization

$$S_4 = \frac{1}{4^{1/2}} \begin{bmatrix} 1 & 1 & 0 & 0 \\ 0 & 0 & \frac{3}{5^{1/2}} & \frac{1}{5^{1/2}} \\ 1 & -1 & 0 & 0 \\ 0 & 0 & \frac{1}{5^{1/2}} & -\frac{3}{5^{1/2}} \end{bmatrix} \begin{bmatrix} 1 & 0 & 0 & 1 \\ 0 & 1 & 1 & 0 \\ 1 & 0 & 0 & -1 \\ 0 & 1 & -1 & 0 \end{bmatrix}. \tag{13}$$

If S_4 is postmultiplied by a column data vector, the first computational pass requires 4 additions and subtractions while the second pass requires a similar number and 4 multiplications. Therefore $A_4 = 8$, $M_4 = 4$.

With some rearrangement of rows and columns, (13) can be applied to S_8 to give

$$S_8 = \frac{1}{2^{1/2}} \left[\begin{array}{cccc|cccc} 1 & 0 & 0 & 0 & & & & \\ 0 & b_8 & a_8 & 0 & & 0 & & \\ 0 & 0 & 0 & 1 & & & & \\ 0 & a_8 & -b_8 & 0 & & & & \\ \hline & & & & 1 & 0 & 0 & 0 \\ & 0 & & & 0 & 1 & 0 & 0 \\ & & & & 0 & 0 & 1 & 0 \\ & & & & 0 & 0 & 0 & 1 \end{array}\right] \left[\begin{array}{cccc|cccc} 1 & 0 & 0 & 0 & 1 & 0 & 0 & 0 \\ 0 & 1 & 0 & 0 & 0 & 1 & 0 & 0 \\ 1 & 0 & 0 & 0 & -1 & 0 & 0 & 0 \\ 0 & 1 & 0 & 0 & 0 & -1 & 0 & 0 \\ \hline 0 & 0 & 1 & 0 & 0 & 0 & 1 & 0 \\ 0 & 0 & 1 & 0 & 0 & 0 & -1 & 0 \\ 0 & 0 & 0 & 1 & 0 & 0 & 0 & -1 \\ 0 & 0 & 0 & 1 & 0 & 0 & 0 & 1 \end{array}\right] \left[\begin{array}{c|c} S_4 & 0 \\ \hline 0 & S_4 \end{array}\right]. \tag{14}$$

This rearrangement orders the rows in increasing order of sequency. If S_8 is postmultiplied by a column data vector, the first and second computational passes execute the multiplication by S_4 on the first and last halves. The third pass requires 8 additions or subtractions and the fourth pass requires 2 additions and 4 multiplications. The total computation is 26 additions or subtractions and 12 multiplications. A flow chart of this computation is shown in Fig. 4.

The decomposition of the general S_N follows from repeated application of (12) [11]. A total of $N \log_2 N + (N/2) - 2$ additions and subtractions together with $2N - 4$ multiplications are required to compute the slant transform of an N-dimensional data vector. For purposes of comparison the $N \times N$ Hadamard transform requires $N \log_2 N$ additions or subtractions.

III. STATISTICAL ANALYSIS OF SLANT TRANSFORM

The development of efficient quantization and coding requires an understanding of the statistical properties of the transform domain samples. This section presents a derivation of the statistical mean and covariance of the slant transform samples and develops models for their probability densities.

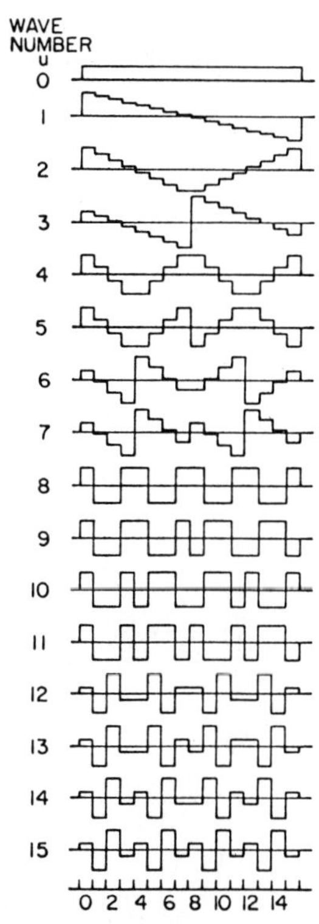

Fig. 3. Slant-transform basis waveforms.

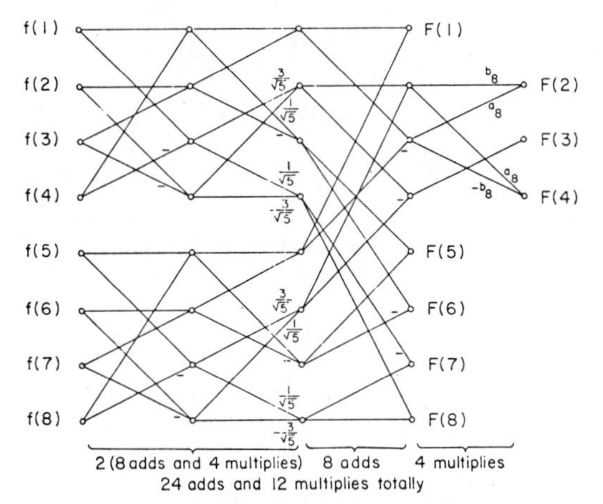

Fig. 4. Slant-transform of order 8 computational flowchart.

A. Mean and Covariance

For the statistical analysis it is assumed that the image array $F(j,k)$ is a sample of a stochastic process with mean

$$E\{F(j,k)\} \equiv \overline{F(j,k)} \tag{15a}$$

and covariance

$$E\{[F(j_1,k_1) - \overline{F(j_1,k_1)}][F(j_2,k_2) - \overline{F(j_2,k_2)}]\} \equiv C_F\{j_1,k_1,j_2,k_2\}. \tag{15b}$$

A convenient, and reasonably accurate, covariance model for an image is the first-order Markov process model for which

$$C_F\{j_1,k_1,j_2,k_2\} = \sigma_R^2\sigma_C^2\rho_R^{|j_1-j_2|}\rho_C^{|k_1-k_2|}$$

where σ_R^2 and σ_C^2 are the row and column variances, and ρ_R and ρ_C are the adjacent pixel correlation factors. From (4), the mean and covariance of the slant transform samples can then be expressed as[1]

$$E\{\mathfrak{F}(u,v)\} \equiv \overline{\mathfrak{F}(u,v)} = \sum_j \sum_k \overline{F(j,k)}\, S(u,j)\, S(k,v) \tag{16}$$

and

$$C_{\mathfrak{F}}\{u_1,u_2,v_1,v_2\} = \sum_{j_1}\sum_{j_2}\sum_{k_1}\sum_{k_2} C_F\{j_1,j_2,k_1,k_2\} \cdot S(u_1,j_1)\,S(k_1,v_1)\,S(u_2,j_2)\,S(k_2,v_2). \tag{17}$$

If the image array is stationary, then as a result of the orthogonality of the kernel,

$$\overline{\mathfrak{F}(0,0)} = \overline{NF(j,k)} \tag{18a}$$

$$\overline{\mathfrak{F}(u,v)} = 0, \; u,v \neq 0. \tag{18b}$$

No closed form expression has been found for the covariance of the slant transform samples, but the covariance may be computed as the two-dimensional slant transform of the function $C\{j_1 - j_2, k_1 - k_2\}$ if the original image field is stationary.

B. Probability Density Models

The probability density of slant transform samples is difficult to obtain since the probability density of the original image array is not usually well defined, and also, the slant transform representation is mathematically complex. However, since the transform operation forms a well behaved weighted sum over all of the pixels in the original image, one can evoke qualitative arguments based upon the central limit theorem [12] to determine reasonable probability density models for the transform domain samples.

The transform domain sample $\mathfrak{F}(0,0)$ is a nonnegative weighted sum of pixel values. Its histogram will generally follow the histogram of pixel values which is often modeled by a Rayleigh density

$$p_{\mathfrak{F}(0,0)}(x) = \frac{x}{\alpha^2}\exp\left\{\frac{-x^2}{2\alpha^2}\right\}, \qquad x \geq 0. \tag{19}$$

All other transform components are bipolar and possess a zero mean. These components generally can be modeled by a Gaussian density

$$p_{\mathfrak{F}(u,v)}(x) = [2\pi\sigma^2(u,v)]^{-1/2}\exp\left\{-\frac{x^2}{2\sigma^2(u,v)}\right\}, \quad x \geq 0;\ (u,v) \neq (0,0) \tag{20}$$

where $\sigma^2(u,v) = C_{\mathfrak{F}}(u,u,v,v)$ is the variance of the transform samples as obtained from (17).

IV. SLANT-TRANSFORM MONOCHROME IMAGE CODING

The basic premise of a monochrome image-transform coding system is that the two-dimensional transform of an image has an energy distribution more suitable to coding

[1] Unless otherwise noted, the summation indices are 1 to N.

than the spatial domain representation. As a result of the inherent pixel-to-pixel correlation of natural monochrome images, the energy in the transform domain tends to be clustered into a relatively small number of transform samples. Low magnitude transform samples can be discarded in an analog transmission system, or grossly quantized in a digital transmission system, without introducing serious image degradation in order to achieve a bandwidth reduction.

Fig. 5 contains a block diagram of the slant-transform coding system for monochrome images. In operation, a two-dimensional slant transform is taken of the image pixels over the entire image, or repeatedly over subsections of the image, called blocks. The transform domain samples are then operated upon by a sample selector that decides which samples are to be transmitted. For an analog communication system, the selected samples are distributed uniformly in time and transmitted by analog modulation, while for a digital communication link, the selected samples are quantized, coded, and transmitted in binary form. At the receiver the incoming data are decoded, and an inverse slant transform is performed to reconstruct the original image.

There are two basic strategies of sample selection: zonal sampling and threshold sampling [1]. In zonal sampling the reconstruction is made with a subset of transform samples lying in certain pre-specified geometric zones—usually the low frequency coefficients. For analog transmission the amplitude of each component in the zone is transmitted. For digital transmission each component in a zone is quantized and assigned a binary code word. The number of quantization levels is usually made proportional to the expected variance of the component, and the number of code bits made proportional to its expected probability of occurrence. In threshold sampling the image reconstruction is made with a subset of the samples which are larger than a specified threshold. Since the locations of the significant samples must be communicated, threshold sampling is usually employed only in digital links.

The following subsections contain an analysis of the slant-transform image-coding process for a zonal sample reduction system utilizing analog transmission and for zonal and threshold coding systems over a digital link. In all instances, a mean-square error performance criterion

$$\mathfrak{E} = \frac{1}{N^2} \sum_j \sum_k E\{[F(j,k) - \hat{F}(j,k)]^2\} \quad (21)$$

is utilized. While it is known that this measure results in some anomalies, it has proven reliable as a performance measure between different transforms and for variants of sampling and coding strategies.

A. Zonal Sampling

The sample selection process for two zones can be analyzed conveniently by defining a transform domain sampling function $T(u,v)$ which takes on the value one for samples to be transmitted and zero for samples to be discarded. The reconstructed image then becomes

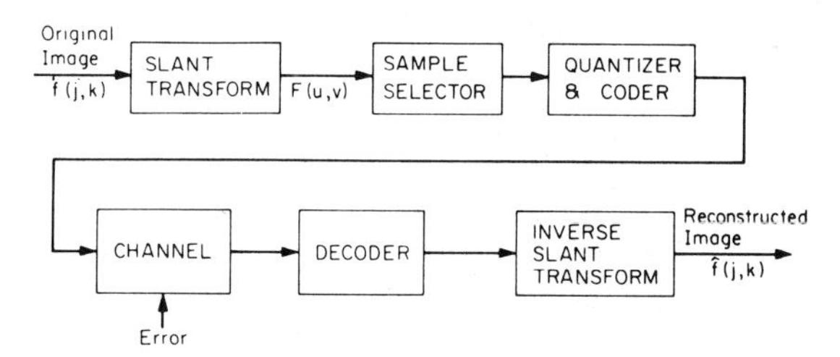

Fig. 5. Slant-transform monochrome-image coding system.

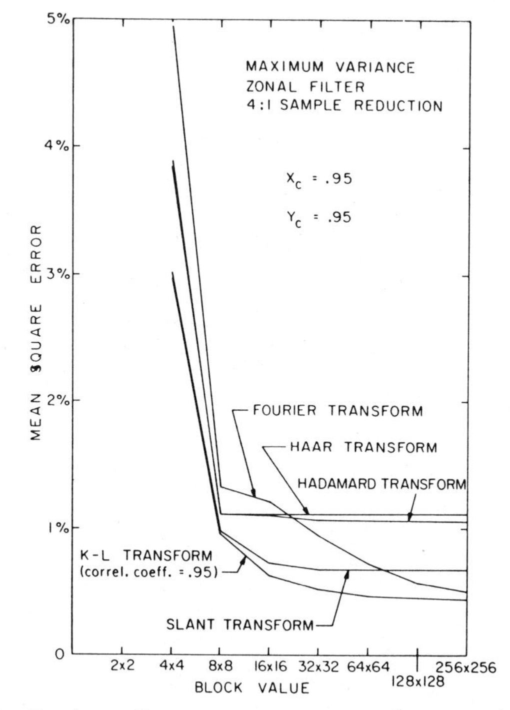

Fig. 6. Zonal sampling mean-square error performance of image transforms as a function of block size.

$$\hat{F}(j,k) = \sum_u \sum_v \mathfrak{F}(u,v)\, T(u,v)\, S(u,j)\, S(k,v). \quad (22)$$

It is then easily shown that the mean-square error expression can be written as

$$\mathfrak{E} = \frac{1}{N^2} \sum_u \sum_v E\{(\mathfrak{F}(u,v)[1 - T(u,v)])^2\}. \quad (23)$$

There are a number of zones that could logically be employed for zonal sampling; for example, a rectangular, elliptical, or triangular zone. Both analytic and experimental studies [11] have indicated that the optimum zone is the so-called maximum variance zone in which $T(u,v)$ is chosen to be unity for those samples having the largest variance as computed by (17) for a given covariance model of the original image. Fig. 6 contains a plot of the mean-square error of an image with a Markov process covariance as a function of block size for various transformations. In this plot the 25 percent of the coefficients with the largest variances were selected, and the remainder discarded. From the figure it is seen that the Karhunen–Loeve transform provides the best mean-square error, while the slant transform results in only a slightly greater error. Also note that the rate of decrease in mean-square error for larger block sizes becomes quite small after a block size of about 16 × 16.

Fig. 7 shows reconstructions of images for maximum variance zonal sampling with the slant transform in 16 × 16 blocks. For purposes of comparison, a series of experiments were performed with the Hadamard, Haar, Fourier, and Karhunen–Loeve transforms, as shown in Fig. 8. It can be seen that the slant transform provides a better subjective quality reconstruction and smaller mean-square error than any other transform except the Karhunen–Loeve transform which does not possess a fast algorithm.

B. Zonal Coding

In the zonal coding system a set of zones is established in each transform block. Transform samples in each zone are then quantized with the same number of quantization levels which is set proportional to the expected variance of the transform coefficients. For a constant word length code, $N_B(u,v)$ code bits are assigned to each coefficient, resulting in

$$L_C(u,v) = 2^{N_B(u,v)} \tag{24}$$

quantization levels. A total of

$$N_B = \sum_u \sum_v N_B(u,v) \tag{25}$$

bits are then required to code the picture. The bit assignment $N_B(u,v)$ for each coefficient is based upon a relation of rate distortion theory [13], [14]. The number of bits is given by

$$N_B(u,v) = \ln [V_F(u,v)] - \ln [D] \tag{26}$$

where $V_F(u,v)$ is the variance of a transform coefficient and D is proportional to the mean-square quantization error. Fig. 9 illustrates a typical bit assignment for coding in 16 × 16 blocks. Quantization decision and reconstruction levels are selected to minimize the mean-square quantization error for the probability density models of Section III using the Max quantization algorithm [15]. With this quantization and coding strategy, it is possible to predict the mean-square coding error for transform coding. Fig. 10 contains a plot of mean-square error as a function of block size for several transforms for coding with an average of 1.5 bits/pixel. The figure indicates that the performance of the slant transform is quite close to the optimal Karhunen–Loeve transform. It is possible to achieve a slightly lower mean-square error for a given channel rate by employing Huffman coding of the quantized coefficients rather than constant word-length coding, but the coder will be much more complex to implement.

Fig. 11 contains simulation results of zonal coding with the slant transform in blocks of 16 × 16 pixels. A comparison with other transforms is shown in Fig. 12.

C. Threshold Coding

In a threshold coding system, each sample whose magnitude is greater than a given threshold level is quantized with a fixed number of levels and its amplitude is coded. It is necessary to code the position of each significant sample in the transform plane. A simple, but quite efficient, technique for position coding is to code the number of nonsignificant samples between significant samples. This scheme, called run length coding, can be implemented as follows.

1) The first sample along each line is coded regardless of its magnitude. A position code word of all zeros or all ones affixed to the amplitude provides a line synchronization code group.

2) The amplitude of the second run length code word is the coded amplitude of the next significant sample. The position code is the binary count of the number of samples of the significant sample from the previous significant sample.

3) If a significant sample is not encountered after scanning the maximum run length of samples, the position and amplitude code bits are set to all ones to indicate a maximum run length.

The advantage of including a line synchronization code group is that it becomes unnecessary to code the line number and, also, it prevents the propagation of channel errors over more than one line.

Fig. 13 contains simulation results for slant-transform threshold coding. As expected, since the coding process is adaptive, its performance is somewhat better than the simpler zonal coding process.

V. SLANT-TRANSFORM COLOR-IMAGE CODING

Fig. 14 contains a block diagram of a typical slant-transform color-image coding system. In the system, the color image is represented by three source tristimulus signals $R(j,k)$, $G(j,k)$, $B(j,k)$ that specify the red, green, and blue content of a pixel at coordinate (j,k), according to the National Television System Commission (NTSC) receiver phosphor primary system [16]. The source tristimulus signals are then converted to a new three-dimensional space, $Y(j,k)$, $I(j,k)$, and $Q(j,k)$, which specify the luminance and the chrominance information of the image pixel according to the NTSC television transmission primary system. The conversion is defined by

$$\begin{bmatrix} Y(j,k) \\ I(j,k) \\ Q(j,k) \end{bmatrix} = \begin{bmatrix} 0.299 & 0.587 & 0.114 \\ 0.596 & -0.274 & -0.322 \\ 0.211 & -0.253 & 0.312 \end{bmatrix} \begin{bmatrix} R(j,k) \\ G(j,k) \\ B(j,k) \end{bmatrix}. \tag{27}$$

The reason for transform coding the $Y\ I\ Q$ signals rather than the $R\ G\ B$ signals is that the $Y\ I\ Q$ signals are reasonably well uncorrelated, and most of the color-image energy is compacted into the Y plane. This permits a more efficient design of the quantizers. Table I compares the energy distribution of the $R\ G\ B$ and $Y\ I\ Q$ color planes. The converted signals then individually undergo a two-dimensional slant transform. This results in three transform domain planes, $\mathfrak{F}_Y(u,v)$, $\mathfrak{F}_I(u,v)$, $\mathfrak{F}_Q(u,v)$, obtained

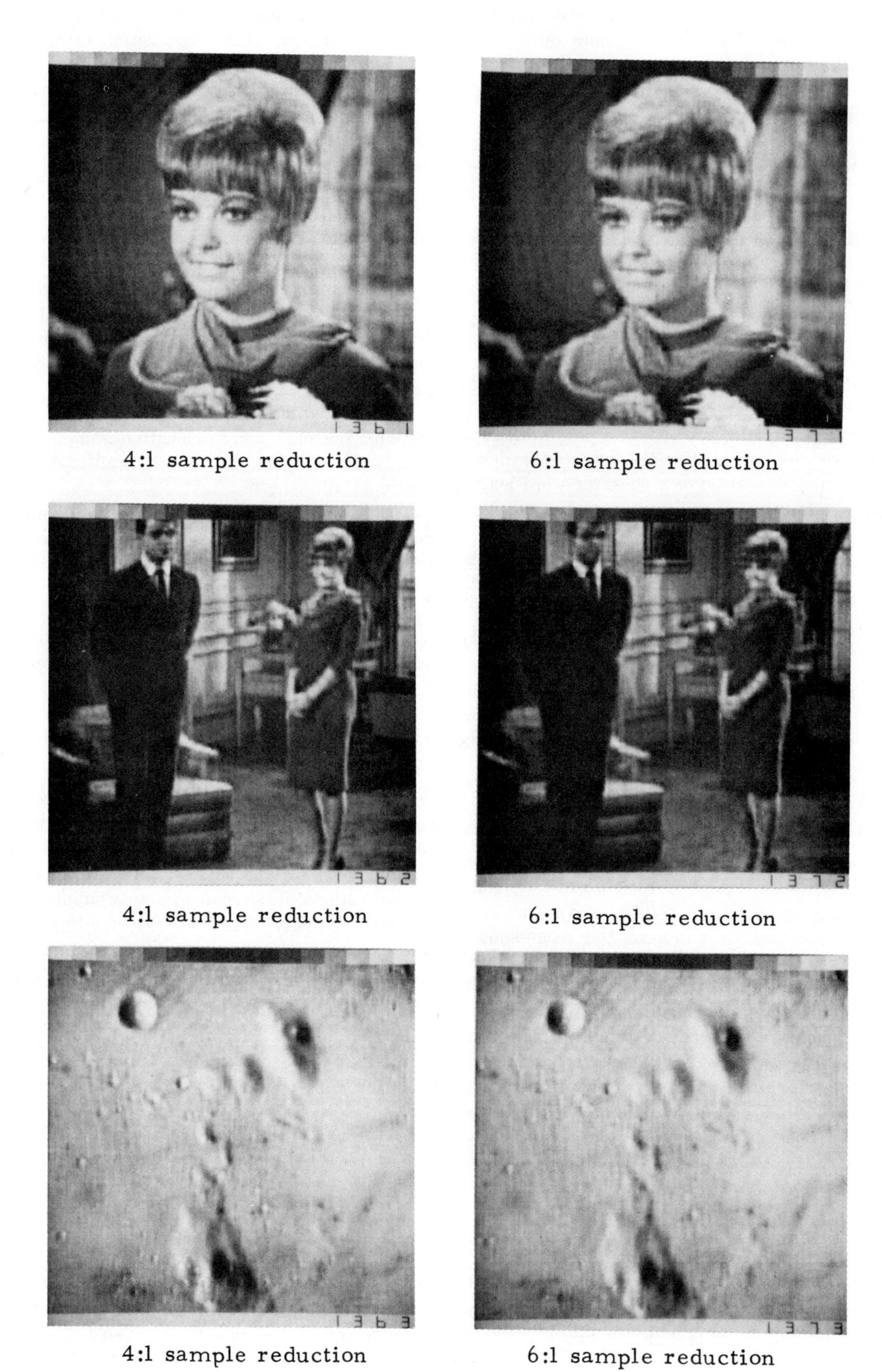

Fig. 7. Slant-transform zonal sampling in 16 × 16 pixel blocks, unquantized transform.

Hadamard transform
4:1 sample reduction

Fourier transform
4:1 sample reduction

Haar transform
4:1 sample reduction

Karhunen-Loeve transform
4:1 sample reduction

Fig. 8. Hadamard, Fourier, Haar, and Karhunen–Loeve transform zonal sampling in 16 × 16 pixel blocks, unquantized transform.

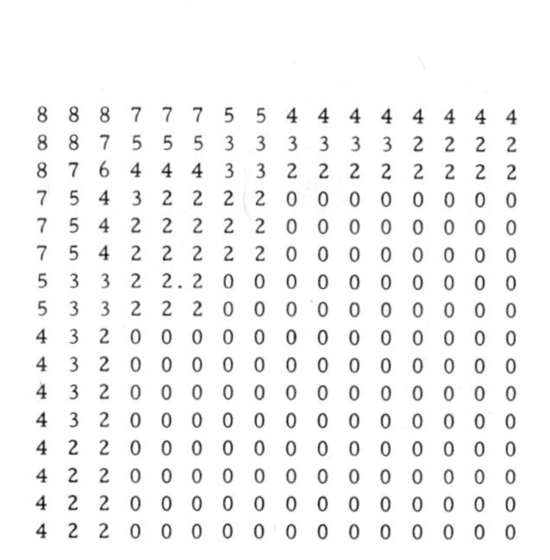

8	8	8	7	7	7	5	5	4	4	4	4	4	4	4	4
8	8	7	5	5	5	3	3	3	3	3	3	2	2	2	2
8	7	6	4	4	4	3	3	2	2	2	2	2	2	2	2
7	5	4	3	2	2	2	2	0	0	0	0	0	0	0	0
7	5	4	2	2	2	2	2	0	0	0	0	0	0	0	0
7	5	4	2	2	2	2	2	0	0	0	0	0	0	0	0
5	3	3	2	2	2	0	0	0	0	0	0	0	0	0	0
5	3	3	2	2	2	0	0	0	0	0	0	0	0	0	0
4	3	2	0	0	0	0	0	0	0	0	0	0	0	0	0
4	3	2	0	0	0	0	0	0	0	0	0	0	0	0	0
4	3	2	0	0	0	0	0	0	0	0	0	0	0	0	0
4	3	2	0	0	0	0	0	0	0	0	0	0	0	0	0
4	2	2	0	0	0	0	0	0	0	0	0	0	0	0	0
4	2	2	0	0	0	0	0	0	0	0	0	0	0	0	0
4	2	2	0	0	0	0	0	0	0	0	0	0	0	0	0
4	2	2	0	0	0	0	0	0	0	0	0	0	0	0	0

Fig. 9. Typical bit assignments for slant-transform zonal coding in 16 × 16 pixel blocks.

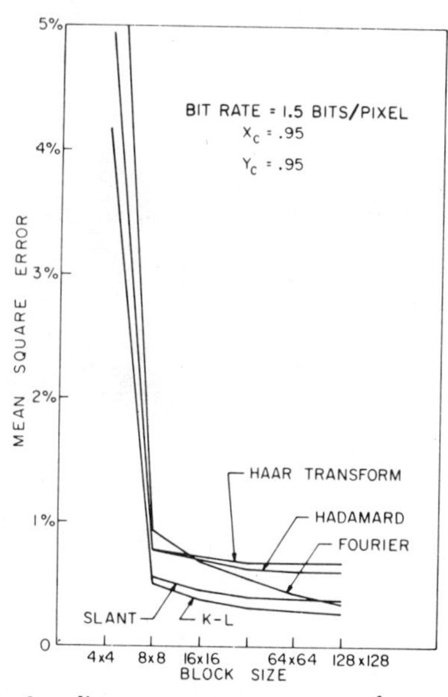

Fig. 10. Zonal coding mean-square error performance of image transforms as a function of block size.

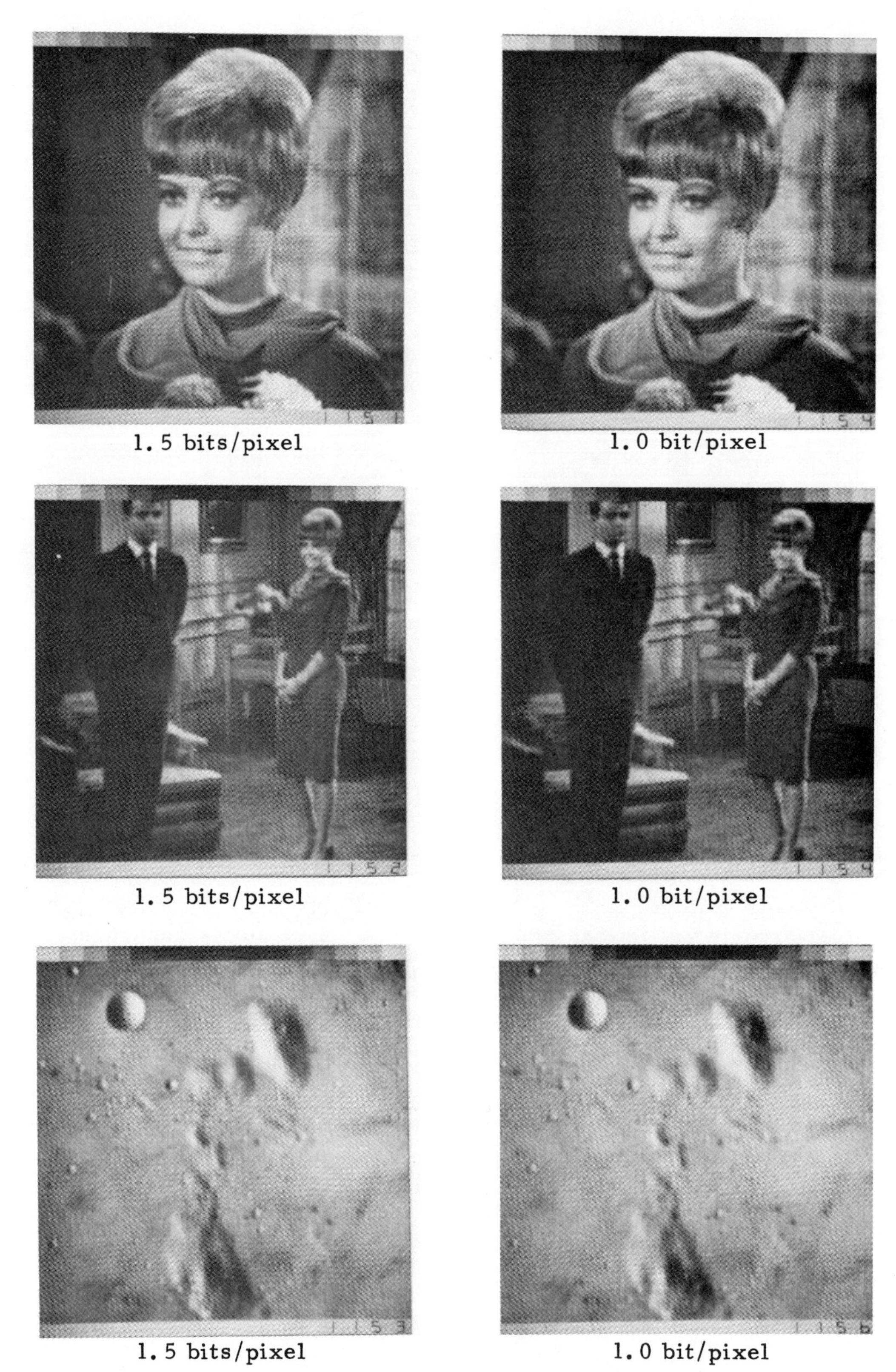

Fig. 11. Slant-transform zonal coding in 16 × 16 pixel blocks, quantized transform.

Hadamard transform
1. 5 bits/pixel

Fourier transform
1. 5 bits/pixel

Haar transform
1. 5 bits/pixel

Karhunen-Loeve transform
1. 5 bits/pixel

Fig. 12. Hadamard, Fourier, Haar, and Karhunen–Loeve transform zonal coding in 16 × 16 pixel blocks, quantized transform.

from

$$[\mathfrak{F}_Y] = [S][Y][S]^T \tag{28a}$$

$$[\mathfrak{F}_I] = [S][I][S]^T \tag{28b}$$

$$[\mathfrak{F}_Q] = [S][Q][S]^T \tag{28c}$$

where $[S]$ is the slant-transform matrix. Note that, since the coordinate conversion and spatial transformation are linear operations, their order may be reversed. Next, the transform samples are quantized with the number of quantum levels made proportional to the expected variance of each pixel, and with the quantization level spacing allowed to be variable to minimize the mean-square quantization error. The quantized samples $\hat{\mathfrak{F}}_Y(j,k)$, $\hat{\mathfrak{F}}_I(j,k)$, and $\hat{\mathfrak{F}}_Q(j,k)$ are then coded and transmitted over a possibly noisy channel. At the receiver the channel output is decoded, and inverse slant transforms are taken to obtain

$$[\hat{Y}] = [S]^T[\hat{\mathfrak{F}}_Y][S] \tag{29a}$$

$$[\hat{I}] = [S]^T[\hat{\mathfrak{F}}_I][S] \tag{29b}$$

$$[\hat{Q}] = [S]^T[\hat{\mathfrak{F}}_Q][S]. \tag{29c}$$

Finally, an inverse coordinate conversion results in the reconstructed tristimulus signals

$$\begin{bmatrix} \hat{R}(j,k) \\ \hat{G}(j,k) \\ \hat{B}(j,k) \end{bmatrix} = \begin{bmatrix} 1.000 & 0.956 & 0.621 \\ 1.000 & -0.272 & -0.647 \\ 1.000 & -1.106 & 1.703 \end{bmatrix} \begin{bmatrix} \hat{Y}(j,k) \\ \hat{I}(j,k) \\ \hat{Q}(j,k) \end{bmatrix}. \tag{30}$$

The coordinate conversion from the $R\ G\ B$ color space to the $Y\ I\ Q$ color space can be considered along with the spatial slant transforms as a three-dimensional transformation. The coordinate conversion provides an energy compaction between color planes and the spatial slant

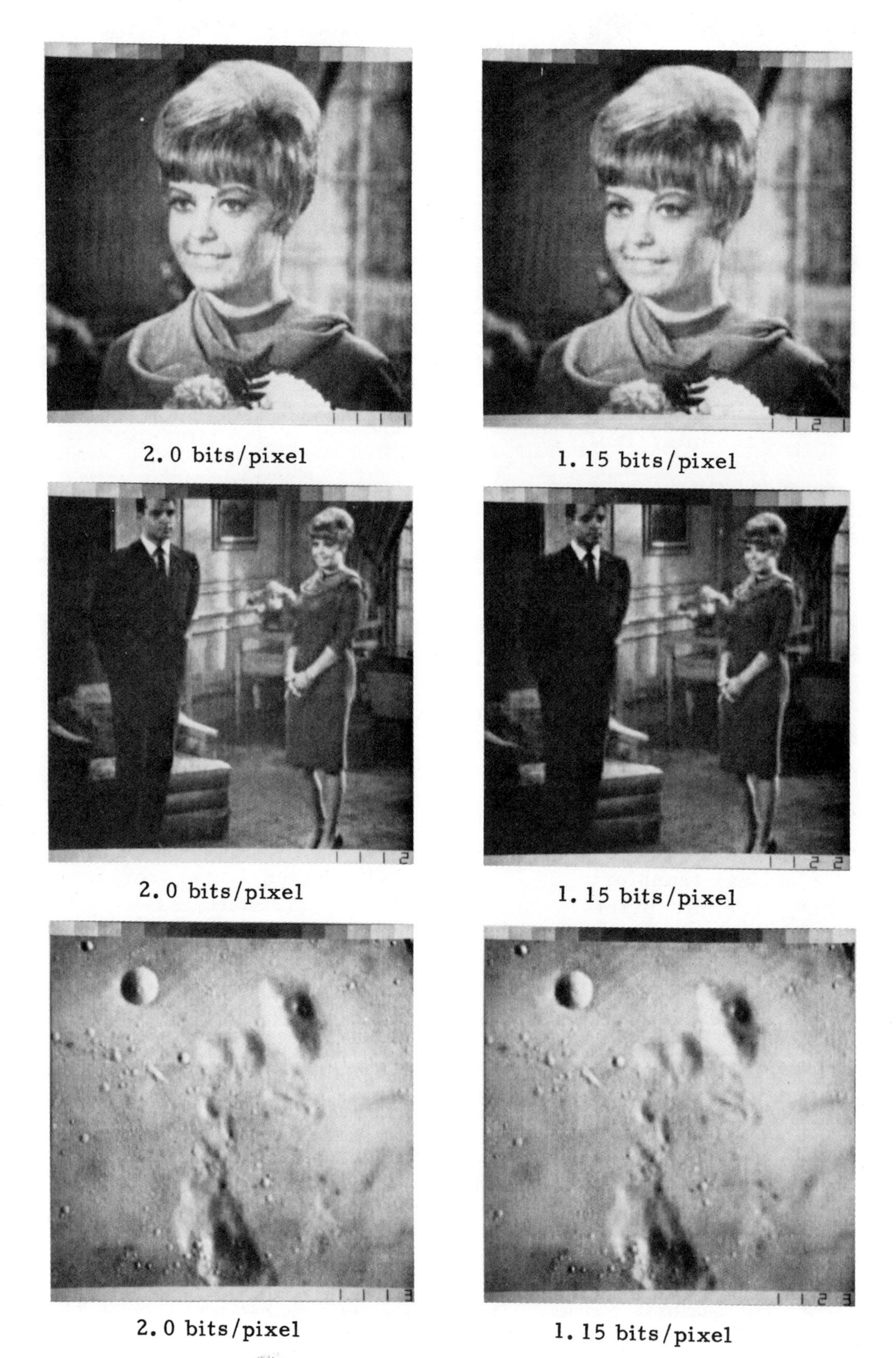

Fig. 13. Slant-transform threshold coding in 16 × 16 pixel blocks, quantized transform.

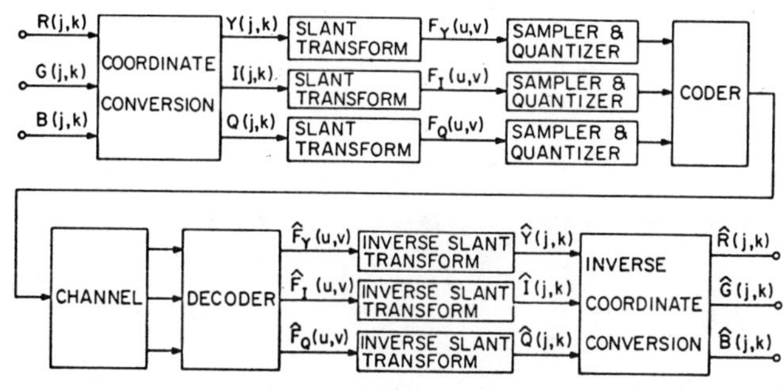

Fig. 14. Slant-transform color-image coding system.

TABLE I
ENERGY COMPACTION OF COORDINATE CONVERSIONS

Test Image	Coordinate System	Percentage of σ_1^2	Percentage of σ_1^2	Percentage of σ_3^2
GIRL	RGB	45.14	35.41	19.45
	YIQ	78.32	17.54	4.14
	$K_1K_2K_3$	85.84	12.10	2.06
COUPLE	RGB	51.55	31.09	17.36
	YIQ	84.84	13.81	1.35
	$K_1K_2K_3$	92.75	6.46	0.79

transforms provide an energy compaction within the color planes. Adopting this philosophy, the optimal three-dimensional transform would be a Karhunen–Loeve transformation which completely decorrelates the $3N^2$ color-image components. It has been shown [17] that the $Y\,I\,Q$ coordinate conversion provides almost as high an energy compaction for color images as does a Karhunen–Loeve color coordinate conversion. This result is verified by the color-image energy distribution of two color images described in Table I.

In order to optimally design the slant-transform image coder, it is necessary to specify some analytic measure of color-image fidelity. Unfortunately, no standard fidelity measures exist. As a rational alternative, the design procedure selected has been to design the transform domain quantization system to minimize the mean-square error between the $Y\,I\,Q$ and $\hat{Y}\,\hat{I}\,\hat{Q}$ color planes as defined by

$$\mathfrak{E} = \frac{1}{3N^2} E\{[Y(j,k) - \hat{Y}(j,k)]^2 + [I(j,k) - \hat{I}(j,k)]^2 + [Q(j,k) - \hat{Q}(j,k)]^2\}. \quad (31)$$

Several variations of the quantization procedure have been investigated [18]. The following has been found to be a highly effective implementation.

1) Model the row and column variance matrices of $R\,G\,B$ as first-order Markov processes and compute the variances of the elements of $Y\,I\,Q$.

2) Model the probability density of the constant basis vector component of F_Y by a Rayleigh density, and the probability densities of the other basis vector components of $F_Y F_I F_Q$ by Gaussian densities with variances computed in 1).

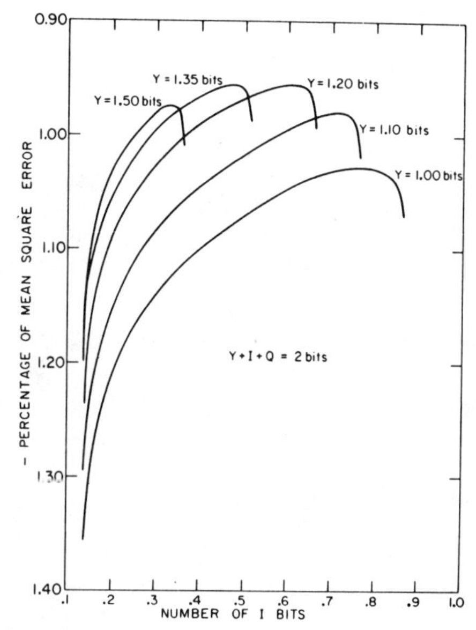

Fig. 15. Quantization error for various color plane bit assignments.

3) Assign

$$L_Y(u,v) = 2^{N_Y(u,v)} \quad (32a)$$

$$L_I(u,v) = 2^{N_I(u,v)} \quad (32b)$$

$$L_Q(u,v) = 2^{N_Q(u,v)} \quad (32c)$$

quantization levels to each transform component where the number of bits allotted to each component is made proportional to the logarithm of its variance computed in 1).

4) The total number of bits allotted to a color image is set at

$$N_B = N_{BY} + N_{BI} + N_{BQ} \quad (33a)$$

where

$$N_{BY} = \sum_{u=1}^{N} \sum_{v=1}^{N} N_Y(u,v) \quad (33b)$$

$$N_{BI} = \sum_{u=1}^{N} \sum_{v=1}^{N} N_I(u,v) \quad (33c)$$

$$N_{BQ} = \sum_{u=1}^{N} \sum_{v=1}^{N} N_Q(u,v). \quad (33d)$$

5) For a given value of N_B, select trial values of N_{BY}, N_{BI}, N_{BQ}, then compute quantization levels from 3), and with the probability density models of 2), perform a variable spacing quantization of each transform component using the Max [15] quantization rule.

6) For representative color images, compute the mean-square error, and, by iterative search techniques, determine optimum bit allocations for transform planes.

It should be noted that the above procedure need not be performed dynamically for every color image to be coded. The optimization need only be performed for typical color images to be coded to obtain a quantization scale which can be designed into the quantizer hardware.

Fig. 15 contains a plot of the mean-square error versus

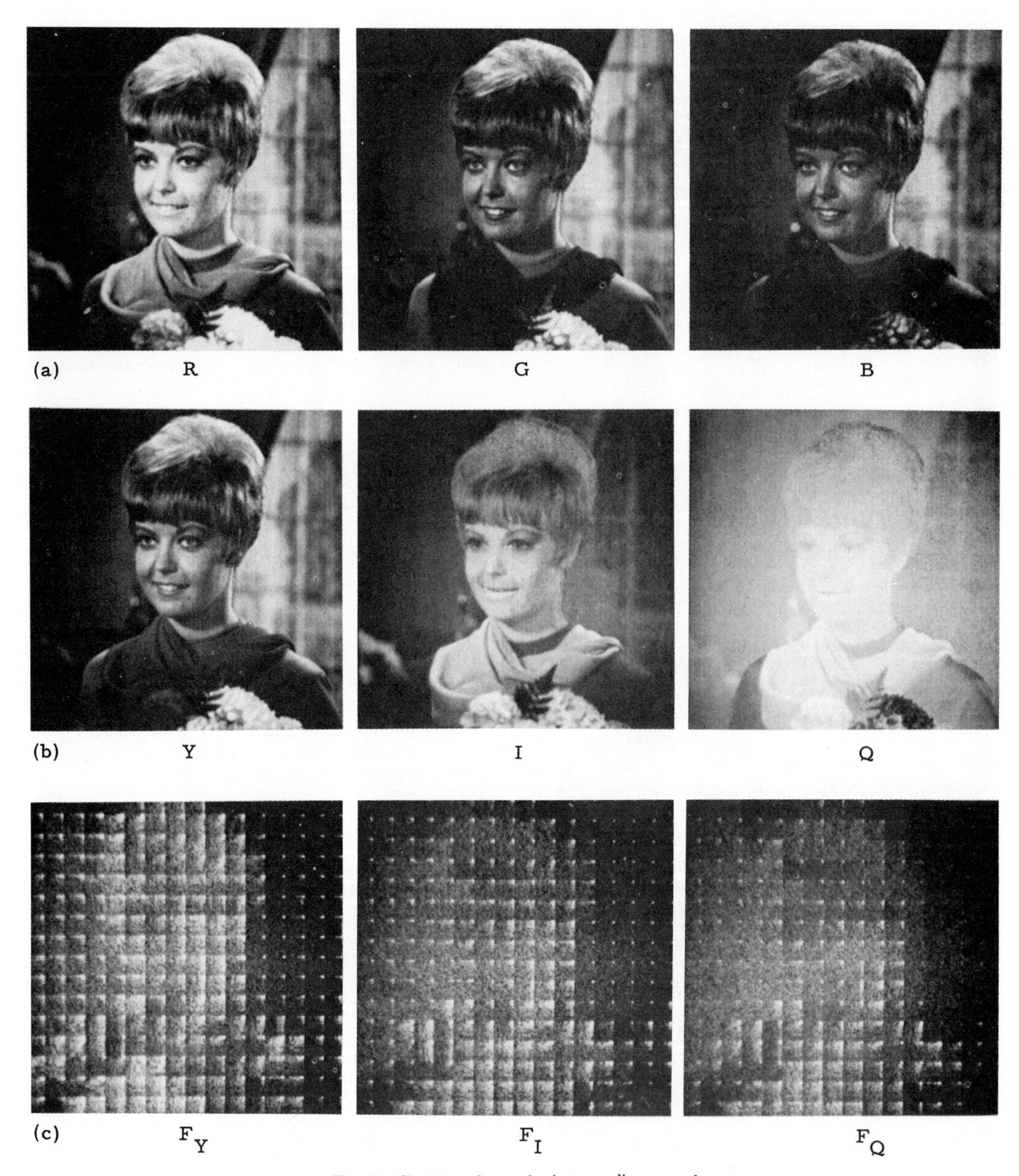

Fig. 16. Slant-transform color-image coding example.

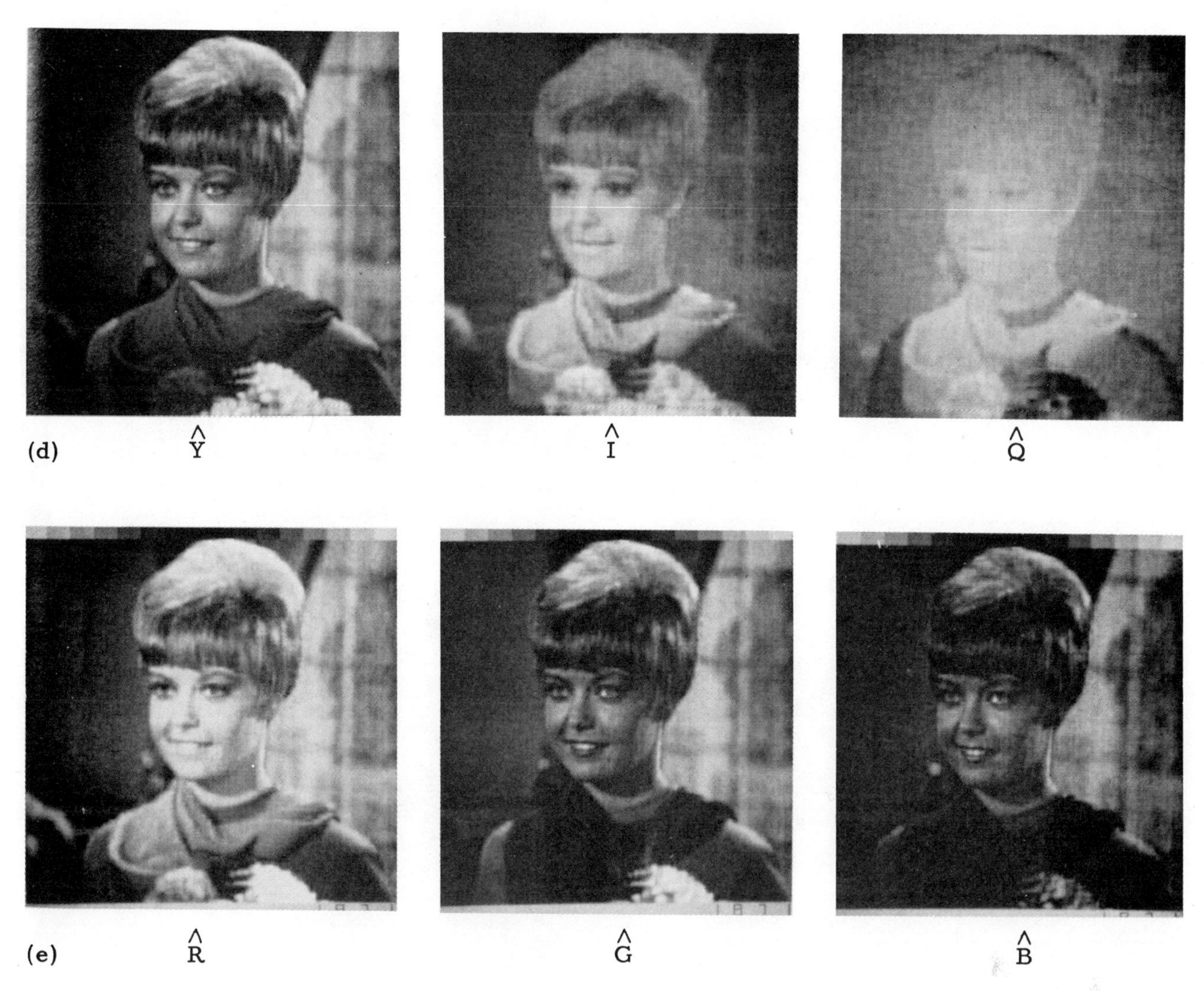

Fig. 16. Cont'd.

N_{BI} for several values of N_{BY} and a fixed value N_B for the girl image. The optimum average bit allocation for this test image is found to be: $N_{BY} = 1.25$, $N_{BI} = 0.55$, $N_{BQ} = 0.20$. The optimum scale does not change appreciably for other images or total bit allotments.

A computer simulation has been performed to subjectively evaluate the performance of the slant-transform color-image coding system. Fig. 16(a) contains monochrome pictures of the red, green, and blue components of an original image of 256 × 256 pixels. Each component of the original is quantized to 255 levels. It should be noted that visually the $R\ G\ B$ components are highly correlated. The corresponding $Y\ I\ Q$ components in Fig. 16(b) appear much less correlated. Fig. 16(c) contains illustrations of the logarithm of the magnitude of each slant-transform plane of the color image for transformation in 16 × 16 pixel blocks to illustrate the spatial energy compaction. In one of the simulation experiments the transform coefficients, $\mathfrak{F}_Y$, $\mathfrak{F}_I$, $\mathfrak{F}_Q$ were assigned code bits such that Y, I, Q were coded with an average of 1.2, 0.54, and 0.26 bits/pixel, respectively. The corresponding reproductions of Y, I, Q and R, G, B are shown in Fig. 16(d) and (e). In this experiment the coding has been reduced from 24 bits/pixel to 2 bits/pixel. The $R\ G\ B$ reconstructions exhibit some degradation as a result of the coding process, but the visual effect of the degradation is much less visible in the color reconstruction because of the spatial frequency limitations of the human visual system. Fig. 17 contains color images slant transform coded with 3 and 2 bits/pixel.

VI. ADDITIONAL TOPICS

The effect of channel errors on slant-transform coded images has been investigated quite extensively [11]. Studies indicate that the coding technique is relatively tolerant to channel errors. With zonal coding, for example, binary symmetric channel errors do not become subjectively noticeable until the error rate is reduced to about 10^{-3}.

Adaptive and semiadaptive quantization and coding

(a) 3 bits/pixel

(b) 3 bits/pixel

(c) 2 bits/pixel

(d) 2 bits/pixel

Fig. 17. Example of slant-transform color-image coding.

techniques have been developed and tested by simulation for the slant transform. These techniques, which are somewhat more complex than the fixed quantization and coding methods described in this paper, provide additional bandwidth reductions of about 2:1 for the same degree of image quality.

VII. SUMMARY

Slant-transform coding of monochrome and color images has proven to provide a substantial bandwidth reduction compared to conventional PCM coding with only a minimal amount of image degradation. It has been found that the slant transform results in a lower mean-square error for moderate size image blocks as compared to other unitary transforms with fast computational algorithms. Subjectively, it has been observed that the image quality of slant-transform coded images is somewhat higher than for images coded with other fast computational unitary transforms.

REFERENCES

[1] P. A. Wintz, "Transform picture coding," *Proc. IEEE*, vol. 60, pp. 809–820, July 1972.

[2] H. C. Andrews and W. K. Pratt, "Fourier transform coding of images," in *Conf. Rec. Hawaii Int. Conf. System Science*, Jan. 1968, pp. 677–679.

[3] G. B. Anderson and T. S. Huang, "Piecewise Fourier transformation for picture bandwidth compression," *IEEE Trans. Commun. Technol.*, vol. COM-19, pp. 133–140, Apr. 1971.
——, "Correction to 'Piecewise Fourier transformation for picture bandwidth compression,'" *IEEE Trans. Commun.* (Corresp.), vol. COM-20, pp. 488–492, June 1972.

[4] W. K. Pratt, J. Kane, and H. C. Andrews, "Hadamard transform image coding," *Proc. IEEE*, vol. 57, pp. 58–68, Jan. 1969.

[5] J. W. Woods and T. S. Huang, "Picture bandwidth compression by linear transformation and block quantization," in *Picture Bandwidth Compression*, T. S. Huang and O. J. Tretiak, Ed. New York: Gordon and Breach, 1972, pp. 555–573.

[6] A. Habibi and P. A. Wintz. "Image coding by linear transformation and block quantization," *IEEE Trans. Commun. Technol.*, vol. COM-19, pp. 50–62, Feb. 1971.

[7] M. Tasto and P. A. Wintz, "Image coding by adaptive block quantization," *IEEE Trans. Commun. Technol.*, vol. COM-19, pp. 957–972, Dec. 1971.

[8] H. C. Andrews, *Computer Techniques in Image Processing.* New York: Academic, 1970.

[9] H. Enomoto and K. Shibata, "Orthogonal transform coding

system for television signals," *IEEE Trans. Electromagn. Compat.*, vol. EMC-13, pp. 11–17, Aug. 1971.
[10] W. K. Pratt, L. R. Welch, and W. H. Chen, "Slant transform for image coding," in *Proc. Symp. Application of Walsh Functions*, Mar. 1972.
[11] W. H. Chen, "Slant transform image coding," University of Southern California, Image Processing Institute, Los Angeles, USCEE Rep. 441, May 1973.
[12] A. Papoulis, *Probability, Random Variables, and Stochastic Processes.* New York: McGraw-Hill, 1965.
[13] J. Pearl, H. C. Andrews, and W. K. Pratt, "Performance measures for transform data coding," *IEEE Trans. Commun.* (Concise Papers), vol. COM-20, pp. 411–415, June 1972.
[14] L. D. Davisson, "Rate-distortion theory and application," *Proc. IEEE*, vol. 60, pp. 800–808, July 1972.
[15] J. Max, "Quantizing for minimum distortion," *IRE Trans. Inform. Theory*, vol. IT-6, pp. 7–12, Mar. 1960.
[16] D. G. Fink, Ed., *Television Engineering Handbook.* New York: McGraw-Hill, 1957.
[17] W. K. Pratt, "Spatial transform coding of color images," *IEEE Trans. Commun. Technol.*, vol. COM-19, pp. 980–992, Dec. 1971.
[18] W. H. Chen and W. K. Pratt, "Color image coding with the slant transform," in *Proc. Symp. Application of Walsh Functions*, Apr. 1, 1973.

23

Reprinted with permission from *IEEE Pattern Anal. Mach. Intell. Trans.* **PAMI-4:**551–555 (1982)

New Algorithm for the Slant Transform

ZHONG-DE WANG

Abstract—Two new algorithms that are more convenient for computation than existing ones for the slant transform are developed. These algorithms reveal the close relationship between the slant transform and the Walsh-Hadamard transform and demonstrate that the slant transform may be approached by a series of steps which gradually change the transform from a Hadamard or Walsh transform to a slant transform.

Index Terms—Fast algorithm, Hadamard transform, orthogonal transform, slant transform, Walsh transform.

I. Introduction

The slant transform was introduced to image coding by Enomoto and Shibata [1] in 1971 and developed by Pratt *et al.* [2], [3]. It has been proven that the slant transform is more efficient and possesses less error in image coding than the Hadamard, Harr, and Fourier transforms [3]. Although the importance of the slant transform has been overshadowed with the discovery of the cosine transform [6] which is widely accepted as the best substitute for the Karhunen–Loêve transform [7], the slant transform still plays an important role for those people who seek the relationship among different orthogonal transforms [8]–[10]. Jones *et al.* [10] have proven that most commonly used orthogonal transforms which possess one-half even and one-half odd vectors may be simply obtained by postprocessing of the Walsh-Hadamard transform coefficients. Among these so-called even/odd transforms, perhaps the slant transform is the one related most closely to the Walsh-Hadamard transform.

The new algorithms discussed in this paper reveal the relationship between the slant transform and the Walsh–Hadamard transform and provide more convient methods for decomposing the slant transform.

II. Composing Slant Matrices

The slant transform, as indicated by its name, possesses a "slant" or a stairlike waveform for its first sequency vector. The matrix of slant transform may be composed by multiplying a series of sparse matrices on the Hadamard matrix. This will be shown as follows.

Let S_N^* denote the slant matrix of order N. The asterisk is used to represent that the sequency order of S_N^* is the same as the Hadamard matrix of the same order N. The subscript denotes the order of the matrix. Then, S_2^* is given by

$$S_2^* = \frac{1}{2^{1/2}} \begin{bmatrix} 1 & 1 \\ 1 & -1 \end{bmatrix}. \tag{1}$$

The slant matrix S_4^* is obtained by

$$S_4^* = \frac{1}{2^{1/2}} \left[\begin{array}{c|c|c|c} 1 & & & \\ \hline & a_4 & -b_4 & \\ & b_4 & a_4 & \\ \hline & & & 1 \end{array}\right] \left[\begin{array}{c|c} S_2^* & S_2^* \\ \hline S_2^* & -S_2^* \end{array}\right] \tag{2}$$

$$= \frac{1}{2}\begin{bmatrix} 1 & 1 & 1 & 1 \\ a_4 - b_4 & -a_4 - b_4 & a_4 + b_4 & -a_4 + b_4 \\ a_4 + b_4 & a_4 - b_4 & -a_4 + b_4 & -a_4 - b_4 \\ 1 & 1 & 1 & 1 \end{bmatrix} \quad (3)$$

where a_4 and b_4 are constants determined uniquely by the orthonomality condition and the step condition for the first sequency row, namely the third row of S_4^*. One may easily obtain

$$a_4 = \frac{2}{5^{1/2}}, \quad b_4 = \frac{1}{5^{1/2}}. \quad (4)$$

Therefore,

$$S_4^* = \frac{1}{2}\begin{bmatrix} 1 \quad 1 \quad 1 \quad 1 \\ \frac{1}{5^{1/2}} \times (1 \quad -3 \quad 3 \quad -1) \\ \frac{1}{5^{1/2}} \times (3 \quad 1 \quad -1 \quad -3) \\ 1 \quad -1 \quad -1 \quad 1 \end{bmatrix}. \quad (5)$$

The slant matrix S_{2N}^* is composed similar to the composing of S_4^*

$$S_{2N} = \frac{1}{2^{1/2}}$$

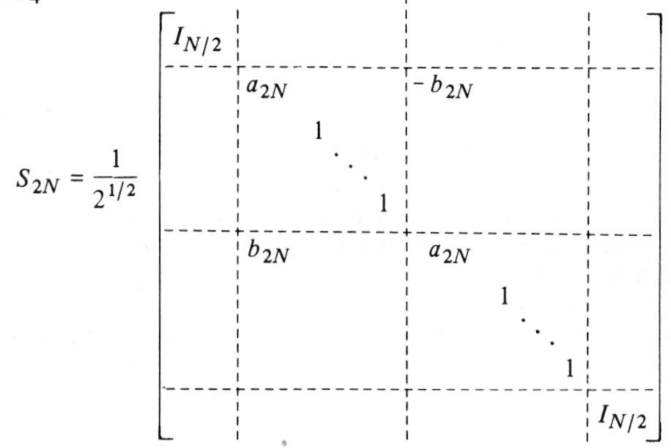

$$\cdot \begin{bmatrix} S_N^* & S_N^* \\ S_N^* & -S_N^* \end{bmatrix} \quad (6)$$

where $I_{N/2}$ is unit matrix of order $N/2$, and a_{2N} and b_{2N} are constants calculated by formulas

$$a_{2N} = \left(\frac{3N^2}{4N^2 - 1}\right)^{1/2}, \quad b_{2N} = \left(\frac{N^2 - 1}{4N^2 - 1}\right)^{1/2}. \quad (7)$$

These have already been given by Pratt [3].

III. Fast Algorithm

The composing method of the slant matrix S_{2N}^*, (6), indicates the fast algorithms for slant transform. But (6) may be modified so as to obtain the simplest fast algorithm. To show this clearly, the slant matrix S_8^* is used as an example. According to (6),

$$S_8^* = \frac{1}{2^{1/2}}\left[\begin{array}{c|cc|cc|c} I_2 & & & & & \\ \hline & a_8 & & -b_8 & & \\ & & 1 & & & \\ \hline & b_8 & & a_8 & & \\ & & & & 1 & \\ \hline & & & & & I_2 \end{array}\right] \left[\begin{array}{c|c} S_4^* & S_4^* \\ \hline S_4^* & -S_4^* \end{array}\right]. \quad (8)$$

The last matrix may be decomposed as

$$\left[\begin{array}{c|c} S_4^* & S_4^* \\ \hline S_4^* & -S_4^* \end{array}\right] = \left[\begin{array}{c|c} I_4 & I_4 \\ \hline I_4 & -I_4 \end{array}\right] \left[\begin{array}{c|c} S_4^* & \\ \hline & S_4^* \end{array}\right]. \quad (9)$$

Let M_4^* denote the first matrix on the right-hand side of (2) and M_8^* denote the first matrix on the right-hand side of (8), namely

$$M_4^* = \left[\begin{array}{c|c|c|c} 1 & & & \\ \hline & a_4 & -b_4 & \\ \hline & b_4 & a_4 & \\ \hline & & & 1 \end{array}\right] \quad (10)$$

and

$$M_8^* = \left[\begin{array}{c|cc|cc|c} I_2 & & & & & \\ \hline & a_8 & & -b_8 & & \\ & & 1 & & & \\ \hline & b_8 & & a_8 & & \\ & & & & 1 & \\ \hline & & & & & I_2 \end{array}\right] \quad (11)$$

Then, S_8^* becomes

$$S_8^* = \frac{1}{2^{1/2}} M_8^* \left[\begin{array}{c|c} I_4 & I_4 \\ \hline I_4 & -I_4 \end{array}\right] \left[\begin{array}{c|c} S_4^* & \\ \hline & S_4^* \end{array}\right]. \quad (12)$$

The last matrix may be decomposed as

$$\left[\begin{array}{c|c} S_4^* & \\ \hline & S_4^* \end{array}\right] = \frac{1}{2^{1/2}}$$

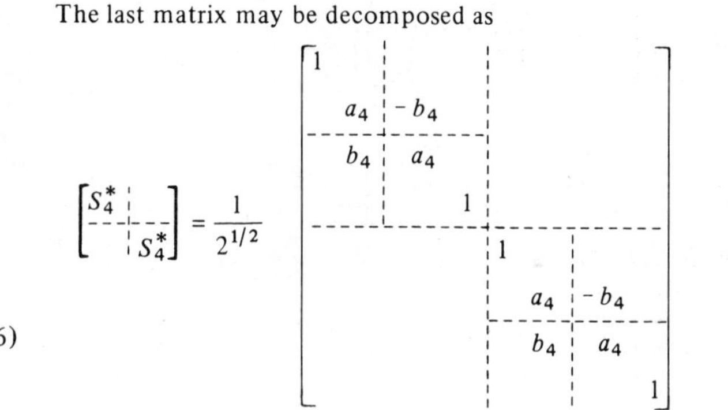

$$\left[\begin{array}{cc|cc} S_2^* & S_2^* & & \\ S_2^* & -S_2^* & & \\ \hline & & S_2^* & S_2^* \\ & & S_2^* & -S_2^* \end{array}\right] = \frac{1}{2^{1/2}} \left[\begin{array}{c|c} M_4^* & \\ \hline & M_4^* \end{array}\right] \cdot \left[\begin{array}{cc|cc} S_2^* & S_2^* & & \\ S_2^* & -S_2^* & & \\ \hline & & S_2^* & S_2^* \\ & & S_2^* & -S_2^* \end{array}\right] \quad (13)$$

Since Hadamard matrices H_2 of order two and H_4 of order four are [4]

$$H_2 = S_2^* = \frac{1}{2^{1/2}}\begin{bmatrix} 1 & 1 \\ 1 & -1 \end{bmatrix} \quad (14)$$

and

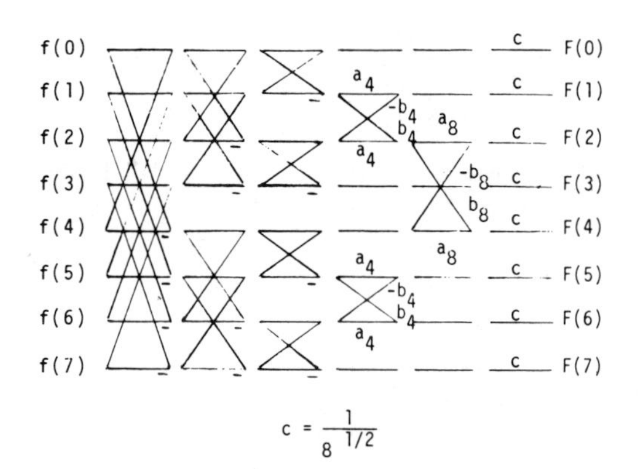

Fig. 1. Fast algorithm flowchart for slant transform of order 8.

$$H_4 = \frac{1}{2^{1/2}}\begin{bmatrix} H_2 & H_2 \\ H_2 & -H_2 \end{bmatrix} = \frac{1}{2^{1/2}}\begin{bmatrix} S_2^* & S_2^* \\ S_2^* & -S_2^* \end{bmatrix} \tag{15}$$

therefore,

$$\begin{bmatrix} S_4^* & \\ & S_4^* \end{bmatrix} = \begin{bmatrix} M_4^* & \\ & M_4^* \end{bmatrix} \begin{bmatrix} H_4 & \\ & H_4 \end{bmatrix}. \tag{16}$$

Substituting (16) into (12),

$$S_8^* = \frac{1}{2^{1/2}} M_8^* \begin{bmatrix} I_4 & I_4 \\ I_4 & -I_4 \end{bmatrix} \begin{bmatrix} M_4^* & \\ & M_4^* \end{bmatrix} \begin{bmatrix} H_4 & \\ & H_4 \end{bmatrix}. \tag{17}$$

But

$$\begin{bmatrix} I_4 & I_4 \\ I_4 & -I_4 \end{bmatrix} \begin{bmatrix} M_4^* & \\ & M_4^* \end{bmatrix} = \begin{bmatrix} M_4^* & \\ & M_4^* \end{bmatrix} \begin{bmatrix} I_4 & I_4 \\ I_4 & -I_4 \end{bmatrix} \tag{18}$$

and

$$\frac{1}{2^{1/2}} \begin{bmatrix} I_4 & I_4 \\ I_4 & -I_4 \end{bmatrix} \begin{bmatrix} H_4 & \\ & H_4 \end{bmatrix} = \frac{1}{2^{1/2}} \begin{bmatrix} H_4 & H_4 \\ H_4 & -H_4 \end{bmatrix} = H_8 \tag{19}$$

where H_8 is Hadamard matrix of order eight [4]. Therefore, S_8^* is obtained as

$$S_8^* = M_8^* \begin{bmatrix} M_4^* & \\ & M_4^* \end{bmatrix} H_8 \tag{20}$$

or

$$S_8^* = M_8^* \cdot M_{8,4}^* \cdot H_8 \tag{21}$$

where

$$M_{8,4}^* = \begin{bmatrix} M_4^* & \\ & M_4^* \end{bmatrix}. \tag{22}$$

The first subscript of $M_{8,4}^*$ denotes the order of the matrix, the second subscript represents that all the submatrices which align in the diagonal of matrix $M_{8,4}^*$ are M_4^*.

Following (20), a flowchart of forward slant transform is shown in Fig. 1 and inverse transform in Fig. 2.

It is easily noticed that the first three steps of Fig. 1 and the last three steps of Fig. 2 compute a Hadamard transform. $N \log_2 N$ additions and subtractions are required. The last two steps of Fig. 1 change the Hadamard transform "gradually" to the slant transform. $N - 2$ additions and subtractions together with $2(N - 2)$ multiplications are required. The same is true for the first two steps of Fig. 2. A total of $N \log_2 N + N - 2$ additions and subtractions together with $2(N - 2)$ multiplications are required to compute the slant transform of an N-dimensional data vector. For hardware implementation, the only requirement is to design a device to accomplish the "modification" transforms ($M_4^*, M_8^*, \cdots$) as a complement to existing Hadamard transform hardware.

IV. A Series of Orthogonal Matrices

Proceeding as in the last section, the slant matrix S_N^* may be obtained as

$$S_N^* = M_N^* \cdot M_{N,N/2}^*, \cdots M_{N,4}^* \cdot H_N \tag{23}$$

where H_N is Hadamard matrix of order N,

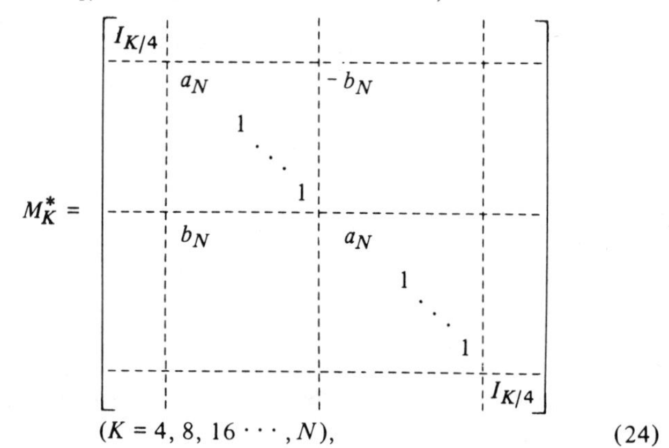

$$M_K^* = \begin{bmatrix} I_{K/4} & & & \\ & a_N & -b_N & \\ & 1 \ \ddots \ 1 & & \\ & b_N & a_N & \\ & & 1 \ \ddots \ 1 & \\ & & & I_{K/4} \end{bmatrix}$$

$$(K = 4, 8, 16 \cdots, N), \tag{24}$$

and

$$M_{N,K}^* = \text{diagonal } [\underbrace{M_K^*, M_K^*, \cdots M_K^*}_{N/K}] \qquad \left(K = 4, 8, \cdots, \frac{N}{2}\right) \tag{25}$$

are modification transforms which change the Hadamard transform gradually into the slant transform. And, a_N and b_N are determined by (7).

It is easy to verify that every M_K^* is a unitary matrix, and so is its "enlarged" matrix $M_{N,K}^*$. Now, a series of unitary matrices may be defined as

$$\left.\begin{aligned} &S_{N,2}^* = H_N, \\ &S_{N,4}^* = M_{N,4}^* \cdot S_{N,2}^*, \\ &\cdots \\ &S_{N,K}^* = M_{N,K}^* \cdot S_{N,K/2}^*, \qquad \left(K = 4, 8, \cdots, \frac{N}{2}\right) \\ &\cdots \\ &S_{N,N}^* = M_N^* \cdot S_{N,N/2} = S_N^* \end{aligned}\right\}. \tag{26}$$

The series of orthogonal matrices $S_{N,K}^*$ begins with the Hadamard matrix H_N when $K = 2$. It becomes more and more like a slant matrix with the doubling of K, and reaches a slant matrix when K reaches N.

V. Fast Algorithm for Slant Transform with Increasing Order of Sequency

It has been shown in the last section that the slant matrix S_N^* may be formed from the Hadamard matrix H_N by multiplying a series of sparse matrices on it. Since the Hadamard matrix and the Walsh matrix are the same matrix except that the Walsh matrix is sequency-ordered but the Hadamard matrix is not, one may reasonably conclude that the slant transform may also be formed from the Walsh transform, keeping the increasing order of sequency unchanged. This will be shown as follows.

Let S_N denote the sequency-ordered slant matrix of order N and W_N denote the Walsh matrix of the same order. Then, S_N may be formed from W_n by multiplying a series of sparse matrices on it:

$$S_N = M_N \cdot M_{N,N/2} \cdots M_{N,4} \cdot W_N \tag{27}$$

where

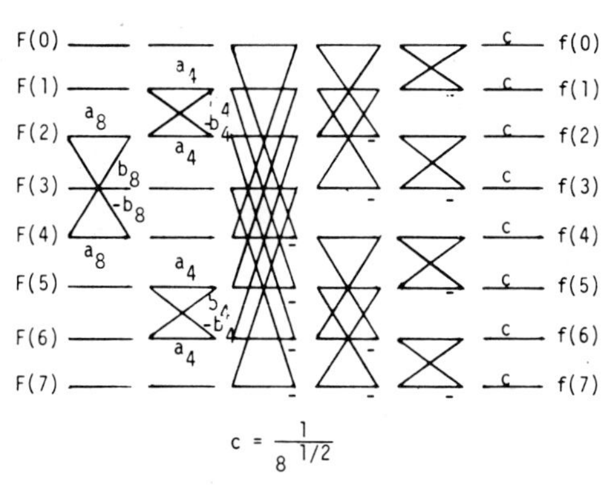

Fig. 2. Fast algorithm flowchart for reverse slant transform of order 8.

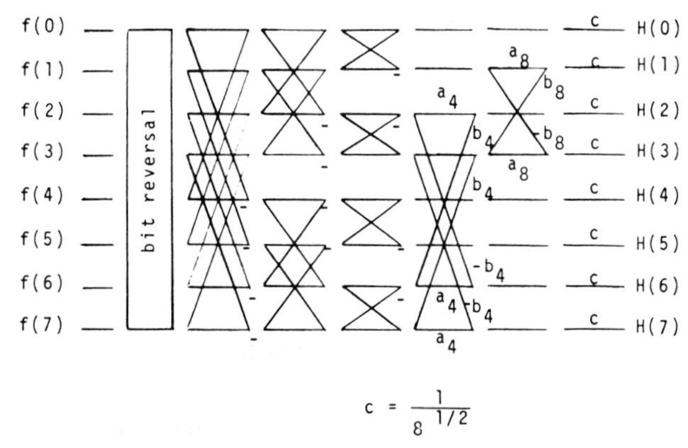

Fig. 3. Fast algorithm flowchart for sequency-ordered slant transform of order 8.

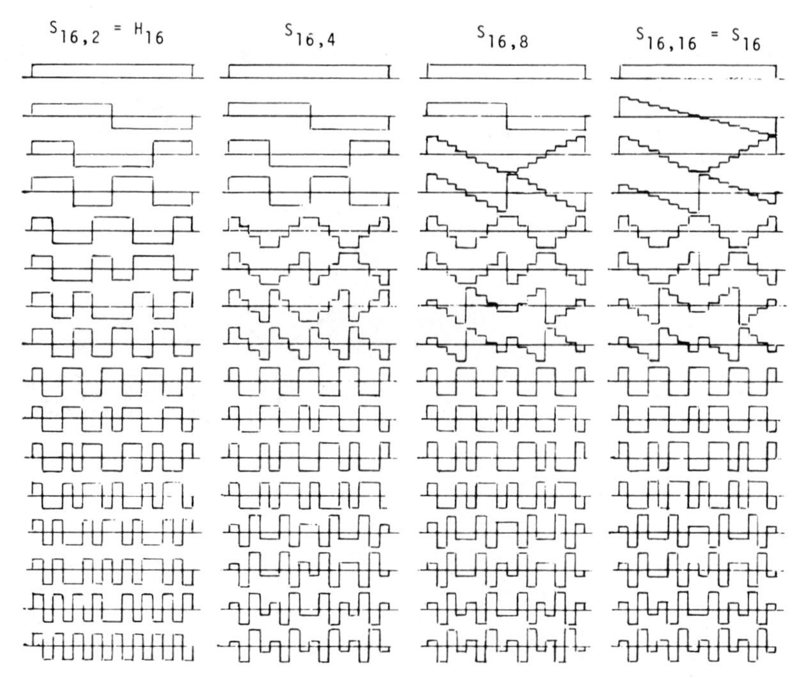

Fig. 4. Waveforms of $S_{N,K}$ for different K.

$$M_{N,4} = \begin{bmatrix} I_{N/4} & & & \\ & a_4 I_{N/4} & & b_4 I_{N/4} \\ & & I_{N/4} & \\ & -b_4 I_{N/4} & & a_4 I_{N/4} \end{bmatrix} \quad (28)$$

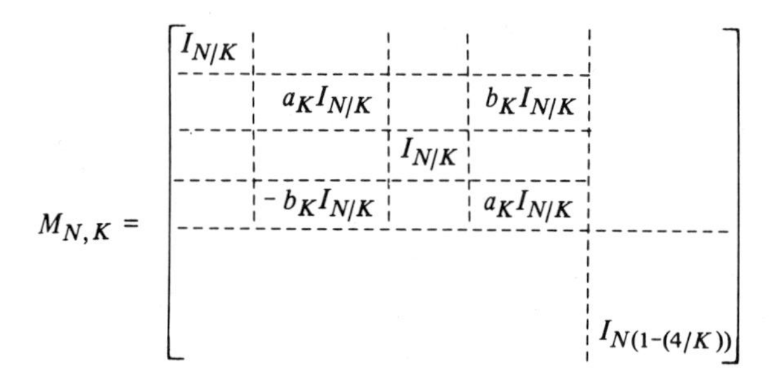

$$\left(K = 4, 8, \cdots, \frac{N}{2}\right) \quad (29)$$

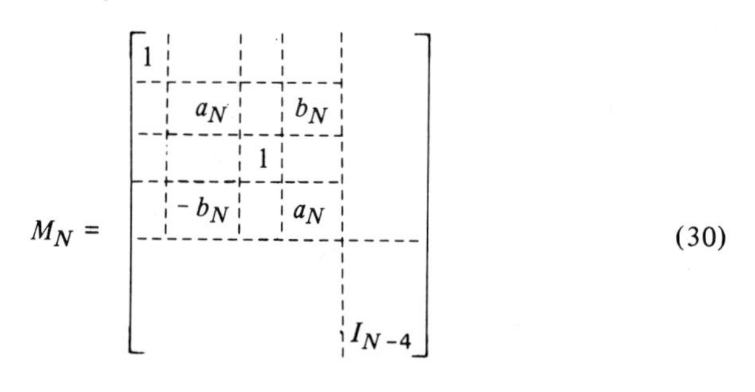

(30)

From (27), another fast algorithm is found which is shown in Fig. 3. The bit reversal and the first three steps compute a Walsh transform which has been given by Manz [5]. The last two steps change the Walsh transform gradually to a slant transform. In this algorithm, an additional bit reversal is required in comparing with the one shown in Fig. 2. For hardware implementation, there is no difference to accomplish $M^*_{N,K}$ or $M_{N,K}$. But, because of its increasing order of sequency, S_N is more preferable to S^*_N.

It is also easy to verify that every matrix $M_{N,K}$ is a unitary matrix. Similar to the last section, a series of unitary matrices may be defined as

$$\left.\begin{aligned} S_{N,2} &= W_N, \\ S_{N,4} &= M_{N,4} \cdot S_{N,2}, \\ &\cdots \\ S_{N,K} &= M_{N,K} \cdot S_{N,K/2}, \\ &\cdots \\ S_{N,N} &= S_N = M_N \cdot S_{N,N/2} \end{aligned}\right\}. \quad (31)$$

It is noticed that $S_{N,K}$ and $S^*_{N,K}$ are the same matrix except that the sequency orders are different. Like the behavior of the series of $S^*_{N,K}$, the series of orthogonal matrices $S_{N,K}$ begins with the Walsh Matrix W_N when $k = 2$, changes gradually with the doubling of k, and reaches S_N when $k = N$. The waveforms of $S_{N,K}$ with different k for $N = 16$ are shown in Fig. 4. That gives an impression of how much change each modification matrix $M_{N,K}$ makes.

ACKNOWLEDGMENT

The author wishes to thank his American sponsor, Prof. B. R. Hunt of the University of Arizona, Tucson, for his invaluable advice and suggestions, Y. Cheng for helpful discussion, and B. Stubbs for her typing.

REFERENCES

[1] H. Enomoto and K. Shibata, "Orthogonal transform coding system for television signals," *IEEE Trans. Electromagn. Compat.*, vol. EMC-13, pp. 11-17, Aug. 1971.

[2] W. K. Pratt, L. R. Welch, and W. H. Chen, "Slant transform for image coding," in *Proc. Symp. Appl. Walsh Function*, March 1972.

[3] W. K. Pratt, W. H. Chen, and L. R. Welch, "Slant transform image coding," *IEEE Trans. Commun.*, vol. COM-22, pp. 1075-1093, Aug. 1974.

[4] N. Ahmed and K. R. Rao, *Orthogonal Transform for Digital Signal Processing*. New York: Springer-Verlag, 1975.

[5] J. W. Manz, "A sequency-ordered fast Walsh transform," *IEEE Trans. Audio Electroacoust.*, vol. AU-20, pp. 204-205, Aug. 1972.

[6] N. Ahmed, T. Natarajan, and K. R. Rao, "Discrete cosine transform," *IEEE Trans. Comput.*, vol. C-23, pp. 90-93, Jan. 1974.

[7] W. K. Pratt, *Digital Image Processing*. New York: Wiley, 1978.

[8] B. J. Fino and V. R. Algazi, "Slant Haar transform," *Proc. IEEE*, vol. 62, pp. 653-654, May 1974.

[9] —, "A unified treatment of discrete fast unitary transform," *SIAM J. Comput.*, vol. 6, no. 4, pp. 700-717, 1977.

[10] H. W. Jones, D. N. Hein, and S. C. Knauer, "The Karhunen-Loevê, discrete cosine, and related transforms obtained via the hadamard transform," presented at the Int. Telemetering Conf., Nov. 1978.

Editor's Comments on Papers 24 and 25

24 JAIN
A Fast Karhunen-Loeve Transform for a Class of Random Processes

25 KITAJIMA and SHIMONO
Some Aspects of the Fast Karhunen-Loeve Transform

KARHUNEN-LOEVE TRANSFORM

Although the Karhunen-Loeve transform (KLT) is an optimal transform in a statistical manner under a variety of criteria, it is rarely used in digital signal processing (Elliott and Rao, 1982). In general, there is no fast algorithm for its implementation. Also, generation of the KLT itself is an involved process. As the KLT is based on the eigenvectors of the covariance matrix of a data sequence, the statistics of the data need to be known. Also, the KLT is not a fixed transform; that is, if the statistics change so does the KLT. However, KLT is used as a benchmark; that is, the other fixed transforms such as the DCT, DST, WHT, ST, etc. are compared to the KLT based on various performance criteria (Paper 5).

KLT not only decorrelates the transform coefficients, it also packs the most energy in the fewest number of transform coefficients; that is, it has the highest EPE of all the discrete transforms. For first order stationary Gauss Markov process, KLT reduces to the DST (Paper 10) provided the boundary values are known (Paper 24). Kitajima and Shimono (Paper 25), however, have shown that, as the sequence has to be modified based on the boundary values, KLT (in this case it reduces to the DST) is not as efficient as the DCT in the bandwidth reduction of the original sequence when the adjacent correlation coefficient ρ is close to unity. Similar conclusions have been drawn by Yip and Rao (Paper 12) and Ahmed and Flickner (1982).

REFERENCES

Ahmed, N., and M. D. Flickner, 1982, *Some Considerations of the Discrete Cosine Transform,* IEEE 16th Asilomar Conference on Circuits, Systems, and Computers, Pacific Grove, Calif., pp. 295–299.
Elliott, D. F., and K. R. Rao, 1982, *Fast Transforms, Algorithms, Analyses and Applications,* Academic Press, New York, 488p.

24

A Fast Karhunen–Loeve Transform for a Class of Random Processes

ANIL K. JAIN, MEMBER, IEEE

***Abstract*–The Karhunen-Loeve transform for a class of signals is proven to be a set of periodic sine functions and this Karhunen-Loeve series expansion can be obtained via an FFT algorithm. This fast algorithm obtained could be useful in data compression and other mean-square signal processing applications.**

I. INTRODUCTION

Various unitary transforms have attracted considerable attention during the past decade for their application in data compression (commonly called "transform coding") and other signal processing applications. If u represents an $N \times 1$ signal vector, then v is its transform given by

$$v = Tu, \tag{1}$$

where T denotes a unitary two-dimensional $N \times N$ matrix. The restriction of T to unitary matrices ensures conservation of signal energy in the transform domain, i.e.,

$$\sum_i |v_i|^2 = \sum |u_i|^2. \tag{2}$$

Examples of such transforms are Fourier, Hadamard, Karhunen-Loeve (KL), Haar, slant, cosine, etc. [1]–[8]. Typically, in the transformed signal v most of the image energy is concentrated

Paper approved by the Editor for Communication Theory of the IEEE Communications Society for publication after presentation at the 1974 National Electronics Conference, Chicago, IL, October 1974. Manuscript received November 26, 1974; revised November 1, 1975 and April 2, 1976. This research was supported in part by NASA under Contract NAS8-31434 and by the Advanced Research Projects Agency of the Department of Defense and monitored by the Air Force Eastern Test Range under Contract F08606-72-C-0008.

The author is with the Department of Electrical Engineering, State University of New York at Buffalo, Amherst, NY 14260.

in relatively few samples (usually the lower "frequency" samples) and these samples are considered sufficient for any subsequent signal processing. Two considerations which become important in selecting a transform for data compression are 1) dimensionality and 2) optimality for compression.

By dimensionality we refer to the computational effort required in implementing (1). For an arbitrary unitary transform T, computation of v takes N^2 mulitplications and about as many additions. When the value of N gets large, the computational load is unbearable so that considerations of dimensionality require the choice of T be restricted to certain classes of fast transforms. One class of such transforms, called Good transforms (after I. J. Good [16]), can be written as a product of several sparse matrices in the form

$$T = T_1 T_2 \cdots T_p, \tag{3}$$

where T_i, $i = 1, \cdots, p (p \ll N)$ are matrices with just a few nonzero entries ($\leqslant r$ entries per row, say, with $r \ll N$). Thus, the multiplication of T with, say an $N \times 1$ vector, is accomplished in about rpN computations. Then, if $N \cong 2^p$, the computations required in (1) reduce to approximately $N \log_2 N$.

By optimality we mean the efficiency of a transform in achieving data compression (or bandwidth compression). Usually this optimality is measured for a class of signals rather than for one signal because conceivably a transform could be optimal for one particular signal and be very poor for others. For a given class of signals having the same second-order statistics, the KL transform is shown to be optimal [9]-[13]. In the past, the success of fast Fourier, cosine transforms, etc. has been noted due to their near optimal performance in many image coding studies [6], [7].

Although the KL transform is optimal, it has dimensionality difficulties. First, the KL transform is unique for a class of signals. Therefore, it has to be computed for that class. Second, even if a closed form analytic expression for the KL transform is known, the transformation calculations of (1) do not, in general, have a fast algorithm available.

In this paper we show that for a class of Markov signals, the KL transform is a set of periodic sine functions, if the boundary values of the signal are fixed (or known). These sine functions are shown to be related to the Fourier transform so that a fast Fourier transform algorithm could be utilized in implementing the KL transform. It is known [11] that the KL transform for the Markov signals considered is given by a set of nonperiodic sinusoids, so that a fast algorithm is unavailable. However, the assumption of known boundary conditions reduces the nonperiodic sine waves to periodic sine waves, thereby leading to a fast computational algorithm.

II. SIGNAL REPRESENTATION

Let $\{x_i\}$ be a finite one-dimensional random process with zero mean and unit variance and an autocorrelation function given by

$$E[x_i x_{i+n}] = \rho^{|n|}, \qquad i = 0, 1, \cdots, N+1. \tag{4}$$

The zero-mean and unity-variance assumptions are nonessential but serve only to present a simplified analysis. It is well known that the sequence x_i can be represented by a first-order stationary Markov process as

$$x_{i+1} = \rho x_i + \epsilon_i, \qquad i \geqslant 0 \tag{5}$$

with

$$E[\epsilon_i] = 0, \qquad E[\epsilon_i \epsilon_j] = (1 - \rho^2)\delta_{ij}, \tag{6}$$

and a suitably chosen initial condition x_0.

It has been shown earlier [14] that the above Markov process, for a fixed N, can also be represented by the following equations:

$$x_i = \alpha(x_{i+1} + x_{i-1}) + \nu_i, \qquad 1 \leqslant i \leqslant N \tag{7}$$

$$x_0 = \rho x_1 + \nu_0 \tag{8}$$

$$x_{N+1} = \rho x_N + \nu_{N+1}, \tag{9}$$

where $\alpha = \rho/(1 + \rho^2)$ and $\{\nu_i, i = 0, 1, \cdots, N, N+1\}$ is a well-defined random process.

Note that the above representation is for a finite process (N = fixed and $i = 0, 1, \cdots, N+1$). The interpretation of this representation is as follows. If $\bar{x}_i$ denotes the best linear mean-square estimate of x_i obtained from a linear combination of the elements of the partial sequence $\{x_j, \forall j \neq i\}$, then the differences $\{\nu_i\}$ given by

$$x_i - \bar{x}_i = \nu_i \tag{10}$$

are such that the variance

$$E[(x_i - \bar{x}_i)^2] \triangleq E[\nu_i^2] \tag{11}$$

is minimum. This will be called the minimum variance representation. In order to find $\bar{x}_i$ we start by writing

$$\bar{x}_i = \sum_{k=1}^{i} a_{ik} x_{i-k} + \sum_{k=1}^{N+1-i} b_{ik} x_{i+k} \tag{12}$$

and find a_{ik} and b_{ik} for each i and k such that $E[(x_i - \bar{x}_i)^2]$ is minimized. Further details of this procedure may be found in [14].

The ν Process: The minimum variance property of the sequence $\{\nu_i\}$ does not guarantee its being uncorrelated. In fact, in our case, the sequence elements ν_i have a nearest neighbor correlation. The correlation properties of the sequence $\{\nu_i\}$ can be expressed as

$$E[\nu_i \nu_j] = \begin{cases} \beta_2^2, & i = j, \\ -\alpha\beta_2^2, & |i - j| = 1, \\ 0, & \text{otherwise}, \end{cases} \tag{13}$$

for $i, j = 1, \cdots, N$ and at the endpoints

$$E[\nu_0^2] = \beta_1^2 \triangleq (1 - \rho^2) = E[\nu_{N+1}^2], \tag{14}$$

$$E[\nu_0 \nu_1] = -\alpha\beta_1^2 = E[\nu_N \nu_{N+1}], \tag{15}$$

$$E[\nu_0 \nu_i] = E[\nu_{N+1} \nu_j] = 0, \qquad i > 1; \qquad j < N. \tag{16}$$

These results are derived in Appendix I.

III. THE KL TRANSFORM OF A CLASS OF MARKOVIAN SIGNALS

If x is a one-dimensional $n \times 1$ vector with autocorrelation matrix R, then the KL transform of x is a matrix Φ, composed of the eigenvectors of R and is defined by the relation

$$\Phi R \Phi^T = \Gamma, \tag{17}$$

where Γ is a diagonal matrix of eigenvalues γ_i^2. If x is a first-order Markov process with autocorrelation given by (4), then the elements of the KL transform are given by [11] (for n = even)

$$\Phi_{ij} = a_j \sin\left[\omega_j\left(i - \frac{n+1}{2}\right) + \frac{j\pi}{2}\right], \tag{18}$$

where

$$\gamma_i^2 = \frac{1-\rho^2}{1 - 2\rho\cos\omega_i + \rho^2},$$

a_i is the normalization constant and $\{\omega_i\}$ are the positive roots of

$$\tan n\omega = -\frac{(1-\rho^2)\sin\omega}{\cos\omega - 2\rho + \rho^2\cos\omega}. \tag{19}$$

Equation (19) is a transcendental equation giving nonharmonic solutions in the sine waves of (18). Now, the KL transform of the vector x may be written as $\hat{x} = \Phi x$ or

$$\hat{x}_i = \sum_{j=1}^{n} \Phi_{ij} x_j = \sum_{j=1}^{n} a_j x_j \sin\left[w_j\left(i + \frac{n+1}{2}\right) + \frac{j\pi}{2}\right]. \tag{20}$$

Due to nonharmonicity of the sine terms, a fast algorithm (like the FFT) is unavailable in computing the series of (20). Therefore, typically n^2 computations are required in computing $\{\hat{x}_i, i = 1, \cdots, n\}$.

From (17) and (20) and using $\Phi^T\Phi = I$, it is easy to see that

$$E[\hat{x}_i \hat{x}_j] = \gamma_i^2 \delta_{ij}. \tag{21}$$

Since the samples $\hat{x}_i$ are uncorrelated, they can be quantized independently. Another advantage of the KL transform is its minimum mean-square error property which makes it optimal for data compression [13].

IV. THE FAST KL TRANSFORM ALGORITHM

From (13) to (16), if we define a vector ν^* with elements $\nu_0, \nu_1, \cdots, \nu_N, \nu_{N+1}$, then its autocorrelation matrix can be written as

$$C^* \triangleq E[\nu^* \nu^{*T}] = \begin{bmatrix} \beta_1^2 & -\rho\beta_2^2 & 0 & \cdots & \cdots & \cdots & 0 \\ -\rho\beta_2^2 & \beta_2^2 & -\alpha\beta_2^2 & & & & \vdots \\ 0 & -\alpha\beta_2^2 & \beta_2^2 & \ddots & \bigcirc & & 0 \\ \vdots & & \ddots & \ddots & \ddots & & \vdots \\ \vdots & \bigcirc & & \ddots & \ddots & -\alpha\beta_2^2 & 0 \\ \vdots & & & & -\alpha\beta_2^2 & \beta_2^2 & -\rho\beta_2^2 \\ 0 & \cdots & \cdots & 0 & & -\rho\beta_2^2 & \beta_1^2 \end{bmatrix} \tag{22}$$

(The inner dashed block is $N \times N$.)

The matrix C^* above is a $(N+2) \times (N+2)$ matrix with $\beta_1^2 = (1-\rho^2)$, $\beta_2^2 = (1-\rho^2)/(1+\rho^2)$, and $\alpha = \rho/(1+\rho^2)$. First, the elements of the sequence $\{\nu_i\}$ are not uncorrelated. Second, except at the boundary points (i.e., for ν_0 and ν_{N+1}), the rest of the elements ν_k, $k = 1, \cdots, N$ have identical properties. If, for the time being, we drop the endpoints and only consider the partial sequence $\{\nu_k, k = 1, \cdots, N\}$, and represent this by a vector ν, then the $N \times N$ autocorrelation matrix C is given as

$$C = E[\nu\nu^T] = \beta_2^2 \begin{bmatrix} 1 & -\alpha & & & \bigcirc \\ -\alpha & 1 & -\alpha & & \\ & \ddots & \ddots & \ddots & \\ & & & & -\alpha \\ \bigcirc & & & -\alpha & 1 \end{bmatrix} \tag{23}$$

$$\triangleq \beta_2^2 Q, \tag{24}$$

where Q is the $N \times N$ matrix in (22). The matrix Q is a symmetric tridiagonal Toeplitz matrix.

Theorem 1: The KL transform of the partial ν sequence $\{\nu_k, k = 1, \cdots, N\}$ is given by

$$\psi_{ij} = \sqrt{\frac{2}{N+1}} \sin\frac{ij\pi}{N+1}. \tag{25}$$

Proof: The eigenvectors ψ_{ij} and the eigenvalues λ_i, of the $N \times N$ symmetric tridiagonal Toeplitz matrix Q are given by [14]

$$\psi_{ij} = \sqrt{\frac{2}{N+1}} \sin\frac{ij\pi}{N+1} \tag{26}$$

and

$$\lambda_i = 1 - 2\alpha\cos\frac{i\pi}{N+1} \qquad \text{for} \quad i, j = 1, \cdots, N. \tag{27}$$

Clearly, the matrix C in (23) has its eigenvectors also given by $\{\psi_{ij}\}$ (since β_2^2 is a scalar constant), so that the matrix $\{\psi_{ij}\}$ is the KL transform of ν.

Theorem 2 [11]: For the first-order stationary finite, Gauss-Markov sequence $\{x_i, i = 0, 1, \cdots, N, N+1\}$, if the boundary conditions x_0 and x_{N+1} are given, then the KL transform of the partial sequence $\{x_k, k = 1, \cdots, N\}$ conditioned on x_0, x_{N+1} is a matrix ψ with elements ψ_{ij} given by (26).

Proof: Let the given boundary conditions be

$$x_0 = c, \quad x_{N+1} = d. \tag{28}$$

If x and ν are defined as $N \times 1$ vectors of components $\{x_1, \cdots, x_N\}$ and $\{\nu_1, \cdots, \nu_N\}$, respectively, then (7) can be written as

$$Qx = \nu + b, \tag{29}$$

where Q is the $N \times N$ tridiagonal matrix in (23) and b is a $N \times 1$ vector containing only the information at the end-points; viz.,

$$b_1 = \alpha c, \quad b_N = \alpha d, \quad b_k = 0, \qquad 2 \leqslant k \leqslant N-1. \tag{30}$$

Since c and d are given and ν and b are uncorrelated (see (5) in Appendix I), we have from (29)

$$\mu \triangleq E[x \mid c,d] = Q^{-1}(b + E[\nu/c,d]) = Q^{-1}b \triangleq x_b \tag{31}$$

and

$$R_b = E[(x-\mu)(x-\mu)^T \mid c,d]$$
$$= Q^{-1}E[\nu\nu^T]Q^{-1} = \beta_2{}^2 Q^{-1}. \tag{32}$$

Hence, the covariance of x given end conditions is simply $\beta_2{}^2 Q^{-1}$. Since $\beta_2{}^2$ is a scalar, the eigenvectors of $\beta_2{}^2 Q^{-1}$ are given by $\{\psi_{ij}\}$, defined above.

Equation (29) above can be rewritten as

$$x = Q^{-1}(\nu + b) = y + x_b \tag{33a}$$

with

$$E[y x_b{}^T] = 0, \tag{33b}$$

where we have defined $y = Q^{-1}\nu$ and $\mu = x_b = Q^{-1}b$. Hence, (33a) is an orthogonal decompositon of the finite random process $\{x_i, 1 \leqslant i \leqslant N\}$ in terms of a zero-mean random process y whose KL transform is a fast transform (viz., the sine transform) and an orthogonal process x_b (to be called "boundary response") is completely determined by the two boundary conditions of $\{x_i\}$ viz., x_0 and x_{N+1}. Denoting $\hat{x} = \psi x$ and similarly $\hat{\nu}$, $\hat{b}$, $\hat{y}$, etc., and realizing from Theorem 1 that $\psi Q \psi = \Lambda$, (33a) is transformed to yield

$$\hat{x}_i = \frac{\hat{\nu}_i}{\lambda_i} + \frac{\hat{b}_i}{\lambda_i} = \hat{y}_i + \hat{\mu}_i. \tag{34}$$

The definition of b in (30) gives

$$\hat{b}_i = \sqrt{\frac{2}{N+1}}\, \alpha(c - (-1)^i d) \sin \frac{i\pi}{N+1} \tag{35}$$

and application of the result in (32) shows $\hat{y}_i{}^0$ are uncorrelated, i.e.,

$$E[\hat{y}_i \hat{y}_j] = \frac{\beta_2{}^2}{\lambda_i}\, \delta_{ij}. \tag{36}$$

From (26), the eigenvectors are independent of the correlation parameter ρ and only the eigenvalues λ_i depend on the statistics of the random process x. This is in contrast with the eigenfunctions ϕ_{ij}, (18), which depend on ρ through ω_i and $\gamma_i{}^2$. Moreover, the eigenvectors of (26) are harmonic sine waves, so that a fast computational algorithm is possible and is developed below.

In order to use the fast KL transform algorithm for data compression, say, of a sequence $\{x_0, x_i, \cdots, x_N, x_{N+1}\}$, first the boundary response vector x_b is calculated according to (31) using the boundary values x_0 and x_{N+1}. The $N \times 1$ vector y is then simply determined as $x - x_b$, where x is the $N \times 1$ vector of the partial sequence $\{x_i, 1 \leqslant i \leqslant N\}$. The elements of y can now be compressed via its $N \times N$ fast KL transform ψ. The two boundary values x_0, x_{N+1} are orthogonal to y, and are compressed independently as a 2×1 vector by its 2×2 KL transform. The reconstructed vector x^* is obtained first generating $x_b{}^*$ from the reconstructed boundary values $x_0{}^*$, $x_{N+1}{}^*$ and adding it to the reconstructed vector y^* from the compressed vector y. Then $\{x_0{}^*, x_1{}^*, \cdots, x_N{}^*, x_{N+1}{}^*\}$ is the reconstructed sequence. Note that this procedure is different from conventional transform method of data compression. For example, if the orthogonal matrix ψ, i.e., the sine transform, is used for conventional transform domain data compression of $\{x_i, 0 \leqslant i \leqslant N+1\}$, then if x_E denotes the $(N+2) \times 1$ vector of these elements, the sine transformation is performed on the entire sequence according to $\hat{x}_E = \psi x_E$ and the elements of $\hat{x}_E$ are coded for transmission (or storage).

Finally, note that if x is non-Gaussian, then the results of Theorem 1 are still true except that μ then is not the conditional mean of x given c, d, but is defined as $Q^{-1}b$ and R_b is not the conditional covariance of x given c, d; but is the covariance of the random process $(x - \mu)$.

Fast Implementation of the Transform

While (26) is directly related to the FFT, the elements ψ_{ij} are not the "imaginary" term in the conventional Fourier transform. It is easy to check that the terms ψ_{ij} form a complete orthonormal set of vectors in constrast with the "sine" terms of the Fourier transform.

Now

$$\hat{x}_k \triangleq \sum_{m=1}^{N} \psi_{km} x_m = \sqrt{\frac{2}{N+1}} \sum_{m=1}^{N} x_m \sin \frac{mk\pi}{N+1}. \tag{37}$$

Equation (37) can be implemented via an FFT algorithm by writing

$$\hat{x}_k = 2\,\frac{1}{\sqrt{2(N+1)}} \sum_{m=0}^{2N+1} x_m \sin \frac{2km\pi}{2N+2} = 2\,\mathrm{Im}\,\{\mathrm{DFT}[x_m]\} \tag{38}$$

for $1 \leqslant k \leqslant N$, where Im implies imaginary terms,

$$\mathrm{DFT}[x_m] \triangleq \frac{1}{\sqrt{M}} \sum_{m=0}^{M-1} x_m \exp\left\{\sqrt{-1}\,\frac{2\pi km}{M}\right\}, \qquad M = 2N+2$$

and $x_m \triangleq 0$ for $m = 0$ and for $m \geqslant N+1$.

Comparison with the Discrete Cosine Transform (DCT)

The DCT of a sequence $\{x_m\}$ is defined as

$$g_0 = \frac{\sqrt{2}}{N} \sum_{m=0}^{N-1} x_m, \qquad g_k = \frac{2}{N} \sum_{m=0}^{N-1} x_m \cos \frac{(2m+1)k\pi}{2N}$$
$$1 \leqslant k \leqslant N-1. \tag{39}$$

In experimental studies reported by Ahmed *et al.* [6], the performance of the DCT (for first-order Markov signals with values of ρ near 0.9) has been shown to be close to the KL transform. The terms g_k in (39) can also be calculated via the

DFT as

$$g_k = \frac{\sqrt{2}}{M} R_e \left\{ \exp\left(\sqrt{-1}\,\frac{k\pi}{2N}\right) \sum_{m=0}^{2N-1} \exp\left(\sqrt{-1}\,\frac{2\pi k m}{2N}\right) \right\},$$
$$1 \leqslant k \leqslant N-1, \tag{40}$$

where $R_e\{\cdot\}$ implies the real part of the terms in { }. Although both the fast KL and the DCT's have obvious relationship with DFT, the DCT does not satisfy the KL transform equation (17) for either R [the original autocorrelation of (4)] or for R_b the [the conditional autocorrelation of (32)]. These calculations are straightforward and therefore, are not included here. The relationship between the KL transform of (18) and the fast KL transform of (26) is more easily seen; the eigenvectors in the latter being periodic with ω_i of (18) being replaced by $i\pi/n + 1$. Although in (38), a $2(N+1)$ step FFT is needed, it can be shown that the $(N+1)$ step sine transform of (37) can also be implemented via an $(N+1)$ step FFT algorithm [17]. However, since the fast KL transform is real, subsequent transform domain calculations for coding or other processing are less for KL transform as compared to the DFT.

Finally, even though Q is a symmetric Toeplitz matrix, Q^{-1} is not Toeplitz in general. (This can be easily seen by inverting a 3×3 Q matrix, as an example.) This implies while $\{\nu_k\}$ is a stationary sequence, the sequence $\{x_k \mid c,d\}$, i.e., $\{x_k\}$ given boundary conditions is not stationary; and the periodic sine functions of the fast KL transform correspond to a nonstationary autocorrelation matrix.

V. EXTENSION TO TWO DIMENSIONS

The one-dimensional results proven in the previous sections are easily extended to two dimensions, when the covariance of the two-dimensional sequence (denoted u_{ij}) is assumed separable. For a zero-mean sequence $\{u_{ij}(0 \leqslant (i,j) \leqslant N+1)\}$ with autocorrelation of the form $E[u_{ij}u_{i+n,j+m}] = \rho_1^{|n|}\rho_2^{|m|}$ (a model often used for images [15]). Equations (4) to (9) of Appendix II give a representation of this sequence (where $\rho_1 = \rho_2$ has been assumed for simplicity). If U, V, ν represent $N \times N$ matrices of elements $\{u_{ij}\}$, $\{v_{ij}\}$, $\{\nu_{ij}\}$; $1 \leqslant (i,j) \leqslant N$, then from (5) and (6) or Appendix II we can write $QU = V + \alpha B_1$, $VQ = \nu + \alpha B_2$, which gives

$$QUQ = \nu + \alpha B_1 Q + \alpha Q B_2 - \alpha^2 B_3 \triangleq \nu + B, \tag{41}$$

where B_1, B_2, B_3 are $N \times N$ matrices containing only the boundary conditions. Specifically, the first and last (Nth) rows of B_1 contain $\{u_{0j}; 1 \leqslant j \leqslant N\}$ and $\{u_{N+1,j}; 1 \leqslant j \leqslant N\}$, respectively, and the rest of the elements are zero. Likewise, the first and the last (Nth) columns of B_2 contain $\{u_{i,0}, 1 \leqslant i \leqslant N\}$ and $\{u_{i,N+1}, 1 \leqslant i \leqslant N\}$, respectively, and the rest are zero. The matrix B_3 has only four nonzero terms viz., $B_3(1,1) = u_{0,0}$, $B_3(N,1) = u_{N+1,0}$, $B_3(1,N) = u_{0,N+1}$, and $B_3(N,N) = u_{N+1,N+1}$. The other elements of B_3 are zero. If $\bar{U}$, $\bar{\nu}$, and $\bar{B}$ are now defined as $N^2 \times 1$ vectors corresponding to a lexicographic (dictionary) order of the matrices U, ν, and B, respectively, (41) can be rewritten as

$$(Q \circledast Q)\bar{U} = \bar{\nu} + \bar{B}, \tag{42}$$

where $\circledast$ denotes Kronecker product. From Appendix II, the vector $\bar{\nu}$ has zero mean and $E[\bar{\nu}\bar{\nu}^T] = \beta_2{}^4(Q \circledast Q)$. Clearly, from Theorem 1 in the last section, $(\psi \circledast \psi)$ is the $N^2 \times N^2$ KL transform matrix of $\bar{\nu}$ and

$$\bar{\mu} \triangleq E[\bar{U} \mid \bar{B}] = \alpha(Q \circledast Q)^{-1}\bar{B} \tag{43}$$

and

$$R_y \triangleq E[(\bar{U} - \bar{\mu})(\bar{U} - \bar{\mu})^T \mid \bar{B}] = \beta_2{}^4(Q \circledast Q)^{-1}. \tag{44}$$

From (44), the matrix of eigenvectors of R_y also is $(\psi \circledast \psi)$, so that the KL transform of $\{u_{ij}$ given boundary conditions $u_{kl}, k,l = 0, N+1\}$ is given by the matrix ψ defined in (26).

Following the development of one dimension, once again it can be shown that

$$E[B_{ij}\nu_{kl}] = 0, \tag{45}$$

i.e., the boundary values and the process $\{\nu_{ij}\}$ are uncorrelated giving the two-dimensional orthogonal decomposition via (41) of the $N \times N$ image U as

$$U = Q^{-1}\nu Q^{-1} + Q^{-1}BQ^{-1} \triangleq Y + U_b, \tag{46a}$$

where Y is a zero-mean two-dimensional random process whose KL transform is the fast sine transform and the boundary response U_b is completely determined by the boundary values of U. Hence, if we define $\hat{U} = \psi U \psi$ and similarly $\hat{Y}$, $\hat{U}_b$ etc., then the following relations are easily verified

$$\hat{U}_{ij} = \frac{\hat{\nu}_{ij}}{\lambda_i\lambda_j} + \frac{\hat{B}_{ij}}{\lambda_i\lambda_j} = \hat{Y}_{ij} + \hat{U}_b(i,j), \tag{46b}$$

where

$$E[\hat{\nu}_{ij}\hat{\nu}_{kl}] = \beta_2{}^4\lambda_i\lambda_j\delta_{ik}\delta_{jl} \tag{46c}$$

$$E[\hat{Y}_{ij}\hat{Y}_{kl}] = \frac{\beta_2{}^4}{\lambda_i\lambda_j}\delta_{ik}\delta_{jl}. \tag{46d}$$

If $\{r_{0j}, 1 \leqslant j \leqslant N\}$ is the one-dimensional sine transform of the $1 \times N$ row $\{u_{0,k} 1 \leqslant k \leqslant N\}$, i.e.,

$$\hat{r}_{0,j} = \sum_{k=1}^{N} \psi_{jk} u_{0,k}$$

and $\hat{r}_{N+1,j}$, $\hat{c}_{i,0}$ and $\hat{c}_{i,N+1}$ are similarly defined as the sine transforms of $\{u_{N+1,k}\}$, $\{u_{k,0}\}$, and $\{u_{k,N+1}\}$, respectively, then the transform domain elements B_{ij} in (46b) are given by

$$\begin{aligned}\hat{B}_{ij} &= \alpha\lambda_j(\hat{r}_{0,j}\psi_{1,i} + \hat{r}_{N+1,j}\psi_{N,i}) + \alpha\lambda_i(\hat{c}_{i,0}\psi_{j,i} + \hat{c}_{i,N+1}\psi_{j,n}) \\ &= -\alpha^2\psi_{1,i}(u_{0,0}\psi_{1,j} + u_{0,N+1}\psi_{N,j}) \\ &\quad -\alpha^2(u_{0,N+1}\psi_{j,1}\psi_{i,N} + u_{N+1,N+1}\psi_{j,N}\psi_{i,N}).\end{aligned}$$

To implement the two-dimensional algorithm, first the boundary response U_b is calculated using the boundary values and then the $N \times N$ matrix Y, which equals $U - U_b$ is calculated and processed by the sine transform ψ. The boundary conditions in B are processed separately. Fig. 1 shows the block diagram for the fast KL transform algorithm implementation for image coding.

VI. DISCUSSION AND CONCLUSIONS

It was shown that the KL transform of finite first-order stationary Gauss-Markov processes conditioned on given boundary conditions becomes a (fast) sine transform. This result implies that a finite first-order Markov process can be modified by its "boundary response" to yield a new process whose KL transform is a fast transform. Extension to two dimensions with reference to images with separable convariance function was shown. Application in image data compression and mean square filtering seem possible. Preliminary experiments on data compression have shown compression ratios around 8 : 1 are possible. Details of these and other experiments concerning Wiener filtering of images may be

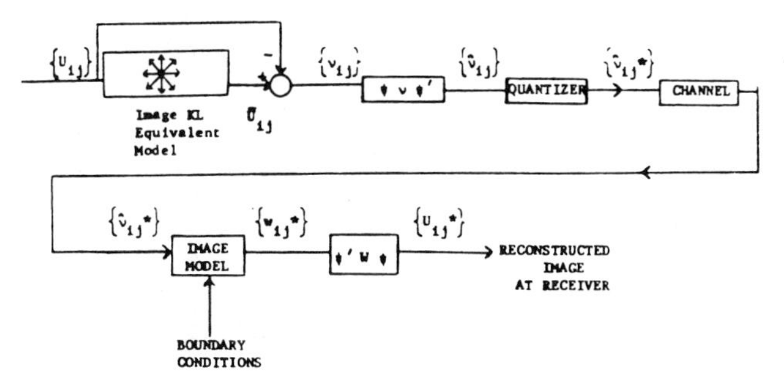

Fig. 1. Fast KL transform algorithm for image data compression.

found in [18], [19]. Other extensions are discussed in [20]–[23].

APPENDIX I

CORRELATION PROPERTIES OF THE MINIMUM VARIANCE REPRESENTATION

The equations of representation of the first-order stationary sequence $\{x_i\}$ with zero mean and autocorrelation

$$E[x_i x_j] = \rho^{|i-j|} \tag{1}$$

are given by

$$x_i = \alpha(x_{i+1} + x_{i-1}) + \nu_i, \qquad i = 1, \cdots, N \tag{2}$$

$$x_0 = \rho x_1 + \nu_0, \quad x_{N+1} = \rho x_N + \nu_{N+1}, \tag{3}$$

where $\alpha = \rho/(1+\rho^2)$. Mulitplying (2) by x_{i+k}, taking expectations and using (1) we get

$$\rho^{|k|} = \alpha(\rho^{|k-1|} + \rho^{|k+1|}) + [\nu_i x_{i+k}] \tag{4}$$

which for $|k| \geq 1$ and with $\alpha = \rho/(1+\rho^2)$, gives

$$E[\nu_i x_{i+k}] = 0, \qquad k \neq 0, \quad 1 \leq i \leq N. \tag{5}$$

Similarly for $k = 0$,

$$E[x_i \nu_i] = (1-\rho^2)/(1+\rho^2) \triangleq \beta_2{}^2; \qquad 1 \leq i \leq N. \tag{6}$$

Multiplication of (2) by ν_{i+k}, taking expectations and use of (5) and (6) yields

$$E[\nu_i \nu_{i+k}] = \beta_2{}^2(\delta_{k,0} - \alpha\delta_{k,-1} - \alpha\delta_{k,1}), \qquad 1 \leq i, \quad i+k \leq N, \tag{7}$$

where δ_{ij} is the Kronecker delta function. Similar procedure when applied to (3) gives

$$E[\nu_0 x_k] = (1-\rho^2)\delta_{k,0};$$

$$E[\nu_0 \nu_k] = (1-\rho^2)\delta_{k,0} - \alpha(1-\rho^2)\delta_{k,1} \tag{8}$$

$$E[\nu_{N+1} x_k] = (1-\rho^2)\delta_{k,N+1};$$

$$E[\nu_{N+1}\nu_k] = (1-\rho^2)\delta_{k,N+1} - \alpha(1-\rho^2)\delta_{k,N}. \tag{9}$$

APPENDIX II

A TWO-DIMENSIONAL IMAGE REPRESENTATION

Let $\{u_{ij}, i,j = 0,1,\cdots,N,N+1\}$ be a two-dimensional stationary sequence with zero mean and a separable autocorrelation function given by

$$E[u_{ij}u_{i+n,j+m}] = \rho^{|n|}\rho^{|m|}. \tag{1}$$

Let $\bar{u}_{ij}$ be the best linear mean-square estimate of u_{ij} obtained from all u_{mn} not including the point u_{ij}. This is obtained by first writing

$$\bar{u}_{ij} = \sum_{k}\sum_{\substack{l \\ k+l\neq 0}} a_{kl}(u_{i+k,j+l} + u_{i+k,j-l} + u_{i-k,j-l} + u_{i-k,j+l}), \tag{2}$$

and minimizing $E[(u_{ij} - \bar{u}_{ij})^2]$ over the coefficients a_{kl}. Differentiation of this expression with respect to a_{kl}, setting it equal to 0 and after some algebraic manipulations we get for $1 \leq (i,j) \leq N$

$$\rho^{|k|}\rho^{|l|} = \sum_{k'}\sum_{l'} a_{k'l'}(\rho^{|k+k'|} + \rho^{|k-k'|}) \cdot (\rho^{|l+l'|} + \rho^{|l-l'|}), \tag{3}$$

for all $k + l \neq 0$, $k' + l' \neq 0$. From (3) it can be proven $a_{01} = \rho/(1+\rho^2) = a_{10}$; $a_{11} = -a_{01}{}^2$ and $a_{kl} = 0$ for $k^2 + l^2 \geq 4$, so that the two-dimensional representation equation defined as $u_{ij} = \bar{u}_{ij} + \nu_{ij}$ becomes

$$u_{ij} = \alpha(u_{i+1,j} + u_{i,j+1} + u_{i,j-1} + u_{i-1,j}) - \alpha^2(u_{i+1,j-1} + u_{i+1,j+1} + u_{i-1,j+1} + u_{i-1,j-1}) + \nu_{ij}, \tag{4}$$

where $\alpha = \rho/(1+\rho^2)$. Equation (4) can be rearranged to give

$$u_{ij} - \alpha(u_{i+1,j} + u_{i-1,j}) = v_{ij} \tag{5}$$

$$v_{ij} - \alpha(v_{i,j+1} + v_{i,j-1}) = \nu_{ij}, \qquad \text{for } 1 \leq (i,j) \leq N. \tag{6}$$

Following Appendix I, the statistical properties of v_{ij} and ν_{ij} are obtained, for $1 \leq (i,j) \leq N$, as

$$E[\nu_{ij}] = 0 = E[v_{ij}] \tag{7}$$

$$E[v_{ij}v_{kl}] = \beta_2{}^2\rho^{|j-l|}(\delta_{ik} - \alpha\delta_{i+1,k} - \alpha\delta_{i-1,k}) \tag{8}$$

$$E[\nu_{ij}\nu_{kl}] = \beta_2{}^4(\delta_{ik} - \alpha\delta_{i+1,k} - \alpha\delta_{i-1,k}) \cdot (\delta_{jl} - \alpha\delta_{j+1,l} - \alpha\delta_{j-1,l}), \tag{9}$$

where $\beta_2{}^2 = ((1-\rho^2)/(1+\rho^2))$.

At boundary points ($i = 0, N+1; j = 0, N+1$) the minimum variance representation equations are somewhat different. These equations are not given since they are not required for the derivation of the fast KL transform result. Fig. 2 shows the structure of the minimum variance representation model for the two-dimensional case.

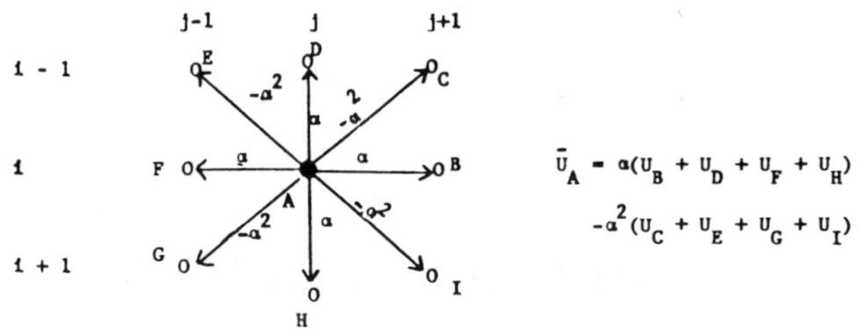

Fig. 2. Minimum variance representation model.

REFERENCES

[1] A. K. Jain, "A fast Karhunen-Loeve transform for finite discrete images," in *Proc. Nat. Electron. Conf.*, Chicago, IL, Oct. 1974.

[2] H. C. Andrews, *Computer Techniques in Image Processing.* New York: Academic Press, 1970.

[3] W. H. Chen, "Slant transform image coding," Ph.D. dissertation, Univ. of Southern California, Los Angeles, May 1973.

[4] W. K. Pratt, "A comparison of digital image transforms," presented at the Mervin J. Kelly Communication Conf., Univ. of Missouri, Rolla, MO, Oct. 1970.

[5] A. Habibi and P. A. Wintz, "Image coding by linear transformation and block quantization," *IEEE Trans. Commun. Tech.*, vol. COM-19, Feb. 1971.

[6] N. Ahmed, T. Natarajan, and K. R. Rao, "Discrete cosine transform," *IEEE Trans. Comput.*, vol. C-23, pp. 90–93, Jan. 1974.

[7] P. A. Wintz, "Transform picture coding," *Proc. IEEE*, vol. 60, pp. 809–820, July 1972.

[8] W. K. Pratt, "Karhunen-Loeve transform coding of images," in *Proc. 1970 IEEE Int. Symp. Inform. Theory*, 1970.

[9] H. P. Kramer and M. V. Mathews, "A linear coding for transmitting a set of correlated signals," *IRE Trans. Inform. Theory*, vol. IT-2, pp. 41–46, Sept. 1956.

[10] H. Hotelling, "Analysis of a complex of statistical variables into principal components," *J. Educ. Psychology*, vol. 24, pp. 417–441, 498–520, 1933.

[11] W. D. Ray and R. M. Driver, "Further decomposition of the Karhunen-Loeve series representation of a stationary random process," *IEEE Trans. Inform. Theory*, vol. IT-16, pp. 663–668, Nov. 1970.

[12] V. R. Algazi and D. J. Sakrison, "On the optimality of the Karhunen-Loeve expansion," *IEEE Trans. Inform. Theory* (Corresp.), pp. 319–321, Mar.1969.

[13] S. Watanabe, "Karhunen-Loeve expansion and factor analysis, theoretical remarks and applications," in *Trans. 4th Prague Conf. Inform. Theory, Statist. Decision Functions and Random Processes*, Prague, The Netherlands, pp. 635–660, 1965.

[14] A. K. Jain, "Image coding via a nearest neighbors image model," *IEEE Trans. Commun.*, vol. COM-23, pp. 318–331, Mar. 1975.

[15] A. K. Jain and E. Angel, "Image restoration, modelling and reduction of dimensionality," *IEEE Trans. Comput.*, vol. C-23, May 1974.

[16] I. J. Good, "The interaction algorithm and practical Fourier analysis," *J. Royal Statistical Soc.* (London), vol. *B20*, p. 361, 1958.

[17] E. O. Brigham, *The Fast Fourier Transform.* Englewood Cliffs, NJ: Prentice Hall, pp. 167–169, 1974.

[18] A. K. Jain, "A fast Karhunen-Loeve transform for recursive filtering of images corrupted by white and colored noise," to be published.

[19] —, "Computer program for fast Karhunen Loeve transform algorithm," Dep. Electr. Eng., SUNY, Buffalo, NY, Final Rep. NASA Contract NAS8-31434, Feb. 1976.

[20] A. K. Jain and D. G. Lainiotis, "A fast biorthogonal expansion for optimum feature extraction of a class of image data," presented at the Johns Hopkins Conf. on Inform. & Syst. Sc., Baltimore, MD, Apr. 1975.

[21] A. K. Jain, "Image restoration by operator factorization of point spread function and image covariance function," Dep. Electr. Eng., SUNY, Buffalo, NY, Tech. Rep. 75-001, Oct. 1975.

[22] A. K. Jain, S. H. Wang, and Y. Z. Liao, "Fast KL transform data compression studies," in *Proc. Nat. Telecommun. Conf.*, Dallas, TX, to be published.

[23] A. K. Jain, *Multidimensional Techniques in Computer Image Processing*, to be published.

25

Some Aspects of the Fast Karhunen–Loeve Transform

HIDEO KITAJIMA AND TETSUO SHIMONO

Abstract—**A few important properties of the fast Karhunen–Loeve transform (FKLT) proposed by Jain [3] are discussed. The purpose here is to emphasize the role of the boundary conditions upon which the entire theory is based. The effects of the zero boundary conditions are also discussed. It is argued that the transform is not as efficient as it might appear when used in bit rate reduction.**

INTRODUCTION

In various signal processing applications, orthogonal transforms play important roles. The Karhunen–Loeve transform is known to be optimal in data compression applications in the sense that it compacts most of the signal energy into low-frequency components of the transform; it also generates uncorrelated data. Mathematically, the two functions are equivalent. Lack of efficient algorithms to implement the KLT has promoted the use of suboptimal transforms such as the DFT, DCT [1], etc. These transforms approximately diagonalize covariance matrices of the $\rho^{|i-j|}$ type [2]. Jain [3], [4] has proposed a sine transform as the fast KLT for a random process which is a modified version of the given process. He has shown that the sine transform exactly diagonalizes the covariance matrix of the modified process. However, the effectiveness of the transform is open to question since it does not directly deal with the original signal. In data compression applications, what counts is not apparent decorrelation, but the saving in bit rate made possible by the use of a particular transform. As mentioned above, the maximum bit rate reduction and complete decorrelation by the use of an orthogonal transform are equivalent. But the resultant transform here, i.e., the modification of the signal sequence by linear combinations of its first and last elements followed by an orthogonal sine transform, is not orthogonal at all. Therefore, the complete diagonalization of the covariance matrix of the modified signal vector does not guarantee the maximum reduction in bit rate; we shall show that the fast KL transform (FKLT) does not attain the maximum bit rate reduction.

We shall also point out implicit assumptions on the boundary conditions. This will enable us to discuss the effects of zero boundary conditions.

FAST KARHUNEN–LOEVE TRANSFORM

The proposed FKLT is outlined here for ease of the succeeding discussion. Let $\{x_i,\ i = 0, 1, \cdots, N+1\}$ be a finite random process with zero mean and unit variance. Also, assume that its covariance is given by

$$E\{x_i x_j\} = \rho^{|i-j|}, \qquad 0 \leqslant \rho < 1,\ 0 \leqslant i,\ j \leqslant N+1. \qquad (1)$$

This process can be represented by the following equations:

$$x_i = \alpha(x_{i+1} + x_{i-1}) + \nu_i, \qquad 1 \leq i \leq N, \tag{2}$$

$$x_0 = \rho x_1 + \nu_0, \tag{3}$$

$$x_{N+1} = \rho x_N + \nu_{N+1} \tag{4}$$

where $\alpha = \rho/1 + \rho^2$. The partial sequence $\{\nu_1, \nu_2, \cdots, \nu_N\}$ of $\{\nu_0, \nu_1, \cdots, \nu_{N+1}\}$ satisfies

$$E\{\nu\nu^t\} = \beta_2{}^2 Q$$

where

$$\nu = (\nu_1 \nu_2 \cdots \nu_N)^t,$$

$$Q_{ij} = \begin{cases} 1, & i = j, \\ -\alpha, & |i - j| = 1, \\ 0, & \text{elsewhere}, \end{cases} \tag{5}$$

$\beta_2{}^2 = 1 - \rho^2/1 + \rho^2$.

With the $N \times N$ tridiagonal Toeplitz matrix defined by (5), a new random process is introduced as

$$y = x - x_b \tag{6}$$

where

$$x = (x_1 x_2 \cdots x_N)^t,$$

$$x_b = \alpha Q^{-1}(x_0 0 0 \cdots 0 x_{N+1})^t. \tag{7}$$
$$| \leftarrow N \rightarrow |$$

We see that y is a vector representation of the partial sequence $\{x_1, x_2, \cdots, x_N\}$ modified by linear combinations of the boundary conditions, i.e., x_0 and x_{N+1}. It can be shown that for the new process y, the KL transform is given by a sine transform ψ defined by

$$\psi_{ij} = \sqrt{\frac{2}{N+1}} \sin \frac{ij\pi}{N+1}, \qquad i, j = 1, 2, \cdots, N. \tag{8}$$

Since ψ is fast implementable, it is called "fast KLT."

We remark that the FKLT is for the N-dimensional vector y, while the original data can be represented by an $(N + 2)$-dimensional vector. If $\{x_0, x_{N+1}\}$ and y are known, $x = (x_1 x_2 \cdots x_N)^t$ is recovered by using (6) and (7); hence, the original sequence $\{x_i\}$ is completely reconstructed if the FKLT of y and the boundary conditions are known. Although the transform ψ is orthogonal, steps (6) and (7) used in the modification of x can never be considered as an orthogonal transform. Hence, the reverse operation, i.e., the reconstruction of x from y and $\{x_0, x_{N+1}\}$, must be carried out with care; small errors with which x_0 and x_{N+1} are possibly accompanied may have considerable effects. We shall return to this point later.

DISCUSSION

Effectiveness of the FKLT

We shall now investigate the effectiveness of the FKLT in data compression applications. Emphasis must be put on the fact that the sine transform is not the FKLT for the original sequence $\{x_i\}$, but for y. It is an open question whether it is optimal in coding $\{x_i\}$. Let us compare the FKLT with the DCT by using the minimum bit rate as a performance measure:

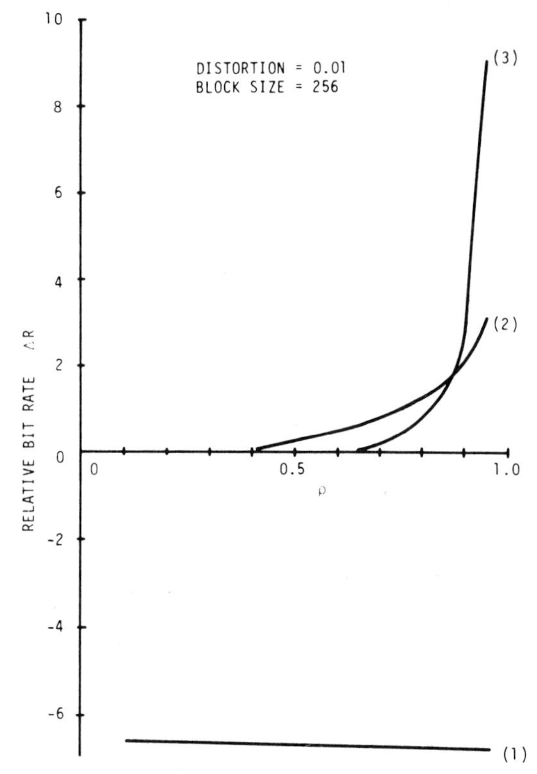

Fig. 1. Bit rate relative to that of DCT.

$$R(D) = \frac{1}{2} \sum_{i=1}^{M} \log_2 \frac{\sigma_i{}^2}{D} \tag{9}$$

where M = block size, $\sigma_i{}^2$ = variances of the transform coefficients, and D is the mean-square error or distortion which we assume satisfies

$$D < \min_i \sigma_i{}^2.$$

The performance measure defined by (9) has been proposed by Pearl *et al.* [5].

The bit rate for the FKLT is shown by graph 1 in Fig. 1 as a function of ρ. The relative bit rate is defined as

$$\Delta R = R - R_{\text{DCT}}.$$

We see that the FKLT appears to be superior to the DCT since R_{FKLT} is smaller than R_{DCT} for all ρ. However, R_{FKLT} is for y and the difference in bit rate is offset by the number of bits required for coding the two boundary conditions x_0 and x_{N+1} which are indispensable to the reconstruction of x.

As we have suggested at the end of the last section, coding errors of x_0 and x_{N+1} are not conserved through the reconstruction procedure. Suppose ϵ_0 and ϵ_{N+1} are coding errors of x_0 and x_{N+1}, respectively. These errors affect the reconstruction of x_b defined by (7), which then affects the reconstruction of x. The resultant mean-square error in the reconstruction of the entire signal sequence x_i can be computed as

$$e^2 = 2\sigma_\epsilon^2 + \frac{2\rho^2}{N+1} \sum_{i=1}^{N} \left[\frac{\sin \dfrac{i}{N+1}\pi}{1 + \rho^2 - 2\rho \cos \dfrac{i}{N+1}\pi} \right]^2 \cdot 2\sigma_\epsilon^2 \tag{10}$$

where we have assumed

$$E\{\epsilon_0\} = E\{\epsilon_{N+1}\} = 0,$$

$$E\{\epsilon_0{}^2\} = E\{\epsilon_{N+1}{}^2\} = \sigma_\epsilon{}^2,$$

$$E\{\epsilon_0 \epsilon_{N+1}\} = 0.$$

The first term in (10) corresponds to ϵ_0 and ϵ_{N+1} directly appearing in x_0 and x_{N+1}. The second term is associated with the proliferation of ϵ_0 and ϵ_{N+1} to other elements of $\{x_i\}$. For $N = 256$ and $\rho = 0.95$, for example, (10) yields

$$e^2 = 10.26 \cdot 2\sigma_\epsilon{}^2,$$

which demonstrates that the coding errors are magnified, far from being conserved.

Using (9), we can obtain an estimate of the number of bits for x_0 and x_{N+1}. There are two important points to note here. First, the error magnification given by (10) must be considered. In the case of the above example, we should replace D in (9) by $D/10.26$. For other combinations of N and ρ, similar steps should be taken. Another thing is that the KLT of x_0 and x_{N+1} is useless since $E\{x_0 x_{N+1}\} = \rho^{N+1} \cong 0$ for most cases of practical importance. Now we can compute the total bit rate for the FKLT:

total bit rate = (bit rate for y) + (bits for x_0 and x_{N+1}).

The total bit rate thus computed for the FKLT is indicated by graph 2 of Fig. 1. We remark that FKLT coding needs more bits than DCT coding in spite of the complete decorrelation in the FKLT domain. The distribution of mean-square errors for FKLT coding is plotted in Fig. 2. Observe that mean-square errors at the end points are extremely small; otherwise, their amplified effects would have considerable effects on x_1 through x_N as we have pointed out. It is rather annoying to assign many bits to the end points while little attention will be paid to them in most applications.

Zero Boundary Conditions

Since the boundary conditions are cumbersome to some extent, one might be tempted to set them to zero. It was actually done in [6] under similar circumstances. However, it is almost obvious that the boundary conditions should not be arbitrary; from the definition of the sequence $\{x_i\}$, x_0 and x_{N+1} must be zero-mean random variables satisfying (1). If they are fixed at zero, i.e., $x_0 = x_{N+1} = 0$, (1) is not satisfied for them. The process $\{0, x_1, \cdots, x_N, 0\}$ cannot be represented by (2)–(4) on which the entire theory of the FLKT is based. If we set $i = 0$ and $x_0 = 0$ in (2), we have

$$x_1 = \alpha x_2 + \nu_1.$$

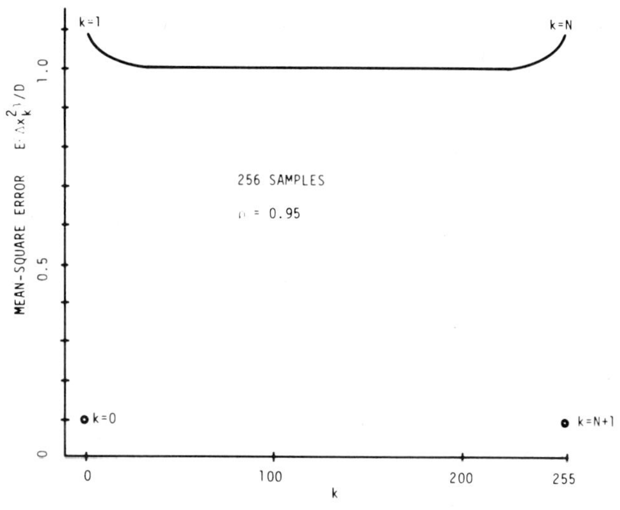

Fig. 2. Error distribution of FKLT coding in the signal sequence.

In the above relation, $\alpha = \rho$ is a suitable choice. But this contradicts the definition of α, $\alpha = \rho/1 + \rho^2$. Equations (3) and (4) with deterministically fixed boundary conditions also lead to contradictions.

If (6) is used with $x_0 = x_{N+1} = 0$, we end up with

$$y = x = (x_1 \cdots x_N)^t. \tag{11}$$

It is now evident that the sine transform ψ defined by (8) is not the KLT for y. We can show that the sine transform generates

$$\hat{x} = \psi y = \psi x = (\hat{x}_1 \cdots \hat{x}_N)^t$$

for zero-boundary conditions with

$$E\{\hat{x}_i \hat{x}_j\} = \begin{cases} 0, \quad i + j = \text{odd}, \\[1ex] \dfrac{1}{N+1} \cdot \dfrac{4\rho^2 \sin \dfrac{i}{N+1}\pi \sin \dfrac{j}{N+1}\pi}{\left(1 + \rho^2 - 2\rho \cos \dfrac{i}{N+1}\pi\right)\left(1 + \rho^2 - 2\rho \cos \dfrac{j}{N+1}\pi\right)}, \\[1ex] \quad i + j = \text{even}. \end{cases}$$

Thus, we observe

$$E\{\hat{x}_i \hat{x}_j\} \to 0 \qquad \text{for } N \to \infty, \quad i \neq j.$$

Although the sine transform is asymptotically equivalent to the KLT as seen above, it is a poor substitute for the KLT when it is compared with the DCT. The relative bit rate for this case appears in Fig. 1 (graph 3). Observe the degraded performance of the sine transform for ρ close to unity. It is interesting, however, to note that for $\rho < 0.85$, the simple sine transform is more efficient than the FKLT in bit rate reduction.

CONCLUDING REMARKS

The effectiveness of the FKLT has been discussed. It has been shown that the extra bit rate reduction made possible by the FKLT is offset by the number of bits assigned to the boundary conditions. While its performance in bit rate reduction may be rated excellent, its theoretical optimality has been discredited. The error amplification has been analyzed by Jain *et al.* [7]; some optimization efforts have been made there in allocation of bits between transform coefficients and boundary values. But the effectiveness of the proposed schemes is uncertain.

It has also been pointed out that with zero-boundary conditions, the FKLT reduces to a simple sine transform of the original data sequence less the boundary conditions. The sine transform is inefficient in bit rate reduction for ρ close to unity.

REFERENCES

[1] N. Ahmed, T. Natarajan, and K. R. Rao, "Discrete cosine transform," *IEEE Trans. Comput.*, vol. C-23, pp. 90–93, Jan. 1974.

[2] H. Kitajima, T. Saito, and T. Kurobe, "Comparison of the discrete cosine and Fourier transforms as possible substitutes for the Karhunen–Loeve transform," *Trans. IECE Japan*, vol. E60, pp. 279–283, June 1977.

[3] A. K. Jain, "A fast Karhunen–Loeve transform for a class of random processes," *IEEE Trans. Commun.*, vol. COM-24, pp. 1023–1029, Sept. 1976.

[4] ——, "A fast Karhunen–Loeve transform for finite discrete images," in *Proc. Nat. Electron. Conf.*, Oct. 1974, pp. 323–328.

[5] J. Pearl, H. C. Andrews, and W. K. Pratt, "Performance measures for transform data coding," *IEEE Trans. Commun. Technol.*, vol. COM-20, pp. 411–415, June 1972.

[6] A. K. Jain, "Image coding via a nearest neighbors image model," *IEEE Trans. Commun.*, vol. COM-23, pp. 318–331, Mar. 1975.

[7] A. K. Jain, S. H. Wang, and Y. Z. Liao, "Fast Karhunen–Loeve transform data compression studies," in *Proc. Nat. Telecommun. Conf.*, Dec. 1976, pp. 615-1–615-5.

Editor's Comments on Papers 26 Through 29

HYBRID TRANSFORMS

The discrete transforms described thus far can be combined in several ways to generate literally an infinite number of hybrid transforms. Examples of such transforms are slant-Haar (SHT) (Paper 26; Rao, Kuo, and Narasimhan, 1979), and Hadamard-Haar (HHT) (Paper 27; Rao et al., 1974). As an example, HHT results from the Kronecker product of WHT and HT matrices. Needless to say, other hybrid transforms such as slant-Hadamard, cosine-Haar, etc., can be generated. The Kronecker product preserves the fast algorithmic properties of the individual transforms. The performances of SHT (Paper 26) and HHT (Paper 27) based on variance distributions or scalar Wiener filtering and data compression have been evaluated.

Hein and Ahmed (Paper 3) originally showed that the DCT can be expressed via the WHT and a conversion matrix that has a sparse block diagonal structure. This concept was later generalized by Jones, Hein, and Knauer (Paper 28), who proved that any even-odd transform (EOT) can be expressed through any other EOT and a conversion matrix having a block diagonal structure. As described earlier, an EOT is one whose matrix has one-half even vectors (v_1, v_2, v_3, v_4 v_4, v_3, v_2, v_1, is an even vector) and one-half odd vectors (v_1, v_2, v_3, v_4, $-v_4$, $-v_3$, $-v_2$, $-v_1$). By intergerizing the conversion matrix for the DCT, they have developed a transform called the C-matrix transform for $N = 8$. This was later extended to $N = 16$ by Srinivasan and Rao (Paper 29). Further extension to $N = 32$ has also been accomplished (Kwak, Srinivasan, and Rao, 1983). CMT also has been applied to image coding (Srinivasan and Rao, 1983).

Venkatraman et al. (1983) have developed the conversion matrices for various EOT such as the DCT, DST, CMT, ST, etc. using the WHT as the base. The computational complexity involved in implementing these transforms and their comparative performances are outlined. It is possible that a single processor can implement all of these transforms with WHT at the intermediate stage.

REFERENCES

Kwak, H. S., R. Srinivasan, and K. R. Rao, 1983, C-Matrix Transform, *IEEE Acoust., Speech, Signal Process. Trans.* **ASSP-31:**1304–1307.

Rao, K. R. et al., 1974, *Hadamard-Haar Transform,* 6th Annual Southeastern Symposium on System Theory, Session FA-5, Baton Rouge, La.

Rao, K. R., J. G. K. Kuo, and M. A. Narasimhan, 1979, Slant-Haar Transform, *Int. J. Comput. Math.,* sec. B, **7:**73–83.

Srinivasan, R., and K. R. Rao, 1983, *Digital Image Coding by C-Matrix Transform,* IFAC Symposium on Real Time Digital Control Applications, Guadalajara, Mexico.

Venkataraman, S., et al. 1983, *Discrete Transforms via the Walsh-Hadamard Transform,* 26th Midwest Symposium on Circuits and Systems, Puebla, Mexico, pp. 74–78.

26

Reprinted with permission from *IEEE Proc.* **62**:653–654 (1974)

Slant Haar Transform

BERNARD J. FINO AND V. RALPH ALGAZI

Abstract—**The slant Haar transform (SHT) is defined and related to the slant Walsh-Hadamard transform (SWHT). A fast algorithm for the SHT is presented and its computational complexity computed. In most applications, the SHT is faster and performs as well as the SWHT.**

INTRODUCTION

The slant Walsh-Hadamard transform (SWHT) (originally called slant transform) has been proposed by Enomoto and Shibata [1] for the order 8 and used in TV image encoding. Pratt *et al.* [2] and Chen [3] have generalized this transform to any order 2^n and compared its performance with other transforms. In [4], we have given a simpler definition of the SWHT as a particular case of a unified treatment of fast unitary transforms and computed the number of elementary operations required by its fast algorithm.[1] The interesting feature of the SWHT is the presence of a slant vector with linearly decreasing components in its basis. On the other hand, we have found that locally dependent basis vectors, such as in the Haar transform (HT), are of interest [5]. In this letter, we define a composite fast unitary transform: the slant Haar transform (SHT). We show that its relations to the SWHT parallel the relations between the HT and the Walsh-Hadamard transform (WHT) [6]. This previous work leads us to expect that the SHT has an advantage over the SWHT because of its speed and comparable performance.

DEFINITION

The generalized Kronecker product of the set $\{\mathcal{A}\}$ of n matrices $[A^j]$ $(j=0, \cdots, n-1)$ of order m and the set $\{\mathcal{B}\}$ of m matrices $[B^k]$ $(k=0, \cdots, m-1)$ of order n is the matrix $[C]$ of order mn such that $C_{vm+w,u'm+w'} = A_{uu'}{}^{w} B_{vw'}{}^{u'}$ when $u, u' < n$ and $w, w' < m$. $[C]$ can be factorized and has a fast algorithm [4]. The generalized Kronecker product provides a simple way to recursively define fast unitary transforms. Consider the matrix of order 2:

$$[T_2] = [F_2(\pi/4)] = \frac{1}{\sqrt{2}}\begin{bmatrix} 1 & 1 \\ 1 & -1 \end{bmatrix}.$$

Then the matrix of order 2^n, denoted $[T_{2^n}]$, is obtained from the matrix of order 2^{n-1} by $\{\mathcal{A}\} \otimes \{[T_{2^{n-1}}], [T_{2^{n-1}}]\}$ where $\otimes$ denotes a generalized Kronecker product. With this recursive notation, the HT is obtained for $\{\mathcal{A}\} = [F_2(\pi/4)], [I_2], \cdots, [I_2]$ and the WHT for $\{\mathcal{A}\} = [F_2(\pi/4)], \cdots, [F_2(\pi/4)]$. For the slant transforms SHT and SWHT, the recursive definitions are as previously given except for a supplementary rotation of rows 1 and 2^{n-1} by the matrix

$$[F_2(\theta_n)] = \begin{bmatrix} \sin\theta_n & \cos\theta_n \\ \cos\theta_n & -\sin\theta_n \end{bmatrix}$$

with $\theta_n < \pi/2$ given by

$$\cos\theta_n = \frac{2^{n-1}}{\sqrt{\dfrac{2^{2n}-1}{3}}}.$$

This rotation introduces the "slant vector," the components of which are linearly decreasing (see [7] for a complete description of slant vectors).

[1] Note that the number of operations given in [3] seems in error and that fewer operations may in fact be needed in a different organization [4].

$$[SHT_8] = \frac{1}{\sqrt{8}}\begin{bmatrix} 1 & 1 & 1 & 1 & 1 & 1 & 1 & 1 \\ 7 & 5 & 3 & 1 & -1 & -3 & -5 & -7 \times 1/\sqrt{21} \\ 3 & 1 & -1 & -3 & -3 & -1 & 1 & 3 \times 1/\sqrt{5} \\ 7 & -1 & -9 & -17 & 17 & 9 & 1 & -7 \times 1/\sqrt{105} \\ 1 & -1 & -1 & 1 & 0 & 0 & 0 & 0 \times \sqrt{2} \\ 0 & 0 & 0 & 0 & 1 & -1 & -1 & 1 \times \sqrt{2} \\ 1 & -3 & 3 & -1 & 0 & 0 & 0 & 0 \times \sqrt{2/5} \\ 0 & 0 & 0 & 0 & 1 & -3 & 3 & -1 \times \sqrt{2/5} \end{bmatrix} \begin{matrix} [MSH_8^{0,0}] \\ [MSH_8^{0,1}] \\ [MSH_8^{1,0}] \\ [MSH_8^{1,1}] \\ \left.\right\} [MSH_8^{2,0}] \\ \\ \left.\right\} [MSH_8^{2,1}] \\ \\ \end{matrix}$$

(a)

$$[SWHT_8] = \frac{1}{\sqrt{8}}\begin{bmatrix} 1 & 1 & 1 & 1 & 1 & 1 & 1 & 1 \\ 7 & 5 & 3 & 1 & -1 & -3 & -5 & -7 \times 1/\sqrt{21} \\ 3 & 1 & -1 & -3 & -3 & -1 & 1 & 3 \times 1/\sqrt{5} \\ 7 & -1 & -9 & -17 & 17 & 9 & 1 & -7 \times 1/\sqrt{105} \\ 1 & -1 & -1 & 1 & 1 & -1 & -1 & 1 \\ 1 & -1 & -1 & 1 & -1 & 1 & 1 & -1 \\ 1 & -3 & 3 & -1 & -1 & 3 & -3 & 1 \times 1/\sqrt{5} \\ 1 & -3 & 3 & -1 & 1 & -3 & 3 & -1 \times 1/\sqrt{5} \end{bmatrix} \begin{matrix} [MSWH_8^{0,0}] \\ [MSWH_8^{0,1}] \\ [MSWH_8^{1,0}] \\ [MSWH_8^{1,1}] \\ \left.\right\} [MSWH_8^{2,0}] \\ \\ \left.\right\} [MSWH_8^{2,1}] \\ \\ \end{matrix}$$

(b)

Fig. 1. Ordered slant transforms of order 8. (a) SHT. (b) SWHT.

ORDERING OF THE BASIS VECTORS

In signal processing practice, the vectors of the HT are used in rank order [4] and those of the WHT are used in sequency (number of sign changes) order. There is an ordering of the slant transforms consistent with these orderings; the basis vectors obtained by the previously given recursive definitions are ordered by: a) decreasing number of nonzero elements (from globally to locally dependent basis vectors); b) increasing number of zero crossings; and c) from left to right. The ordered matrices $[SHT_8]$ and $[SWHT_8]$ are shown in Fig. 1.

RELATIONS BETWEEN SHT AND SWHT

The SHT and SWHT have relations similar to the relations between the HT and the WHT [6]. Partition the ordered SHT (SWHT) into $2n$ rectangular submatrices, denoted $[MSH_{2^n}{}^{k,i}]$ ($[MSWH_{2^n}{}^{k,i}]$), $k=0, \cdots, n-1$ and $i=0, 1$. For $k, i=0, 1$, $[MSH_{2^n}{}^{0,0}]$, $[MSH_{2^n}{}^{0,1}]$, $[MSH_{2^n}{}^{1,0}]$, and $[MSH_{2^n}{}^{1,1}]$ are the four first rows of the ordered matrix $[SHT_{2^n}]$. For $k, i > 1$, $[MSH_{2^n}{}^{k,i}]$ is formed from the rows of ranks $2^{k-1} + i2^{k-2} \le r < 3 \times 2^{k-2} + i2^{k-2}$. The matrices $[MSWH_{2^n}{}^{k,i}]$ are similarly defined. The submatrices for the order 8 are shown in Fig. 1. It can be shown, following the proof given in [6], that

$$[MSWH_{2^n}{}^{k,i}] = [S_{2^{k-2}}]^i \, [WHT_{2^{k-2}}][MSH_{2^n}{}^{k,i}] \qquad (1)$$

where

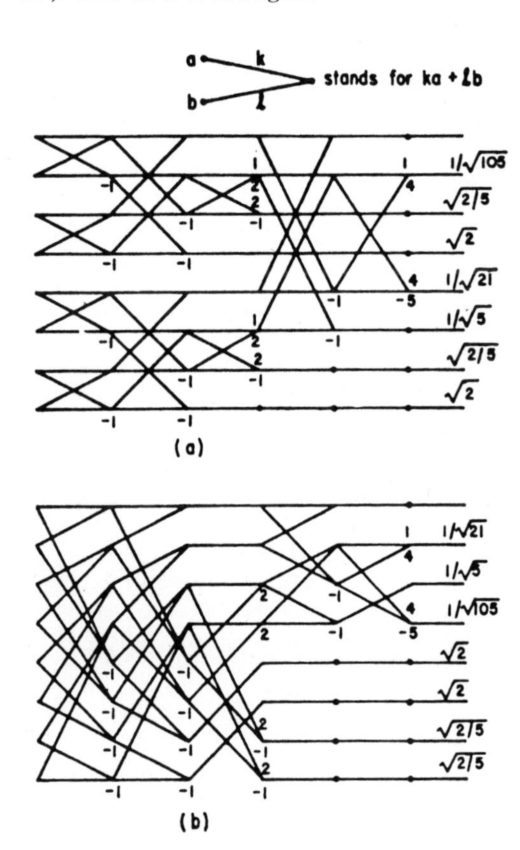

Fig. 2. Fast algorithm for the SHT of order 8.
(a) Unordered rows. (b) Ordered rows.

$$[S_m] = \begin{bmatrix} & 0 & & 1 \\ & & \cdot & \\ & 1 & & \\ \cdot & & 0 & \\ 1 & & & \end{bmatrix}$$

and $[WHT_{2^n}]$ is the ordered WH matrix of order 2^n.

As for the HT and WHT, these relations imply "zonal" relations between the components of the representation of a vector by the SHT and SWHT and, in particular, an identical energy partition according to these zones.

Fast Algorithm

The framework developed in [4] gives the fast algorithm presented in Fig. 2(a) for the order 8. Note that the step-by-step normalizations in the rotations by the matrix $[F_2(\theta_n)]$ can be delayed so that the rows 1 and 2^{n-1} are in fact rotated by the matrix

$$\begin{bmatrix} 1 & 2^{n-1} \\ 2^{n-1} & -\dfrac{(2^{2n-2}-1)}{3} \end{bmatrix}$$

which requires only 2 shifts, 2 additions, and 1 multiplication. This algorithm can be reorganized as shown in Fig. 2(b) to give the ordered transform.

The relations (1) also yield a decomposition of the SWHT algorithm into an SHT and WHT of lower orders.

Computational Complexity

The required number of each elementary operation can be computed with general formulas [4] and the previous definition of the SHT. We find that the computation of the SHT of a vector of order 2^n requires $2^{n+2}-6$ additions, $2^{n-2}-1$ multiplications, 2^n-2 shifts, and $3(2^{n-2})$ normalizations[2] at the last stage of computation.

By another algorithm, the SHT of order 4, which is also the SWHT of the same order, can be performed with 8 additions and 2 multiplications [3], compared with 10 additions and 2 shifts as given by the above formulas. This order-4 algorithm can be introduced in the recursive definition to trade 2^{n-1} additions and 2^{n-1} shifts for 2^{n-1} multiplications in the previously given results.

The SHT has the same number of multiplications and shifts than the SWHT, but has $(n-3)2^n+4$ fewer additions and 1 more normalization; therefore the SHT is faster than the SWHT.

Other Slant Transforms

The two slant transforms considered in this letter are only two members of a large family of slant transforms; this family has been studied in some detail in [7].

Conclusions

We have defined the SHT and presented some properties of this transform, mainly as compared to the SWHT, which has been successfully considered for image encoding. The SHT is faster and preserves some local properties of the signal and we believe it should be preferred over the SWHT.

References

[1] H. Enomoto and K. Shibata, "Orthogonal transform coding system for television signals," in *Proc. 1971 Symp. Applications of the Walsh Functions*, AD-727 000, pp. 11–17.

[2] W. K. Pratt, L. R. Welch, and W. H. Chen, "Slant transform for image coding," in *Proc. 1972 Symp. Applications of the Walsh Functions*, AD-744 650, pp. 229–234.

[3] W. H. Chen, "Slant transform image coding," Electron. Science Lab., Univ. Southern California, Tech. Rep. 441, May 1973.

[4] B. J. Fino and V. R. Algazi, "A unified treatment for fast unitary transforms," to be published.

[5] B. J. Fino, "Experimental study of image coding by Haar and complex Hadamard transforms" (in French), *Ann. Telecommun.*, pp. 185–208, May–June 1972.

[6] ——, "Relations between Haar and Walsh/Hadamard transforms," *Proc. IEEE* (Lett.), vol. 60, pp. 647–648, May 1972.

[7] ——, "Recursive definition and computation of fast unitary transforms," Ph.D. dissertation, Univ. California, Berkeley, Nov. 1973.

[2] We assume that the transform coefficients are scaled to the most commonly encountered normalization factor; thus the transform is unitary within a scale factor.

Image Data Processing by Hadamard-Haar Transform

K. R. RAO, SENIOR MEMBER, IEEE, M. A. NARASIMHAN, AND KRISHNAIAH REVULURI

***Abstract*—A hybrid version of the Haar and Walsh–Hadamard transforms (HT and WHT) called Hadamard–Haar transform (HHT)$_r$ is defined and developed. Efficient algorithms for fast computation of the (HHT)$_r$ and its inverse are developed. (HHT)$_r$ is applied to digital signal and image processing and its utility and effectiveness are compared with other discrete transforms on the basis of some standard performance criteria.**

***Index Terms*—Data compression, digital time processing, feature selection, Hadamard–Haar transform (HHT)$_r$, Wiener filtering.**

I. INTRODUCTION

DIGITAL signal and image processing has come into prominence in recent years. This requires, in many cases, utilization of discrete orthogonal transforms [1], [2]. Fourier [3], slant [4], Walsh–Hadamard [5], Haar [1], [6], [7], discrete linear basis [8], rapid [9], [10], slant Haar [11], and discrete cosine [12] have already been utilized in these areas. This utilization is stimulated in part by the rapid development of digital hardware. Efficient algorithms for fast implementation of the orthogonal transforms have further accelerated their effectiveness, leading to the design and development of special-purpose digital processors tailored for specific transforms. Since the linear transformation of image data results in compaction of its energy into fewer coefficients [13], image processing by transform techniques can lead to lower transmission rates with negligible image degradation [4], [14].

Manuscript received August 1974; revised March 1975.
K. R. Rao is with the Department of Electrical Engineering, University of Texas at Arlington, Arlington, Tex. 76019.
M. A. Narasimhan is with Texas Instruments, Inc., Dallas, Tex.
K. Revuluri is with the Department of Information Engineering, University of Illinois, Chicago Circle, Ill. 60680.

II. HADAMARD–HAAR TRANSFORM [15]

The objectives of this paper are to develop a hybrid version of the well-known Walsh–Hadamard (WHT) and Haar transforms (HT) such that the advantages of both of them can be utilized and to test its utility and effectiveness in image processing in terms of standard performance criteria, such as mean-square error (mse) subjective image quality, and variance distribution. The rth-order Hadamard–Haar transform $(\mathrm{HHT})_r$ and its inverse $(\mathrm{HHT})_r$ can be respectively defined as

$$[\mathbf{L}_r(n)] = \frac{1}{N}[HH_r(n)][\mathbf{x}(n)]$$

$$\text{and} \quad [\mathbf{x}(n)] = [HH_r(n)]^T[\mathbf{L}_r(n)] \quad (1)$$

where

$$[\mathrm{HH}_r(n)] = [G_0(r)] \otimes [H_0(n-r)] \quad (2)$$

is the rth-order HHT matrix of size $(2^n \times 2^n)$. $x(m)$ and $L_r(m)$, $m = 0,1,\cdots,N-1$ are the data and rth-order transform sequences, respectively. $n = \log_2 N$ and $[G_0(m)]$ and $[H_0(m)]$ are WHT and HT matrices of sizes $(2^m \times 2^m)$, respectively. All other notation is described elsewhere [15]. Based on the matrix factoring of $[G_0(m)]$ and $[H_0(m)]$, $[HH_r(n)]$ can be factored into a number of sparse matrices which result in the fast algorithms. As an example for $r = 1$ the transform matrix can be factored as

$$[HH_1(n)] = [G_0(1)] \otimes [H_0(n-1)] = \left[\begin{array}{c|c} [I(n-1)] & [I(n-1)] \\ \hline [I(n-1)] & -[I(n-1)] \end{array}\right] \cdot \prod_{j=1}^{n-1} \left[\begin{array}{c|c} [H_0^{(j)}(n-1)] & 0 \\ \hline 0 & [H_0^{(j)}(n-1)] \end{array}\right] \quad (3)$$

where $[I(m)]$ is identity matrix of size $(2^m \times 2^m)$ and $[H_0^{(j)}(n-1)]$, $j = 1,2,\cdots,n-1$ are the matrix factors of $[H_0(n-1)]$. For $n = 3$,

$$[HH_1(3)] = [G_0(1)] \otimes [H_0(2)] = \begin{bmatrix} 1 & 1 & 1 & 1 & 1 & 1 & 1 & 1 \\ 1 & 1 & -1 & -1 & 1 & 1 & -1 & -1 \\ \sqrt{2} & -\sqrt{2} & 0 & 0 & \sqrt{2} & -\sqrt{2} & 0 & 0 \\ 0 & 0 & \sqrt{2} & -\sqrt{2} & 0 & 0 & \sqrt{2} & -\sqrt{2} \\ 1 & 1 & 1 & 1 & -1 & -1 & -1 & -1 \\ 1 & 1 & -1 & -1 & -1 & -1 & 1 & 1 \\ \sqrt{2} & -\sqrt{2} & 0 & 0 & -\sqrt{2} & \sqrt{2} & 0 & 0 \\ 0 & 0 & \sqrt{2} & -\sqrt{2} & 0 & 0 & -\sqrt{2} & \sqrt{2} \end{bmatrix} \begin{matrix} \text{base functions} \\ \phi_0 \\ \phi_2^1 + \phi_2^2 \\ \phi_3^1 + \phi_3^3 \\ \phi_3^2 + \phi_3^4 \\ \phi_1 \\ \phi_2^1 - \phi_2^2 \\ \phi_3^1 - \phi_3^3 \\ \phi_3^2 - \phi_3^4 \end{matrix} \quad (4)$$

The base functions of $(\mathrm{HHT})_r$ are the linear combinations of Haar functions ϕ_m^l. The signal flow graphs for $(\mathrm{HHT})_1$ when $N = 8$ and $N = 16$, based on (1), (3), and (4), are shown in Figs. 1 and 2, respectively. The number of additions (or subtractions) required for fast implementation of HT, WHT, and $(\mathrm{HHT})_r$ are $2(N-1)$, $N \log 2\, N$, and $2^{r+1}(N/2^r - 1) + rN - [(r+2)N - 2^{r+1}]$, respectively.

III. APPLICATIONS

The applications of $(\mathrm{HHT})_r$ in feature selection, Wiener filtering, and image processing are investigated.

A. Feature Selection

The technique developed by Andrews [6] for selecting certain features in the rotated (transform) space based on the variance criterion [4], [12]–[14] is utilized here (Fig. 3). The transform that packs the most energy in the fewest number of its coefficients is the best in terms of feature selection, data compression, and mse. It is well known that the Karhunen–Loêve transform is optimal as it generates uncorrelated coefficients, compacts most of the energy in few coefficients, and results in minimum mse [13]. However, generation and implementation of this transform is tedious although some simplifications have been suggested [16] and some fast algorithms have been described for certain class of signals [17]. To test the effectiveness of $(\mathrm{HHT})_r$, the variance distributions for a first-order Markov process [18] with a correlation coefficient $\rho = 0.9$ and signal-to-noise ratio K_0 of unity for $N = 16$, for $(\mathrm{HHT})_{1,2}$ are compared with those of HT and WHT in Table I. Inspection of Table I indicates that these transforms can be ranked as WHT > $(\mathrm{HHT})_2$ > $(\mathrm{HHT})_1$ > HT.

B. Wiener Filtering [7]

Wiener filtering has been extended to transform processing (Fig. 4). The object here is to select the $(2^n \times 2^n)$ filter matrix $[G(n)]$ such that $[\hat{\mathbf{x}}(n)]$ the estimate of $[\mathbf{x}(n)]$ is the best in the mse sense when the signal $[\mathbf{x}(n)]$ has been corrupted by an additive noise $[\mathbf{w}(n)]$.

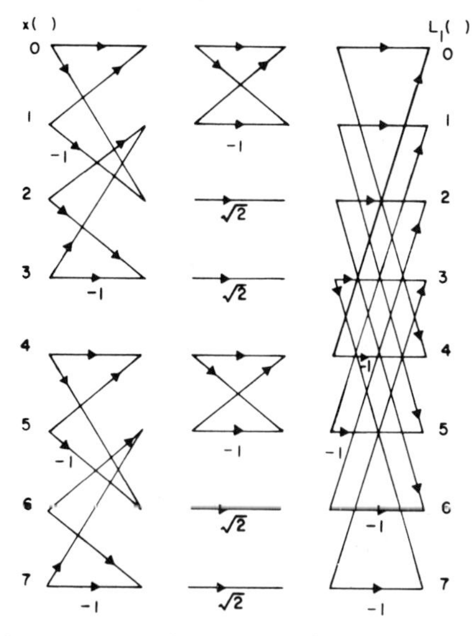

Fig. 1. Signal flow graph computing $(\mathrm{HHT})_1$, $N = 8$ (the multiplier 1/8 is not shown).

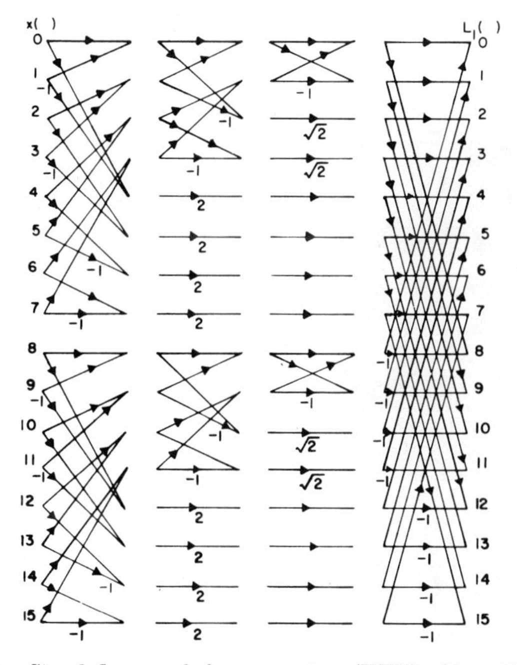

Fig. 2. Signal flow graph for computing $(\mathrm{HHT})_1$, $N = 16$ (the multiplier 1/16 is not shown).

Pearl [18] has developed the expressions for estimating the mse due to scalar filtering when $[G(n)]$ is constrained to be a diagonal matrix. This expression is utilized for computing the mse (Table II) for various transforms when the random process is that described in the feature selection.

C. Image Processing

The standard digitized (girl) data supplied by the Image Processing Institute of the University of Southern California, Los Angeles, have been utilized for transform image processing. The University of Southern California

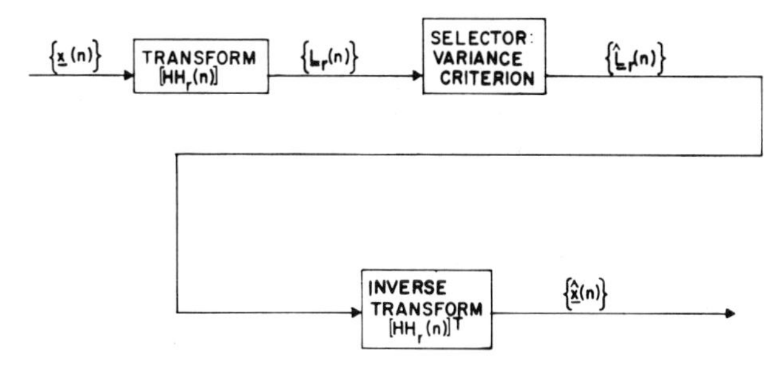

Fig. 3. Maximum variance zonal sampling.

TABLE I
NORMALIZED VARIANCE FOR FIRST-ORDER MARKOV PROCESS WHEN $\rho = 0.9$ AND $K_0 = 1$, FOR $N = 16$

Transform / l	HT	WHT	$(\mathrm{HHT})_1$	$(\mathrm{HHT})_2$
1	9.8346	9.8346	9.8346	9.8346
2	2.5364	2.5364	2.5364	2.5364
3	0.8638	1.020	1.0209	1.0209
4	0.8638	0.78	0.7061	0.7061
5	0.2755	0.706	0.2946	0.3066
6	0.2755	0.307	0.2946	0.3031
7	0.2755	0.303	0.2562	0.2864
8	0.2755	0.283	0.2562	0.2059
9	0.1	0.206	0.1024	0.1038
10	0.1	0.105	0.1024	0.1038
11	0.1	0.105	0.1024	0.1034
12	0.1	0.104	0.1024	0.1034
13	0.1	0.104	0.0976	0.1013
14	0.1	0.103	0.0976	0.1013
15	0.1	0.102	0.0976	0.0913
16	0.1	0.098	0.0976	0.0913

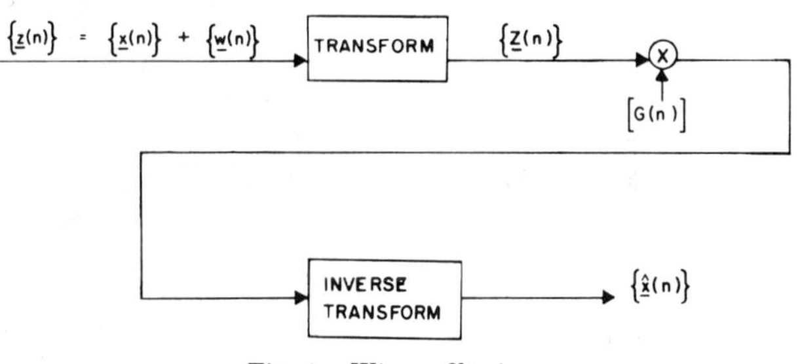

Fig. 4. Wiener filtering.

image data consist of (256×256) pixels with each pixel quantized uniformly to 256 gray levels. The transform processing (Fig. 5) is carried out on subimages of (16×16) block size [7], [13]. The selector performs threshold sampling on the transformed data, i.e., picks out the l percent of the transform coefficients having the largest magnitudes in each subimage (l is variable) and sets all the other to zero [13]. The original and the reconstructed images are reduced to 13 equiprobable levels [2], [14] to obtain the line printer plots, which are then photographed. The mse between the original and the reconstructed images [4] is computed for various data compressions and is presented in Table III. In increasing order of mse the transforms are as follows:

TABLE II
MSE FOR FIRST-ORDER MARKOV PROCESS WHEN $\rho = 0.9$ AND $K_0 = 1$ FOR SCALAR WIENER FILTERING

N \ Transform	HT	WHT	$(HHT)_1$	$(HHT)_2$	$(HHT)_3$
4	0.2942	0.2942	0.2942	0.2942	
8	0.2650	0.2649	0.26487	0.2649	
16	0.2589	0.2582	0.2584	0.2582	0.2582
32	0.2582	0.2582	0.257582	0.25677	0.25656
64	0.2581	0.2559	0.257987	0.25711	0.25620

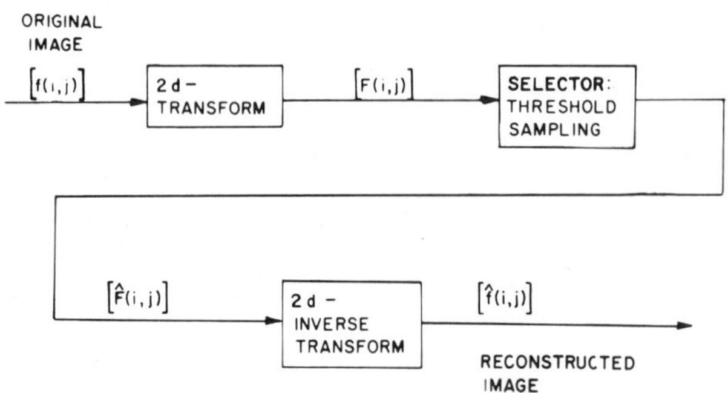

Fig. 5. Transform image processing.

TABLE III
MSE IN PERCENT BETWEEN THE ORIGINAL GIRL AND THE RECONSTRUCTED IMAGE BASED ON THRESHOLD SAMPLING AS A FUNCTION OF DATA COMPRESSION FOR HT, $(HHT)_1$, $(HHT)_2$, AND WHT

Transform \ Data Compression	4:1	6:1	8:1	12:1
HT	0.224655	0.401599	0.570233	0.883826
$(HHT)_1$	0.291588	0.500686	0.690139	1.022737
$(HHT)_2$	0.292468	0.498852	0.687942	1.022565
WHT	0.285804	0.492695	0.684362	1.025216

Fig. 6. Original monochrome image girl.

Data Compression	Transforms arranged in increasing order of mse
4:1	HT, WHT, $(HHT)_1$, $(HHT)_2$
6:1	HT, WHT, $(HHT)_2$, $(HHT)_1$
8:1	HT, WHT, $(HHT)_2$, $(HHT)_1$
12:1	HT, $(HHT)_2$, $(HHT)_1$, (WHT)

Fig. 6 shows the original monochrome image (girl). Figs. 7–10 show the reconstructed images using 4:1, 6:1, 8:1, 12:1 data compressions, respectively. A comparison of these images shows that good quality reconstruction is achieved even up to data compression of 12:1. In all of these threshold sampling processes, the effects of quantization and/or coding have not been taken into account, the objective being the energy compaction property of the orthogonal transforms which can be utilized in decreasing the channel capacity with negligible degradation in image quality.

IV. CONCLUSIONS

$(HHT)_r$, the hybrid version of the WHT and the HT, is defined and developed. Sparse matrix factoring of these later transforms is utilized in developing the efficient algorithms for fast implementation of $(HHT)_r$. Some of

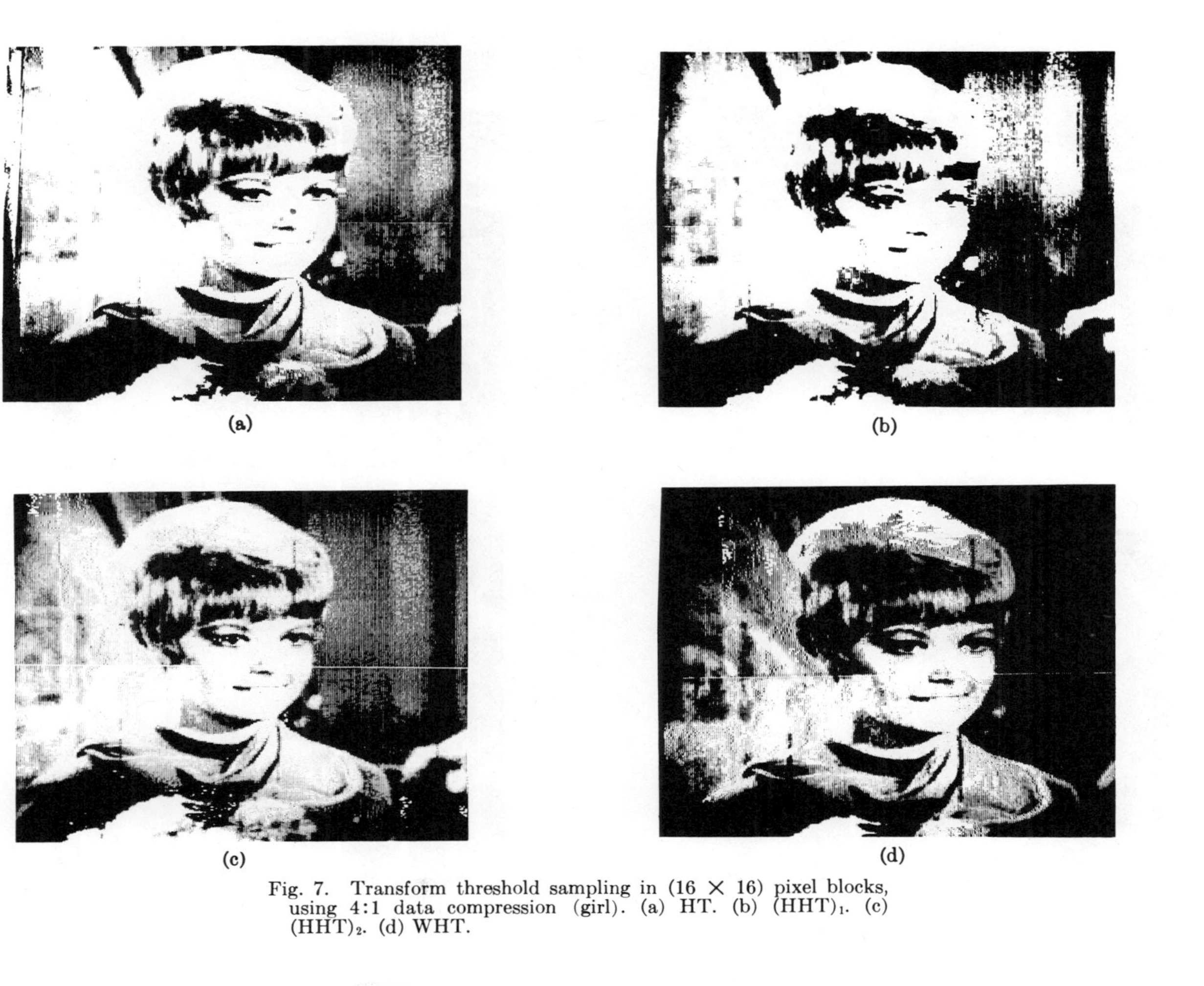

Fig. 7. Transform threshold sampling in (16 × 16) pixel blocks, using 4:1 data compression (girl). (a) HT. (b) $(HHT)_1$. (c) $(HHT)_2$. (d) WHT.

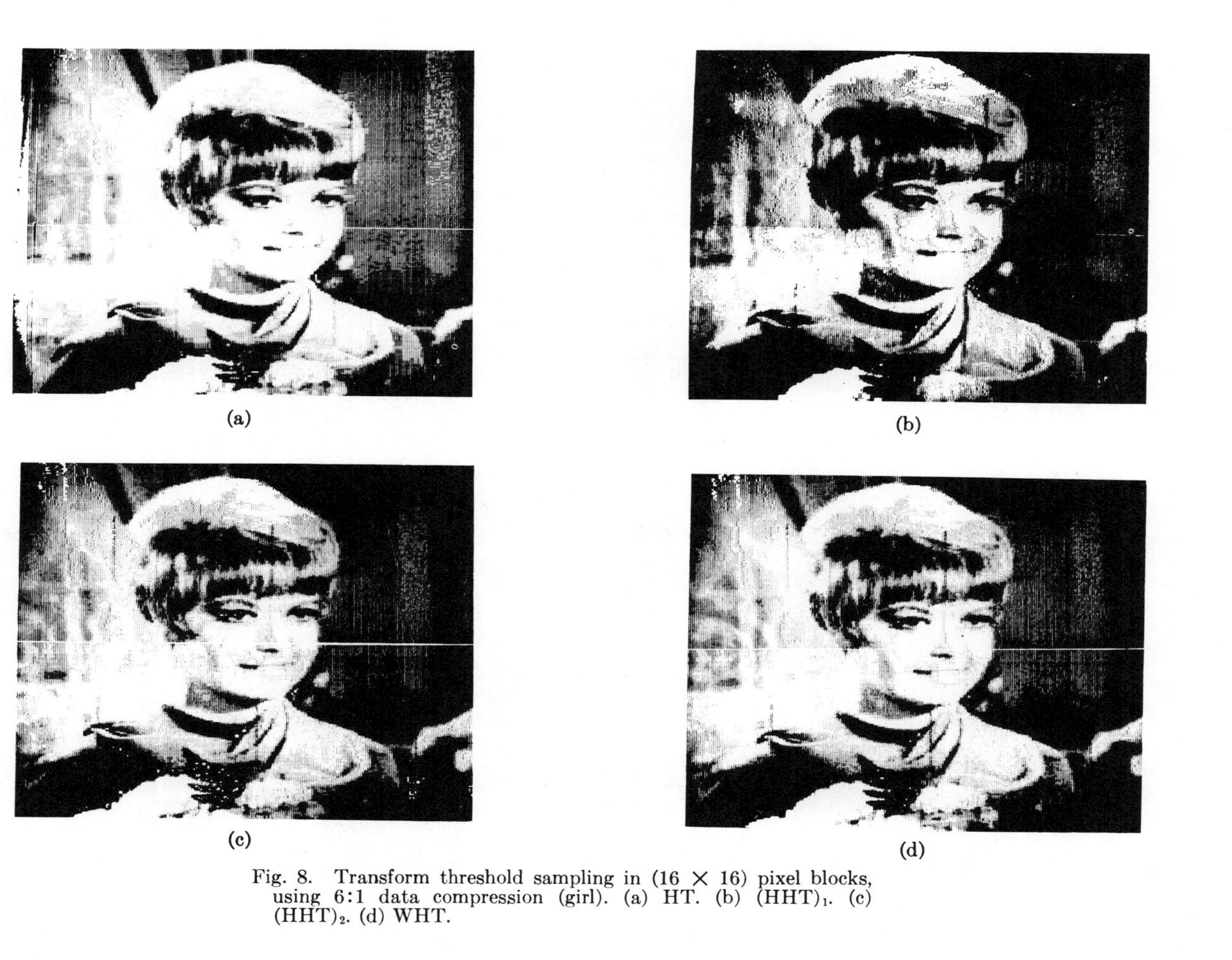

Fig. 8. Transform threshold sampling in (16 × 16) pixel blocks, using 6:1 data compression (girl). (a) HT. (b) $(HHT)_1$. (c) $(HHT)_2$. (d) WHT.

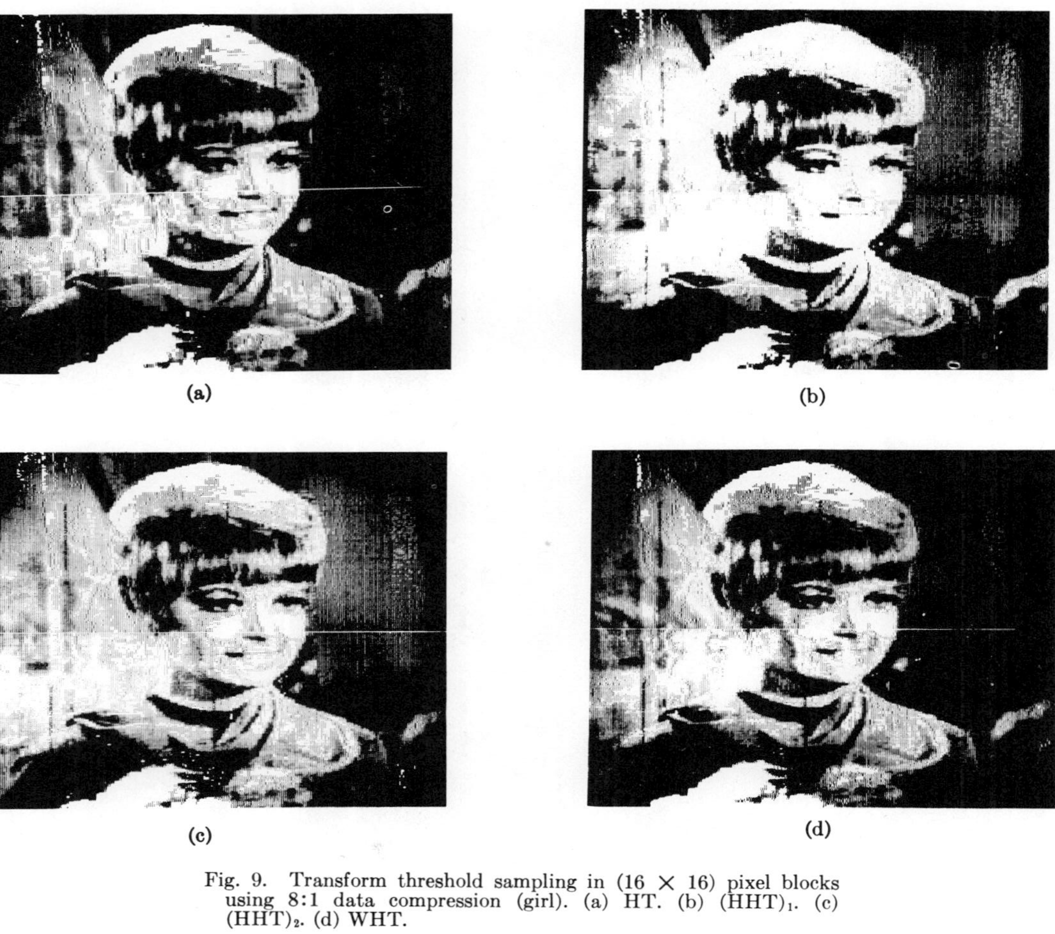

(a) (b) (c) (d)

Fig. 9. Transform threshold sampling in (16 × 16) pixel blocks using 8:1 data compression (girl). (a) HT. (b) $(HHT)_1$. (c) $(HHT)_2$. (d) WHT.

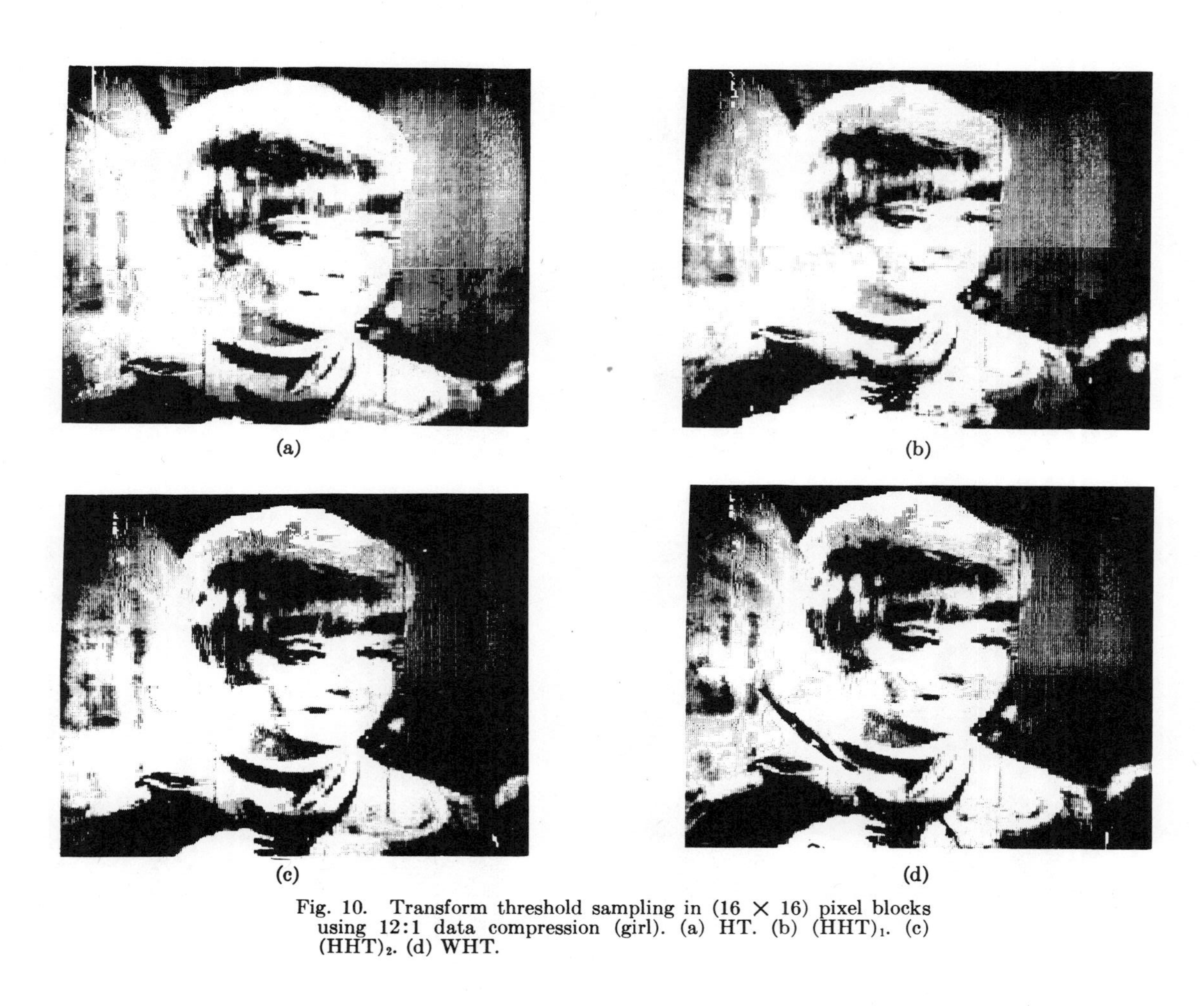

(a) (b) (c) (d)

Fig. 10. Transform threshold sampling in (16 × 16) pixel blocks using 12:1 data compression (girl). (a) HT. (b) $(HHT)_1$. (c) $(HHT)_2$. (d) WHT.

the applications of $(HHT)_r$ such as feature selection in pattern recognition, Wiener filtering, and transform image processing are illustrated. HT is superior to WHT in terms of computational complexity and data compression, whereas WHT performs better than HT for variance distribution and Wiener filtering. $(HHT)_r$ serves as a compromise between HT and WHT for all these applications except for low data compressions where both HT and WHT are superior to $(HHT)_r$.

REFERENCES

[1] H. C. Andrews, *Computer Techniques in Image Processing.* New York: Academic, 1970.

[2] N. Ahmed and K. R. Rao, *Orthogonal Transforms for Digital Signal Processing.* Berlin, Germany: Springer-Verlag, 1975.

[3] E. O. Brigham, *The Fast Fourier Transform.* Englewood Cliffs, N. J.: Prentice-Hall, 1974.

[4] W. K. Pratt, W. H. Chen, and L. R. Welch, "Slant transform image coding," *IEEE Trans. Commun.*, vol. COM-22, pp. 1075–1093, Aug. 1974.

[5] W. K. Pratt, J. Kane, and H. C. Andrews, "Hadamard transform image coding," *Proc. IEEE*, vol. 57, pp. 58–68, Jan. 1969.

[6] H. C. Andrews, "Multidimensional rotations in feature selection," *IEEE Trans. Comput.* (*Special Issue on Feature Extraction and Selection in Pattern Recognition*) (Short Notes), vol. C-20, pp. 1045–1051, Sept. 1971.

[7] W. K. Pratt, "Generalized Wiener filtering computation techniques," *IEEE Trans. Comput.* (*Special Issue on Two-Dimensional Signal Processing*), vol. C-21, pp. 636–641, July 1972.

[8] R. M. Haralick and K. Shanmugam, "Comparative study of a discrete linear basis for image data compression," *IEEE Trans. Syst., Man, Cybern.*, vol. SMC-4, pp. 16–27, Jan. 1974.

[9] H. Reitboeck and T. P. Brody, "A transformation with invariance under cyclic permutation for applications in pattern recognition," *Inform. Contr.*, vol. 15, pp. 130–154, 1969.

[10] P. P. Wang and R. C. Shiau, "Machine recognition of printed Chinese characters via transformation algorithms," *Pattern Recognition*, vol. 5, pp. 303–321, 1973.

[11] B. J. Fino and V. R. Algazi, "Slant Haar transform," *Proc. IEEE* (Lett.), vol. 62, pp. 653–654, May 1974.

[12] N. Ahmed, T. Natarajan, and K. R. Rao, "Discrete cosine transform," *IEEE Trans. Comput.* (Corresp.), vol. C-23, pp. 90–93, Jan. 1974.

[13] P. A. Wintz, "Transform picture coding," *Proc. IEEE* (*Special Issue on Digital Picture Processing*), vol. 60, pp. 809–820, July 1972.

[14] N. Ahmed *et al.*, "On the orthogonal transform processing of image data," in *Proc. 1973 Nat. Electronics Conf.*, vol. 28, pp. 274–279.

[15] K. R. Rao *et al.*, "Hadamard–Haar transform," presented at the Sixth Annu. Southeastern Symp. Syst. Theory, Baton Rouge, La., Feb. 21–22, 1974.

[16] K. Shanmugam and R. M. Haralick, "A computationally simple procedure for imagery data compression by the Karhunen–Loêve method," *IEEE Trans. Syst., Man, Cybern.* (Corresp.), vol. SMC-3, pp. 202–204, Mar. 1973.

[17] A. K. Jain, "A fast Karhunen–Loêve transform for finite discrete images," in *Proc. 1974 Nat. Electronics Conf.*, vol. 29, Chicago, Ill., pp. 323–326.

[18] J. Pearl, "Walsh processing of random signals," *IEEE Trans. Electromagn. Compat.* (*Applications of Walsh Functions 1971 Symp. Proc.*), vol. EMC-13, pp. 137–141, Aug. 1971.

28

Reprinted from *Int. Found. Telemetering Conf. Proc.* **14:**87-98 (1978)

THE KARHUNEN-LOEVE, DISCRETE COSINE, AND RELATED TRANSFORMS OBTAINED VIA THE HADAMARD TRANSFORM*

H. W. Jones, D. N. Hein, and S. C. Knauer

ABSTRACT

The Karhunen-Loeve transform for stationary data, the discrete cosine transform, the Walsh-Hadamard transform, and most other commonly used transforms have one-half even and one-half odd transform vectors. Such even/odd transforms can be implemented by following a Walsh-Hadamard transform by a sparse matrix multiplication, as previously reported by Hein and Ahmed for the discrete cosine transform. The discrete cosine transform provides data compression nearly equal to that of the Karhunen-Loeve transform, for the first order Markov correlation model. The Walsh-Hadamard transform provides most of the potential data compression for this correlation model, but it always provides less data compression than the discrete cosine transform. Even/odd transforms can be designed to approach the performance of the Karhunen-Loeve or discrete cosine transform, while meeting various restrictions which can simplify hardware implementation. The performance of some even/odd transforms is compared theoretically and experimentally. About one-half of the performance difference between the Walsh-Hadamard and the discrete cosine transforms is obtained by simple post-processing of the Walsh-Hadamard transform coefficients.

INTRODUCTION

It is well known that the Karhunen-Loeve, or eigenvector transform (KLT), provides decorrelated vector coefficients with the maximum energy compaction, and that the discrete cosine transform (DCT) is a close approximation to the KLT for first-order Markov data (1). We will show that the general class of even/odd transforms includes this particular KLT, as well as the DCT, the Walsh-Hadamard transform (WHT), and other familiar transforms. The more complex even/odd transforms can be computed by combining a simpler even/odd transform with a sparse matrix multiplication. A theoretical performance measure is computed for some even/odd transforms, and two image compression experiments are reported.

EVEN/ODD TRANSFORMS

Orthogonal transforms are frequently used to compress correlated sampled data. Most commonly used transforms, including the Fourier, slant, DCT, and WHT have one-half even and one-half odd transform vectors. Several properties of such even/odd transforms are given in this section. The even vector coefficients are uncorrelated with the odd vector coefficients for a data correlation class which includes stationary data. The KLT is an even/odd transform for this class of data correlation. A conversion from one even/odd transform to another requires only multiplication by a sparse matrix, having one-half of its elements equal to zero.

If N, the number of data points, is an even number, a vector

$$V = (v_1 v_2 \cdot \cdot \cdot v_{N-1} v_N)^T$$

is said to be even if

$$v_i = v_{N+1-i} \qquad i = 1, \cdot \cdot \cdot, N/2$$

and is odd if

*Portions of this research were performed under NASA contract NAS2-9703 and NASA Interchange Agreement No. NCA2-OR-363-702.

$$v_i = -v_{N+1-i} \qquad i = 1, \cdots, N/2$$

For a data vector of length N

$$X = (x_1 x_2 \cdots x_{N-1} x_N)^T$$

the NxN correlation matrix is given by

$$\Sigma_x = E(XX^T)$$

Since Σ_x is a symmetric matrix, it can be partitioned into four N/2 x N/2 submatrices in the following manner:

$$\Sigma_x = \begin{vmatrix} A & B \\ B^T & C \end{vmatrix}$$

where

$$A = A^T \quad \text{and} \quad C = C^T$$

The general form for a transform matrix with one-half even and one-half odd basis vectors (called an even/odd transform), can be written as a partitioned matrix

$$H = \begin{vmatrix} E & \tilde{E} \\ D & -\tilde{D} \end{vmatrix}$$

where E and D are N/2 x N/2 orthogonal matrices, and $\tilde{E}$ and $\tilde{D}$ are formed by reversing the order of the columns in E and D, that is,

$$\tilde{E} = E\tilde{I} \quad \text{and} \quad \tilde{D} = D\tilde{I}$$

where the permutation matrix $\tilde{I}$ is the opposite diagonal identity matrix. The matrix H can then be factored into the product of two matrices

$$H = \begin{vmatrix} E & 0 \\ 0 & D \end{vmatrix} \begin{vmatrix} I & \tilde{I} \\ I & -\tilde{I} \end{vmatrix}$$

It is next shown that the even and odd vector coefficients of an even/odd transform are uncorrelated, for a general class of data correlation matrices. The correlation matrix for the transformed data vector, Y = HX, is given by the similarity transform.

$$\Sigma_y = H \Sigma_x H^T$$

$$= \begin{vmatrix} E(A + B\tilde{I} + \tilde{I}B^T + \tilde{I}C\tilde{I})E^T & E(A - B\tilde{I} + \tilde{I}B^T - \tilde{I}C\tilde{I})D^T \\ D(A + B\tilde{I} - \tilde{I}B^T - \tilde{I}C\tilde{I})E^T & D(A - B\tilde{I} - \tilde{I}B^T + \tilde{I}C\tilde{I})D^T \end{vmatrix}$$

The even and odd vector coefficients are uncorrelated when the opposite diagonal submatrices are identically zero. This is obviously true in the special case where

$$A = \tilde{I}C\tilde{I} \quad \text{and} \quad B = \tilde{I}B^T\tilde{I}$$

These equations state that the data correlation matrix Σ_x is symmetric about both the main diagonal and the opposite diagonal. This condition is satisfied if $E(x_i x_j)$ is a function of the magnitude of $i - j$, that is, if the process is stationary. For stationary data, the correlation matrix is a symmetric Toeplitz matrix (1, 2).

This decorrelation property of even/odd transforms is used to show that the KLT is an even/odd transform. For K, a reordered matrix of the KLT vectors, the transformed vector is Z, Z = KX. The correlation matrix for the transformed vector is given by

$$\Sigma_z = K \Sigma_x K^T$$

Suppose that the data are first transformed by the matrix H, above, and that the data are such that the even and odd coefficients are uncorrelated by this even/odd transform:

$$\Sigma_y = H \Sigma_x H^T$$

$$= 2 \begin{vmatrix} E(A + BI)E^T & 0 \\ 0 & D(A - BI)D^T \end{vmatrix}$$

$$= \begin{vmatrix} Y_1 & 0 \\ 0 & Y_2 \end{vmatrix}$$

Since H is invertible, $K = AH$, for $A = KH^T$.

$$\Sigma_z = AH \Sigma_x H^T A^T$$

$$\Sigma_z = A \Sigma_y A^T$$

$$\Sigma_z = A \begin{vmatrix} Y_1 & 0 \\ 0 & Y_2 \end{vmatrix} A^T$$

Suppose that

$$A = \begin{vmatrix} A_1 & A_2 \\ A_3 & A_4 \end{vmatrix}$$

$$\Sigma_z = \begin{vmatrix} A_1Y_1A_1^T + A_2Y_2A_2^T & A_1Y_1A_3^T + A_2Y_2A_4^T \\ A_3Y_1A_1^T + A_4Y_2A_2^T & A_3Y_1A_3^T + A_4Y_2A_4^T \end{vmatrix}$$

Since the KLT produces N fully decorrelated coefficients, Σ_z is a diagonal matrix. Either both A_2 and A_3, or both A_1 and A_4 must be identically zero. For $A_2 = A_3 = 0$, the first N/2 vectors remain even, while for $A_1 = A_4 = 0$, the even and odd vectors are interchanged in K.

$$K = AH$$

$$= \begin{vmatrix} A_1 & 0 \\ 0 & A_4 \end{vmatrix} \begin{vmatrix} E & \tilde{E} \\ D & -\tilde{D} \end{vmatrix}$$

$$= \begin{vmatrix} A_1E & A_1\tilde{E} \\ A_4D & -A_4\tilde{D} \end{vmatrix}$$

$$= \begin{vmatrix} A_1E & A_1E\tilde{I} \\ A_4D & -A_4D\tilde{I} \end{vmatrix}$$

The KLT is an even/odd transform for the class of correlation matrices for which even/odd transforms decorrelate the even and odd vector coefficients.

If H and J are two even/odd NXN transforms, the multiplication matrix for conversion between them is sparse:

$$H = \begin{vmatrix} E & \tilde{E} \\ D & -\tilde{D} \end{vmatrix}$$

$$J = \begin{vmatrix} F & \tilde{F} \\ G & -\tilde{G} \end{vmatrix}$$

The conversion is defined by

$$H = SJ$$

$$S = HJ^T$$

$$= \begin{vmatrix} E & \tilde{E} \\ D & -\tilde{D} \end{vmatrix} \begin{vmatrix} F^T & G^T \\ \tilde{F}^T & -\tilde{G}^T \end{vmatrix}$$

$$= \begin{vmatrix} EF^T + \tilde{E}\tilde{F}^T & EG^T - \tilde{E}\tilde{G}^T \\ DF^T - \tilde{D}\tilde{F}^T & DG^T + \tilde{D}\tilde{G}^T \end{vmatrix}$$

However,

$$\tilde{E}\tilde{G}^T = E\tilde{I}(G\tilde{I})^T = E\tilde{I}\tilde{I}G^T = EG^T$$

$$\tilde{D}\tilde{F}^T = DF^T$$

It follows that

$$S = 2\begin{vmatrix} EF^T & 0 \\ 0 & DG^T \end{vmatrix}$$

The conversion between any two even/odd transforms requires $N^2/2$ rather than N^2 multiplications.

We have shown that the class of even/odd transforms has no correlation between the even and odd vector coefficients, for a class of data correlation matrices including the stationary data matrix. The KLT for this data correlation class, and many familiar transforms, are even/odd transforms. The coefficients of any even/odd transform can be obtained by a sparse matrix multiplication of the coefficients of any other even/odd transform. This observation was the basis of a previous implementation of the DCT and suggested the investigation of even/odd transforms described below.

THE DISCRETE COSINE TRANSFORM OBTAINED VIA THE HADAMARD TRANSFORM

Hein and Ahmed have shown how the DCT vectors can be obtained by a sparse matrix multiplication on the WHT vectors (3, 4). Since the DCT, unlike the general KLT, has a constant vector and a shifted square-wave vector in common with the WHT, the number of matrix multiplications is less than $N^2/2$. The A matrix, which generates the DCT vectors for $N = 8$ from the WHT vectors, is given by Hein and Ahmed, and is reproduced here as Figure 1. Although this implementation of the DCT requires more operations for large N than the most efficient DCT implementation (5), it is very satisfactory for N equal to 8.

If a transform has even and odd vectors and has a constant vector, as is typical, it can be obtained via the WHT in the same way as the DCT. The slant transform is an example (1, 6). A hardware implementation of the DCT via the WHT is being constructed at Ames Research Center, using $N = 8$ and the matrix of Figure 1. Since this implementation contains the matrix multiplication factors in inexpensive read-only memories, it will be possible to consider the real-time quantization design and evaluation of a large class of transforms. Transforms with suboptimum performance are acceptable only if they can be implemented with reduced complexity. Transform performance can be determined theoretically from the vector energy compaction, while the implementation complexity can be estimated from the number and type of operations added after the WHT.

COMPARISON OF TRANSFORMS USING THE FIRST-ORDER MARKOV CORRELATION MODEL

It is generally accepted that the sample-to-sample correlation of an image line scan is approximated by the first-order Markov model (7).

$$c(x_i, x_j) = c(|i - j|) = r^{|i-j|}$$

The correlation of adjacent samples, r, varies from 0.99 for low detail images to 0.80 for high detail images, with an average of about 0.95 (8). The correlation matrix, $\sum_x$, was generated using the first-order Markov model, for various r, and the corresponding KLT's and vector energies were numerically computed. (The analytic solution is known (9).) In addition, the matrix Σ_x was used to compute the transform vector energies and correlations for the WHT, DCT, and other transforms.

As is well known, the KLT vectors for $r = 0.95$ are very similar to the DCT vectors and have nearly identical vector energies (1, 3). The most apparent difference between the DCT and the KLT is that the KLT vector corresponding to the constant DCT vector is not exactly constant, but weights the central samples in a fixed transform block more than samples near the edge of the block. As r approaches 1.00, this KLT vector approaches the constant vector, and all the KLT vectors approach the corresponding DCT vectors. The vector energies of the KLT and the DCT are nearly identical for r greater than 0.90, and differ only slightly for r greater than 0.50. The KLT and DCT vector energies for $N = 8$ and $r = 0.50$ are plotted in Figure 2. The energy compaction at $r = 0.5$ is much less than at the typical $r = 0.95$.

The rate-distortion performance of a transform depends on the transform energy compaction. If the distortion d is less than the coefficient variance σ_i^2 for all i, all N transform vectors are quantized and transmitted. The number of bits required is (10):

$$b = \sum_{i=1}^{N} \frac{1}{2} \log_2 (\sigma_i^2/d)$$

$$= \frac{1}{2} \sum_{i=1}^{N} \log_2 \sigma_i^2 - \frac{N}{2} \log_2 d$$

The first term of b can be used as a figure of merit for a transform.

$$f = \frac{1}{2} \sum_{i=1}^{N} \log_2 \sigma_i^2$$

The figure of merit f is a negative number; the larger its magnitude, the greater the rate reduction achieved by the transform. Table I gives f for the KLT, DCT, WHT, and two even/odd transforms that will be described below. At correlation $r = 0.95$, the KLT gains 0.014 bits more than the DCT and 1.183 bits more than the WHT. The WHT achieves most of the available data compression, and the DCT achieves nearly all. As this rate reduction is obtained for all N vectors, the increased compression of the DCT over the WHT, for $r = 0.95$, is 1.169/8, or 0.15 bits per sample.

EVEN/ODD TRANSFORMS OBTAINED VIA THE WALSH-HADAMARD TRANSFORM

The sequency of a transform vector is defined as the number of sign changes in the vector. The vector sequencies of the vectors corresponding to the matrix of Figure 1 are in bit-reverse order, as indicated (0, 4, 2, 6, 1, 5, 3, 7). The energy compaction of the WHT and DCT for $r = 0.95$ and $N = 8$ is shown in Figure 3. In the conversion from WHT to DCT, the two-by-two matrix operation on vectors 2 and 6 transfers energy from 6 to 2. The four-by-four matrix operation on the vectors of sequency 1, 5, 3, and 7 reduces the energy of 3, 5, and 7 and increases the energy of 1. These operations remove most of the residual correlation of the WHT vectors. The matrix multiplication requires 20 multiplications by 10 different factors (15 factors including sign differences).

We first consider a simplified operation on the 2 and 6 and the 1 and 3 sequency vectors. This operation consists of multiplying the WHT vectors by matrix B (Figure 4). This further transform is designed to reduce correlation and to generate new transform vectors in a way somewhat similar to the A matrix multiplication which produces the DCT. There are two identical two-by-two operations, and a total of eight multiplications by two different factors (three including sign). The energy compaction of the B-matrix transform is shown in Figure 3, with the energies of the WHT and DCT. As the B-matrix transform vectors of sequency 0, 4, 5, and 7 are identical to the WHT vectors, they have identical energy. The B-matrix transform vectors of sequencies 0, 1, 2, 3, 4, and 6 are identical to the corresponding DCT vectors (0, 4) or very similar. For example, the B-matrix vector of sequency 1 is a slanted vector of step width 2 and step size 2 (3, 3, 1, 1, -1, -1, -3, -3). The performance of the B-matrix transform, in terms of the figure of merit, is given in Table I above. The B-matrix transform has something more than one-half of the gain of the DCT over the WHT, with something less than one-half of the multiplications, and less than one-fourth the hardware if the two-by-two transformer is used twice.

As a second example, suppose that it is desired to approximate the DCT by adding integer products of the WHT vectors. For small integers, this operation can be implemented by digital shifts-and-adds, and requires fewer significant bits to be retained. The matrix C, given in Figure 5, is an orthonormal transform matrix that is similar to the DCT. The two-by-two matrix, operating on the vectors of sequency 2 and 6, is a specialization of the general two-by-two matrix having orthogonal rows with identical factors. The four-by-four operation on the vectors of odd sequency is a specialization of the general four-by-four matrix with orthogonal rows, identical factors, and the additional requirement of a positive diagonal.

The specializations of the general matrices were made by requiring that the two-by-two matrix integers have approximately the ratios found in the second (and third) rows of the A matrix, and that the four-by-four matrix integers have approximately the ratios found in the fifth (and eighth) rows of the A matrix. Since the A-matrix transform is the DCT, it is ensured that the C transform vectors of sequency 2, 6, 1, and 7 will approximate the corresponding DCT vectors.

The energy compaction results of the C transform, with the results of the WHT and DCT, are given in Figure 6, for $r = 0.95$ and $N = 8$. The energy of the vectors of sequency 2, 6, 1, and 7 is very similar to the energy of the DCT vectors, but the vectors of sequency 3 and 5 are different. The energy correspondence could be improved by matching the four-by-four matrix factors to the average of the fifth and sixth rows in the A matrix, but there is little potential data compression remaining. The theoretical performance of the C matrix, in terms of the figure of merit, is given in Table I. The C-matrix transform obtains nearly all the gain of the DCT over the WHT. If the rational form, instead of the integer form, of the C-matrix transform were used, the computation would require 16 multiplications by 4 different factors (7 factors including sign differences). There is some reduction in complexity from the implementation of matrix A.

EXPERIMENTAL IMAGE COMPRESSION RESULTS

Experimental results were obtained for two-dimensional, 8X8 sample block implementations of the transforms considered above. Four video test images — Harry Reasoner, two Girls, two Men, and band — were used in all tests. These images have correlation of 0.97 to 0.98 between elements in the scan line, and fit the first order Markov model, except for the very detailed band image, which deviates from the Markov model and has an average in-line correlation of 0.85 (11). Two different compression experiments were made.

The test images were first compressed by representing either thirty-two or sixteen of the sixty-four 8X8 transform vectors, using an eight-bit uniform, full-range quantizer. The other vectors were neglected. The patterns of the vectors transmitted and neglected are given in Figure 7. The vectors are in sequency order, with the lowest sequency average vector in the upper left corner of the pattern. The mean-square error for this compression method and the four transforms is given in Table II. The B-matrix transform error is intermediate between the WHT and DCT errors, and the C-matrix error is very close to the DCT error. This is consistent with the Markov model energy compaction results above.

To obtain the greatest transform compression, the transmitted bits should be assigned to the vectors according to the equation given above, and the coefficient quantizers should be designed for minimum error given the coefficient energy and amplitude distributions. The optimum theoretical bit assignments and quantizers depend on the particular transform used. The test images, and most typical images, contain low-contrast, high-correlation background areas, and edges where correlation is low. The bit assignments and quantizer designs based on the stationary Markov model ignore this nonstationarity, and designs that consider low-contrast areas and edges give improved mean-square error and subjective performance. Such improved designs have been devised for the WHT (11), and have been tested with the DCT, B-matrix, and C-matrix transforms. The transmission rate and mean-square error results are given in Figure 8, for the test images compressed in the video field. The DCT gives improved error performance, and the B and C matrix transforms are intermediate, but the B and C matrix results are relatively poorer than those in Table II. The DCT gives more rate reduction than the WHT — about 0.2 to 0.5 bits per sample. As a two-dimensional transform has twice the gain of a one-dimensional transform (10), the theoretical gain of the DCT over the WHT, for $r = 0.95$, should be twice the 0.15 bits per sample of Table I, or 0.30 bits per sample.

The lower error of the DCT, B-matrix, and C-matrix transforms does indicate subjective improvement in the compressed images. This subjective improvement is larger at lower total bit rates, due to the relative increase of larger, more noticeable errors at the lower rates, and due to the more objectionable, blocky nature of large WHT errors. The B and C matrix errors are subjectively more similar to the DCT errors than to WHT errors, because the higher energy vectors approximate the DCT vectors.

It is not surprising that a design optimized for the WHT gives good results for the DCT and similar transforms. The transform compression introduces errors in three ways: by not transmitting vector coefficients, by using quantizers that are too narrow, and by quantization errors within the quantizer ranges. The DCT, because of its superior energy compaction, reduces the first two sources of error. Although the quantizers used are nearly uniform, they do have smaller quantization steps for low coefficient values, so the third source of error is also reduced. Any compression design will give better performance with the DCT. From the similarity in energy compaction, a good design for the WHT should be reasonably effective for the DCT. However, further performance gains can be made with the DCT and with the B-matrix and C-matrix transforms, by optimizing the compression designs for the transform used.

The error statistics show that the lower mean-square error of the DCT is due both to fewer large errors, which nearly always occur at edges, and to fewer small errors, which occur in flat areas and edges. The subjective appearance of the compressed image confirms that the DCT produces both smoother low contrast areas and less distorted edges. Since the low contrast areas have very high correlation, and since the edges — though not noise-like — can be approximated by a low-correlation Markov model, the mean-square error and subjective results agree with the theoretical result that the DCT is superior to the WHT for all values of correlation (see Table I).

CONCLUSION

The Karhunen-Loeve transform for data with stationary correlation, the discrete cosine transform, the Walsh-Hadamard transform, and other familiar transforms are even/odd vector transforms whose coefficients can be obtained by sparse matrix multiplications of the coefficients of other even/odd transforms. Of the familiar transforms, the Walsh-Hadamard transform is the simplest to implement, but has the smallest compression gain. Using the Walsh-Hadamard transform followed by a sparse matrix multiplication allows implementation of any even/odd transform. The discrete cosine transform has a difficult implementation, but very closely approaches the optimum performance for first-order Markov data. As the form of the vectors is modified to approach that of the discrete cosine vectors, the vector energy compaction and the theoretical and experimental image compression results approach those of the discrete cosine transform. The theoretical data compression reliably indicates the difference in experimental performance for these transforms. About one-half of the performance difference between the Walsh-Hadamard and the discrete cosine transforms can be achieved by simple post processing of the Walsh-Hadamard coefficients.

ACKNOWLEDGMENTS

The authors are grateful to D. R. Lumb and L. B. Hofman of Ames Research Center, and to N. Ahmed of Kansas State University, for their assistance and encouragement. The authors are also grateful for the use of the Ames Research Video Compression Facility.

REFERENCES

1. Ahmed, N. and Rao, K. R., Orthogonal Transforms for Digital Signal Processing. Springer-Verlag, Berlin, 1975.
2. Berger, T., Rate Distortion Theory. Prentice Hall, Englewood Cliffs, New Jersey, 1971.
3. Ahmed, N., Natarajan, T., and Rao, K. R., "Discrete Cosine Transform," IEEE Trans. Computers. Vol. C-23, Jan. 1974, pp. 90-93.
4. Hein, D. and Ahmed, N., "On a Real-Time Walsh-Hadamard/Cosine Transform Image Processor," IEEE Trans. Electromagnetic Compatibility. Vol. EMC-20, Aug. 1978, pp. 453-457.
5. Chen, W. H., Smith, C. H., and Fralick, S. C., "A Fast Computational Algorithm for the Discrete Cosine Transform," IEEE Trans. Communications. Vol. COM-25, Sept. 1977, pp. 1004-1009.
6. Pratt, W. K., Chen, W. H., and Welch, L. R., "Slant Transform Image Coding," IEEE Trans. Communications. Vol. COM-22, Aug. 1974, pp. 1075-1093.
7. Franks, L. E., "A Model for the Random Video Process," Bell Syst. Tech. J., April 1966, pp. 609-630.
8. Connor, D. J. and Limb, J. O., "Properties of Frame-Difference Signals Generated by Moving Images," IEEE Trans. Communications. Vol. COM-22, Oct. 1974, pp. 1564-1575.
9. Jain, A. K., "A Fast Karhunen-Loeve Transform for a Class of Random Processes," IEEE Trans. Communications. Vol. COM-24, Sept. 1976, pp. 1023-1029.
10. Davisson, L. D., "Rate-Distortion Theory and Application," Proc. IEEE. Vol. 60, July 1972, pp. 800-808.
11. Jones, H. W., "A Comparison of Theoretical and Experimental Video Compression Designs," to appear in IEEE Trans. Electromagnetic Compatibility, Feb. 1979.

TABLE I The Figure of Merit, $f = \sum \log_2 \sigma_i$, for Different Transforms at N = 8 and Various Correlations.

	Transform				
Correlation, r	KLT	DCT	WHT	B matrix	C matrix
0.99	-19.817	-19.775	-18.489	-19.205	-19.597
0.95	-11.743	-11.729	-10.560	-11.206	-11.558
0.90	-8.379	-8.341	-7.311	-7.875	-8.180
0.80	-5.162	-5.092	-4.317	-4.731	-4.954
0.70	-3.402	-3.328	-2.765	-3.056	-3.214
0.50	-1.453	-1.396	-1.136	-1.261	-1.333
0.00	0.00				

TABLE II The Mean-Square Error for the WHT, DCT, B Matrix and C Matrix Transforms with a Subset of Vectors Retained.

Mean-square error for 32 vectors retained

	Reasoner	Two Girls	Two Men	Band
WHT	0.558	0.806	1.694	3.948
B matrix	0.500	0.738	1.581	3.628
C matrix	0.442	0.666	1.536	3.310
DCT	0.446	0.660	1.535	3.056

Mean-square error for 16 vectors retained

	Reasoner	Two Girls	Two Men	Band
WHT	1.619	2.206	4.801	12.322
B matrix	1.507	2.093	4.557	12.056
C matrix	1.427	2.029	4.447	11.897
DCT	1.430	2.031	4.406	11.828

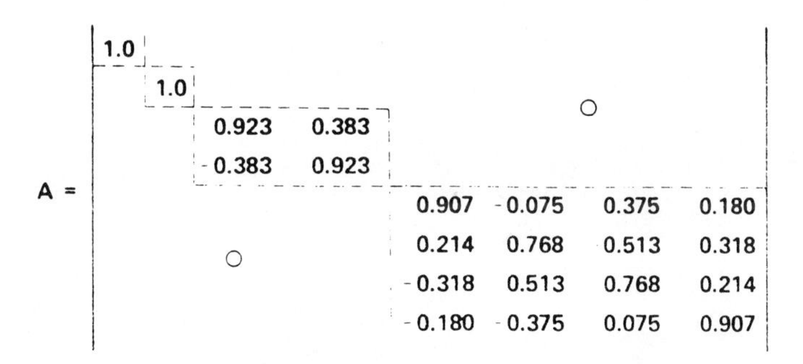

Figure 1 - The A Matrix Used to Obtain the DCT From the WHT.

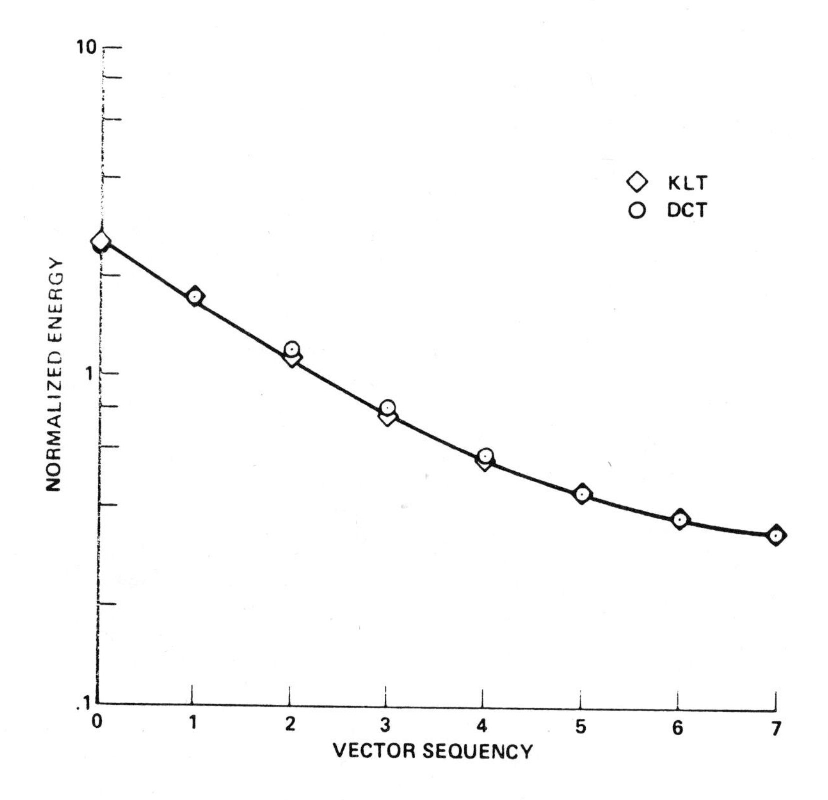

Figure 2 · KLT and DCT Vector Energies for N = 8 and r = 0.5.

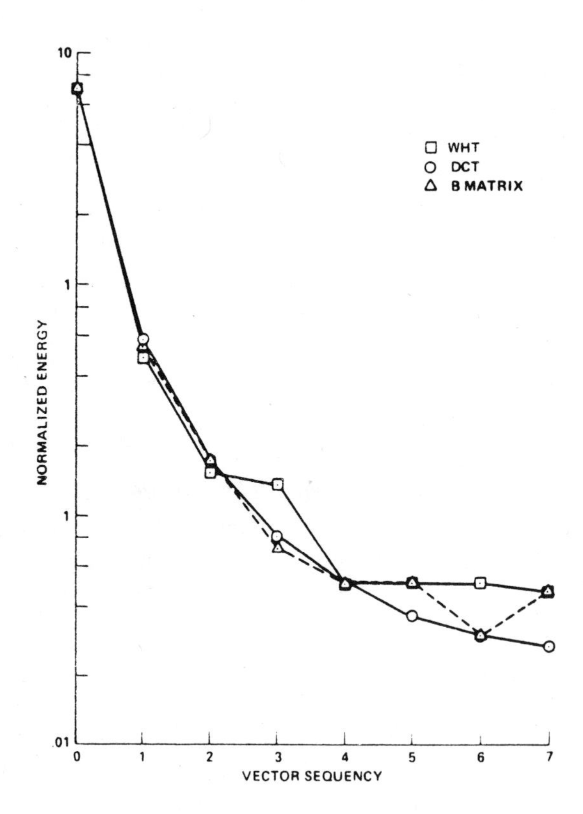

Figure 3 · The Energy Compaction of the DCT, WHT, and B Matrix Transforms for N = 8 and r = 0.95.

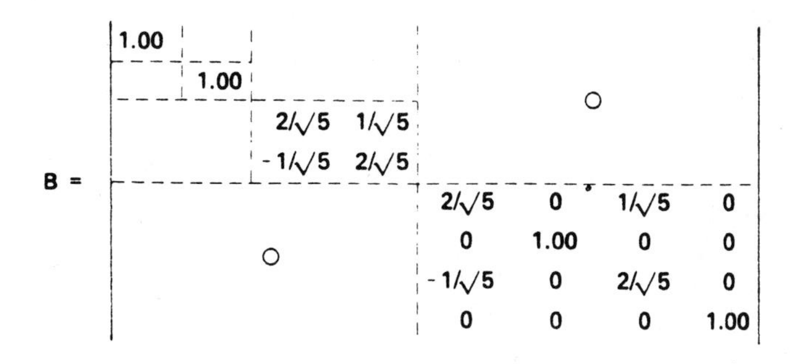

Figure 4 - The B Matrix.

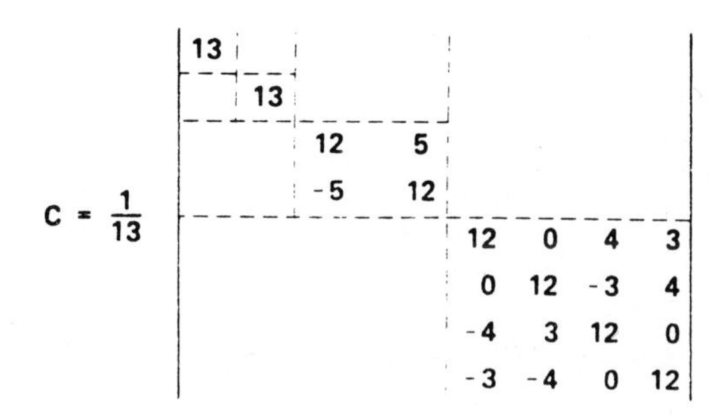

Figure 5 - The C Matrix.

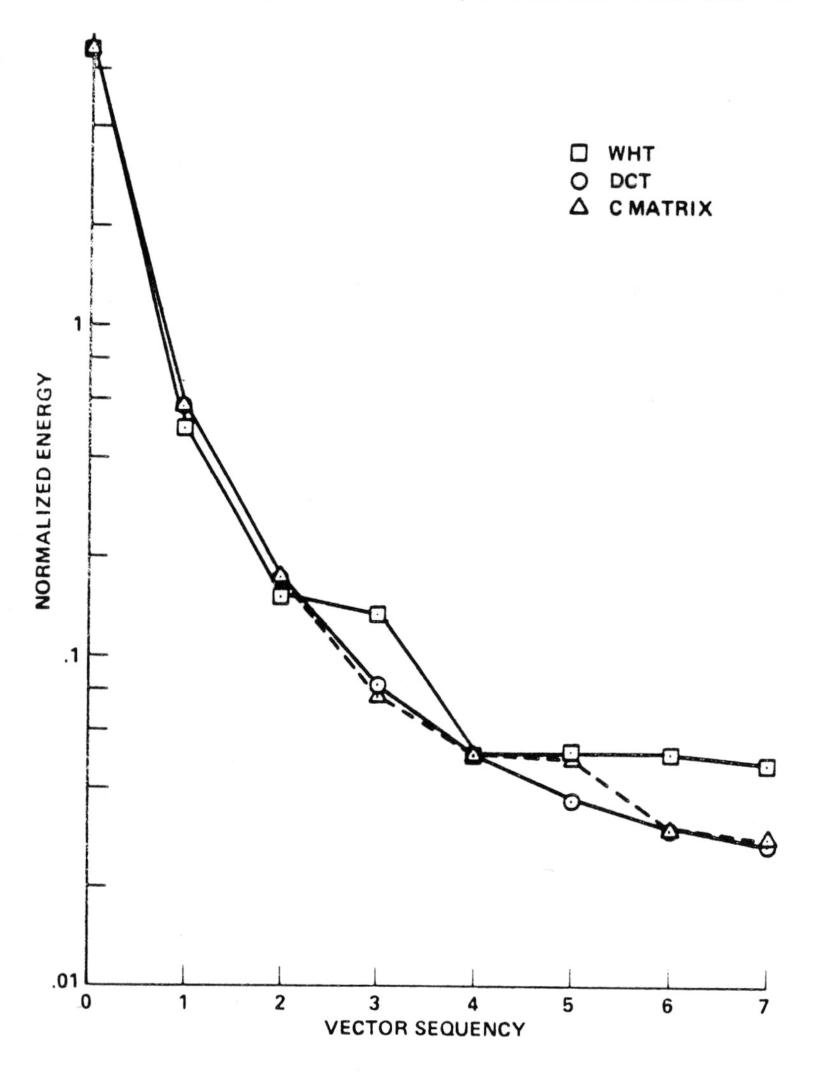

Figure 6 - The Energy Compaction of the WHT, DCT, and C Matrix Transforms for N = 8 and r = 0.95.

8	8	8	8	8	8	8	8
8	8	8	8	8	8	8	8
8	8	8	8	0	0	0	0
8	8	8	8	0	0	0	0
8	8	0	0	0	0	0	0
8	8	0	0	0	0	0	0
8	8	0	0	0	0	0	0
8	8	0	0	0	0	0	0

32 vector pattern

8	8	8	8	8	8	8	8
8	8	0	0	0	0	0	0
8	0	0	0	0	0	0	0
8	0	0	0	0	0	0	0
8	0	0	0	0	0	0	0
8	0	0	0	0	0	0	0
8	0	0	0	0	0	0	0
8	0	0	0	0	0	0	0

16 vector pattern

Figure 7 - Patterns of Vector Coefficients Retained and Neglected.

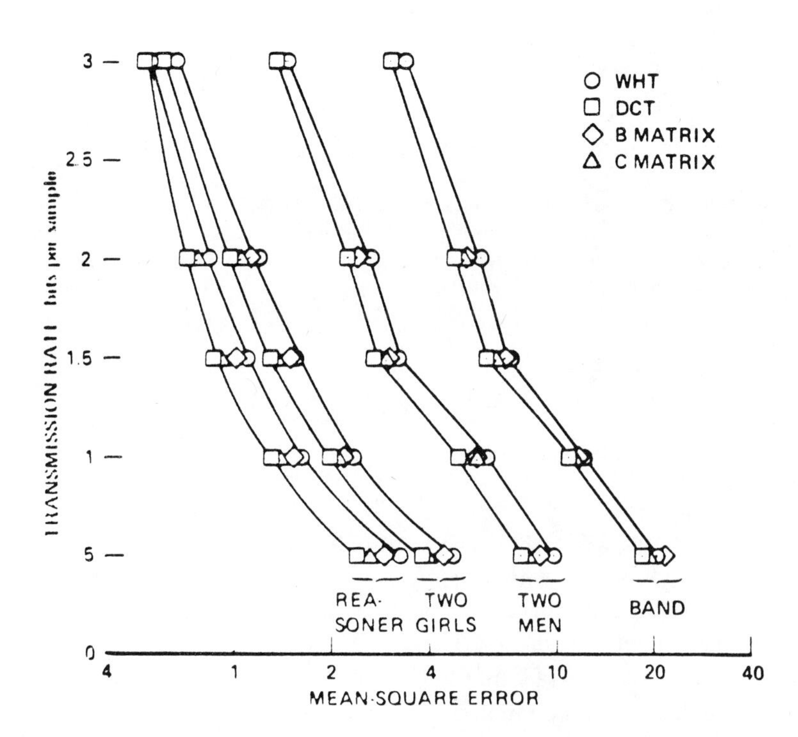

Figure 8 Transmission Rate Versus Error for the Four Test Images.

Reprinted from *Signal Process.* **5**:81–85 (1983)

AN APPROXIMATION TO THE DISCRETE COSINE TRANSFORM FOR $N = 16$*

R. SRINIVASAN and K.R. RAO
Electrical Engineering Department, The University of Texas at Arlignton, P.O. Box 19016, Arlington, TX 76019, USA

Received 9 November 1981
Revised 5 April 1982

Abstract. An approximation to the discrete cosine transform (DCT) called the *C*-matrix transform (CMT) has been developed by Jones *et al.* [3] for $N = 8$. This is now extended to $N = 16$ and its performance is compared with the DCT based on some standard criteria. CMT is computationally simpler as it involves only integer arithmetic. It has potential in signal processing applications because of its closeness to the DCT.

Zusammenfassung. Eine Näherung für die Diskrete Kosinus-Transformation (DCT) unter der Bezeichnung '*C*-Matrix-Transformation' (CMT) wurde für $N = 8$ von Jones *et al.* [3] entwickelt. Sie wird nun auf $N = 16$ erweitert und bezüglich einiger üblicher Kriterien mit der DCT verglichen. Die CMT ist einfacher zu realisieren, da sie nur Integer-Arithmetik benötigt. Wegen ihrer Nähe zur DCT liegen mögliche Anwendungen in der Sprachverarbeitung.

Résumé. Une approximation de la transformation de cosinus discrète (TCD), appelée transformation de matrice *C* (TMC), a été développée par Jones *et al.* [3] pour $N = 8$. Ceci est maintenant étendu à $N = 16$ et ses performances sont comparées avec celles de la TCD sur la base d'un critère standard. La TMC est plus simple du point de vue du calcul et implique seulement de l'arithmétique entière. Elle a un potentiel dans les applications du traitement des signaux à cause de sa parenté avec la TCD.

Keywords. Discrete transforms, fast algorithms.

Introduction

The discrete cosine transform (DCT) [1] has been widely used in digital signal and image processing because of its nearly optimal performance. Hein and Ahmed [2] have shown that the DCT can be implemented via the Walsh–Hadamard transform (WHT) [4] through a conversion matrix which has block diagonal structure. Jones *et al.* [3] have generalized this process and they have shown that any even/odd transform (this transform has one-half even and one-half odd vectors) can be expressed in terms of any other even/odd transform and the conversion matrix has sparse block diagonal format. During this development they have obtained an approximation to the DCT for $N = 8$ (N is the length of the data sequence). This transform called *C*-matrix transform (CMT) has the following properties:

(i) It is an even/odd transform.

(ii) The conversion matrix has only integers as its elements. Hence, only integer arithmetic is involved.

* This paper is based on the research by Mr. R. Srinivasan for the M.S. Thesis as a partial requirement for the M.S. degree from the University of Texas at Arlington. A paper based on this research was presented at the 19th Annual Allerton Conference on Communication, Control and Computing, Monticello, IL, Sept. 30–Oct. 2, 1981.

(iii) Its performance is very close to the DCT.

In the present paper the CMT is extended to $N = 16$. CMT for $N = 32$ is being developed. (The complexity of developing the CMT increases exponentially as N increases.) The effectiveness of the CMT is compared with other transforms based on variance distribution [8], rate distortion, mean-square error for scalar Wiener filters and computational complexity. The inherent advantage of the CMT appears to be the simplicity in its implementation (integer arithmetic plus adders and subtractors) yet negligible performance degradation compared to the DCT.[1]

C-matrix transform

Deriving DCT or CMT through WHT is by means of a conversion matrix. In the case of DCT it is given by:

$$[\widetilde{T(N)}] = [A(N)][\tilde{H}(N)] \tag{1}$$

where the symbol $\tilde{}$ stands for the row bit-reversed order of the matrix. $[T(N)]$ and $[H(N)]$ are the DCT and WHT matrices respectively of size $(N \times N)$.

The conversion matrix for the DCT, hereafter referred to as the A-matrix is reproduced here for $N=8$ for the sake of clarity in Fig. 1. Jones *et al.* [3] tried to manipulate this matrix by approximating the ratios found in certain rows of the A-matrix, thereby getting them as integers. The requirements were that the resulting structure have orthogonal rows, sparse block-diagonal structure and also a positive diagonal. Such a complexity in manipulating the A-matrix to get the C-matrix becomes exponential as N increases. In this process it should not deviate much from the corresponding DCT-vectors lest the closeness in performance be lost. The resulting C-matrix transform can be expressed as

$$[\tilde{A}_c(N)] = [C(N)][\widetilde{H(N)}] \tag{2}$$

where $[\tilde{A}_c(N)]$ is the transform matrix and $[C(N)]$ is the conversion matrix.

The C-matrix for $N = 16$ was derived by trial and error with certain guidelines listed below:

(1) The rows are to be orthogonal to one another.

(2) The structure should be block-diagonal.

(3) Maintain the ratios of elements in a row as far as possible, to those of the corresponding A-matrix. This helps to better approximate the CMT basis vectors with those of the DCT having the same sequences.

(4) The 2×2, 4×4 and 8×8 non-sparse sections of the conversion matrix be of the form

$$\left[\begin{array}{c|c} [A_1(N)] & [A_2(N)] \\ \hline -J(N)[A_2(N)]J(N) & J(N)[A_1(N)]J(N) \end{array}\right] \tag{3}$$

where $N = 1, 2, 4$ denote 1×1, 2×2, and 4×4 matrix subsections and J is the opposite diagonal unit matrix i.e.,

$$J(2) = \begin{bmatrix} 0 & 1 \\ 1 & 0 \end{bmatrix}.$$

[1] Transform image processing, in general is implemented based on $N = 4$, 8, and 16 for which the assumption of stationary statistics is reasonably valid. As the performance of the DCT does not improve significantly for $N > 16$, [10] it is conjectured that the CMT also has similar property.

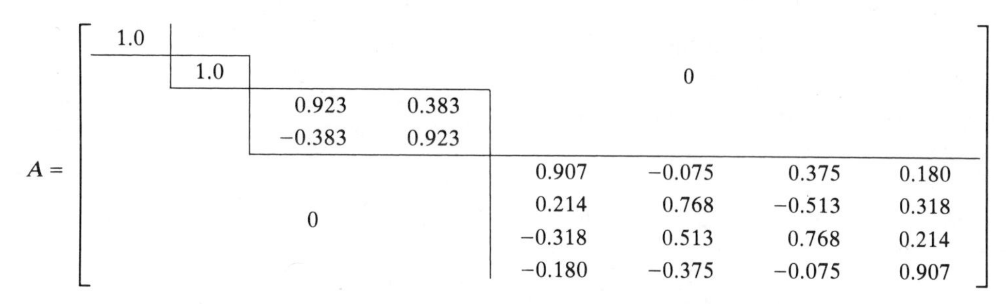

Fig. 1. The A-matrix used to obtain the DCT from the WHT.

Fig. 2 shows $[A_1(1)]$, $[A_2(1)]$, $[A_1(2)]$ and $[A_2(2)]$ for the 2×2 and 4×4 matrix blocks of the conversion matrix for $N=8$.

$$C=\frac{1}{13}\begin{bmatrix} 13 & 0 & 0 & 0 & 0 & 0 & 0 & 0 \\ 0 & 13 & 0 & 0 & 0 & 0 & 0 & 0 \\ 0 & 0 & 12 & 5 & 0 & 0 & 0 & 0 \\ 0 & 0 & -5 & 12 & 0 & 0 & 0 & 0 \\ 0 & 0 & 0 & 0 & 12 & 0 & 4 & 3 \\ 0 & 0 & 0 & 0 & 0 & 12 & -3 & 4 \\ 0 & 0 & 0 & 0 & -4 & 3 & 12 & 0 \\ 0 & 0 & 0 & 0 & -3 & -4 & 0 & 12 \end{bmatrix}$$

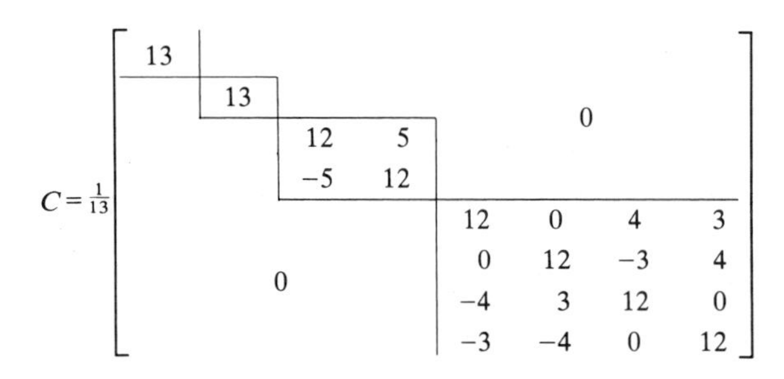

$A_1(1)=12;\quad A_2(1)=5;$

$$A_1(2)=\begin{bmatrix} 12 & 0 \\ 0 & 12 \end{bmatrix};\quad A_2(2)=\begin{bmatrix} 4 & 3 \\ -3 & 4 \end{bmatrix}$$

Fig. 2. The C-matrix for $N=8$ and its sub-matrices.

(5) The upper 8×8 block is the same as that of the C-matrix for $N=8$ (Fig. 2). Fig. 3 shows the lower 8×8 diagonal block of the C-matrix obtained for $N=16$.

$$\frac{1}{13}\begin{bmatrix} 12 & 0 & 0 & -1 & 4 & 0 & 2 & 2 \\ 0 & 12 & -1 & 0 & 0 & 4 & -2 & 2 \\ 0 & 1 & 12 & 0 & -2 & 2 & 4 & 0 \\ 1 & 0 & 0 & 12 & -2 & -2 & 0 & 4 \\ -4 & 0 & 2 & 2 & 12 & 0 & 0 & 1 \\ 0 & -4 & -2 & 2 & 0 & 12 & 1 & 0 \\ -2 & 2 & -4 & 0 & 0 & -1 & 12 & 0 \\ -2 & -2 & 0 & -4 & -1 & 0 & 0 & 12 \end{bmatrix}$$

Fig. 3. The lower 8×8 diagonal block for the C-matrix for $N=16$.

No claim to optimality is made by the authors for this conversion matrix. Using (2), the CMT for $N=16$ can be completed. The basis vectors for $N=16$ for the DCT and CMT are shown in Fig. 4. It is seen that but for a few, most of the CMT basis vectors are identical or nearly identical to the DCT basis vectors.

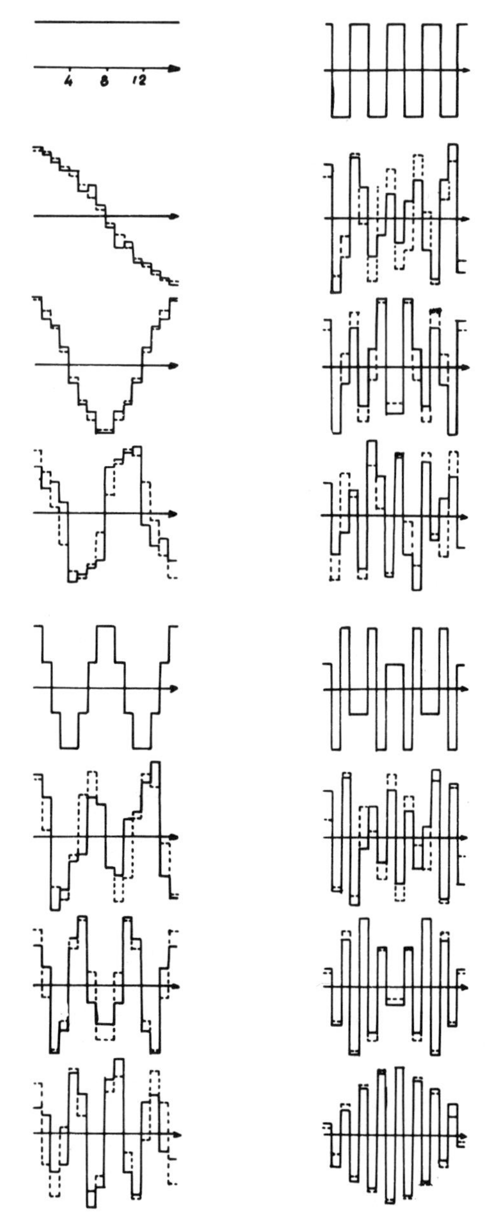

Fig. 4. Basis vectors for $N=16$ for CMT and DCT, ——CMT, - - - DCT.

Evaluation criteria

The C-matrix transform for $N=8$ and $N=16$ has been compared with the DCT from various points of view like mean square error [4], variance distribution [3], rate distortion [4], residual correlation [5], sum of the absolute values of the off-diagonal elements [6] and figure of merit [7]. The statistics of the data sequence are assumed to be governed by a first-order Markov process (see Table 1). Based on this criteria, it can be shown easily that the performance of the CMT is very close to that of the DCT.

Table 1

Sum of the absolute values of off-diagonal elements of the transformed covariance matrix for a Markov-I process for $\rho = 0.9$

	$N = 8$	$N = 16$
DCT	0.901	3.334
WHT	2.371	10.294
CMT	1.218	5.695

Computational requirements

An efficient method reported for DCT is the algorithm proposed by Chen *et al.* [9]. But the method proposed by Hein and Ahmed [2] also yields WHT coefficients in the process. Table 2 lists the computational requirements for $N = 8$ and 16. A comparison of CMT with DCT via WHT shows that the former is computationally much simpler for $N = 16$ and is of the same order for $N = 8$. However, the DCT algorithm of Chen *et al.* [9] requires less number of arithmetic operations compared to the CMT. In any case the latter can be implemented with integer arithmetic.

Table 2

Computational requirements for the DCT and CMT for $N = 8$ and 16

	$N = 8$		$N = 16$	
	Adds	Mults	Adds	Mults
FDCT [9]	26	16	74	44
DCT via WHT	38	20	134	84
CMT	34	16	106	56

Hardware implementation

The CMT coefficients for $N = 8$ can be computed using a similar structure proposed by Hein and Ahmed [2] but with the ease of integer arithmetic and much smaller ROM size. The ROM size for CMT is only two-thirds the size for the DCT. The table-lookup used 12-bit input words for the DCT in this method. For the CMT the same accuracy can be achieved with lesser number of bits. Table 3 presents a comparison for 12-bit input words for both the transforms.

Table 3

Hardware requirements for implementation of the CMT and DCT via WHT [2]

	$N = 8$	$N = 16$
DCT via WHT	2.88 K bytes	12.096 K bytes
CMT	2.304 K bytes	8.064 K bytes

Conclusions

A new transform for $N = 16$ is presented that appears to be comparable and a cheaper substitute for the DCT. The various performance measures prove its closeness to the DCT. The hardware requirements and implementation are also presented for this new transform. Because of the integer arithmetic, computationally it is much simpler compared to the DCT via WHT.

References

[1] N. Ahmed, T. Natarajan and K.R. Rao, "Discrete cosine transform," *IEEE Trans. Comput.*, Vol. C-23, Jan. 1974, pp. 90–93.

[2] D. Hein and N. Ahmed, "On a real-time Walsh–Hadamard cosine transform image processor," *IEEE Trans. Electromag. Compat.*, Vol. EMC-20, Aug. 1978, pp. 453–457.

[3] H.W. Jones, D.N. Hein and S.C. Knauer, "The Karhunen–Loeve discrete cosine and related transforms obtained via the Hadamard transform," *Int. Telemetering Conf.*, Los Angeles, Nov. 14–16, 1978.

[4] N. Ahmed and K.R. Rao, "Orthogonal transforms for digital signal processing," Springer-Verlag, New York, 1975.

[5] M. Hamidi and J. Pearl, "Comparison of cosine and Fourier transforms of Markov-I signals," *IEEE Trans. Acoust. Speech Signal Process*, Vol. ASSP-24, Oct. 1976, pp. 428–429.

[6] H.B. Kekre and J.K. Solanki, "Comparative performance of various trigonometric unitary transforms for transform image coding, *Int. J. Electronics*, Vol. 44, 1978, pp. 305–315.

[7] H.C. Andrews, "Picture processing and digital filtering," Springer-Verlag, New York, 1975, Chap. 2.

[8] K.R. Rao, John G.K. Kuo, and M.A. Narasimhan, "Slant-Haar transform," *Int. J. Computer Math.*, Sec. B, Vol. 7, 1979, pp. 73–83.

[9] W.H. Chen, C.H. Smith and S.C. Fralick, "A fast computational algorithm for the discrete cosine transform," *IEEE Trans. Commun.*, Vol. COM-25, Sept. 1977, pp. 1004–1009.

[10] A.N. Netravali and J.O. Limb, "Picture coding: a review," *Proc. IEEE*, Vol. 68, Mar. 1980, pp. 366–406.

Part II

APPLICATIONS

Editor's Comments on Papers 30, 31, and 32

PATTERN RECOGNITION

By mapping a pattern from the spatial to the transform domain, it is possible to enhance the features of the pattern. Then, only a few, but significant, features need be used in such a pattern recognition scheme. The classifier needs to be trained in the transform domain. If the transform implementation can be simplified, then the pattern classification, recognition, or identification can be made a viable scheme.

Several transforms such as the WHT, HT, DFT, and RT have been utilized in character recognition. In particular, because of its several invariant properties, RT has been utilized in alphanumeric character recognition (Papers 7 and 30). With the WHT, the RT also has been used in scene matching by Schütte, Frydrychowicz, and Schröder (Paper 31). Also, HT (Paper 32), with WHT and hybrid transforms (Wendling, Gagneux, and Stamon, 1978), has been utilized in handwritten alphabet recognition. The circular shift invariant power spectral property of the WHT has been utilized in signature verification (Nemcek and Lin, 1974). RT has been utilized in machine print alphanumeric recognition (Milson and Rao, 1976). Barger and Rao (1979) have compared the effectiveness of WHT, DFT, and RT in speech recognition. The various shift invariant power spectra developed by Hama and Yamashita (Paper 20) have applications in pattern matching and recognition. Burkhardt and Muller (1980) have developed a class of fast translation invariant transforms which have applications in position independent pattern classification and recognition problems (Muller and Burkhardt 1982; Muller, Schutte and Burkhardt, 1980).

REFERENCES

Barger, H. A., and K. R. Rao, 1979, Evaluation of Discrete Transforms for Use in Digital Speech Recognition, *Comput. Electr. Eng.* **6:**183–197

Burkhardt, H., and X. Muller, 1980, On Invariant Sets of Certain Class of Fast Translation-Invariant Transforms, *IEEE Acoust., Speech, Signal Process. Trans.* **ASSP-28:**517–523.

Milson, T. E., and K. R. Rao, 1976, A Statistical Model for Machine Print Recognition, *IEEE Syst. Man, Cybern. Trans.* **SMC-6:**671–678.

Muller, X., and H. Burkhardt, 1982, *Two-Dimensional, Fast Translation Invariant Transforms with Improved Mapping Properties,* IEEE 6th International Conference on Pattern Recognition, Munich, West Germany, pp. 427–430.

Muller, X., H. Schutte, and H. Burkhardt, 1980, *Two Dimensional Fast Transforms with High Degree of Completeness for Translation Invariant Pattern Recognition Problems,* IEEE 5th International Conference on Pattern Recognition, Miami Beach, Florida, pp. 170–173.

Nemcek, W. F., and W. C. Lin, 1974, Experimental Investigation of Automatic Signature Verification, *IEEE Syst., Man Cybern Trans.* **SMC-4:**121-126.

Wendling, S., G. Gagneux, and G. Stamon, 1978, A Set of Invariants within the Power Spectrum of Unitary Transformations, *IEEE Comput. Trans.* **C-27:**1213-1216.

Simulation of Alphanumeric Machine Print Recognition

M. A. NARASIMHAN, MEMBER, IEEE, VENKAT DEVARAJAN, AND K. R. RAO, SENIOR MEMBER, IEEE

***Abstract*—The rapid transform (RT) is used in the recognition of printed alphanumeric characters. The position invariant property of the transformation is explored by recognizing characters shifted in both x and y directions. The recognition algorithm is simulated on the digital computer. The test pattern is represented by two gray levels (binary 0 and 1) with dark background and bright characters. A two-dimensional rapid transform of the test pattern is computed. Feature selection is carried out in the transform domain based on maximum variance zonal sampling. These features are used with the corresponding features of the template patterns in computing the Euclidian distance function and the decision is made based on this minimum distance criterion. Test patterns with the gray levels interchanged can be classified with the same accuracy because significant features remain invariant to the interchange of gray levels. These features also remain invariant to cyclic shifts of the test pattern in both x and y directions. Test patterns corrupted by uniformly distributed random noise from 5 percent to 12 percent are also used for the recognition. Higher accuracy is achieved by smoothing the test pattern so that the stray bits that are on the digitized image are eliminated. It is shown through the examples that the reversed gray levels do not affect the prominent features for gray levels other than the binary case.**

Introduction

Rapid transform techniques are being utilized in the feature selection stage of character recognition [1]–[8]. The procedure generally involves selecting fewer but relevant features in the rapid transform domain based on some criteria such as the variance or magnitude of the transform component. This can result in simplified classifier design with negligible degradation in the overall recognition process. Fast algorithms resulting in reduced computational and memory requirements and corresponding reduced round-off error have further enhanced the utility of this transform. Reitboeck and Brody [1] have applied the fast Fourier transform (FFT) and rapid transform [1] (RT) to an alphabet recognition scheme where letters having different positional distortions, inclination, rotation up to 15°, and size variation up to 1:3 have been recognized with 80–100 percent accuracy. Though they mention data compression–feature selection in the transform space—the recognition process involves all the features of the prototype and the test set. Wang and Shiau [2] have utilized the RT among other transforms, recognizing single-font Chinese characters based on the feature selection by the variance criterion.

Manuscript received July 12, 1978; revised June 7, 1979 and January 30, 1980.

M. A. Narasimhan is with the Corporate Engineering Center, M.S. 452, Texas Instruments, Incorporated, P.O. Box 5621, Dallas, TX 75222.

V. Devarajan is with the Unit 1-68200, Vought Corporation, P.O. Box 225907, Dallas, TX 75265.

K. R. Rao is with the Department of Electrical Engineering, University of Texas at Arlington, Arlington, TX 76019.

Rapid Transform

The RT, which has some very attractive features, was originally developed by Reitboeck and Brody [1]. The RT results from a minor modification of the Walsh–Hadamard transform (WHT) [5]. The signal flow graph for the RT is identical to that of the WHT, except that the absolute value of the output of each stage of the iteration is taken before feeding it to the next stage (Fig. 1). This is not an orthogonal transform, as no inverse exists. With the help of additional data, however, the signal can be recovered from the transform sequence [14]. RT has some interesting properties such as invariance to cyclic shift, reflection of the data sequence, and the slight rotation of a two-dimensional pattern. It is applicable to both binary and analog inputs and it can be extended to multiple dimensions. The algorithm based on the flow graph shown in Fig. 1 can be implemented in $N\log_2 N$ additions/subtractions (N the dimension of the input data is an integer power of two) and has the "in-place" structure [5]. Improved algorithms for computation of the RT have been developed by Ulman [9] and Kunt [15]. Various properties of the RT have been developed by Wagh and others [10], [13].

Objectives

The objective of this correspondence is to utilize RT for recognizing machine print alphanumeric characters (Fig. 2). These characters obtained from the U.S. Postal Service include different types of degradation such as mixed fonts, rotation, scaling, and mutation. Initially prototypes are generated by averaging over, say, 50 of each alphanumeric characters. The statistical properties of the character set are assumed to be governed by a first-order Markov process. Subsequently 50 test patterns for each class (36 classes for the entire alphanumeric set) are utilized to determine the variance distribution for the test data. The feature selection is based on the variance criterion, i.e., retaining the transform components with large variances. Various data compression ratios such as 2:1, 4:1, are utilized in the selection process. As an example, for 4:1 data compression only 25 percent of the transform components with large variances are selected as relevant features. Noise filtering prior to transformation is achieved by smoothing the corrupted characters so that any stray bits in the digitized image are eliminated. The test data is then mapped onto the RT domain and the mean square error (mse) between the test and transformed prototype characters is computed. The decision for classifying the test character is based on this minimum mse. As the RT involves only additions and subtractions and as the data have only binary levels, both computer simulation and hardware realization are simplified. The results of the recognition scheme are outlined and the classification accuracy as a function of the data compression ratio is discussed.

Recognition System

The recognition system (Fig. 2) is simulated on the digital computer. The pattern class consists of the entire upper case and the scheme can be described in stages as follows.

1) Each character is represented by a 32×32 matrix with digitized data. Only two gray levels (binary 0 and 1) with the

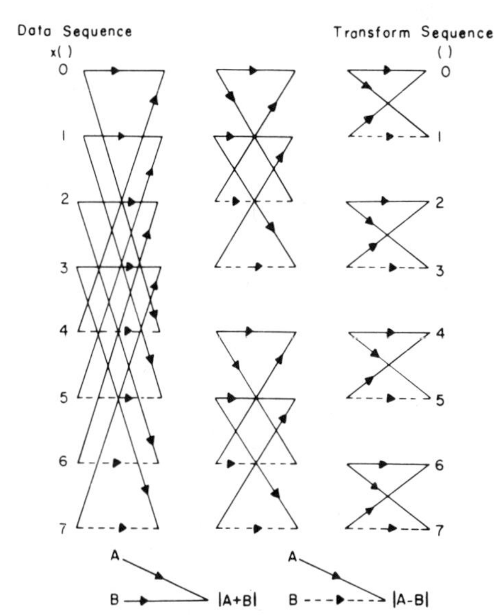

Fig. 1. Signal flow graph for the rapid transform $N=8$.

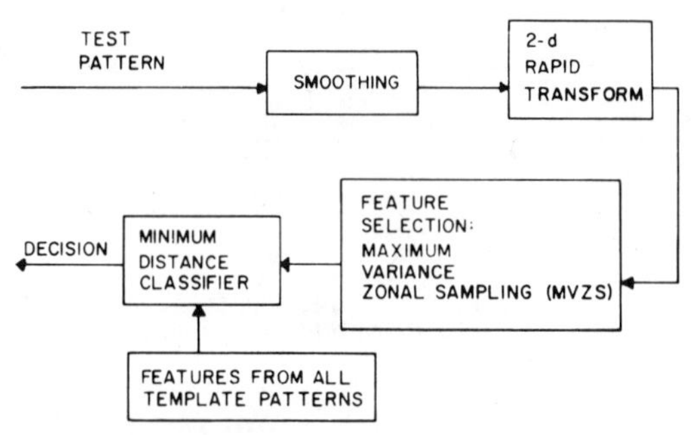

Fig. 2. Flow graph describing the character recognition by rapid transform.

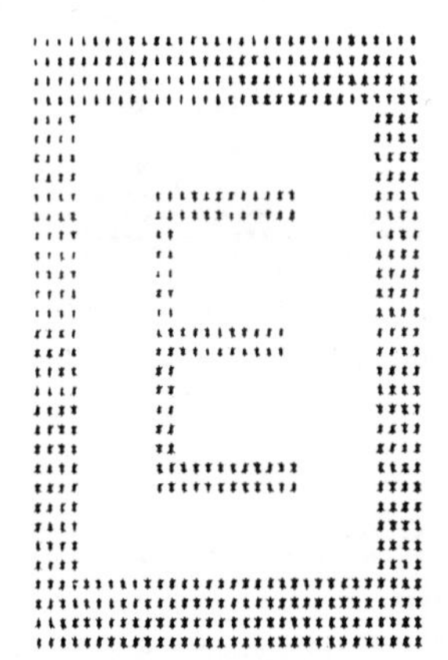

Fig. 3. Prototype character E represented by a matrix in a binary field.

background being white and the pattern being dark are allowed for all the characters.

2) A two-dimensional RT of all the prototypes is taken. This involves application of the one-dimensional RT sequentially, first by rows (columns) and then by columns (rows).

3) Feature selection in the transform domain initially is carried out on the basis of maximum variance zonal sampling (MVZS). The covariance matrix Σ of the characters is assumed to be that of a first-order Markov process with correlation coefficient $\rho=0.9$; i.e., $\Sigma=[\rho^{|k-l|}]$, $k,l=0,1,\cdots,N-1$. The corresponding covariance matrix in the transform domain $\tilde{\Sigma}$ is the two-dimensional RT of Σ. The variance distribution [5] of the patterns then is $[\tilde{\Sigma}(0,0)\tilde{\Sigma}(N-1,N-1)]^T[\tilde{\Sigma}(0,0)\tilde{\Sigma}(1,1)\cdots\tilde{\Sigma}(N-1,N-1)]$ where $\tilde{\Sigma}(j,j)$ is the jth diagonal element of $\tilde{\Sigma}$. For each prototype, the transform components are arranged in order of decreasing variance and are stored in the memory. Depending on the data compression, the corresponding transform components are selected as the relevant features for the recognition scheme. As the nonnull space of the characters is limited to (16×16) matrix, the periodicity property of RT reduces the variance distribution, feature selection, and classification based on any of (16×16) submatrices of the (32×32) transformed matrix.

4) Test characters are drawn from the data base of the U.S. Postal Service. Details of the data base and tape format are furnished in [8].

5) Noise filtering is achieved by smoothing the corrupted character before transform processing. The binary level of a data point is interchanged (0 to 1 or vice versa) if all its immediate neighbors are different from itself.

6) After filtering the noise, two-dimensional RT and feature selection based on MVZS is carried out on the filtered character.

7) The Euclidean distance between the test character and all the prototypes based on the selected features in the transform domain is computed and a decision is made based on the minimum mean square distance, i.e., $\min\Sigma_i\Sigma_j(Z(i,j)-W_k(i,j))^2$ where $Z(i,j)$ is the two-dimensional transform component of the character being recognized, $W_k(i,j)$ is the two-dimensional transform component of the kth pattern class and the summation is over all the features selected for classification purposes.

To improve the feature selection and the recognition accuracy, prototypes were selected based on the average of 50 samples for each test character. Statistical mean and variance distribution for the test set is developed on the basis of 50 samples for each character (total population size is 1800). The variance distribution representative of the test data is utilized for the selection matrix instead of the first-order Markov process described in step 3). Results are presented for both situations, a) when the alphanumeric set is treated as a homogeneous single set and b) when the alphanumeric set is split into alphabets and numerals. The recognition scheme is implemented on the IBM 370/155 digital computer.

Computer Simulation

The results of the simulation of the recognition process are described here. The various steps of the recognition system outlined earlier can be followed through with an example. The prototype E represented by a (32×32) matrix digitized to two levels and its two-dimensional RT are shown in Figs. 3 and 4 respectively. The feature selection process based on the MVZS is shown in Fig. 5. The test pattern E corrupted with 5 percent random noise and the filtered character are shown in Fig. 6 (see also Fig. 9). The same test pattern with 7.5 percent random noise and the filtered version shifted along both x and y are shown in Fig. 7. The results of the decision scheme are shown in Table I and Table II.

Gray level interchange for the prototype E is shown in Fig. 8. The two-dimensional RT of this character differs from that of

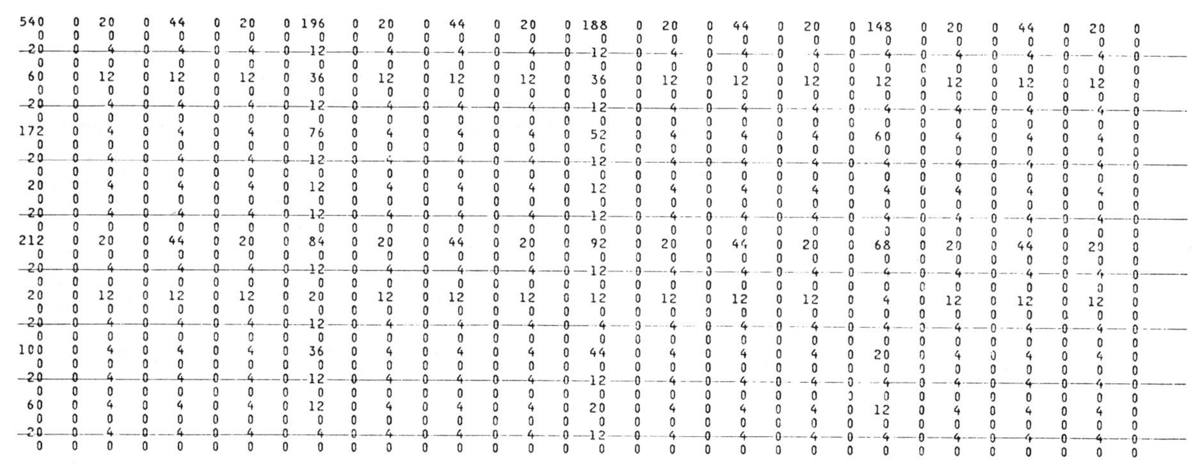

Fig. 4. Two-dimensional RT of prototype character E shown in Fig. 3.

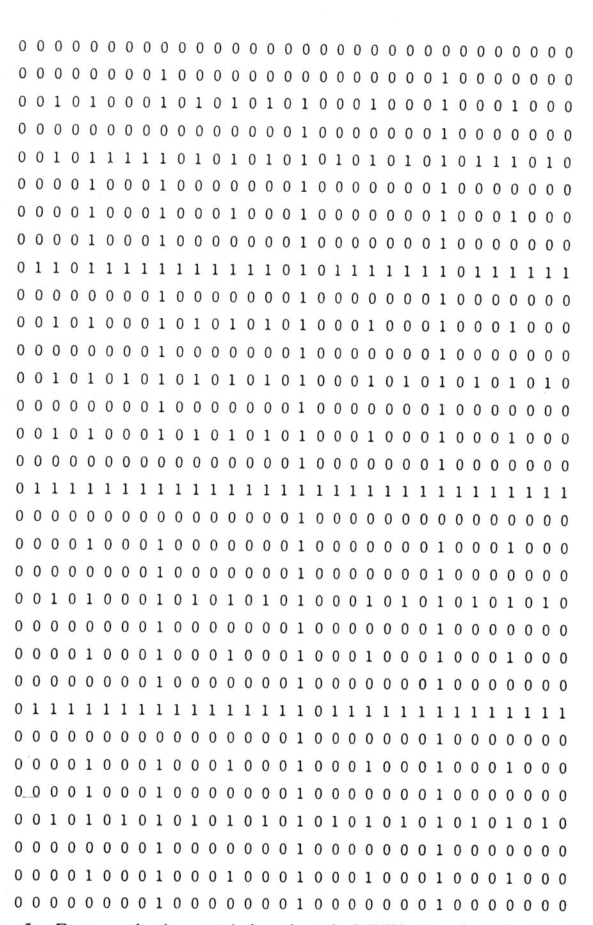

Fig. 5. Feature selection matrix based on the MVZS. Element 1 signifies the corresponding transform components of a character retained for the recognition process and element 0 denotes the component discarded. This matrix is based on 4:1 data compression.

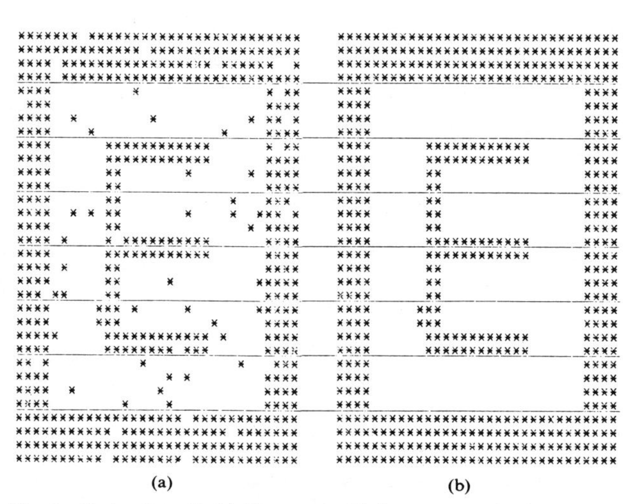

(a) (b)

Fig. 6. Test pattern E. (a) Corrupted with 5 percent randomly generated noise. (b) After filtering the noise.

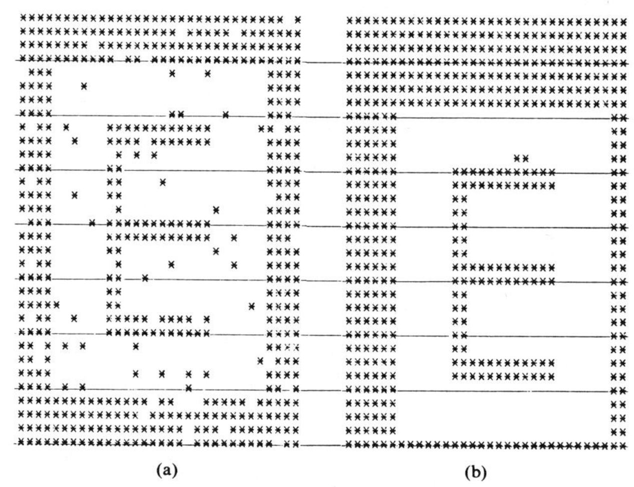

(a) (b)

Fig. 7. Test pattern E. (a) Corrupted with 7.5 percent random noise. (b) Test pattern E with x and y shifts after smoothing.

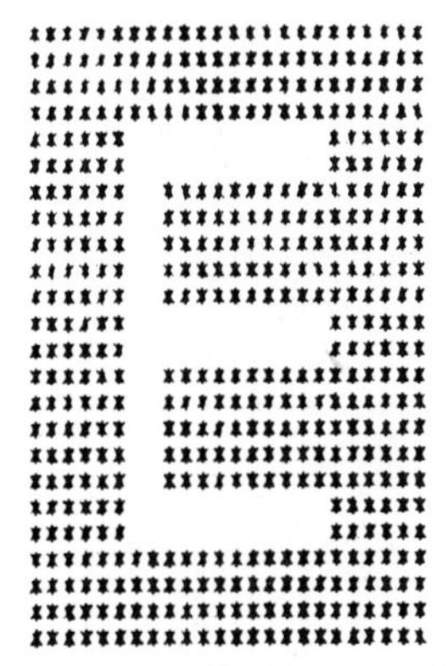

Fig. 8. Prototype character E with gray level interchange.

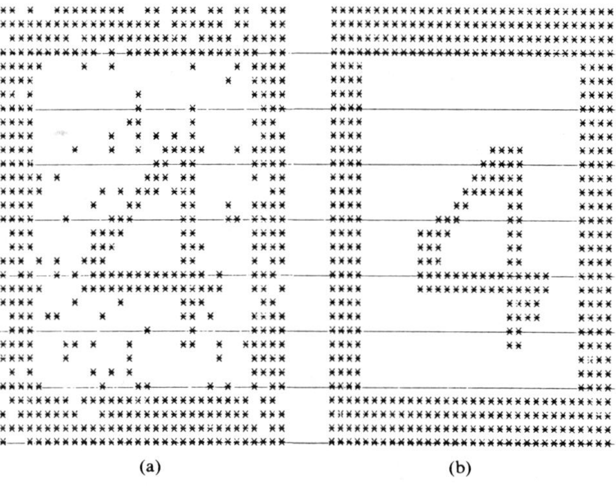

(a) (b)

Fig. 9. Test character 4. (a) 10 percent noise. (b) After filtering noise.

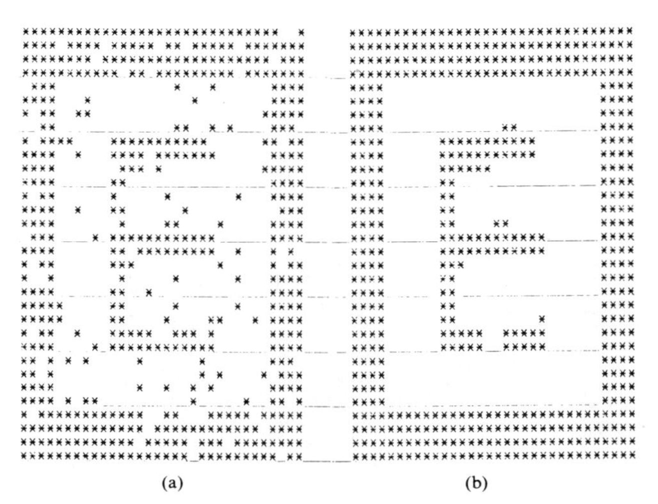

(a) (b)

Fig. 10. Test pattern E. (a) 12 percent noise. (b) Smoothed pattern.

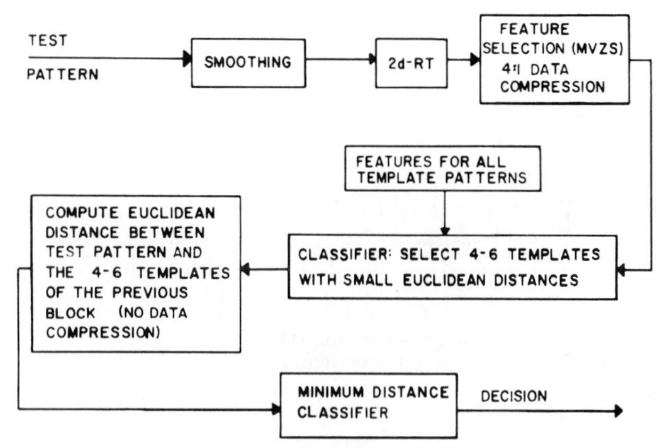

Fig. 11. Flow graph describing the sequential recognition scheme using RT.

TABLE I
EUCLIDIAN DISTANCE BETWEEN THE CHARACTER DESCRIBED IN FIG. 6(B) AND THE PROTOTYPES

TEMPLATE	
E	6192
F	13124
L	17200
C	14976
THE TEST PATTERN IS AN E	

TABLE II
EUCLIDIAN DISTANCE BETWEEN THE CHARACTER DESCRIBED IN FIG. 7 AND THE PROTOTYPES

TEMPLATE	EUCLIDIAN DISTANCE
E	3136
F	13124
L	17200
C	14976
THE TEST PATTERN IS AN E	

TABLE III
EUCLIDIAN DISTANCE BETWEEN THE CHARACTER SHOWN IN FIG. 9(B) AND THE PROTOTYPES

TEMPLATE	EUCLIDIAN DISTANCE
E	22784
F	20032
L	20992
C	23160
T	23936
I	32000
H	32000
0	32000
1	28480
7	13440
4	6312
THE TEST PATTERN IS A 4	

TABLE IV
EUCLIDIAN DISTANCE BETWEEN THE CHARACTER SHOWN IN FIG. 10(B) AND THE PROTOTYPES

SMOOTHED PATTERN TEMPLATE	CORRELATION DIFFERENCE
E	7520
F	12992
L	18720
C	17920
T	22048
I	32000
H	32000
0	32000
1	20344
7	15472
4	15816
THE TEST PATTERN IS AN E	

TABLE V

CONFUSION MATRIX FOR THE ENTIRE ALPHANUMERIC FIELD

	A	B	C	D	E	F	G	H	I	J	K	L	M	N	O	P	Q	R	S	T	U	V	W	X	Y	Z	0	1	2	3	4	5	6	7	8	9
A	15	0	0	0	0	0	0	0	1	1	0	0	0	0	0	0	3	0	0	0	0	0	0	0	0	0	0	0	0	0	1	1	3	0	0	0
B	0	10	0	0	1	0	0	0	0	0	1	0	1	0	0	0	1	5	0	0	0	0	0	1	0	0	0	1	0	0	0	0	0	0	3	1
C	0	0	11	0	0	0	1	0	0	5	0	0	0	0	0	0	1	0	1	0	0	0	0	0	2	1	0	0	0	0	0	0	3	0	0	0
D	1	1	0	5	0	0	0	0	1	2	1	0	0	0	4	0	0	2	0	0	1	0	0	0	0	0	3	0	1	1	0	0	0	0	2	0
E	0	0	0	0	16	2	0	0	0	0	0	0	0	0	0	2	0	1	0	0	0	0	1	0	0	2	0	1	0	0	0	0	0	0	0	0
F	0	0	0	0	1	11	0	0	0	0	0	5	0	0	0	2	1	0	0	0	0	0	0	0	0	1	0	1	0	0	0	0	0	2	1	0
G	0	0	0	1	0	0	10	0	0	4	0	0	0	2	0	0	0	0	0	0	0	0	0	2	2	0	0	0	0	0	0	2	0	2	0	0
H	0	0	0	0	0	0	1	17	0	0	0	0	1	0	0	0	0	2	0	0	1	0	0	0	0	0	0	0	0	0	1	0	0	2	0	0
I	0	0	0	0	0	0	0	1	19	0	0	1	0	0	0	0	0	0	0	0	0	0	0	0	0	0	0	2	1	0	0	0	0	0	1	0
J	0	0	3	0	0	0	0	0	0	12	0	3	0	0	0	0	0	0	0	0	1	1	0	0	2	0	0	0	0	0	0	0	0	2	0	1
K	0	4	0	0	1	0	0	0	0	0	8	0	0	0	1	0	1	0	0	0	0	3	0	0	1	0	5	0	0	0	0	0	0	0	1	0
L	0	0	0	0	0	0	0	0	0	0	0	12	0	0	0	0	0	0	0	2	1	0	0	0	1	0	0	0	0	0	3	0	0	6	0	0
M	0	0	0	0	0	0	0	1	0	0	0	0	8	0	0	0	0	2	0	0	0	0	8	2	0	1	0	0	0	0	1	0	0	0	2	0
N	0	0	0	0	0	0	0	0	0	0	0	0	1	8	1	0	0	0	0	0	0	0	3	0	0	0	1	0	0	3	0	1	0	0	6	1
O	0	2	0	2	0	0	0	0	0	0	1	0	0	0	6	0	2	1	0	0	1	0	0	0	1	0	3	2	0	1	0	0	1	0	2	0
P	0	0	0	0	3	1	3	0	0	1	0	0	0	0	0	7	2	0	0	0	0	0	1	1	0	0	0	0	0	0	0	5	0	1	0	0
Q	0	3	0	4	0	0	0	0	0	0	1	0	0	0	1	0	11	0	0	0	1	0	0	1	0	0	0	1	0	0	0	1	0	0	0	1
R	1	6	0	0	0	0	0	0	0	1	1	0	1	1	0	1	0	8	0	0	0	0	1	0	0	0	0	0	0	0	0	0	0	0	4	0
S	1	0	3	0	0	3	0	0	1	1	0	0	0	0	1	1	0	0	2	0	0	0	0	1	0	1	0	0	0	3	0	4	2	0	0	1
T	0	0	0	0	0	0	0	1	0	0	0	14	0	0	0	0	0	0	0	4	1	0	0	0	0	0	0	3	0	0	0	0	0	2	0	0
U	0	0	1	1	0	0	0	0	0	2	0	0	0	0	1	0	0	1	0	0	16	0	0	0	1	0	0	2	0	0	0	0	0	0	0	0
V	0	0	0	0	0	0	0	2	2	0	0	0	0	1	1	0	0	0	0	0	0	13	0	1	0	0	3	0	0	0	0	0	1	0	1	0
W	0	0	1	0	0	0	0	0	0	0	0	0	5	1	0	0	0	0	0	0	1	0	15	0	0	0	0	0	0	0	1	0	0	0	1	0
X	0	2	1	0	0	0	0	0	0	2	5	0	3	0	0	0	1	0	0	0	0	0	0	6	0	2	2	0	0	0	1	0	0	0	0	0
Y	0	0	2	0	0	0	0	1	1	4	0	0	0	0	0	0	0	0	0	0	3	0	0	0	10	0	0	0	3	1	0	0	0	0	0	0
Z	0	0	2	0	0	1	0	0	2	0	0	2	1	0	0	1	0	0	0	1	0	0	4	0	0	5	0	0	1	1	2	0	1	1	0	0
0	0	1	1	1	0	0	0	0	0	0	3	1	0	0	5	0	2	1	0	0	0	0	0	3	0	0	2	1	0	1	0	0	0	0	3	0
1	0	0	0	0	0	0	0	0	5	0	0	3	0	0	0	1	0	0	0	0	1	0	0	0	0	0	0	9	0	1	1	0	0	3	1	0
2	0	0	1	0	2	0	0	0	5	0	0	1	0	0	0	0	0	0	2	0	0	0	0	0	0	3	0	1	7	0	0	0	2	1	0	0
3	0	0	1	0	0	0	0	0	0	0	0	1	0	0	0	0	0	0	2	0	3	0	0	1	0	1	0	0	1	12	0	0	1	1	0	1
4	1	0	0	0	0	0	0	0	0	0	0	2	0	0	0	0	0	0	0	0	0	1	2	0	0	2	0	0	0	0	16	0	0	1	0	0
5	0	4	0	0	1	1	0	0	0	0	0	1	1	0	0	1	0	0	0	0	0	0	0	0	1	1	0	0	1	0	0	12	1	0	0	0
6	0	0	2	1	0	0	0	0	3	1	0	0	0	0	0	1	2	0	1	0	0	0	0	0	0	2	0	2	1	1	0	2	6	0	0	0
7	0	0	1	0	0	0	0	0	0	0	0	8	0	0	0	0	0	0	0	1	1	0	0	0	0	0	0	2	1	0	2	0	0	9	0	0
8	0	3	0	3	0	0	0	0	2	0	2	0	1	0	0	0	1	0	0	0	0	1	0	0	0	0	1	0	0	0	0	0	0	0	11	0
9	0	0	0	0	1	0	0	0	3	3	0	0	0	0	0	0	3	0	0	0	0	0	0	0	0	1	0	0	2	1	0	0	7	1	1	2

Prototype is developed from the test set, i.e., average 50 samples for each character. No smoothing 4:1 compression.

TABLE VI

CONFUSION MATRIX FOR THE ALPHABET CHARACTER RECOGNITION

	A	B	C	D	E	F	G	H	I	J	K	L	M	N	O	P	Q	R	S	T	U	V	W	X	Y	Z
A	17	0	0	0	0	0	1	0	1	1	0	0	1	0	0	1	2	0	0	0	0	1	0	0	0	0
B	1	13	0	0	2	0	0	0	0	0	2	0	1	0	0	0	1	3	0	0	0	1	0	1	0	0
C	0	0	11	0	0	1	0	0	5	0	0	0	0	0	0	0	0	0	4	0	0	0	0	0	2	1
D	2	1	0	8	0	0	0	0	0	1	2	0	0	0	5	0	0	1	0	0	2	1	0	2	0	0
E	1	0	0	0	15	3	0	0	0	0	1	0	0	0	0	2	0	0	0	0	0	0	2	0	0	1
F	0	1	0	0	1	13	0	0	0	1	0	4	0	0	0	2	0	0	2	0	0	0	1	0	0	0
G	0	0	1	1	0	0	6	0	0	5	0	0	0	4	0	1	1	2	1	0	0	0	0	1	2	0
H	0	0	0	0	0	0	1	19	0	1	0	0	1	0	0	0	0	2	0	0	1	0	0	0	0	0
I	0	0	0	0	0	0	0	1	20	0	0	1	0	0	0	0	0	0	1	0	2	0	0	0	0	0
J	0	0	3	0	0	0	0	0	0	11	1	3	0	0	0	1	0	0	1	0	1	1	0	0	4	0
K	0	4	0	0	1	0	1	0	0	0	8	0	0	1	2	0	2	1	0	0	0	4	0	0	1	0
L	1	0	0	0	0	0	0	0	0	2	0	15	0	0	0	0	0	0	0	3	1	0	0	0	2	1
M	1	1	0	0	0	0	0	1	0	0	2	0	6	0	0	0	0	1	0	0	0	0	10	1	0	2
N	0	2	0	0	0	0	0	1	0	0	1	0	1	11	2	0	0	1	3	0	0	0	3	0	0	0
O	1	3	0	2	0	0	0	0	2	1	1	0	0	0	10	0	3	1	0	0	1	0	0	0	0	0
P	2	3	0	0	3	1	3	0	0	1	0	0	0	0	0	8	2	0	0	0	0	0	1	0	0	1
Q	1	4	0	3	0	0	0	0	0	1	3	0	0	0	1	0	9	0	1	0	0	0	0	2	0	0
R	1	8	0	0	0	0	0	0	0	1	0	0	1	1	0	1	0	10	0	0	0	0	1	1	0	0
S	0	1	1	0	1	2	2	0	1	2	0	0	0	1	2	1	0	1	7	0	0	0	0	1	0	2
T	1	0	0	0	0	0	0	4	0	0	0	15	0	0	0	0	0	0	0	4	1	0	0	0	0	0
U	0	0	1	1	0	0	0	0	1	2	0	0	0	0	2	0	0	1	0	0	16	0	0	0	1	0
V	0	0	0	0	0	0	0	2	2	0	0	0	0	1	0	1	0	0	0	0	0	18	0	1	0	0
W	0	1	1	0	0	0	0	0	0	2	0	0	4	1	0	0	0	0	0	0	0	0	16	0	0	0
X	0	2	1	0	0	0	0	0	0	2	5	0	2	1	2	0	1	0	0	0	0	0	0	6	0	3
Y	0	0	2	0	0	0	0	2	1	5	0	0	0	1	0	0	0	0	1	0	3	1	0	0	9	0
Z	0	0	2	0	0	1	0	0	2	0	0	3	1	0	0	1	0	0	3	2	0	0	4	0	0	6

Prototype is developed from the test set, i.e., average 50 samples for each character. Both the prototype and test set are smoothed. 4:1 data compression.

TABLE VII

CONFUSION MATRIX FOR THE ALPHABET CHARACTER RECOGNITION

	A	B	C	D	E	F	G	H	I	J	K	L	M	N	O	P	Q	R	S	T	U	V	W	X	Y	Z
A	17	0	0	0	0	0	0	1	0	1	0	0	0	1	0	0	0	3	0	0	0	0	0	0	1	1
B	1	13	0	0	2	0	1	0	0	0	0	0	1	0	0	0	1	5	0	0	0	0	0	1	0	0
C	0	0	11	0	0	0	1	0	0	5	0	0	0	0	0	0	1	0	4	0	1	0	0	0	2	1
D	1	0	1	7	0	0	0	0	1	2	2	0	0	0	5	0	0	3	0	0	1	2	0	0	0	0
E	0	1	0	0	16	2	0	0	0	0	0	0	0	0	0	2	1	0	1	0	0	0	0	0	0	1
F	0	1	0	0	1	13	0	0	0	0	0	5	0	0	0	2	1	0	1	0	0	0	0	0	0	1
G	0	1	0	1	0	0	10	0	0	5	0	0	0	2	0	0	0	1	0	0	0	0	0	2	3	6
H	0	0	0	0	0	0	1	19	0	0	0	0	1	0	0	0	0	2	0	1	1	0	0	0	0	0
I	0	0	0	0	0	0	0	1	20	0	0	1	0	0	2	0	0	0	2	0	1	0	0	1	0	0
J	0	0	3	0	0	0	1	0	0	11	0	3	0	0	0	1	0	0	1	0	1	1	0	0	0	0
K	0	4	0	0	1	0	0	0	0	0	14	0	0	1	2	0	3	0	0	0	1	2	0	1	0	0
L	1	0	0	0	0	0	0	0	0	1	0	14	0	0	0	0	0	0	0	5	1	0	0	0	2	1
M	0	1	0	0	0	0	0	1	0	0	0	0	8	0	0	0	0	3	0	0	0	0	8	2	0	2
N	0	2	0	0	0	0	0	0	0	0	1	0	1	14	2	0	1	0	1	0	0	0	3	0	0	0
O	0	3	0	3	0	0	0	0	0	0	1	0	0	0	10	0	3	1	2	1	1	0	0	0	1	0
P	1	3	0	0	4	1	2	0	0	1	0	0	0	0	0	8	2	0	0	0	1	0	1	2	0	0
Q	1	3	0	4	0	0	0	0	0	0	1	0	0	0	1	1	11	0	1	0	1	0	0	1	0	0
R	1	8	0	0	0	0	0	0	0	1	1	0	1	1	0	1	0	10	0	0	0	0	1	0	0	0
S	1	1	2	0	0	3	0	0	1	1	0	0	0	0	3	2	0	0	9	0	0	0	0	1	0	1
T	1	0	0	0	0	0	0	3	0	0	0	14	0	0	0	0	0	0	0	6	1	0	0	0	0	0
U	0	0	1	1	0	0	0	0	1	2	0	3	0	0	1	0	0	1	0	0	17	0	0	0	1	0
V	0	0	0	0	0	0	0	2	2	0	0	0	0	2	1	0	0	0	0	0	0	17	0	1	0	0
W	0	1	1	0	0	0	0	0	0	0	0	0	4	1	0	0	0	0	0	0	1	0	15	0	1	0
X	0	2	1	0	0	0	0	0	0	2	5	0	3	0	2	0	1	0	0	0	0	0	0	6	0	3
Y	0	0	2	0	0	0	0	1	1	3	0	0	1	1	0	0	0	0	1	0	3	1	0	0	11	0
Z	0	0	3	0	0	1	0	0	2	0	0	3	1	0	0	1	0	0	3	1	0	0	4	0	0	6

Prototype developed from the test set, i.e., average 50 samples for each character. No smoothing. 2:1 data compression.

TABLE VIII
CONFUSION MATRIX FOR THE ALPHABET CHARACTER RECOGNITION

	A	B	C	D	E	F	G	H	I	J	K	L	M	N	O	P	Q	R	S	T	U	V	W	X	Y	Z
A	18	0	0	0	0	0	0	0	1	0	0	1	1	0	0	1	3	0	0	0	0	0	0	0	0	0
B	2	13	0	0	1	0	0	0	0	1	0	0	1	0	0	0	1	5	0	0	0	0	0	1	0	0
C	0	0	11	0	0	0	1	0	0	0	0	5	0	0	0	0	1	0	4	0	0	0	0	0	2	1
D	1	1	1	7	0	0	0	0	1	2	0	2	0	0	5	0	0	2	0	0	1	2	0	0	0	0
E	0	1	0	0	16	2	0	0	0	0	0	0	0	0	0	2	0	1	0	0	0	0	1	0	0	2
F	0	1	0	0	1	12	0	0	0	0	5	0	0	0	0	2	1	0	1	1	0	0	0	0	0	1
G	0	1	0	1	0	0	10	0	0	0	0	5	0	3	0	0	0	0	0	0	0	0	0	2	3	0
H	0	0	0	0	0	0	1	19	0	0	0	0	1	0	0	0	0	2	0	1	1	0	0	0	0	0
I	0	0	0	0	0	0	0	1	20	0	1	0	0	0	0	0	0	0	2	0	1	0	0	0	0	0
J	0	0	3	0	0	0	0	0	0	0	3	12	0	0	0	1	0	0	1	0	1	1	0	0	3	0
K	0	4	0	0	1	0	0	0	0	10	0	0	0	0	2	0	3	0	0	0	1	3	0	0	1	0
L	1	0	0	0	0	0	0	0	0	0	14	1	0	0	0	0	0	0	0	5	1	0	0	0	2	1
M	0	1	0	0	0	0	0	1	0	0	0	0	8	0	0	0	0	3	0	0	0	0	8	2	0	2
N	0	2	0	0	0	0	0	0	0	0	0	0	1	14	2	0	1	1	1	0	0	0	3	0	0	0
O	1	3	0	3	0	0	0	0	0	1	0	0	0	0	9	0	3	1	2	0	1	0	0	0	1	0
P	1	3	0	0	4	1	3	0	0	0	0	1	0	0	0	8	2	0	0	0	0	0	1	1	0	0
Q	1	3	0	4	0	0	0	0	0	1	0	0	0	0	1	1	11	0	1	0	1	0	0	1	0	0
R	1	8	0	0	0	0	0	0	0	1	0	1	1	1	0	1	0	10	0	0	0	0	1	0	0	0
S	1	0	3	0	1	3	0	0	1	0	0	1	0	0	3	2	0	0	8	0	0	0	0	1	0	1
T	1	0	0	0	0	0	0	3	0	0	14	0	0	0	0	0	0	0	0	6	1	0	0	0	0	0
U	0	0	1	1	0	0	0	0	1	0	0	2	0	0	1	0	0	1	0	0	17	0	0	0	1	0
V	0	0	0	0	0	0	0	2	2	0	0	0	0	2	1	0	0	6	0	0	0	17	0	1	0	0
W	0	1	1	0	0	0	0	0	0	0	0	1	5	1	0	0	0	0	0	0	1	0	15	0	0	0
X	0	2	1	0	0	0	0	0	0	5	0	2	3	0	2	0	1	0	0	0	0	0	0	6	0	3
Y	0	0	2	0	0	0	0	1	1	0	0	4	1	1	0	0	0	0	1	0	3	1	0	0	10	0
Z	0	0	3	0	0	1	0	0	2	0	3	0	1	0	0	1	0	0	3	1	0	0	4	0	0	6

Prototype is developed from the test set i.e., average of 50 samples for each character. No smoothing. 4:1 data compression.

TABLE IX
CONFUSION MATRIX FOR THE NUMERICAL CHARACTER RECOGNITION

	N0	N1	N2	N3	N4	N5	N6	N7	N8	N9
N0	15	1	1	1	0	1	0	1	5	0
N1	0	12	1	2	1	0	0	6	3	0
N2	0	2	11	0	0	1	6	3	2	1
N3	0	2	3	17	1	0	2	2	1	1
N4	0	0	1	0	20	1	0	0	0	0
N5	0	0	2	1	0	16	1	3	2	0
N6	1	3	3	2	1	2	8	2	2	1
N7	0	2	1	2	2	0	0	18	0	0
N8	4	0	0	0	2	1	1	0	18	1
N9	1	0	3	3	2	1	8	3	4	3

Both prototypes and characters to be recognized are from the test set. No smoothing. 4:1 data compression.

TABLE X
CONFUSION MATRIX FOR THE NUMERICAL CHARACTER RECOGNITION

	N0	N1	N2	N3	N4	N5	N6	N7	N8	N9
N0	11	1	0	2	1	1	4	6	3	2
N1	0	11	0	2	1	0	0	7	3	0
N2	0	1	13	5	2	1	0	1	1	1
N3	0	0	1	14	0	4	2	1	2	2
N4	0	0	8	0	1	2	1	12	0	3
N5	0	1	6	4	0	10	0	2	2	1
N6	0	0	10	9	0	2	2	6	2	0
N7	0	3	2	2	1	0	0	17	0	0
N8	1	0	0	1	0	0	2	1	20	0
N9	0	2	1	8	0	2	2	1	3	0

Both test set and prototypes are from the trained patterns. No smoothing. 4:1 data compression.

the original character (Fig. 3) in only the dc (0,0) component. All other transform components remain invariant to the binary level interchange (positive to negative effect). The results of the decision scheme are shown in Table III. Results of increasing this noise level further to 12 percent are shown in Fig. 10 and Table IV.

RESULTS AND CONCLUSION

The recognition results for the alphanumeric test set are shown in Tables V through X. Feature selection based on the variance distribution developed for the test set rather than on the Markov process gave lower misclassification errors. Also smoothing of both the prototype and the test set increased the recognition accuracy. Dividing the entire alphanumeric family into two groups, alphabet and numerals, further simplified the classification algorithm, thereby improving the recognition process. On the basis of the above results a sequential decision procedure is suggested wherein the test set is initially associated with four to six classes based on the Euclidean distances and then final classification is made considering all the features with no data compression (Fig. 11).

The recognition system developed here differs from that of Wang and Shiau [2] in the following aspects. (a) Test characters are filtered from noise (smoothing) before recognition. (b) The recognition is invariant to gray level interchange of the characters (positive to negative effect). (c) The relationship between classification accuracy and amount of noise introduced is developed.

REFERENCES

[1] H. Reitbroeck and T. P. Brody, "A transformation with invariance under cyclic permutation for applications in pattern recognition," *Information and Control*, vol. 15, pp. 130–154, July 1969.

[2] P. P. Wang and R. C. Shiau, "Machine recognition of printed chinese characters via transformation algorithms," *Pattern Recognition*, vol. 5, pp. 303–321, Dec. 1973.

[3] H. C. Andrews, "Multidimensional rotations in feature selection," *IEEE Trans. Comput.*, vol. C-20, pp. 1045–1051, Sep. 1971.

[4] W. F. Nemcek and W. C. Lin, "Experimental investigation of automatic signature verification," *IEEE Trans. Syst. Man, and Cybern.*, vol. SMC-4, pp. 121–126, Jan. 1974.

[5] N. Ahmed and K. R. Rao, *Orthogonal Transforms for Digital Signal Processing*. New York/Berlin: Springer-Verlag, 1975.

[6] H. C. Andrews, *Introduction to Mathematical Techniques in Pattern Recognition*. New York: Wiley-Interscience, 1972.

[7] M. A. Narasimhan and K. R. Rao, "Printed alphanumeric character recognition by rapid transform," presented at Ninth Annual Asilomar Conf. on Circuits, Systems, and Computers, Pacific Grove, CA, Nov. 3–5, 1975, in *Proc.*, pp. 558–563.

[8] T. E. Milson and K. R. Rao, "A statistical model for machine print recognition," *IEEE Trans. Syst. Man, and Cybern.*, vol. SMC-6, pp. 671–678, Oct. 1976.

[9] L. J. Ulman, "Computation of the Hadamard transform and the R-transform in ordered form," *IEEE Trans. Comput.*, vol. C-19, pp. 359–360, Apr. 1970.

[10] M. D. Wagh and S. V. Kanetkar, "A multiplexing theorem and generalization of R-transform," *Intern. J. Comput.*, vol. C-24, pp. 1120–1121, Nov. 1975.

[11] M. D. Wagh, "Periodicity in R-transformation," *J. Inst. of Electron. and Telecommuni. Engrs.*, vol. 21, pp. 560–561, 1975.

[12] ——, "An extension of R-transform to patterns of arbitrary lengths," *Intern. J. Comput. Math.*, vol. 7, Sec. B, pp. 1–12, 1977.

[13] M. D. Wagh and S. V. Kanetkar, "A class of translation invariant transforms," *IEEE Trans. Acoust., Speech, Signal Processing*, vol. ASSP-25, pp. 203–205, Apr. 1977.

[14] V. Vlasenko, K. R. Rao, and V. Devarajan, "Unified matrix treatment of discrete transforms," presented at Tenth Annual Southeastern Symp. on System Theory, Mississippi State, MS, Mar. 13–14, 1978, in *Proc.* pp. 11, B-18–11, B-29.

[15] M. Kunt, "On computation of the Hadamard transform and the R-transform in ordered form," *IEEE Trans. Comput.*, vol. C-24, pp. 1120–1121, Nov. 1975.

31

Reprinted with permission from pages 195-198 of *5th International Conference of Pattern Recognition Proceedings,* Institute of Electrical and Electronics Engineers, New York, 1980

SCENE MATCHING WITH TRANSLATION INVARIANT TRANSFORMS

Holger Schütte, Stephan Frydrychowicz, Johannes Schröder

DORNIER GmbH, Postfach 1420, 7990 Friedrichshafen 1
West Germany

SUMMARY

The problem of locating a reference image within a larger image is discussed.

The method most widely used for automatic determination of local similarity between two structured data sets is the correlation, which can be calculated by various methods.

In this paper a new method is introduced, which calculates the similarity with few characteristic reference coefficients in a modified sequency domain (R-Transform). Most important to the new method is the property to yield translation invariant coefficients.

Because of this property, we can compute the coarse location with few wide steps of the scanning window. In the detected area we can determine the exact position using another set of reference coefficients in the original sequency domain (Walsh-Hadamard-Transform).

The real time computation (TV-field rate) is possible because of the simple operators (+ and |-|) and the high degree of parallelism of the method.

I. Introduction

The problem of locating a reference image within a larger image using a correlation technique is discussed.

Because of the large computation costs, the application of correlation methods are related to limited data sets. A small search area will be used for tracking and a reduction of quantisation (for example 1 bit for 1 picture element) will be used, if the search area is as big as the complete picture each time.

This paper will demonstrate, that the application of translation invariant transforms for similarity detection will yield efficient and accurate results, in spite of large data sets and high quantisation rate.

In section II the translational displacement problem and the correlation methods for solving them will be introduced.

Section III will describe the properties of transforms in the sequency domain.

In section IV the complete method will be described and some results of scene matching will be presented.

In section V a conclusion is given.

II. The Translational Displacement Problem

Finding the location of a reference image within the field of view (FOV) of a video sensor in real time is considered using a correlation technique in the sequency domain. Real time means that the location of the reference image is found within the TV-frame-rate [1], [2].

The pictures do not differ in magnitude and rotation und their digital picture elements are within the same grey scale range.

In the following we will explain the
- direct correlation
- FFT-correlation
- sequency domain-correlation

A. Direct correlation

Let two images, S the search area and R the reference window be defined as shown in Fig. 1

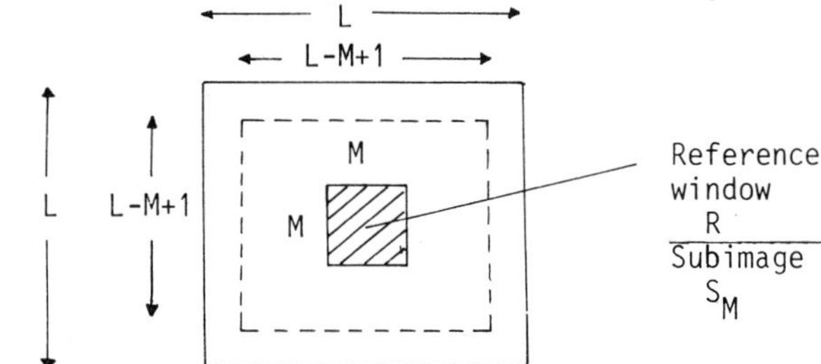

Fig. 1: Search space

S is an L x L array of digital picture elements and R is an M x M, M smaller than L, array of digital picture elements. To compare the two images, we have to compare each of the $(L-M+1)^2$ subimages (array size M x M) with the reference window (array size M x M).

This shall be done with the equation most widely used, the normalised product-correlation [1]:

$$K^2(i,j) = \frac{\left(\sum_{l=1}^{M}\sum_{m=1}^{M} R(l,m)\cdot S_M^{i,j}(l,m)\right)^2}{\left[\sum_{l=m}^{M}\sum_{m=1}^{M} R^2(l,m)\right]\left[\sum_{l=1}^{M}\sum_{m=1}^{M} S_M^{2\,i,j}(l,m)\right]} \qquad (1)$$

$$1 \le i,\ j \le L - M + 1$$

B. FFT-Correlation

The basic idea of the FFT-Correlation [3] is to replace the shift of the subimages in the original domain by multiplications in the frequency domain.

This manner of correlation can be done by comparing the complete L x L video array with an enlarged reference array (from M x M to L x L). Or this can be done with the more realistic concept of smaller reference arrays (e.g. from M x M to 2 M x 2 M) and 50 % overlapping of the subimages.

C. Sequency domain correlation

The basic idea is the application of translation invariant transforms ([4], R-transform) for fast computation of the coarse displacement of two pictures. This can be explained as follows:

a) The translation invariant transforms have the following property: the cyclic permutation of the input pattern has no influence on the transformation pattern (Figure 2)

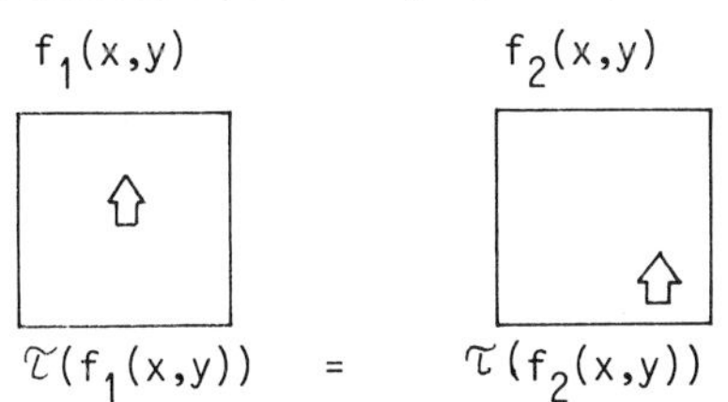

Fig. 2: Property of translation invariant transforms

If the pattern drifts out of the input array or if a strange pattern drifts into the input array, the transformation pattern $\tau(f(x,y))$ will change continuously [4].

b) The differences of transformed patterns (from one array to another) can be calculated for classification by the mean absolute difference (MAD), by the product correlation or by the Euclidean distance [4].
The graphical representation of the correlation of a transformation window (M x M) along a row inside of a search area (L_1 x L_2) can be seen in Figure 3.

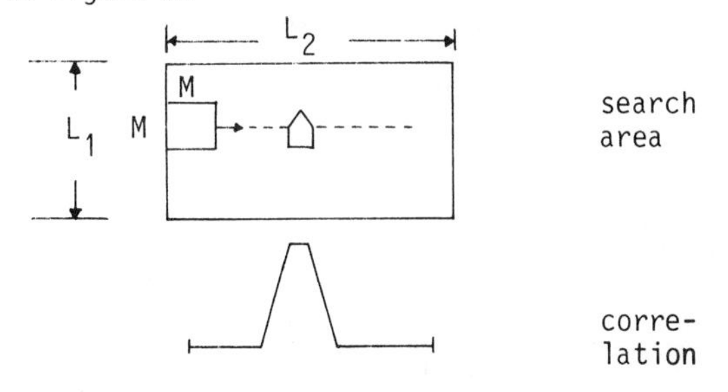

Fig. 3: Graphical representation of the correlation

c) This characteristic peculiarity implies, that the shift of the M x M-search-window along the row can be made with large steps, without missing the top of the correlation peak (Figure 4).

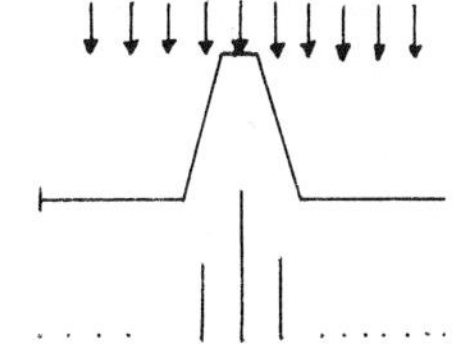

Fig. 4: Coarse correlation

d) The width of the correlation peak (at the same level) and the width of the scanning steps of the M x M window are dependent on the selection of the reference coefficients in the modified sequency domain.

Having found the coarse position by determining the maximum value of the correlation and threshold level, the correct position can be found by denser sampling of the search window.

The exact location of the reference window in the search area will be obtained by operations in the original sequency domain (Walsh-Hadamard-Transform).

III. Properties of Transforms in the Sequency Domain

In this section we are going to describe some properties of the Rapid-Transform [4] , the Modified-Rapid-Transform [5] and the Walsh-Hadamard-Transform.

These transforms have been defined recursively, by connecting four elements simultaneously. We define the input- and output-elements of the transforms, which have the same structure but different operators:

$$\begin{pmatrix} A & B \\ C & D \end{pmatrix} \Rightarrow \boxed{\tau} \Rightarrow \begin{pmatrix} \tilde{A} & \tilde{B} \\ \tilde{C} & \tilde{D} \end{pmatrix}$$

R-Transform [4] (RT)

$$\begin{aligned} \tilde{A} &= A + B + C + D \\ \tilde{B} &= |(A + B) - (C + D)| \\ \tilde{C} &= |A - B| + |C - D| \\ \tilde{D} &= ||A - B| - |C - D|| \end{aligned} \quad (2)$$

MR-Transform [5] (MRT)

$$\begin{aligned} \tilde{A} &= A + B + C + D \\ \tilde{B} &= |(A + B) - (C + D)| \\ \tilde{C} &= |(A - B) + (C - D)| \\ \tilde{D} &= |(A - B) - (C - D)| \end{aligned} \quad (3)$$

Walsh-Hadamard-Transform (WHT)

$$\begin{aligned} \tilde{A} &= A + B + C + D \\ \tilde{B} &= (A + B) - (C + D) \\ \tilde{C} &= (A - B) + (C - D) \\ \tilde{D} &= (A - B) - (C - D) \end{aligned} \quad (4)$$

As demonstrated in [4] , the nonlinear R-Transform is invariant under cyclic permutation. This property can be demonstrated for the MR-Transform by analogous conclusions. The Walsh-Hadamard-Transform has, on the contrary, only linear operators.

For special changes of the gray scale value we can describe the effect in the sequency domain.

We use a short-hand notation for the gray scale values of the field:

$$x = (x_{i,j}) \quad i = 1,\ldots N, \quad j = 1,\ldots N$$

and we define the change factors $\lambda x + C$ with $\lambda \geq 0$ and C real.

The following is valid:

$$RT(\lambda x + C) = \lambda \cdot RT(x) + RT(C)$$
$$MRT(\lambda x + C) = \lambda \cdot MRT(x) + MRT(C)$$
$$WHT(\lambda x + C) = \lambda \cdot WHT(x) + WHT(C) \qquad (5)$$

It can be shown, that the constant additive factor C influences only the first element of the transformed array

$$RT(C)_{i,j} = MRT(C)_{i,j} = WHT(C)_{i,j} = \begin{cases} N^2 \cdot C & i,j = 0 \\ 0 & \text{elsewhere} \end{cases}$$

The constant multiplicative factor λ will result in a spread, with the same factor, in the sequency domain.

Because of these properties, we can classify normalized reference images in the modified sequency domain as well as in the original domain.

The computation costs for normalization are lower in the sequency domain (few selected coefficients).

IV. The New System Concept and Results

A.) System Concept

Finding the location of a reference image in a sensed image is realized in three loops using the R-Transform or the Walsh-Hadamard-Transform.

In the first loop, the search window is shifted over the sensed image in large steps. The decision, that the reference image is contained in the sensed image, follows from a comparison of the characteristic features of the sensed and the reference image using the R-Transform.

In the second loop a determination of the position can be made in the same manner with fine steps in the limited region.

In the third loop the determination of the exact position can be realized by shift steps of one picture element using the Walsh-Hadamard-Transform.

The selection of the reference features for the first loop will be made in the following form:

Having defined a target area in a larger image, the transformation window will be shifted near the target systematically in such a form that a kernel area is contained in each transformation window. Those coefficients with the least differences during the shift of the transformation window will be chosen for reference.

The size of the kernel area in relation to the transformation window is important to the shift steps of the search window and important to the signal/noise ratio. Using a weighting function, the selection of the reference features in the modified sequency domain can be influenced decisively.

To extract the reference features in the second loop (fine steps) a set of coefficients of the R-Transform is used and in the third loop (pixel steps) a set of coefficients of the Walsh-Hadamard-Transform is used. The chosen set of coefficients contains only 2-3 % of the complete set.

The equation (6) calculates the normalized similarity:

$$x^2 = \left(\sum_{i=1}^{K} S_{ki\,li}\, \tau_{ki\,li}\right)^2 \Big/ \sum_{i=1}^{K} \left(S_{ki\,li}\, \tau_{ki\,li}\right)^2 \qquad (6)$$

The $S_{ki\,li}$ are the K present transformation coefficients and the $\tau_{ki\,li}$ are the K reference transformation coefficients, which have been weighted in advance (offline).

B.) Computation cost

The amount of computation required for one search-window is summarized in Table I:

	Transformation (RT, MRT, WHT) M x M - window	Classification K feature coefficient
ADD	$M^2 (2 \log_2 M)$	2 K - 1
MULT	-	2 K
DIV	-	1

Table I: Cost of correlation with R-Transform

The complete computational load of the similarity detection in three loops is dependent on the overlap (D) of the search windows (M x M) in a search area (L x L). The number of comparisons (α) in the first loop is:

$$\alpha = \left(\frac{L - M}{M - D} + 1\right)^2 \qquad (7)$$

With increasing M, the saving of computation costs as opposed to the correlation function ($D = 1 \rightarrow \alpha = (L-M+1)^2$) becomes more and more important.

C.) Results

In the following section we describe some results using digitized TV-pictures (256 x 384 picture elements).

In picture 1 and 2 are two examples of computing a modified Euclidean distance (reciprocal similarity) by the R-Transform and comparing the result with a threshold value.

To detect the road crossing we have selected 100 translation invariant coefficients for reference. In picture 1 the smaller road crossing field as kernel area is recognized within the larger search window (64 x 64), because the computed distance is smaller than the threshold distance (the bar on the left is the threshold distance and the other one is the present distance). In picture 2, the wrong road crossing in the search window is rejected.

Picture 1: Detection of the road crossing

Picture 2: Rejection of the wrong road crossing

For a better understanding of the properties of the R-Transform we have computed the distances of all 320 transformation windows (64 x 64) along a row inside of the search area (256 x 384).

In picture 3 we can see the normalized distance graphically.

Picture 3: Normalized distance graph along a line

The broad peak of the distance in the surrounding of the signed target is obvious and has been described in section II.

V. Conclusions

A general class of algorithms for similarity detection in the sequency domain has been introduced, applicable to the specific problem of translational image registration.

Experimental results have been presented to show the efficiency of the new method.

The structure of the new algorithm is especially suited for digital similarity detections. There is a saving of computation time of two orders of magnitude on a multi purpose computer and it is planned to realize the algorithm in a special purpose computer for real time computation (TV-field-rate).

Acknowledgement

The research reported in this paper was supported by the German Bundesministerium für Verteidigung (BMVg).

Literature

1 Barnea, Daniel.I.; Silverman, Harvey, F.:
A Class of Algorithms for Fast Digital Image Registration
IEEE Transactions on Computers, Vol.C-21,No.2 February 1972

2 Boland, J.S. et al
Design of a Correlation for Real-Time Video Comparisons
IEEE Transactions on Aerospace and Electronic Systems, Vol. AES-15, No. 1, January 1979

3 Brigham, E.O.
The Fast Fourier Transform
Prentice Hall, Inc., Englewood Cliffs, N.J. 1974

4 Reitboeck, H.; Brody, T.P.
A Transformation with Invariance Under Cyclic Permutation for Applications in Pattern Recognition
Information and Control, 15, 130-154, 1969

5 Müller, X.; Schütte, H.; Burkhardt, H.
Two Dimensional Fast Transforms with High Degree of Completenes for Translation Invariant Pattern Recognition Problems
"Short paper" for the 5-ICPR

Reprinted with permission from pages 844-848 of *3rd International Conference of Pattern Recognition Proceedings,* Institute of Electrical and Electronics Engineers, New York, 1976

USE OF THE HAAR TRANSFORM AND SOME OF ITS PROPERTIES IN CHARACTER RECOGNITION

S. Wendling, G. Gagneux, and G. Stamon

Summary

Haar transform is defined and several of its properties are developed. A fast computing algorithm for Haar coefficients is given.

Haar power spectrum is introduced and it is shown that the obtained results can be generalized to any transform defined by an orthogonal matrix whose first row vector contains ones only. A character recognition experiment using both invariants within the power spectrum and features within the transform domain is described.

Introduction

Properties of global transforms such as Fourier transform or Hadamard transform and, for each of these a fast computation algorithm, account for the widespread use of these in picture (transmission, recognition,...) or speech processing.

This paper will introduce Haar transform which has over the former two (which are globally sensitive) the advantage of being both locally and globally sensitive. Once the definition of the haar transform recalled, some of its properties will be established and an algorithm for fast and efficient computation of Haar coefficients developed. Then the Haar transform efficiency will be tested in a character recognition experiment.

Haar functions

Let E be the vector space of the numerical functions, defined within $[0,1]$, bounded, accepting on any point a limit on the left and on the right, and continuous to the right. With the usual inner product of L^2 ($[0,1]$), E is a pre-Hilbert space in which it is possible to define a complete orthonormal system (f_n), called Haar system, and defined in the following way :

1) $f_0(x) = 1 \quad \forall x \in [0,1]$

2) $\forall n > 0$, integer, let m be the greatest integer such as $2^m \leq n$ and $n = 2^m + k$

Then $f_n(x) = 2^{m/2}$ for $\frac{2k}{2^{m+1}} \leq x < \frac{2k+1}{2^{m+1}}$

$f_n(x) = -2^{m/2}$ for $\frac{2k+1}{2^{m+1}} \leq x < \frac{2k+2}{2^{m+1}}$

$f_n(x) = 0$ elsewhere.

Figure 1a shows these functions for n=0,1,2,...,7

Haar matrices

The sub-set of the Haar system consisting of the $f_k (0 \leq k < n)$ functions where $n = 2^P$, is considered.

The previous considerations entail that in each of the intervals

$$I_j = \left[\frac{j}{2^P}, \frac{j+1}{2^P}\right] \text{ with } j \in \left[0,1,2,\ldots,2^P-1\right]$$

the f_k $(0 \leq k \leq 2^P-1)$ functions are constant.

$H(2^P)$, the $2^P x 2^P$ matrix, called Haar matrix, is then obtained in the following way :

$H(2^P) = \left[h_{ij}\right]$ where h_{ij}, the generic term, located at the intersection of the ith row and the jth column, equals $f_i(I_j)$ (i.e. the value taken by f_i on the I_j interval ; see figure 1b).

The row vectors of $H(2^P)$ then establish an orthogonal system of R^{2^P}, the orthogonality with the first row vector implying that

$$\sum_{j=0}^{j=2^P-1} h_{ij} = 0 \text{ for } i = 1,2,\ldots,2^P-1 \qquad (1)$$

Moreover, $H(2^P)$ is such as $H(2^P)\,H(2^P)^t = 2^P I_{2^P}$ where I_{2^P} is the $2^P \times 2^P$ identity matrix.

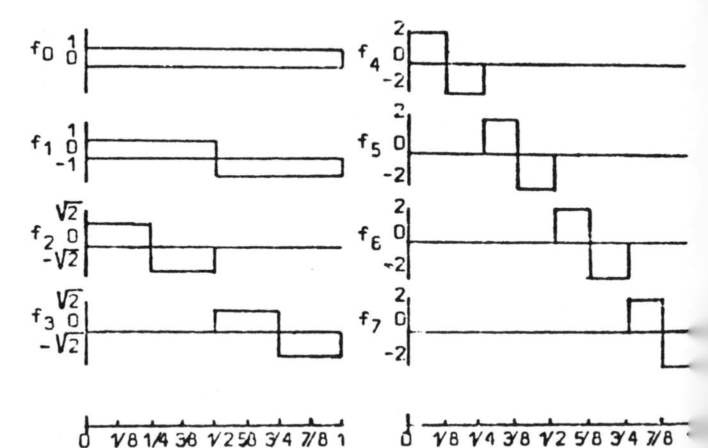

fig. 1a. Haar functions for n = 0,1,..., 7.

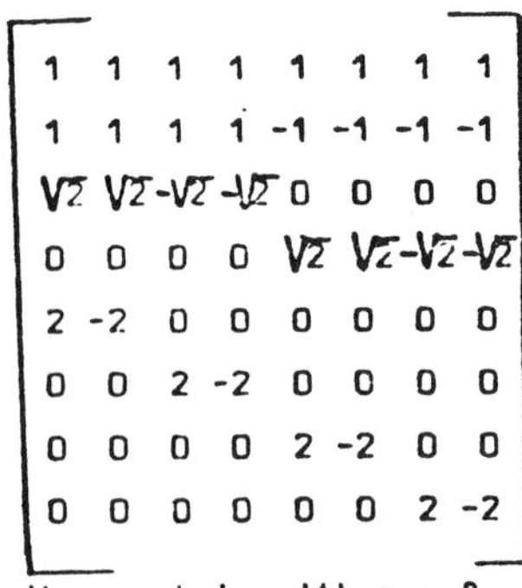

1	1	1	1	1	1	1	1
1	1	1	1	-1	-1	-1	-1
√2	√2	-√2	-√2	0	0	0	0
0	0	0	0	√2	√2	-√2	-√2
2	-2	0	0	0	0	0	0
0	0	2	-2	0	0	0	0
0	0	0	0	2	-2	0	0
0	0	0	0	0	0	2	-2

fig. 1b. Haar matrix with n = 8

Haar transform

The Haar transform of an n periodic sequence, $X = \{x_0, x_1, \ldots, x_{n-1}\}$ is defined as

$$[T(X)] = 1/n\,[H][X] \qquad (2)$$

where $[X]$ is the vector representation of sequence X, $[H]$ the n x n Haar matrix and

$[T(X)]^t = \{t_0(X), t_1(X), \ldots, t_{n-1}(X)\}$

the coefficients of the Haar transform.

As $[H][H]^t = nI$, the inverse transform will be defined as $[X] = [H]^t\,[T(X)]$ (3)

Fast Haar transform

Using matrix factoring or matrix partitioning similar to those used for FFT (Fast Fourier Transform) and FHT (Fast Hadamard Transform) computation, fast and efficient algorithms for the Haar transform are obtained.

Given the particular structure of the n X n = $2^P \times 2^P$ Haar matrices (fig. 1b for n = 2^3), the total number of iterations needed to compute the transform coefficients is P. Each iteration reduces by half the number of values still to be computed.

Thus, the first iteration gives the n/2 coefficients:

$$t_k(X) \qquad n/2 \leq k < n$$

the second iteration gives the n/4 coefficients

$$t_k(X) \quad ; \quad n/4 \leq k < n/2$$

$$\vdots \qquad\qquad \vdots$$

the Pth iteration gives the 2 coefficients :

$$t_0(X) \text{ and } t_1(X)$$

For example, with n = 8, the signal flow graph of fig. 2 is obtained. Apart from the multipliers, the total number of arithmetic operations (additions or subtractions) required for computing the Haar coefficients is 8 + 4 + 2 = 14.

Generalization

The signal flow graph in fig. 2 can be straightforwardly generalized for n = 2^P. The following remarks can be made :

1. the total number of iterations is given by $P = \log_2 n$
2. the 1st iteration requires n arithmetic operations (additions or subtractions). At each iteration this number is divided by 2, which yields the total number of n + n/2 + n/4 + ... + 2 = 2(n-1) operations. In the same way, the number of multiplications can be evaluated as n/2 + n/4 + ... + 2 = (n-1). A single evaluation of (2) would have required n^2 operations.
3. Compared to Hadamard transform which can be evaluated in n $\log_2 n$ additions or subtractions, Haar transform requires nearly $(\log_2 n)/2$ less operations. As an example, with n = 2^{10} = 1024, FHT will need 10240 additions or subtractions whereas Haar transform will need only 2046, that is to say about five times less.

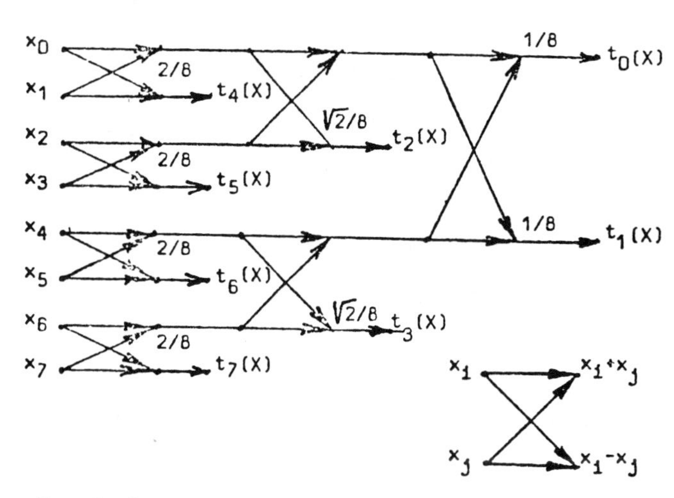

fig. 2. Signal flow graph illustrating the computation of the Haar transform coefficients for n = 8.

Parseval's formula (Power Spectrum)

The transpose of (2) yields :

$$[T(X)]^t = 1/n\,([H][X])^t = 1/n\,[X]^t[H]^t \qquad (4)$$

From (2) and (4) it follows that :

$$[T(X)]^t\,[T(X)] = 1/n^2\,[X]^t[H]^t[H][X]$$

$$[T(X)]^t\,[T(X)] = 1/n\,[X]^t[X]$$

or

$$\sum_{i=0}^{n-1} t_i^2(X) = 1/n \sum_{i=0}^{n-1} x_i^2 \qquad (5)$$

Thus, if $X = \{x_0, x_1, \ldots, x_{n-1}\}$ represents the sampled values of a signal, (5) shows that the power of this signal is conserved within the transform domain. The set $\{t_i^2(X)\}$, however, does not represent the individual spectral points as it is not invariant to the shift of the sampled data.

Spectrum development

Now a set of invariants is being looked for. The case where n = 8 is first considered.

Let X^l denote X shifted to the left by l positions.

$$X^l = \{x_l, x_{l+1}, \ldots, x_{l-2}, x_{l-1}\} \quad l=1,2,\ldots,7$$

The Haar transform of X^l is :

$$[T(X^l)] = 1/8\,[H][X^l]$$

$= 1/8\,[H][S^l][X]$ where $[S^l]$ is the l positions to the left translation matrix $(1 \leq l \leq 7)$

$$= 1/64\,[H][S^l][H]^t[H][X]$$

$[T(X^l)] = 1/8\,[A^l][T(X)]$ where $[A^l] = 1/8[H][S^l][H]^t$

Now $[A^l]$ will always be orthogonal since

$$[A^l]^t[A^l] = 1/64\,[[H]^t]^t[S^l]^t[H]^t[H][S^l][H]^t$$

$$= I_8 \text{ for } [S^l]^t[S^l] = I_8$$

Moreover $[A^l]$ can always be written

$$A^l = \begin{bmatrix} 1 & 0 \cdots 0 \\ 0 & \\ \vdots & \tilde{A}^l \\ 0 & \end{bmatrix}$$

where $[\tilde{A}^l]$ is a 7X7 matrix such as $[\tilde{A}^l][\tilde{A}^l]^t = I_7$

Indeed : $a^1_{oo} = 1$ since $\forall 1$,

$$a^1_{oo} = 1/8 \begin{bmatrix}1111111\end{bmatrix} \begin{bmatrix}1\\1\\1\\1\\1\\1\\1\\1\end{bmatrix}$$

$a^1_{oj} = 0$ for $1 \leq j \leq 7$ since in the Haar matrix, the sum of the elements in each row is zero (except in the first one)

$a^1_{io} = 0$ for $1 \leq i \leq 7$ since the row vectors of the Haar matrix are orthogonal to vector $[1111111]$ (except the first row) which is invariant to shifts.

So, $[\tilde{A}^1]$ is necessarily orthogonal.

It follows that :

$$(t_o\ (X^1))^2 = (t_o\ (X))^2$$

and
$$\sum_{i=1}^{7} (t_i (X^1))^2 = \sum_{i=1}^{7} (t_i\ (X))^2 \qquad (6) \qquad l=0,1,\ldots,7$$

Results (6), obtained by considering the particular case where n=8, can be straight-forwardly generalized for $nXn=2^P X 2^P$ Haar matrices.
The following equations are then obtained :

$$(t_o\ (X^1))^2 = (t_o\ (X))^2$$

and
$$\sum_{i=1}^{n-1} (t_i\ (X^1))^2 = \sum_{i=1}^{n-1} (t_i\ (X))^2 \qquad (7) \qquad l=0,1,\ldots,(n-1)$$

Generalization

The results in (5) and (7) obtained for Haar matrices can be generalized for other orthogonal matrices whose first row vector consists of ones only. Moreover for the transformations defined by such matrices, there always exists at least two invariants in the power spectrum.

Indeed, let $[M]$ be a nXn matrix, whose generic term is m_{ij}, and verifying :

$$m_{oj} = 1 \quad o \leq j \leq n-1 \qquad (I)$$

$$[M]^t [M] = nI_n \qquad (II)$$

This second hypothesis implies in particular that

$$\sum_{j=0}^{n-1} m_{ij} = 0 \quad 1 \leq i \leq n-1 \qquad (8)$$

Let the transform of a vector X by $[M]$ be defined as :

$$[T\ (X)] = 1/n\ [M][X] \qquad (9)$$

(II) then necessarily implies that :

$$[T\ (X)]^t\ [T\ (X)] = 1/n^2\ [X]^t [M]^t [M][X]$$

$$[T\ (X)]^t\ [T\ (X)] = 1/n\ [X]^t [X]$$

or else

$$\sum_{i=0}^{n-1} (t_i\ (X))^2 = 1/n \sum_{i=0}^{n-1} x_i^2 \qquad (10)$$

Under the condition of a l positions left shift of X $(1 \leq l \leq n-1)$, (8) entails that :

$[T\ (X^1)] = 1/n\ [M][S^1][X]$ where $[S^1]$ is the l positions to the left translation matrix

$$[T\ (X^1)] = 1/n^2 [M][S^1][M]^t [M][X]$$

$[T\ (X^1)] = 1/n\ [A^1][T\ (X)]$ where $[A^1] = 1/n[M][S^1][M]^t$

Given (I) and (II), it is easy to show that the matrix $[A^1]$ will always be written :

$$A^1 = 1/n \begin{bmatrix} n & 0 \cdots 00 \\ 0 & \\ \vdots & \tilde{A}^1 \\ 0 & \end{bmatrix}$$

where $[\tilde{A}^1]$ is such as $[\tilde{A}^1][\tilde{A}^1]^t = I_{n-1}$

Hence the following conclusion :

Under the hypotheses (I) and (II) there will always exist two invariants at least, viz. :

$$(t_o\ (X^1))^2 = (t_o\ (X))^2$$

$$\sum_{i=1}^{n-1} (t_i\ (X^1))^2 = \sum_{i=1}^{n-1} (t_i\ (X))^2 \qquad (11) \qquad 1 \leq l \leq n-1$$

Particular cases illustrating property (11) are given by the Haar, Fourier or Hadamard transforms, with however, additional properties for the last two, which result into the breaking of $(\tilde{A}^1)$ into orthogonal "block sub-matrices". This accounts for the fact that more than two invariants are obtained (n/2 + 1 for Fourier and P + 1 for Hadamard when $n = 2^P$).

Application of Haar transform

Given, on the one hand, the velocity of Haar transform computing and, on the other hand, the two invariants obtained in the power spectrum (7), it may seem interesting to apply it to character recognition.

The characters used are those given by MUNSON'S data basis, which consists of 49 X 3 alphabets, each corresponding to the 46 non-blank FORTRAN characters. The alphabets have been hand-written by 49 authors, each of these having produced three alphabets (so 3 X 46 characters each). The obtained images are binary and held in a 24 X 24 matrix. To obtain a 2^P X 2^P matrix, each image has been transcribed into a 32 X 32 matrix by bounding it with zeroes so the earlier described algorithm could be used.

As for the experiment itself, repeated independently for each author, it was carried out in the following way :

Considering that the three hand-written alphabets for a given author were representative of his hand-writing characteristic features for each of the 46 characters were selected.

For a given character, two types of features were extracted :

. features within the power spectrum (2 invariants)

.. features within the Haar transform.

These characteristics obtained, each of the 138 characters (3 X 46) for one author was tried to be recognized. For this purpose, in a first clustering stage, the two invariants in the power spectrum were used, then in a second stage the transform features were used to complete the recognition process.

Features within the power spectrum

Feature extraction

For each of the i samples ($1 \leq i \leq 3$) of a given character, the two invariants defined in (7), which will be called P^i_{W0} and P^i_{W1}, were computed and the maximal and minimal values of these P^i_{W0} and P^i_{W1} ($1 \leq i \leq 3$) retained as representative for this character.

For the k characters in the alphabet ($1 \leq k \leq 46$) two values were twice obtained :

$$PWMIN0(k) = \inf_i(P^i_{W0}(k)) \text{ and } PWMAX0(k) = \sup_i(P^i_{W0}(k))$$

$$PWMIN1(k) = \inf_i(P^i_{W1}(k)) \text{ and } PWMAX1(k) = \sup_i(P^i_{W1}(k)) \qquad (12)$$

$$1 \leq k \leq 46$$

That is to say that for each of the 3 X 2 invariants $P^i_{W0}(k)$ and $P^i_{W1}(k)$, an interval which includes the i ($1 \leq i \leq 3$) samples of the character k ($1 \leq k \leq 46$) is defined.

Use of features

Every time a character is to be recognized, its invariants (LPW0 and LPW1) will be computed, which will allow, by comparing them to the references defined in (12), to sort out of all the possible characters those for wich :

$$LPW0 \notin [PWMIN0(k), PWMAX0(k)]$$

$$LPW1 \notin [PWMIN1(k), PWMAX1(k)] \qquad 1 \leq k \leq 46$$

So, starting with 46 classes, this first clustering stage will considerably reduce the number of possibilites to affect the still unknown character to a class. Moreover, if the samples used to establish the references defined in (12) are representative of the author's hand-writing the character to be recognized will necessarily belong to one of the remaining classes.

Figure 3 shows the average percentage of affectation possibilities eliminated by this first processing (only the 10 most representative authors have been represented).

For the whole of the 49 authors, the average of eliminated possibilities is 51.53%, i.e. starting with 46 possible classes, only 22 are still to be considered.

Features within the Haar transform

Feature extraction

When Haar transform is applied to a given image, 32 X 32 = 1024 coefficients are obtained. Obviously, the same method as that used for the power spectrum (definition for each of the coefficients of an interval containing all the possible values of this coefficient) cannot be used since 1024 X 2 = 2048 values would have to be conserved for each character. At most it would be possible for a given character to keep only the p intervals which differ most from those chosen for the other characters. However, this method has not been applied for two reasons : first because the coefficients obtained within the transform domain are not invariant to a shift of the input data so the intervals would be too large and second because the selection of references similar to those defined in (12) requires a great memory which was not avalaible for our application. Hence the following solution :

Let $t^i_j(k)$ be the jth Haar transform coefficient of the ith sample ($1 \leq i \leq n$; here n = 3) of the kth character ($1 \leq k \leq 46$).

The average value of

$$t_j(k) = 1/n \sum_{i=1}^{n} t^i_j(k) \qquad \begin{matrix} 1 \leq j \leq 1024 \\ 1 \leq k \leq 46 \end{matrix} \quad \text{is computed.}$$

That is, for character k, the value of the jth coefficient is chosen equal to the mean of the values it takes in the set of character k samples.

Concurrently, $m_j = 1/46 \sum_{k=1}^{46} t_j(k)$ i.e. the mean of all the $t_j(k)$ in the alphabet is computed.

This being done, $d_j(k) = (t_j(k) - m_j)^2 \quad 1 \leq j \leq 1024$ is computed for each character, and the p values $t_j(k)$ for which $d_j(k)$ is maximal are taken as characteristic features of character k. That means that, for a given character, the coefficients that deviate most from the average of the whole alphabet are considered as characterizing this character. Note however that this method is far from being the most efficient since the $t_j(k)$ are obtained by a mean which does not convey the variations between one sample and another of the same character. Nevertheless, the aim of this experiment was not to find relevant characteristics within the transform domain.

Use of the features

For each character X to be recognized, its Haar transform and its two invariants are computed. Coefficients t_j and PW0 and PW1 are obtained. Then, using only PW0 and PW1, as above indicated, the 1st sorting is done. Going on with the n remaining characters, the p characteristics retained for the transform itself are used, and the distance between character X and each of the n classes is evaluated :

$$d_k(X) = \sum_{i=1}^{P} (t_l - r_i(k))^2 \quad 1 \leq k \leq n \text{ where } l \text{ is the index}$$

of the ith characteristic coefficient of class k, and $r_i(k)$ is the characteristic value retained for this coefficient in class k.

Character X is then affected to the class for which $d_k(X)$ is minimal.

Figure 4 shows the recognition percentages obtained by taking 1,2,...,15 features in Haar transform.

When 15 features are considered, the recognition average for all the authors above mentioned amounts to 97.75%, that is to say that among the 138 (3X46) tested characters for an individual author, only 3 are badly classed. This, if the preceeding remark about the choice of features in Haar transform in taken into account, leads one to surmising that a complete recognition can be achieved when more relevant characteristics are chosen in the transform domain.

	NUMBER OF THE AUTHOR									
	4	10	12	14	15	18	21	24	35	44
% of elimi-nation	65.12	39.04	62.38	65.97	59.74	36.01	60.49	60.30	59.74	63.04

fig. 3 Average percentage of eliminated possibilities in the first phase of recognition.

Number of Features	NUMBER OF THE AUTHOR									
	4	10	12	14	15	18	21	24	35	44
1	49.28	29.71	52.17	44.93	44.20	31.16	44.20	39.96	42.75	54.35
2	64.49	54.35	78.99	66.67	66.67	63.77	70.29	63.77	77.54	65.94
3	73.19	66.67	84.06	71.01	80.43	79.71	83.33	81.16	86.96	80.43
4	79.71	72.46	90.58	77.54	87.68	83.33	92.03	88.41	92.03	85.51
5	82.51	81.16	92.03	78.26	92.75	87.68	93.48	92.03	93.48	92.03
6	87.68	82.61	95.65	81.16	95.65	89.86	94.93	92.75	94.20	92.75
7	89.86	86.96	97.10	86.96	97.10	92.03	97.83	95.65	98.55	92.03
8	91.30	88.41	97.10	89.86	97.83	93.48	98.55	96.38	99.28	95.65
9	92.03	92.75	97.10	91.30	97.10	95.65	99.28	97.10	99.28	96.38
10	94.20	94.93	97.83	92.75	96.38	97.10	100.00	96.38	100.00	96.38
11	94.93	94.20	97.83	94.20	94.93	97.10	100.00	95.65	100.00	98.55
12	95.65	92.75	97.83	94.20	96.38	97.10	100.00	95.65	100.00	98.55
13	95.65	94.20	98.55	96.38	97.10	95.65	100.00	97.10	100.00	99.28
14	97.10	93.48	98.55	97.10	97.10	96.38	99.28	98.55	99.28	96.55
15	96.38	94.93	98.55	97.10	97.10	96.38	100.00	98.55	99.28	99.28

fig. 4 Recognition percentages according to the number of features.

Conclusion

Haar orthonormal system and matrices were first introduced. Haar transform was then defined and some of its properties studied. A fast computation algorithm was thus obtained. After the particular case of Haar transform power spectrum was studied, it was demonstrated that, under certain conditions (orthogonal matrix whose 1st row vector consists of ones only), any transform satisfied Parseval's formula and possessed at least two invariants by translation. Finally, to bring to light the benefit of using invariants thus defined, a character recognition experiment was described.

References

1. WILLIAM K. PRATT, JULIUS KANE, C. ANDREWS "Hadamard Transform Image Coding" IEEE, vol.57, N°1, January,1969
2. NASIR AHMED, K.R. RAO, A.L.ABDUSSATTAR "BIFORE or Hadamard Transform" IEEE, vol.Au-19, n° 3, September, 1971
3. PAUL P. WANG, ROBERT C. SHIAU "Machine Recognition of printed Chinese Characters via Transformation Algorithms" Pattern Recognition, vol.5, pp.303-321, 1973
4. Laveen KANAL "Patterns in Pattern Recognition: 1968-1974" IEEE, vol.IT-20, n° 6, November, 1974
5. A. ROSENFELD "Picture Processing by Computer" Academic Press, 1969
6. G. STAMON, S. WENDLING "Etude Comparative de deux algorithmes de traitement global" A.F.C.E.T. (RABAT), 1975
7. H. ANDREWS "Introduction to Mathematical Techniques in Pattern Recognition" New-York : Wiley, 1972

Editor's Comments on Papers 33 Through 37

33 **NARASIMHA and PETERSON**
Design of a 24-Channel Transmultiplexer

34 **AHMED, NATARAJAN, and RAINBOLT**
On Generating Walsh Spectrograms

35 **COX and CROCHIERE**
Real-Time Simulation of Adaptive Transform Coding

36 **MALAH, CROCHIERE, and COX**
Performance of Transform and Subband Coding Systems Combined with Harmonic Scaling of Speech

37 **FRANGOULIS and TURNER**
Hadamard-Transformation Technique of Speech Coding: Some Further Results

SPEECH CODING

Digital speech coding by discrete transforms for achieving bandwidth compression was originally proposed by Boesswetter (1970). Bit rate reduction was achieved by transmitting only a few of the WHT coefficients. A similar technique was proposed by Shum and Elliott (1973) by transmitting only 4 or 8 of 64 dominant coefficients in the WHT domain. Frangoulis (1977) extended this concept by adaptively selecting these coefficients taking into account the nonstationarity of speech. He has also evaluated the sensitivity of these coefficients to sampling rate, male or female speakers, and so forth. Some techniques were used for reducing the overhead bits, that is, those required for identifying which of the WHT coefficients have been processed and transmitted. He claims that good quality speech can be reconstructed from only 6 to 8 of 64 WHT coefficients. By using block quantization and variance distribution of the transform coefficients, Campanella and Robinson (1971) have compared the KLT, DFT, and WHT for processing digital speech at reduced bit rates.

As well as in the bandwidth reduction of speech, discrete transforms have also been utilized in transmultiplexers. Specifically, a bandpass filter bank was realized by cascading a DCT processor with a weighting network (Paper 33). This transmultiplexer using the DCT is available commercially. Also, similar to Fourier spectrograms (time-frequency-amplitude plots) of speech, Walsh spectrograms (time-sequency-amplitude plots) have been developed (Paper 34).

Zelinski and Noll (1977) analyze the transform coding of speech based on both fixed and adaptive schemes. An important development here is that, by using the short-term basis spectrum as the side information, both the bit allocation and quantizer levels are adaptively controlled over each speech segment. They have shown that this adaptive transform coding (ATC) performs better than the adaptive DPCM technique. In a subsequent paper (1979), they have developed techniques for modifying the ATC when applied to speech coding at low bit rates (12 KBPS or less). Perceptual quality of speech is not lowered by changing the bit assignment scheme and by substitution of discarded transform coefficients with noise. The superior performance of the DCT compared to any other fixed transforms has been demonstrated.

For speech coding at low bit rates (9.6–16 KBPS) by transform techniques, new developments are suggested by Cox and Crochiere (Paper 35). These developments include symmetrical DFT that results in the reduction in end effects in the blocks, homomorphic model as the side information, and real-time simulation on an array processing computer. These developments have led the authors to conduct real telephone conversations as opposed to stored speech utterances over a transform coder. By combining time domain harmonic scaling (TDHS) with subband coding (SBC) and ATC, Malah, Crochiere, and Cox (Paper 36) show that speech can be transmitted at high communications quality at 9.6 KBPS and 7.2 KBPS respectively. These rates represent savings of 7 KBPS and 4 KBPS respectively over the SBC and ATC alone. They emphasize that the TDHS-SBC is particularly suitable for real-time hardware implementation for speech coding at 9.6 KBPS as this scheme is much less complex compared to the TDHS-ATC.

REFERENCES

Boesswetter, C., 1970, *Analog Sequency Analysis and Synthesis of Voice Signals*, Proceedings of the Symposium on Applications of Walsh Functions, Washington, D.C., pp. 220–229.

Campanella, S., and G. S. Robinson, 1971, A Comparison of Orthogonal Transformations for Digital Speech Processing, *IEEE Commun. Tech. Trans.* **COM-19:**1045–1049.

Ying, F. Y. Y., A. R. Elliott, and W. O. Brown, 1973, Speech Processing with Walsh-Hadamard Transforms, *IEEE Audio Electroacoust. Trans.* **AU-21:**174–179.

Zelinski, R., and P. Noll, 1977, Adaptive Transform Coding of Speech Signals, *IEEE Acoust., Speech, Signal Process. Trans.* **ASSP-25:**299–309.

Zelinski, R., and P. Noll, 1979, Approaches to Adaptive Transform Speech Coding at Low-bit Rates, *IEEE Acoust., Speech, Signal Process. Trans.* **ASSP-27:**89–95.

Design of a 24-Channel Transmultiplexer

MADIHALLY J. NARASIMHA MEMBER, IEEE, AND ALLEN M. PETERSON, FELLOW, IEEE

***Abstract*—The design of a transmultiplexer capable of performing the bilateral conversion between one 1544 kbit/s digital signal (which represents 24 PCM coded voice channels) and two analog group signals (each one containing 12 voice channels in the 60–108 kHz band) is investigated. It is shown that an FIR filter bank required as part of such a transmultiplexer can be realized efficiently by cascading a discrete cosine transform processor and a weighting network. Fast convolution algorithms are derived for evaluating the cosine transform. A method of using the symmetry conditions to reduce the computation rate in the weighting network and an elegant hardware configuration for implementing it are also discussed.**

I. Introduction

Two types of multiplexing techniques are used in the present telephone network for the transmission of several channels over a common link—the older and predominant frequency division multiplexing (FDM) and the recent and fast-growing time division multiplexing (TDM). The increasing use of PCM terminals and digital switching exchanges, coupled with the slow pace of replacing existing analog carrier equipment, has created a need for the FDM and TDM systems to coexist and communicate with each other. As a result, a method of translating between these two systems is required in the mixed analog-digital (MAD) telephone network. Such an interface box will be referred to as a transmultiplexer. The present technique for accomplishing this translation uses a tandem connection of digital and analog channel banks. The cost and inefficiency associated with this scheme has emphasized the need for a more direct (all-digital) translation procedure.

Freeny *et al.* [1]–[3] investigated a digital signal processing system for emulating an FDM channel bank which, in addition, would also perform TDM-FDM translation. Their scheme for effecting the necessary single-sideband (SSB) modulation and demodulation used the classical Weaver technique on a per-channel basis, and did not make use of the computational savings attainable by treating the frequency translation of a group of 12 channels taken jointly. Darlington [4] first suggested that algorithms reminiscent of fast Fourier transform (FFT) techniques can be used to reduce the computational burden in modulating and demodulating a group of channels. The paper by Bellanger and Daguet [5] showed that the TDM-FDM translation problem can be solved by the combination of a discrete Fourier transform (DFT) processor and a polyphase network. This decomposition arises from the fact that the individual filters in the filter bank required for SSB modulation and demodulation are frequency-shifted versions of a low-pass prototype. Extensions and modifications of this decomposition to real as well as complex signals, and to FIR as well as IIR filters, are described in [6]–[8]. There have been several U.S. Patents [9]–[11] on these concepts. Other papers on this subject appeared in the May 1978 issue of IEEE Transactions on Communications [12]–[14]. A paper by Tsuda *et al.* [15] in the same issue does not use DFT processors, but relies on efficient multirate filtering techniques to reduce the computational burden.

The polyphase network [5] can be viewed as a set of N filters (where N denotes the number of channels involved) obtained from phase shifting and undersampling the low-pass prototype filter. If the prototype is an FIR filter, it is appropriate to call the polyphase network a weighting (or window) network, thereby rendering the terminology familiar to those using the DFT for spectral analysis. In fact, the demodulation of an FDM signal is equivalent to a spectral analysis operation. If an N-point DFT is used to accomplish this with conventional windows, such as Hamming or Hanning, the required passband characteristics and crosstalk rejection cannot be obtained. In order to achieve these performance objectives, the window, which corresponds to the impulse response of a prototype low-pass filter, will have to be over many N-point data frames instead of a single frame. The methods described in the above-cited references describe how it is possible to extend the window over several (or even an infinite number of) frames and still keep the transform length to N. An alternative method of viewing this process is that if a spectral analysis has to be carried over many N-point frames of data using a window, and only N of the spectral components uniformly spaced in frequency are desired, then the analysis can be carried out with an N-point DFT with some preprocessing.

The motivation behind decomposing the frequency translation problem into a DFT and a weighting network is that the efficient FFT algorithm can be employed to reduce the computation rate in the DFT. With the introduction of new high-speed convolution algorithms for computing the DFT [16], the efficiency of the translation scheme can be further improved. This was shown by Peled and Winograd [17].

It should be pointed out that the use of a DFT processor in implementing a bandpass filter bank has been studied under several different contexts. Notable among these are speech analysis and synthesis using phase vocoders [18], [19] and

Manuscript received January 15, 1979; revised July 15, 1979. This work was supported by NASA under Grant NGL 05-020-014.
M. Narasimha is with Granger Associates, Santa Clara, CA 95051.
A. M. Peterson is with the Department of Electrical Engineering, Stanford University, Stanford, CA 94305.

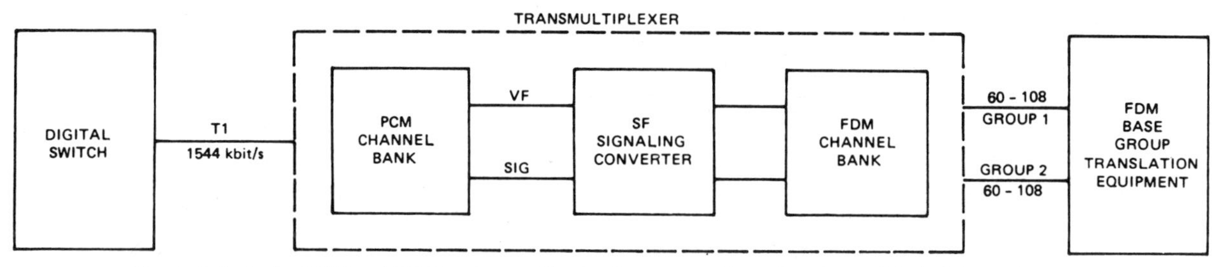

Fig. 1. Interfacing TDM and FDM systems with a tandem connection of analog and digital channel banks.

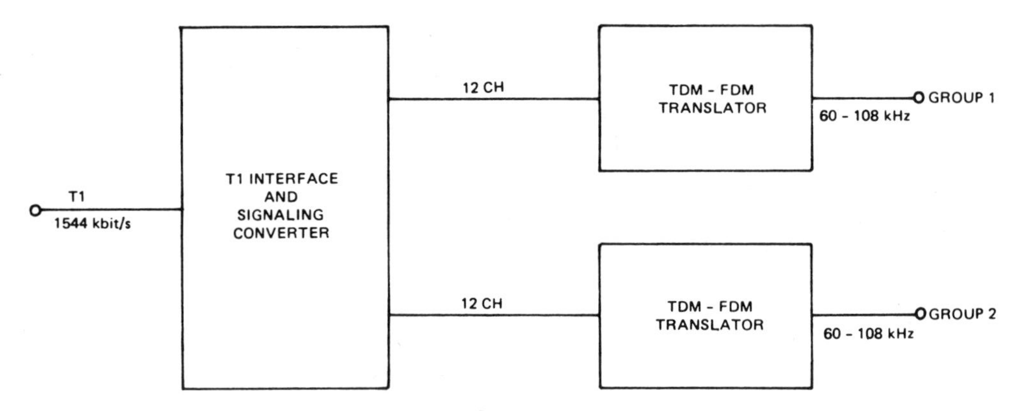

Fig. 2. An all-digital 24-channel transmultiplexer.

spectral analysis [20]. A paper by Narasimha and Peterson [21] has a unified treatment of this topic.

This paper describes a 12-channel TDM-FDM translator, required in a transmultiplexer between 24-channel PCM systems and 2-FDM groups, with the following characteristics.

1) An FIR bandpass filter bank for achieving both channel filtering and sample-rate conversion. The voice frequency sampling rate is 8 kHz (corresponding to PCM systems) and the FDM sequence is digitized at 112 kHz.

2) Recognizing the transpose relation between the modulator and demodulator networks allows the solution obtained in one direction to be used in the other.

3) The required FIR filter bank is decomposed into a combination of a 14-point discrete cosine transform (DCT) and a 561-tap weighting network.

4) A fast algorithm for computing the 14-point DCT (16 multiplies and 76 adds at the 8 kHz rate).

5) A method of using the symmetry conditions in the linear phase FIR prototype filter to reduce the number of multiplies in the weighting network by a factor of two.

6) The total computation rate of the translator in any one direction is about 0.2 million multiplies/s/channel and 0.4 million adds/s/channel.

II. Functions of a Transmultiplexer

At the present time, the setup needed to interconnect a TDM system, such as the $T1$ lines emanating from a digital switch, with the FDM base group translating equipment, is shown in Fig. 1. The functions to be performed by the transmultiplexer are shown enclosed in dotted lines. In Fig. 1, the digital channel banks (D-banks) convert the $T1$ signal to voice frequency and E and M signaling leads. These are combined in the signaling adaptor to get voice frequency signals containing single-frequency (SF) signaling information, which are then fed to two 12-channel FDM channel banks to obtain two basic group signals in the 60–108 kHz band. These groups will then be translated to the supergroup level by the group translation equipment. The main noise contributions in this setup arise from the digital and FDM channel banks.

An all-digital solution to translating a $T1$ signal into two FDM group bands is depicted in Fig. 2. The $T1$ interface and signaling converter takes the place of the D-banks and signaling adaptors. The design of the $T1$ interface part is similar to that of a $D3$ channel bank with one exception: instead of the final analog–digital conversion process, the voice information in μ-255 coding is changed to a 14-bit linear code. Mapping of the $T1$ signaling bit into 2600 Hz in-band SF signaling is accomplished in the signaling converter using digital filters. These filters mimic their analog counterparts used in standard signaling adaptors. A digital processor can be used to accomplish this task. The design of such a processor is fairly conventional and will not be discussed any further.

The constituent parts of a 12-channel TDM-FDM translator are shown in Fig. 3(a). A digital filter bank modulates (and demodulates) the 12 linearly coded channels, each sampled at 8 kHz, containing in-band signaling information, into the 4–52 kHz frequency band, with the FDM sequence sampled at 112 kHz. It will be shown later that this filter bank can be obtained efficiently by the combination of a discrete cosine transform (DCT) and a weighting network (polyphase network). From this FDM sequence, the mirror image band lying in the 60–108 kHz region can be picked off by a digital-to-analog converter and a bandpass filter [Fig. 3(b)] in the TDM to FDM direction, while bandpass filtering followed by digitizing at 112 kHz rate would translate the group signal to the 4–52 kHz band in the opposite direction.

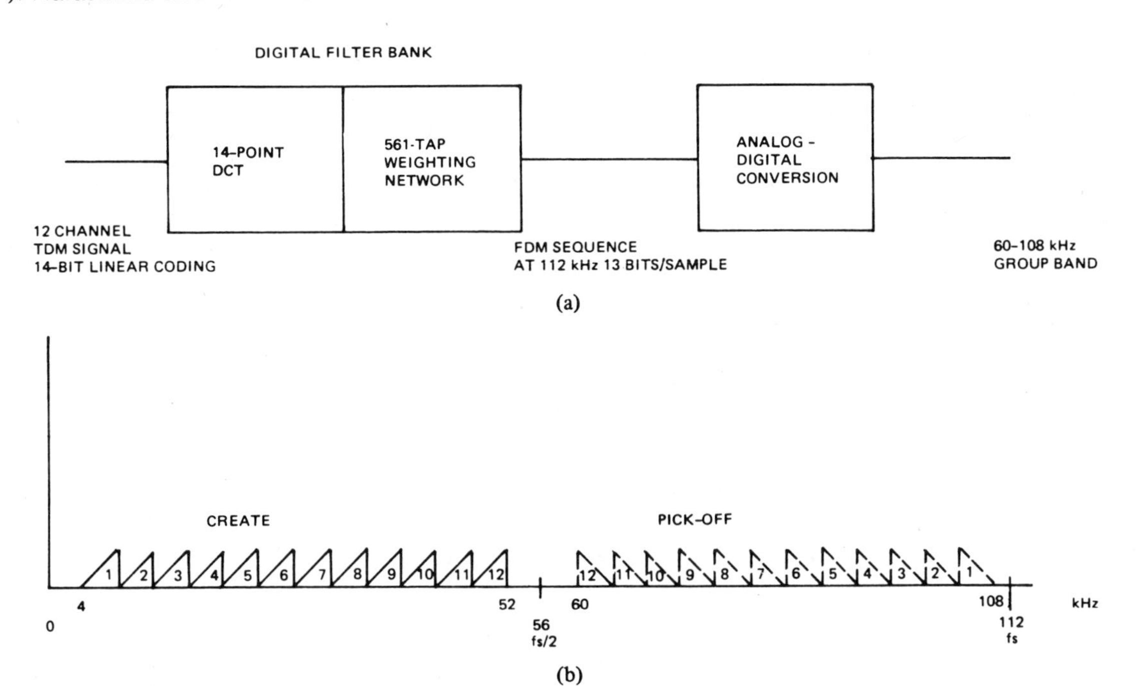

Fig. 3. (a) Block diagram of a 12-channel TDM-FDM translator. (b) Frequency scheme of the multiplex.

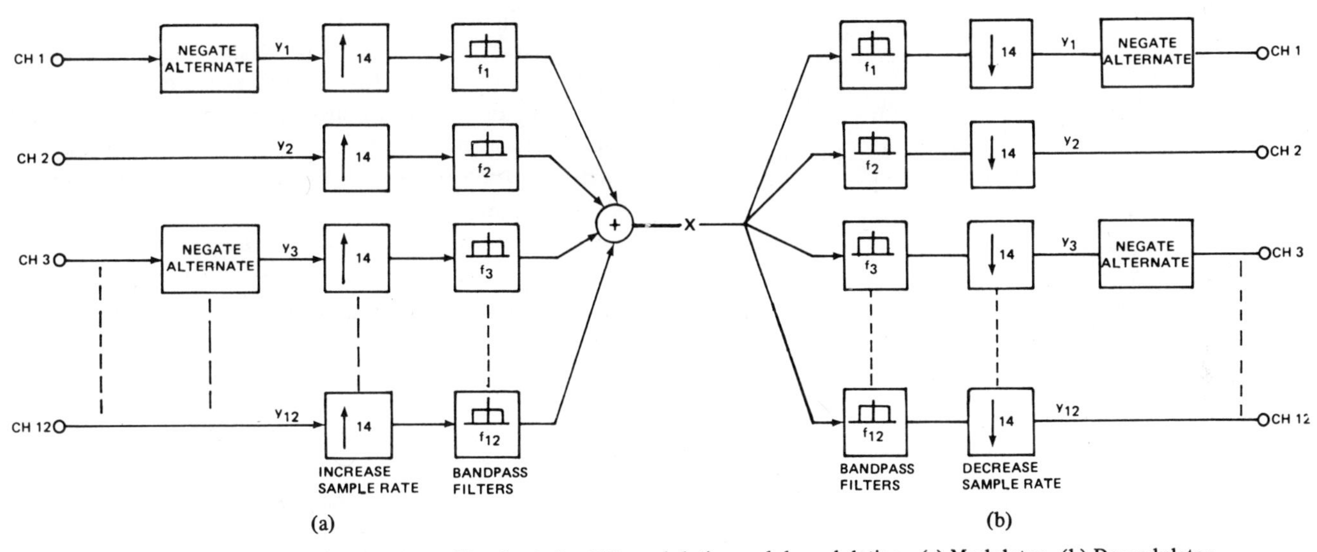

Fig. 4. Block diagram of the bandpass filter bank for SSB modulation and demodulation. (a) Modulator. (b) Demodulator.

III. Transpose Relation Between Modulator and Demodulator

Consider the problem of combining 12 voice channels, each sampled at 8 kHz, on a frequency division multiplex basis according to the frequency scheme shown in Fig. 3(b). In order to simplify the following D-A conversion and bandpass filtering, two guard channels will be included in the FDM system. The composite signal sampled at a 112 kHz rate can be obtained by increasing the sample rate on each of the channels to this rate (which can be done by inserting zeros), bandpass filtering with a set of filters uniformly spaced in frequency, and summing the output of these filters. This is depicted in Fig. 4(a), where the center frequency of the nth BPF is $f_n = (n + \frac{1}{2})4$ kHz. It should be noted that the spectrum of odd-numbered channels needs to be inverted before filtering in order to obtain upright sidebands in all the positions. The "NEGATE ALTERNATE" boxes do this spectral inversion by multiplying the input with the sequence $1, -1, 1, -1, \cdots$.

The demodulation of the wide-band frequency-multiplexed signal into individual channel outputs can be accomplished by the setup depicted in Fig. 4(b). Here, the down conversion is down by deleting samples.

Considering the processes of sample rate increase and decrease as duals, it is clear from Fig. 4 that the demodulator can be obtained from the modulator network by the transposition operation. (The transpose is obtained by simply reversing the direction of all branches in the flowgraph representation of the network.) The importance of this observation is that it is sufficient to solve the bandpass filter-bank problem in only one direction.

In the rest of the paper the digital bandpass filter network

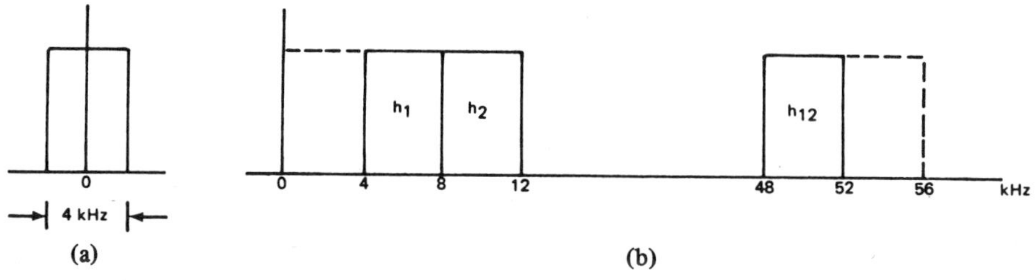

(a) (b)

Fig. 5. (a) Prototype filter. (b) Frequency scheme of the bandpass filter bank.

for only the demodulation process is derived. The corresponding modulator can be obtained by transposition.

IV. Decomposition of the Filter Bank into a DCT and a Weighting Network

Let $x(\cdot)$ be the wide-band FDM signal sampled at 112 kHz. The demodulation of x into the baseband channels $y_1, y_2, \cdots, y_{12}$ can be accomplished by passing it through the filter bank depicted in Fig. 5(b). If $h_n(m)$ denotes the impulse response of the nth filter in this bank, it can be expressed in terms of the response $h(m)$ of a low-pass prototype shown in Fig. 5(a) by the relation

$$h_n(m) = h(m) \cos[\pi(2n+1)m/28], \quad n = 1, 2, \cdots, 12. \quad (1)$$

Considering the demodulation process with input $x(r)$, the output $y_n(r)$ of the nth channel (assuming an FIR prototype filter of length $M+1$) is given by

$$y_n(r) = \sum_{m=0}^{M} x(r-m)\, h(m) \cos[\pi(2n+1)m/28], \quad n = 1, 2, \cdots, 12. \quad (2)$$

In order to make use of the symmetry conditions in the prototype, it is advantageous to choose the filter length to be a multiple of 28, plus one. For this application, a 561-tap filter appears to be sufficient. Making use of the relation $h(m) = h(560-m)$, the output of the nth channel can be rewritten as

$$y_n(r) = x(r-280)\, h(280) + \sum_{m=0}^{279} [x(r-m) + x(r-560+m)]\, h(m) \cos[\pi(2n+1)m/28]. \quad (3)$$

Divide the filter length into blocks of 28 each. That is, let

$$m = 28p + q, \quad p = 0, 1, \cdots, 9; \quad q = 0, 1, \cdots, 27. \quad (4)$$

This substitution in (3) and subsequent simplification results in

$$y_n(r) = \sum_{q=0}^{13} u(r,q) \cos[\pi(2n+1)q/28], \quad n = 1, 2, \cdots, 12. \quad (5)$$

The sequence $u(r,q)$, $q = 0, 1, \cdots, 13$ in the above equation is given by

$$u(r,0) = x(r-280)\, h(280) + v(r,0)$$

$$u(r,q) = v(r,q) - v(r, 28-q), \quad q = 1, 2, \cdots, 13$$

where

$$v(r,q) = \sum_{p=0}^{9} (-1)^p [x(r-28p-q) + x(r-560+28p+q)]\, h(28p+q). \quad (6)$$

From (5) it is clear that the channel outputs $y_1, y_2, \cdots, y_{12}$ (including the guard bands y_0 and y_{13}) can be obtained by computing the inverse DCT of the weighted sequence $u(r,q)$. It should be pointed out that the channel outputs are diminished in bandwidth by a factor of 14, and hence can be undersampled by this ratio. That is, one needs to compute these outputs only once every 14 input samples.

Incorporating this undersampling effect and rewriting the defining equation for $u(r,q)$, $q = 1, 2, \cdots, 13$ in a more compact way [by using the symmetry property $h(m) = h(560-m)$], the demodulated channel outputs are given by

$$y_n(14r) = \sum_{q=0}^{13} u(14r, q) \cos[\pi(2n+1)q/28], \quad n = 1, 2, \cdots, 12 \quad (7)$$

where

$$u(14r, 0) = x(14r-280)\, h(280) + \sum_{p=0}^{9} (-1)^p h(28p)[x(14r-28p) + x(14r-560+28p)]$$

$$u(14r, q) = \sum_{p=0}^{19} (-1)^p h(28p+q)[x(14r-28p-q) + x(14r-560+28p+q)], \quad q = 1, 2, \cdots, 13. \quad (8)$$

Fig. 6 is a flowgraph representation of the demodulator described by (7) and (8). The detailed operations required for only the weighting network part are shown. A method of calculating the inverse DCT will be described later.

The FDM sequence $x(\cdot)$, sampled at 112 kHz, is passed through appropriate delay networks $1, z^{-1}, \cdots, z^{-13}$. It is then undersampled by retaining only 1 out of every 14 sam-

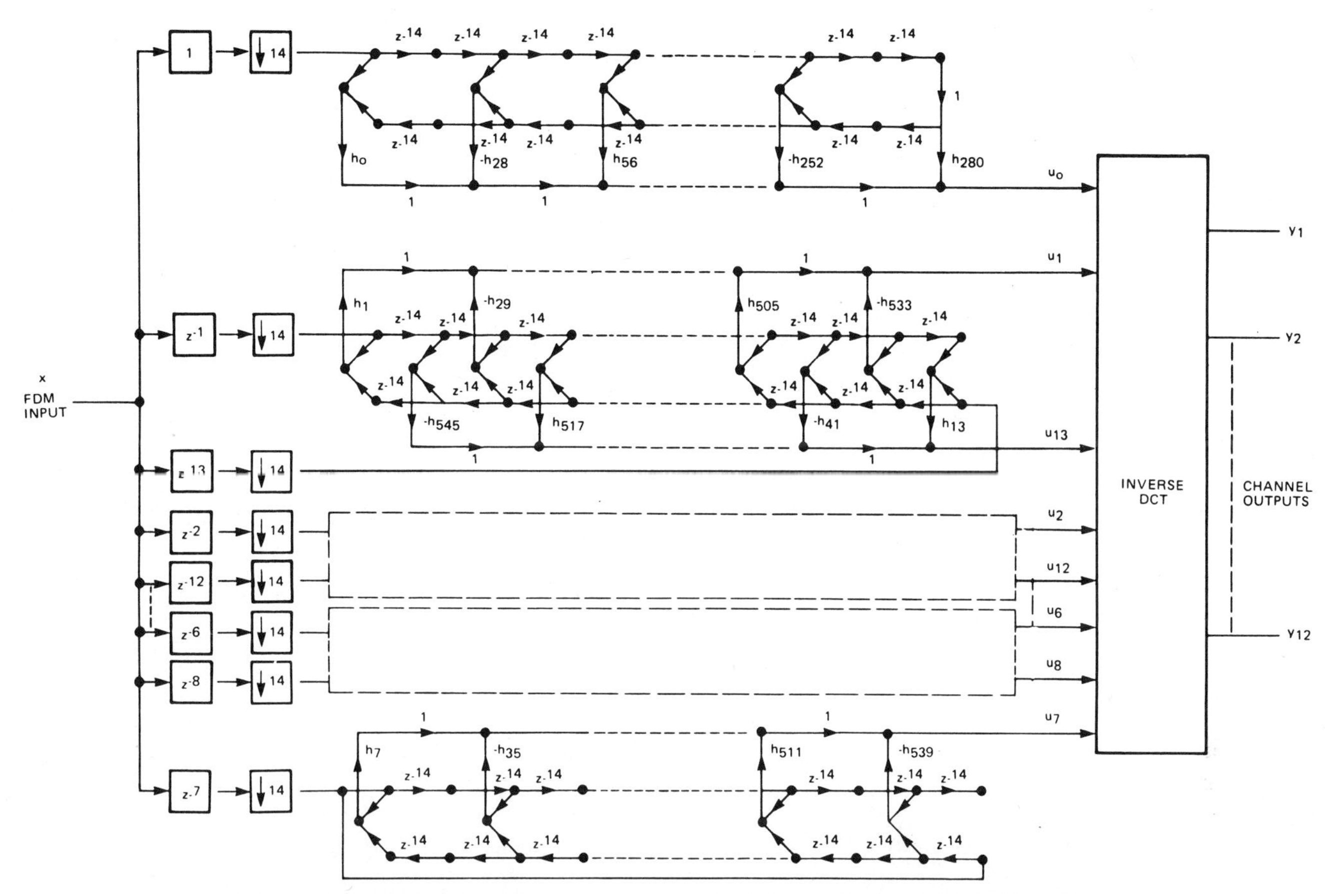

Fig. 6. Demodulator filter bank depicting a flowgraph representation for the weighting network. The z^{-14} branches shown correspond to a unit sample delay at the 8 kHz rate. $u(\cdot, q)$ is denoted by u_q.

ples. This reduces the sampling rate to 8 kHz. The decimated sequences then enter a set of shift registers, each having 40 taps. (A branch denoted by z^{-14} in Fig. 6 corresponds to a unit sample delay at the 8 kHz rate.) After summing the appropriate tap outputs, the signals are weighted by the filter coefficients and accumulated to yield the sequence $u(\cdot, q)$, $q = 0, 1, \cdots, 13$. An inverse DCT is then performed on the weighted sequence to obtain the channel outputs. It should be noted in Fig. 6 that the computation of $u(\cdot, 0)$ needs only 11 multiplies. The data coming out of the delay networks z^{-1} and z^{-13} are involved in the computation of $u(\cdot, 1)$ and $u(\cdot, 13)$, each requiring 20 multiplies. Similarly, $u(\cdot, 2)$ and $u(\cdot, 12)$ are evaluated using the data going through the z^{-2} and z^{-12} delays, etc. Only the z^{-7} path is used for the computation of $u(\cdot, 7)$.

It is appropriate to mention at this point that transposing the flowgraph of Fig. 6 yields the filter bank for the modulator. In this case a forward DCT has to be performed on the channel inputs. The transformed outputs are then passed through the transpose of the weighting network shown in Fig. 6. Thereafter, the sampling rate of the weighted sequences is increased to 112 kHz by inserting zeros. These oversampled sequences should then be appropriately delayed and added together to obtain the FDM sequence. Actually, the combination of inserting zeros, delaying, and summing the sequences corresponds to simply interleaving the points coming out of the weighting network.

V. Hardware Realization of the Weighting Network

The flowgraph of Fig. 6 depicts a set of 14 filters operating on appropriately delayed input points to evaluate the weighted sequence $u(\cdot, q)$. It may be preferable in practice to share the arithmetic unit among these filters in order to reduce the hardware complexity. This section describes an implementation of the weighting network which resembles the direct form realization of a linear phase FIR filter. The unit sample delays of the FIR filter will be replaced by 14-length shift registers performing a certain pattern of permutation.

A slight simplification in the hardware can be achieved by assuming that the last filter coefficient $h(560)$ is zero. Such a filter can be obtained by designing a 559-tap filter and then setting the end tap gains $h(0)$ and $h(560)$ to zero. This last operation does not change the frequency response of the original filter and preserves the linear phase characteristics. In this case, the describing equation for $u(14r, 0)$ can be rewritten as

$$u(14r, 0) = \sum_{p=0}^{19} (-1)^p \, h(28p) \, x(14r - 28p). \tag{9}$$

Fig. 7 shows a hardware configuration suitable for evaluating the weighted sequence $u(\cdot, q)$ in a serial manner. The FDM sequence is first reversed in blocks of 14 by passing it through the box labeled $R14$. That is, if the input sequence to this unit

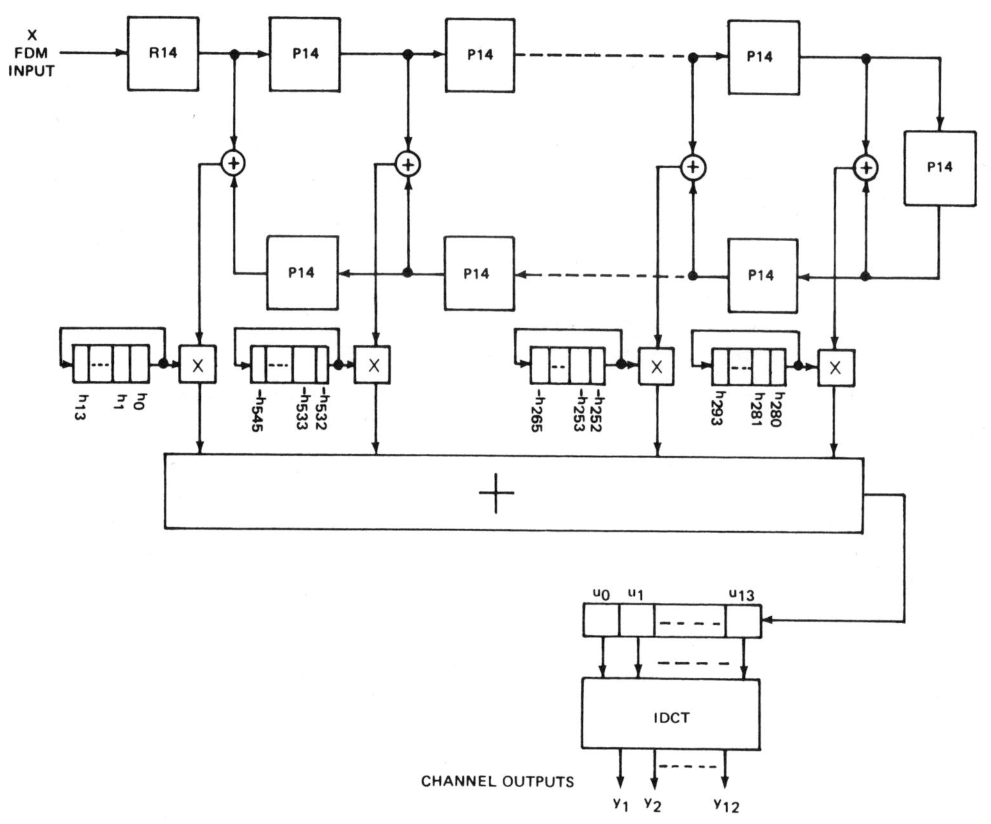

Fig. 7. FDM demodulator showing a suitable hardware configuration for the weighting network.

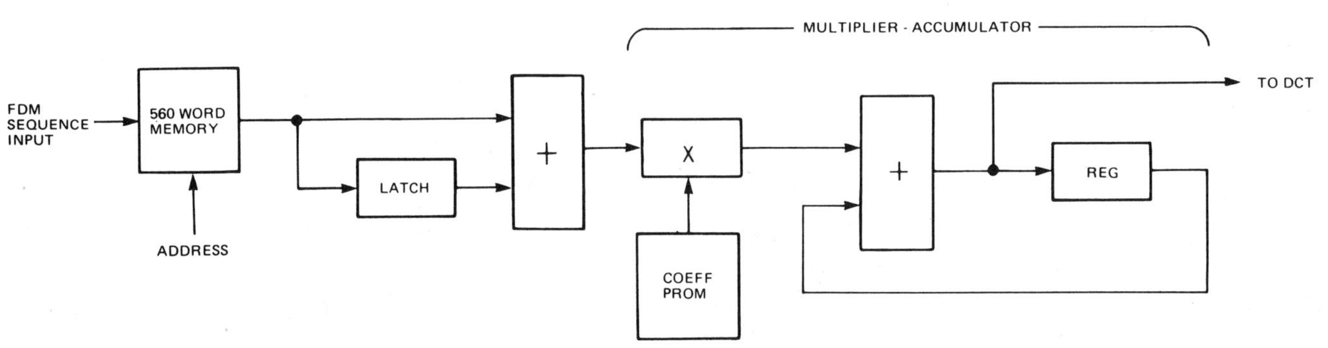

Fig. 8. A simplified hardware structure for the weighting network of the demodulator using a single random access memory for data storage and a high-speed multiplier accumulator.

is $x_0, x_1, \cdots, x_{13}|x_{14}, x_{15}, \cdots, x_{27}| \cdots$, the output with 14 sample delay is $x_{13}, x_{12}, \cdots, x_0|x_{27}, x_{26}, \cdots, x_{14}| \cdots$. This time-reversed sequence is then shifted through permutation shift registers denoted by $P14$. A certain pattern of permutation is performed on the input sequence, in blocks of 14, in addition to shifting it by 14 samples in these units. For the same input sequence $x_0, x_1, \cdots, x_{13}|x_{14}, x_{15}, \cdots, x_{27}| \cdots$, the output of the $P14$ box will be $x_0, x_{13}, \cdots, x_1|x_{14}, x_{27}, \cdots, x_{15}| \cdots$ with a delay of 14 samples. Notice that the permutation involved is one of retaining the first sample and reversing the other 13 in a block of 14. Also, two of these permutations correspond to a simple delay of 28 samples. The outputs from these $P14$ blocks are summed in a manner analogous to that done in linear phase FIR filters, multiplied by filter coefficients stored in 14-length recirculating shift registers, and finally accumulated. The weighted sequence will be obtained in the order $u(\cdot, 0), u(\cdot, 1), \cdots, u(\cdot, 13)$.

In calculating the output $u(\cdot, 0)$, there is no summing before multiplying by the filter coefficients, and only one of the inputs to the two-input adders will be active and the other should be set to zero. A careful examination of (9) will reveal this fact.

The reversing and permutation operations can be easily accomplished by a proper addressing of random access memories. A set of read-only memories driven by a counter will serve the function of recirculating shift registers for storing the filter coefficients.

With the availability of high-speed LSI multipler accumulators (MAC), the hardware required to implement the weighting network can be greatly reduced. Fig. 8 shows a possible organization using a single 560-word memory for data storage and a TRW 16-bit MAC (TDC 1010J). In this implementation the MAC has to complete an operation in about 446 ns. Since the TDC 1010J is twice as fast, one of these devices can be used

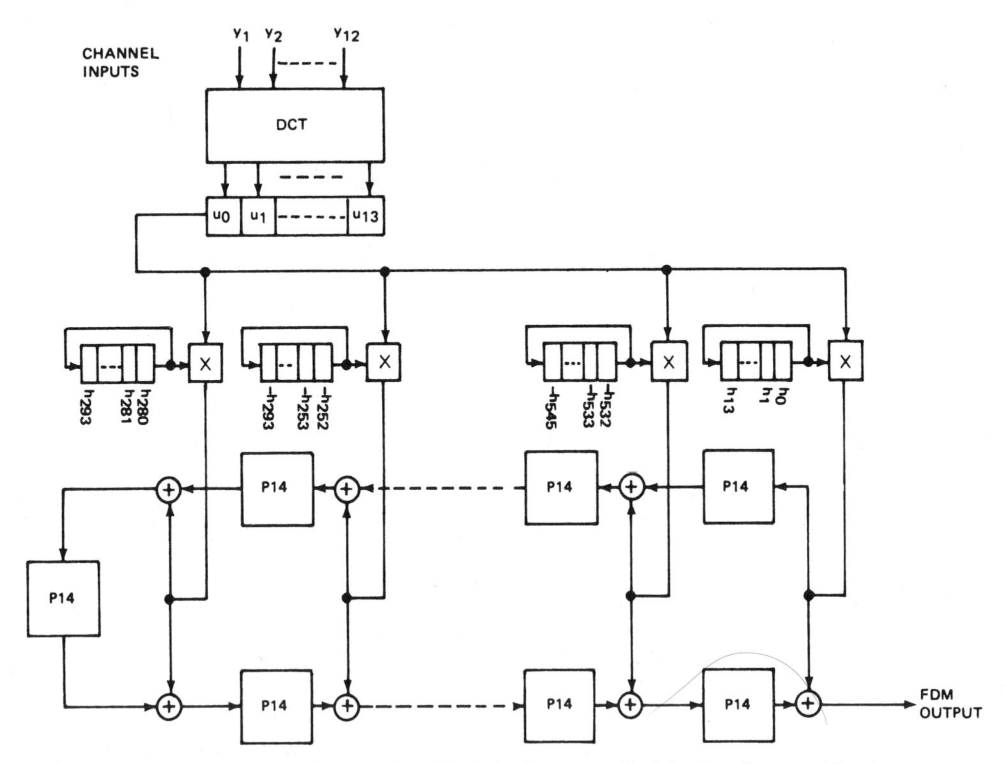

Fig. 9. Modulator filter bank. This is the "transpose" of the one shown in Fig. 7.

for implementing the demodulator weighting networks in both of the groups.

The main control problem in the implementation of Fig. 8 will be that of generating the RAM address. A paper by Narasimha [22] discusses the memory addressing problem in FIR filters using RAM's for data storage. The techniques described there can be extended to cover this case.

For the weighting network in the corresponding modulator, the "transpose" of the hardware in Fig. 7 can be employed. This is shown in Fig. 9. The recirculating registers for the filter coefficients and the $P14$ permutation blocks will remain the same. Reversing of the FDM outputs, however, is not necessary. In order to understand this last phenomenon, one should refer back to the transpose of Fig. 6 and compare it with the "hardware transpose" obtained here. This modulator can also be implemented using a simplified hardware organization analogous to that shown in Fig. 8.

It should be pointed out that, for the demodulator, the initial reversing of the FDM sequence by $R14$ is not required if the weighted sequence is evaluated in the order $u(\cdot, 13)$, $u(\cdot, 12), \cdots, u(\cdot, 0)$. In this case, the permutation performed by the $P14$ blocks will be different and the filter coefficients in the recirculating registers will be stored in a different order. However, if one wants to retain the same permutation blocks and recirculating filter coefficients in the modulator too, the reversing operation will have to be performed at the FDM outputs.

VI. Computation of the Inverse DCT

In order to obtain the channel outputs $y_n(\cdot)$, an inverse DCT has to be performed on the weighted sequence $u(\cdot, q)$. Denoting $u(\cdot, q)$ by u_q, the relation between y_n and u_q is given by

$$y_n = \sum_{q=0}^{13} u_q \cos [\pi(2n + 1)q/28], \quad n = 1, 2, \cdots, 12. \quad (10)$$

A 14-point DFT algorithm can be used to compute this DCT as described in Narasimha and Peterson [23]. This method, using fast convolution algorithms for evaluating the DFT [16], would require about 40 multiplies to compute the above DCT. Alternatively, high-speed convolution methods can be applied directly for computing y_n from u_q. The method presented here for the computation of the inverse DCT requires 16 multiplies and 76 adds. Some of the techniques discussed in Peled and Winograd [17] are used in the derivation of this efficient algorithm.

Notice the symmetry relation

$$y_{14-n-1} = \sum_{q=0}^{13} (-1)^q u_q \cos [\pi(2n + 1)q/28]. \quad (11)$$

Thus, defining two new sequences

$$a_n = \frac{1}{2}(y_n + y_{14-n-1}) = \sum_{q=0}^{6} u_{2q} \cos [\pi(2n + 1)2q/28],$$
$$n = 1, 2, \cdots, 6 \quad (12)$$

$$b_n = \frac{1}{2}(y_n - y_{14-n-1}) = \sum_{q=0}^{6} u_{2q+1}$$
$$\cdot \cos [\pi(2n + 1)(2q + 1)/28], \quad n = 1, 2, \cdots, 6 \quad (13)$$

one can obtain $y_n, n = 1, 2, \cdots, 12$ by adding and subtracting the a_n and b_n values.

A procedure for evaluating the a_n sequence is described below. Let

$$c_k = \cos(\pi k/28), \quad k = 1, 2, 3, \cdots. \tag{14}$$

Then

$$a_3 = u_0 - u_4 + u_8 - u_{12} \tag{15}$$

$$\begin{bmatrix} (a_1 + a_5)/2 \\ (a_2 + a_4)/2 \\ a_6' \end{bmatrix} = \begin{bmatrix} u_0 \\ u_0 \\ u_0 \end{bmatrix} + \begin{bmatrix} c_{12} & -c_4 & -c_8 \\ -c_8 & -c_{12} & c_4 \\ c_4 & c_8 & c_{12} \end{bmatrix} \begin{bmatrix} u_4 \\ u_8 \\ u_{12} \end{bmatrix} \tag{16}$$

$$\begin{bmatrix} (a_1 - a_5)/2 \\ (a_2 - a_4)/2 \\ a_6'' \end{bmatrix} = \begin{bmatrix} c_6 & -c_{10} & -c_2 \\ c_{10} & -c_2 & c_6 \\ -c_2 & -c_6 & -c_{10} \end{bmatrix} \begin{bmatrix} u_2 \\ u_6 \\ u_{10} \end{bmatrix} \tag{17}$$

where a_6' and a_6'' are such that

$$a_6 = a_6' + a_6''. \tag{18}$$

It is clear that $a_1, \cdots, a_6$ can be obtained easily if one solves the two matrix multiplication problems indicated above.

By a suitable reordering of columns and rows (and changing signs), these can be reduced to a standard matrix multiplication problem shown below:

$$\begin{bmatrix} z_0 \\ z_1 \\ z_2 \end{bmatrix} = \begin{bmatrix} x_0 & x_1 & x_2 \\ x_1 & x_2 & x_0 \\ x_2 & x_0 & x_1 \end{bmatrix} \begin{bmatrix} y_0 \\ y_1 \\ y_2 \end{bmatrix}. \tag{19}$$

The evaluation of z_0, z_1, and z_2 in (19) can be accomplished by the following algorithm which requires 4 multiplies and 11 adds. Let

$$\begin{aligned} m_0 &= [(x_0 + x_1 + x_2)/3](y_0 + y_1 + y_2) \\ m_1 &= [(x_2 + x_1 - 2x_0)/3](y_0 - y_2) \\ m_2 &= [(x_0 + x_2 - 2x_1)/3](y_2 - y_1) \\ m_3 &= [(x_0 + x_1 - 2x_2)/3](y_1 - y_0). \end{aligned} \tag{20}$$

Then

$$\begin{aligned} z_0 &= m_0 - m_1 + m_2 \\ z_1 &= m_0 + m_1 - m_3 \\ z_2 &= m_0 - m_2 + m_3. \end{aligned} \tag{21}$$

Note that the factors $(x_0 + x_1 + x_2)/3$, etc., can be precomputed.

The evaluation of b_n, defined in (13), is slightly more difficult. Again letting $c_k = \cos(\pi k/28)$, and using the relations

$$\begin{aligned} c_1 &= c_7(c_6 + c_8) \\ c_3 &= c_7(c_4 + c_{10}) \\ c_5 &= c_7(c_2 + c_{12}) \\ c_9 &= c_7(c_2 - c_{12}) \\ c_{11} &= c_7(c_4 - c_{10}) \\ c_{13} &= c_7(c_6 - c_8) \end{aligned} \tag{22}$$

the equations for obtaining $b_1, \cdots, b_6$ can be written as follows.

Let

$$u_k' = c_7 u_k = u_k/\sqrt{2}, \quad k = 1, 3, 5, 7, 9, 11, 13. \tag{23}$$

Then

$$b_3 = u_7' + (u_1' - u_{13}') - (u_3' + u_{11}') - (u_5' - u_9') \tag{24}$$

$$\begin{bmatrix} (b_1 + b_5)/2 \\ (b_2 - b_4)/2 \\ b_6' \end{bmatrix} = \begin{bmatrix} c_4 & -c_{12} & c_8 \\ c_{12} & c_8 & -c_4 \\ -c_8 & -c_4 & -c_{12} \end{bmatrix} \begin{bmatrix} u_1' - u_{13}' \\ u_3' + u_{11}' \\ u_5' - u_9' \end{bmatrix} - \begin{bmatrix} u_7' \\ u_7' \\ u_7' \end{bmatrix} \tag{25}$$

$$\begin{bmatrix} (b_1 - b_5)/2 \\ (b_2 + b_4)/2 \\ b_6'' \end{bmatrix} = \begin{bmatrix} c_{10} & c_2 & -c_6 \\ c_2 & -c_6 & -c_{10} \\ c_6 & c_{10} & c_2 \end{bmatrix} \begin{bmatrix} u_1' + u_{13}' \\ u_3' - u_{11}' \\ u_5' + u_9' \end{bmatrix} \tag{26}$$

where

$$b_6' + b_6'' = b_6. \tag{27}$$

It is not necessary to perform multiplications by $1/\sqrt{2}$ to obtain u_k' from u_k in the DCT processor since these multiplies can be taken care of by premultiplying the odd coefficients of the weighting network. The above matrix multiplications can be reduced to the standard form as before, and the 4 multiplies, 11 adds algorithm can be used to evaluate them.

A complete flowgraph for the evaluation of the channel outputs y_n from $u_0, u_2, \cdots, u_{12}, u_1', u_3', \cdots, u_{13}'$ is shown in Fig. 10.

In the TDM to FDM direction (modulation), a forward DCT

$$u_q = \sum_{n=1}^{12} y_n \cos[\pi(2n+1)q/28] \quad q = 0, 1, \cdots, 13, \tag{28}$$

has to be evaluated on the channel inputs $y_1, y_2, \cdots, y_{12}$. This is clearly the transpose matrix multiplication problem to the one posed in (10); hence, it can be solved by simply transposing the flowgraph of Fig. 10. (This result also follows from the discussion in Section III.) It should be pointed out that the transposition operation computes u_q', defined in (23), for the odd values of q, which can be taken care of by adjusting the coefficients of the weighting network.

The terms "forward" and "inverse DCT" are used rather loosely in this paper for the sake of convenience. Their definition differs slightly from the standard ones used in the literature [23]. Furthermore, they are not true inverses of each other. There is, however, the transpose relationship between the two.

VII. Computational Requirements

For the hardware configuration depicted in Figs. 7 and 8, the amount of computation to be done in the weighting network, at a 112 kHz rate, is 20 multiplies and 39 adds. The following DCT requires about 16 multiplies and 76 adds. Noting that a DCT has to be evaluated every 125 μs, the total computational burden for the 12-channel demodulator amounts to about 2.3 million multiplies/s and 4.8 million adds/s. This figure compares favorably with that obtained by Peled and

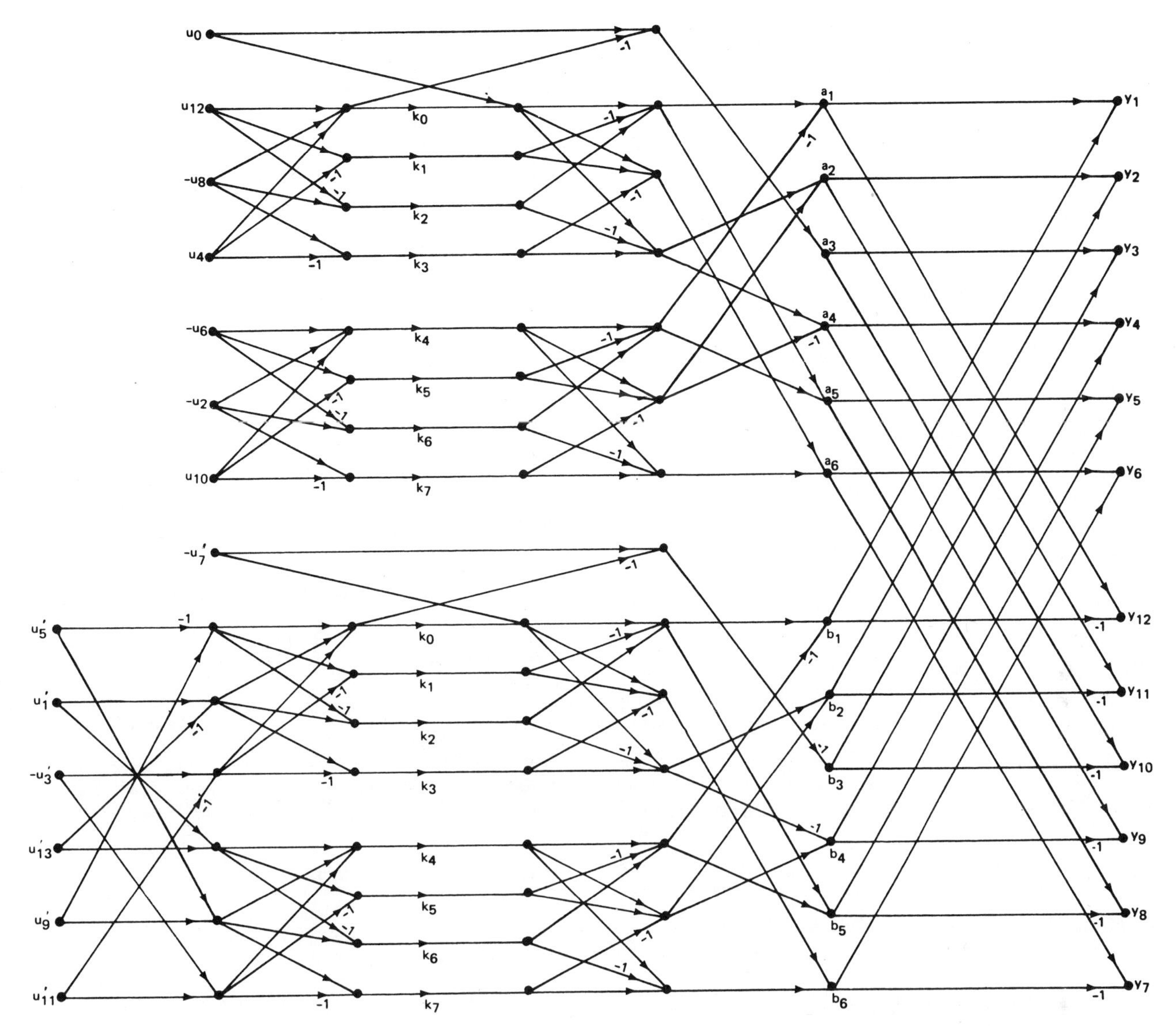

Fig. 10. Flowgraph for the 14-point inverse DCT needed in the demodulator. The outputs y_0 and y_{13} (guard channels) are not shown. The values for the constants $k_0, \cdots, k_7$ are as follows: $k_0 = 0.1666667$; $k_1 = 0.05585543$; $k_2 = 0.7343022$; $k_3 = k_1 + k_2$; $k_4 = 0.4409586$; $k_5 = 0.5339694$; $k_6 = 0.3408729$; $k_7 = k_5 + k_6$.

Winograd [17] using the Weaver scheme for frequency translation. The same amount of computation has to be performed in the modulator. It should be pointed out that since only FIR filters are used in the weighting network, 16-bit arithmetic appears to be sufficient.

VIII. Synchronization Considerations

A major application of the transmultiplexer is to interface between digital switching equipment and analog carrier systems. The special requirement for digital terminals interconnecting with digital switches is that they must have the means to be loop timed. The transmultiplexer should also have this provision, that is, to have the local oscillator phase-locked to the timing of the incoming signal and return this timing to the far-end switch. Since the incoming clock from the switch is highly stable (better than 1 part in 10^{10} per day), the resulting frequency translation error in the outgoing 60–108 kHz group band will be negligible. This time-keeping procedure will be satisfactory in cases where the incoming group band signal does not have much frequency deviation. (According to CCITT Recommendation G.225, the virtual channel carrier frequencies in the group should have an accuracy of $\pm 10^{-6}$. This corresponds to a translation error of about 0.1 Hz for a 108 kHz signal.)

Frequency translation errors can be minimized by synchronizing the TDM side to the incoming 1544 kHz clock and the FDM side to a pilot derived from the analog carrier system. Such a scheme will require elastic buffers in the transmultiplexer and has to allow for slips if the digital and analog networks are not synchronized.

When operating with standard $D3$ channel banks (which have as much as 130 ppm line error), a highly stable internal 1544 kHz clock in the transmultiplexer can provide timing information to the $D3$ terminals. If these terminals operate in a loop time mode, frequency translation errors will be reduced. The accuracy of the clock should be better than 10 ppm in order to limit translation errors to less than 1 Hz in either direction.

IX. Analog-Digital Conversion Requirements

The analog-to-digital converter on the group band side is one of the critical components of the transmultiplexer. Its specifications can be obtained by considering the performance objectives for the total noise in a channel under two conditions: 1) when no signal is present on the group side (idle channel noise), and 2) when the other 11 channels are loaded with conventional telephone signals corresponding to CCITT Recommendation G.227 (loaded channel noise). The method for obtaining the signal-to-noise ratio masks for the two cases are explained in [24]. For the 12-channel FDM signal sampled at 112 kHz, a 13-bit A-D converter appears to be sufficient.

By computing the peak envelope power of a 12-channel FDM signal made up of +3.17 dBm0 limited PCM signals and knowing the allowed contribution of the decoder to the idle noise, the dynamic range of the D-A can be obtained. Again, for this application, a 13-bit D-A seems adequate.

X. Concluding Remarks

This paper discussed the design of a transmultiplexer between 24-channel PCM systems and two FDM base groups. It was shown that the TDM-FDM translator required in such a transmultiplexer can be realized efficiently by cascading a weighting network and a discrete cosine transform processor. This decomposition arises from the fact that the individual filters in the filter bank required for the translation are frequency-shifted versions of a low-pass prototype. It holds good both in the modulation and demodulation processes, since these two are related by the transposition operation. The use of symmetry conditions in the weighting network and employing high-speed convolution algorithms to evaluate the DCT can greatly reduce the computational burden in the system.

An FIR filter bank was employed to accomplish both the channel filtering and sample rate alteration in the TDM-FDM translator. This single-stage approach simplifies the hardware implementation and does not seem to increase the number of arithmetic operations.

Although an IIR design can be used in the filter bank, such an approach does not appear to offer any significant savings in computation because sample rate alterations, by a fairly large factor (14), need to be performed. Moreover, IIR filters require, in general, longer word lengths and do not possess good phase characteristics. However, they have lower absolute delay.

Besides the TDM-FDM translator, a signaling converter is usually required in a transmultiplexer. This maps PCM signaling information to single-frequency tones suitable in carrier systems and vice versa. The particular frequency used for signaling is 2600 Hz in the North American Continent and 3825 Hz in most other places. Consequently, the design of the converter would be quite different for these two systems. The mapping will have to be performed in the digital domain, using digital filters. Presumably a single set of digital filters, which are designed to mimic the filters in the analog versions of the signaling adaptors, can be multiplexed among the various channels.

The question of performance specifications, synchronizing schemes, etc., for transmultiplexers is still under study by the CCITT and other organizations. A 561-tap FIR design for the channel filters was used as an example in this paper to satisfy certain reasonable specifications, such as ±0.5 dB in-band ripple and 65 dB crosstalk rejection in a loopback test. Similarly, the A-D requirements and the word lengths to be employed will depend on the noise performance demanded of this equipment. The use of FIR filters appears to allow 16-bit representation of numbers in the DCT and weighting network.

References

[1] S. L. Freeny *et al.*, "Design of digital filters for an all digital FDM-TDM translator," *IEEE Trans. Circuit Theory*, vol. CT-18, pp. 702-711, Nov. 1971.

[2] S. L. Freeny *et al.*, "System analysis of a TDM-FDM translator/digital *A*-type channel bank," *IEEE Trans. Commun. Technol.*, vol. COM-19, pp. 1050-1059, Dec. 1971.

[3] S. L. Freeny *et al.*, "An exploratory terminal for translating between analog frequency division and digital time division signals," in *Proc. ICC'71*, pp. 22-31-22-36.

[4] S. Darlington, "On digital single-sideband modulators," *IEEE Trans. Circuit Theory*, vol. CT-17, pp. 409-414, Aug. 1970.

[5] M. Bellanger and J. L. Daguet, "TDM-FDM translator: Digital polyphase and FFT," *IEEE Trans. Commun.*, vol. COM-22, pp. 1199-1204, Sept. 1974

[6] M. J. Narasimha *et al.*, "Design of digital filter-banks for voice multiplexing systems," Stanford Electron. Lab., Stanford Univ., Stanford, CA, Internal Rep., Oct. 1975.

[7] P. M. Terrell and P. J. W. Rayner, "A digital block processor for SSB-FDM modulation and demodulation," *IEEE Trans. Commun.*, vol. COM-23, pp. 282-286, Feb. 1975.

[8] M. Tomlinson and K. M. Wong, "Techniques for digital interfacing of TDM-FDM systems," *Proc. Inst. Elec. Eng.*, vol. 123, pp. 1285-1290, Dec. 1976.

[9] J. L. Daguet *et al.*, "Single sideband system for digitally processing a given number of channel signals," U.S. Patent 3 891 803, issued June 24, 1975.

[10] M. G. Bellanger and J. L. Daguet, "Circuit arrangement for digitally processing a given number of channel signals," U.S. Patent 3 971 922, issued July 27, 1976.

[11] M. J. Narasimha, "Digital filter bank," U.S. Patent 4 131 766, issued Dec. 26, 1978.

[12] G. Bonnerot *et al.*, "Digital processing techniques in the 60-channel transmultiplexer," *IEEE Trans. Commun.*, vol. COM-26, pp. 698-706, May 1978.

[13] R. Maruta and A. Tomozawa, "An improved method for digital SSB-FDM modulation and demodulation," *IEEE Trans. Commun.*, vol. COM-26, pp. 720-725, May 1978.

[14] F. Takahata *et al.*, "Development of a TDM/FDM transmultiplexer," *IEEE Trans. Commun.*, vol. COM-26, pp. 726-733, May 1978.

[15] T. Tsuda *et al.*, "Digital TDM-FDM translator with multistage structure," *IEEE Trans. Commun.*, vol. COM-26, pp. 734-741, May 1978.

[16] D. P. Kolba and T. W. Parks, "A prime factor FFT algorithm using high speed convolution," *IEEE Trans. Acoust., Speech, Signal Processing*, vol. ASSP-25, pp. 281-294, Aug. 1977.

[17] A. Peled and S. Winograd, "TDM-FDM conversion requiring reduced computational complexity," *IEEE Trans. Commun.*, vol. COM-26, pp. 707-719, May 1978.

[18] R. W. Schafer and L. R. Rabiner, "Design and simulation of a speech analysis-synthesis system based on short time Fourier analysis," *IEEE Trans. Audio Electroacoust.*, vol. AU-21, pp. 165-174, June 1973.

[19] M. R. Portnoff, "Implementation of the digital phase vocoder using the FFT," *IEEE Trans. Acoust., Speech, Signal Processing*, vol. ASSP-24, pp. 243-248, June 1976.

[20] K. Shenoi, "Design and application of digital filters," Ph.D. dissertation, Dep. Elec. Eng., Stanford Univ., Stanford, CA, May 1977.

[21] M. J. Narasimha and A. M. Peterson, "Design and applications

of uniform digital bandpass filter-banks," presented at the IEEE Int. Conf. on Acoust., Speech, Signal Processing, Tulsa, OK, Apr. 1978.

[22] M. J. Narasimha, "Implementation of FIR filters with random access memories," presented at ELECTRO'79, New York, NY, Apr. 24-26, 1979, paper 8/2.

[23] M. J. Narasimha and A. M. Peterson, "On the computation of the discrete cosine transform," *IEEE Trans. Commun.*, vol. COM-26, pp. 934-936, June 1978.

[24] M. G. Bellanger *et al.*, "Specification of A/D and D/A Converters for FDM telephone signals," *IEEE Trans. Circuits Syst.*, vol. CAS-25, pp. 461-467, July 1978.

34

On Generating Walsh Spectrograms

N. AHMED, T. NATARAJAN, AND H. R. RAINBOLT

***Abstract*—A simple method for generating a class of *time-sequency-amplitude* plots called Walsh spectrograms is presented. A Fortran computer program which enables one to generate such spectrograms readily is provided. Illustrative examples related to generating spectrograms associated with pulse-code modulation and speech data are included.**

I. Introduction

Time-frequency-amplitude plots or spectrograms have found digital signal processing applications with respect to the analysis of nonstationary data. Examples of such applications include transient signal detection and speech processing [1]–[4]. Such spectrograms may be referred to as "Fourier spectrograms," in the sense that the discrete Fourier transform (DFT) plays a central role in generating them.

The problem of generating Walsh spectrograms using the Walsh–Hadamard transform has received relatively little attention, although it has been considered for applications in areas where such spectrograms can be used to good advantage [5]–[8]. A survey of the literature shows that one area where Walsh spectrograms have been used is with respect to marine seismic data processing [9], [10]. However, the quality of such spectrograms can be considerably improved by employing a relatively inexpensive method which has been used effectively to generate Fourier spectrograms [11]. To this end, the main objective of this paper is to present a corresponding procedure for generating Walsh spectrograms using the sequency-ordered Walsh–Hadamard transform $(\mathrm{WHT})_w$.[1] Illustrative examples of Walsh spectrograms pertaining to pulse-code-modulation (PCM) and speech data are provided. A computer program that enables one to generate Walsh spectrograms readily, is also included.

II. Transform Power Spectra

A basic requirement for an orthogonal transform to be useful for generating sequency spectrograms is that it possesses a power spectrum in which each spectral point represents the power in a particular sequency. A $(\mathrm{WHT})_w$ spectrum which has this property was developed by Harmuth [13]. For example, when $N = 8$, the $(\mathrm{WHT})_w$ power spectrum is defined as

$$\begin{aligned} P_w(0) &= W^2(0) \\ P_w(1) &= W^2(1) + W^2(2) \\ P_w(2) &= W^2(3) + W^2(4) \\ P_w(3) &= W^2(5) + W^2(6) \\ P_w(4) &= W^2(7) \end{aligned} \qquad (1)$$

Manuscript received April 2, 1975.
N. Ahmed and T. Natarajan are with the Department of Electrical Engineering, Kansas State University, Manhattan, KS 66506. (913) 532-5600.
H. R. Rainbolt is with the Department of Speech, Kansas State University, Manhattan, KS 66506.
[1] The notation $(\mathrm{WHT})_w$ denotes Walsh or sequency-ordered Walsh–Hadamard transform [12].

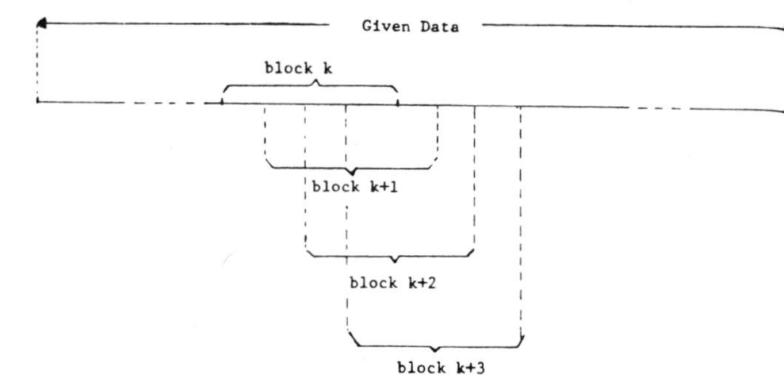

Fig. 1. Illustration of successive overlapping blocks.

where $W(k)$ denotes the kth $(\mathrm{WHT})_w$ coefficient and $P_w(s)$ is the sth power spectral point associated with the sequency s The general form of this spectrum is given by [13]

$$\begin{aligned} P_w(0) &= W^2(0) \\ P_w(s) &= W^2(2s-1) + W^2(2s), \qquad s = 1,2,\cdots,(N/2-1) \\ P_w(N/2) &= W^2(N-1). \end{aligned} \qquad (2)$$

We observe that the above spectrum has $(N/2 + 1)$ points.

III. Outline of Method

The outline given in this section essentially follows from a method used by Rothanser and Maiwald [11] in connection with generating digitized sound spectrographs. The procedure consists of the following steps.

Step 1: The given data sequence is divided into overlapping blocks such that each block consists of N points, as illustrated in Fig. 1 for four successive blocks. The amount of overlap between contiguous blocks can be varied. For the present study, we consider a 75 percent overlap. Such overlapping helps to improve the consistency of the resulting spectral estimates [14].

Step 2: Let $\{W(m)\}_k = \{W_k(0)W_k(1)\cdots W_k(N-1)\}$ denote the sequence obtained by computing the $(\mathrm{WHT})_w$ of the kth data block. Then the corresponding $(\mathrm{WHT})_w$ power spectrum is given by [see (2)]

$$\begin{aligned} P_w^{(k)}(0) &= W_k^2(0) \\ P_w^{(k)}(s) &= W_k^2(2s-1) + W_k^2(2s), \qquad s = 1,2,\cdots,N/2 \\ P_w^{(k)}(N/2) &= W_k^2(N-1). \end{aligned} \qquad (3)$$

Step 3: Average the spectrum computed for each block over adjacent spectral points using the relation

$$\begin{aligned} \bar{P}_w^{(k)}(0) &= P_w^{(k)}(0) \\ \bar{P}_w^{(k)}(s) &= \tfrac{1}{3}[P_w^{(k)}(s-1) + P_w^{(k)}(s) + P_w^{(k)}(s+1)], \\ & \qquad s = 1,2,\cdots,(N/2-1) \\ \bar{P}_w^{(k)}(N/2) &= P_w^{(k)}(N/2). \end{aligned} \qquad (4)$$

The 3-point averaging indicated in (4) serves to smooth the power spectrum and hence improve the visual quality of the corresponding spectrogram.

TABLE I
A BASIC CHARACTER SET

Grey level #	1	2	3	4	5	6	7	8
First print		-	=	+	X	X	0	0
Second print				+	X	X	X	W
Third print						=	*	M
Fourth print								*

Step 4: For each $\bar{P}_w^{(k)}(s)$ obtained in (4), compute the corresponding decibel (dB) value given by

$$L^{(k)}(s) = 10 \log_{10} [\bar{P}_w^{(k)}(s)] \text{ dB}, \qquad s = 0,1,\cdots,N/2. \quad (5)$$

Step 5: Plot the $L^{(k)}(s)$ obtained via (12) on a line printer in the form of a density-modulated string. This step is accomplished in two stages as follows:

i) Divide $L^{(k)}(s)$ values obtained in (5) into 8 grey levels. The decibel spacing (DBS) between the grey levels can be chosen to cover the dynamic range of the power spectrum. It can be obtained as

$$\text{DBS} = (L_{\max} - L_{\min})/8 \quad (6)$$

where $L_{\max}$ and $L_{\min}$ are estimates of the maximum and minimum decibel values of the power spectrum for the given data sequence. In (6), DBS is rounded-off to the nearest integer.

Next, the grey level associated with the sth spectral point $G^{(k)}(s)$ is computed as

$$G^{(k)}(s) = [(L^{(k)}(s) - L_{\min})/\text{DBS}] + 1 \quad (7)$$

where $[\cdot]$ denotes the integer portion of the quantity enclosed. Again, if $G^{(k)}(s)$ in (7) is less than 1 or greater than 8, it is set to 1 and 8, respectively.

ii) By means of superposition (i.e., overprinting) of members of the character set in Table I, obtain a time-sequency-amplitude plot. This resulting plot possesses a perceptually linear gradation in eight shades of grey, and is the desired spectrogram.

The above steps are implemented using Fortran as the programming language. The corresponding computer program is listed in Appendix I. The associated comment statements readily enable one to identify each of the above steps with a specific portion of the program.

IV. ILLUSTRATIVE EXAMPLES

We now apply the steps listed in the previous section to two types of data: i) PCM, and ii) speech.

PCM Data: This data was simulated using a series of independent Gaussian random variables. A total of 5760 points were generated, of which the portion between the points 2170 and 3705 was clipped by setting each value in this range to either 1 or 0, depending upon whether the value of a data point is positive or negative, respectively. This data was processed in blocks consisting of 128 points per block, and an overlap of 75 percent was used [see Fig. 1]. The resulting Walsh and Fourier spectrograms are shown in Fig. 2.

Speech Data: As a second example, we consider speech data which was obtained by sampling the sentence "noon is the sleepy time of day," as spoken by a partially deaf speaker. The sampling interval was approximately 120 ms, which resulted in a Nyquist frequency of approximately 4 kHz. The resulting data was processed in blocks, with 256 points per block. Thus, each block represented approximately 31 ms of speech. The spectrograms that resulted from a 75 percent overlap are shown in Fig. 3. For the purposes of illustration, the spectrogram obtained by means of a conventional analog spectrograph (KAY Electric Sonograph 606 1A), is also included in Fig. 3.

In each of the above examples, an 8:1 reduction was used while photographing the line printer plots.

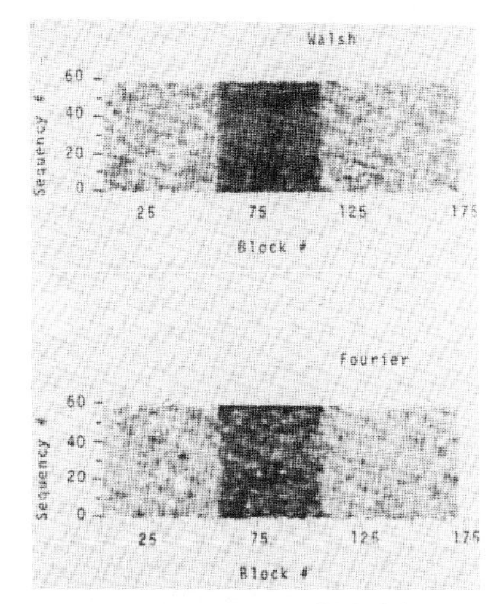

Fig. 2. Spectrograms of PCM data.

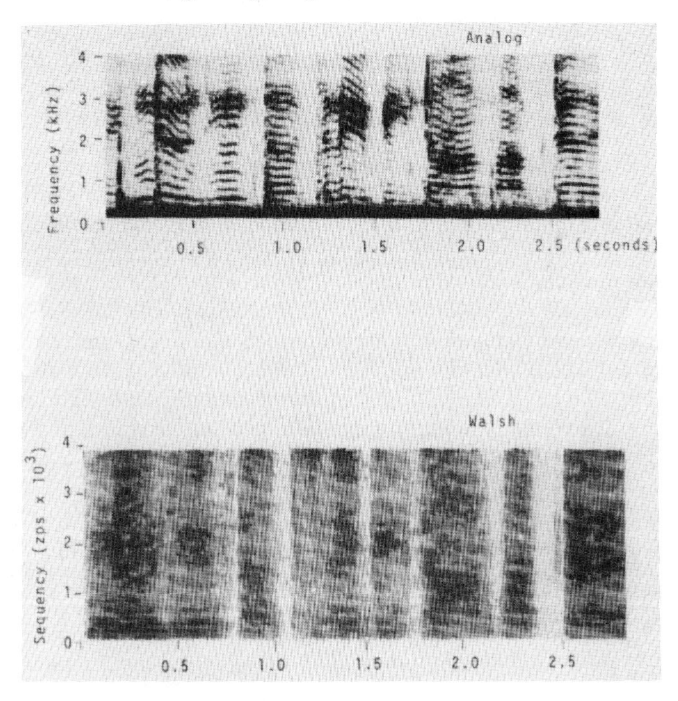

Fig. 3. Spectrograms of "noon is the sleepy time of day."

V. CONCLUSION

A simple method for generating Walsh spectrograms has been presented along with an appropriate computer program. Details pertaining to the method have been summarized in a form which makes it easy to use. By means of illustrative examples, it has been demonstrated that the quality of the spectrograms so obtained is considerably better than those developed in the past [9], [10]. It would be worthwhile comparing such Walsh spectrograms with corresponding ones generated by means of a method which involves overprinting via a dot-cluster technique [15].

APPENDIX I

```
C
C     THIS PROGRAM GENERATES TIME-SEQUENCY-AMPLITUDE PLOT OF A
C     DATA SEQUENCE.
C     THE DATA SECUENCE IS DIVIDED INTO OVERLAPPING SEGMENTS OF
C     N DATA POINTS. THE SEQUENCY POWER SPECTRUM,IN DB.,IS
C     COMPUTED FOR EACH SEGMENT. EACH DB. VALUE IS ASSIGNED
C     A GREY LEVEL BETWEEN 1 AND 8 DEPENDING UPON ITS MAGNITUDE.
C     THE EIGHT LEVELS COVER THE DYNAMIC RANGE OF THE DB
C     VALUES OF THE POWER SPECTRM. CORRESPONDING TO EACH
C     SEGMENT,A LINE OF SUPERPOSED CHARACTERS IS PRINTED. THE
C     NUMBER OF OVERPRINTS AND THE TYPE OF CHARACTER DEPENDS
C     UPON THE GREY LEVEL.
C     DEFINITION OF VARIABLES :
C     NN: TOTAL LENGTH OF THE INPUT DATA SEQUENCE
C     N: NUMBER OF POINTS IN EACH SEGMENT
C     NS: INTERVAL BETWEEN TWO OVELAPPING SEGMENTS
C
      DIMENSION XX(6000),X(256),P(129),IP(129)
    1 FORMAT('1')
      READ 2,NN,N,NS,LOG2N
    2 FORMAT(4I5)
      PRINT 3,NN,N,NS
    3 FORMAT('1',5X,'LENGTH OF THE DATA SEQUENCE =',I5,
     .5X,'NUMBER OF POINTS IN A SEGMENT =',I4,5X,
     .'INTERVAL BETWEEN TWO SEGMENTS =',I3,/,' ',5X,
     .'LOG. N TO THE BASE 2 =',I3)
C     READ THE DATA SEQUENCE
      READ 5,(XX(I),I=1,NN)
    5 FORMAT(16F5.2)
      NB=(NN/N-1)*N/NS+1
C     NB = NUMBER OF OVERLAPPING SEGMENTS
      PRINT 1
      DO 100 IB=1,NB
      M=(IB-1)*NS
      DO 10 I=1,N
   10 X(I)=XX(I+M)
      CALL FWTS(X,N,LOG2N)
      N21=N/2+1
      N1=N-1
C     COMPUTE THE SEQUENCY POWER SPECTRUM
      P(1)=X(1)*X(1)
      P(N21)=X(N)*X(N)
      K=1
      DO 20 I=2,N1,2
      K=K+1
      P(K)=X(I)*X(I)+X(I+1)*X(I+1)
   20 CONTINUE
      DO 25 I=1,N21
      IF ((I.EQ.1).OR.(I.EQ.N21)) GO TO 22
C     SMOOTHING OF THE POWER SPECTRUM
      PA=(P(I-1)+P(I)+P(I+1))/3.
      GO TO 23
   22 PA=P(I)
   23 IF (PA.LT.1.) GO TO 24
C     THE DB. VALUES ARE COMPUTED
      IP(I)=10.*ALOG10(PA)
      GO TO 25
   24 IP(I)=0
   25 CONTINUE
C     OUTPUT A LINE OF OVERPRINTED CHARACTERS ON THE PRINTER
      CALL SPCTGR(IP,N21)
  100 CONTINUE
      STOP
      END
      SUBROUTINE FWTS(X,N,ITER)
C
C     THIS ROUTINE IMPLEMENTS THE FAST WALSH TRANSFORM
C
      DIMENSION X(256),Y(256)
      DO 15 J=1,ITER
      J2=2**J
      N2J=N/J2
      NJM1=2**(J-1)
      DO 10 I=1,N2J
      IN2J=(I-1)*J2
      IS2=IN2J+1
      IS4=IS2+1
      DO 10 K=1,NJM1
      IS1=K+IN2J
      IS3=IS1+NJM1
      Y(IS2) = X(IS1) + X(IS3)
      Y(IS4) = X(IS1) - X(IS3)
      IT=IS2
      IS2=IS4+2
      IS4=IT+2
   10 CONTINUE
      DO 15 M=1,N
   15 X(M) = Y(M)
      RETURN
      END
      SUBROUTINE SPCTGR(IP,N)
C
C     THIS ROUTINE MAPS THE DB. VALUES INTO GREY LEVELS
C
      DIMENSION IP(1),IOUT(132)
      DO 50 I=1,N
      L=IP(I)/3+1
      IF (L.LT.0) L=0
      IF (L.GT.8) L=8
      IOUT(I)=L
   50 CONTINUE
      CALL PRINTR(IOUT,N)
      RETURN
      END
      SUBROUTINE PRINTR(IOUT,N)
C
C     THIS ROUTINE PRINTS A LINE OF OVERPRINTED CHARACTERS
C     TO REPRESENT DIFFERENT SHADES OF GREY.
C
      DIMENSION IOUT(1),LINE(133),CHAR(32)
      INTEGER CHAR,SPACE,PLUS
C
C     CHARACTER SET FOR MULTIPRINT LINE OUTPUT,
C     8 LEVELS, 4 OVERPRINTS
C
      DATA CHAR /' ','-','=','+','X','X','O','C',
     .           ' ',' ',' ','+','X','X','X','W',
     .           ' ',' ',' ',' ',' ','=','*','M',
     .           ' ',' ',' ',' ',' ',' ',' ','*'/
      DATA SPACE,PLUS /' ','+'/

      NP1=N+1
      DO 50 IP=1,4
      DO 10 I=1,N
      L=IOUT(I)
      IND=(IP-1)*8+L
   10 LINE(I+1)=CHAR(IND)
      IF (IP.NE.1) GO TO 20
C     LINE FEED PRINTER CONTROL CHARACTER IS SPECIFIED
      LINE(1)=SPACE
      GO TO 30
C     OVERPRINT CONTROL CHARACTER IS SPECIFIED
   20 LINE(1)=PLUS
   30 PRINT 40,(LINE(I),I=1,NP1)
   40 FORMAT(133A1)
   50 CONTINUE
      RETURN
      END
```

REFERENCES

[1] C. R. Arnold, "Spectral estimation for transient waveforms," *IEEE Trans. Audio Electroacoust.*, pp. 248–257, Sept. 1970.

[2] J. Flanagnan, *Speech Analysis, Synthesis and Perception*. New York: Academic Press, 1965.

[3] B. Gold and C. Rader, *Digital Processing of Signals*. New York: McGraw-Hill, 1969.

[4] A. V. Oppenheim, "Speech spectrograms using the fast Fourier transform," *IEEE Spectrum*, pp. 57–62, August 1970.

[5] S. J. Campanella and G. S. Robinson, "A comparison of Walsh and Fourier transformations for applications of speech," in *Proc. 1971 Symp. Applications of Walsh Functions*, AD-727000, pp. 199–202.

[6] E. J. Claire, "Acoustic signature detection and classification techniques," in *Proc. 1971 Symp. Applications of Walsh Functions*, AD-727000, pp. 125–129.

[7] H. Goetheoffer, "Speech processing with Walsh functions," in *Proc. 1972 Symp. Applications of Walsh Functions*, AD-744 650, pp. 163–168.

[8] C. Boesswetter, "Analog sequency analysis and synthesis of voice signals," in *Proc. 1970 Symp. Applications of Walsh Functions*, AD-707 431, pp. 220–229.

[9] C. Chen, "Walsh domain processing of marine seismic data," *Proc. 1972 Symp. Applications of Walsh Functions*, AD-744 650, pp. 64–67.

[10] ——, "Further results on Walsh domain processing of marine seismic data," in *Proc. 1973 Symp. Applications of Walsh Functions*, AD-763000, pp. 253–256.

[11] E. Rothauser and D. Maiwald, "A digitized sound spectrograph using FFT and multiprint techniques," IBM Research Publication, RZ 295 (no. 11466), Jan. 1969. (Copies may be requested from IBM Thomas J. Watson Research Center, P.O. Box 218, Yorkstown Heights, NY 10598.)

[12] N. Ahmed, H. Schreiber, and P. Lopresti, "On notation and definition of terms related to a class of complete orthogonal functions," *IEEE Trans. Electromagn. Compat.*, vol. EMC-15, pp. 75–80, 1973.

[13] H. F. Harmuth, *Transmission of Information by Orthogonal Functions*. New York: Springer-Verlag, 1969.

[14] P. D. Welch, "The use of fast Fourier transform for the estimation of power spectra: A method based on time averaging over short modified pediograms," *IEEE Trans. Audio Electroacoust.*, pp. 70–76, June 1967.

[15] Kennett, "Short-term spectral analysis and sequency filtering of seismic data," in *Proc. NATO Advanced Study Institute*. Leiden, The Netherlands: Noordhoff, 1975.

Reprinted with permission from *IEEE Trans.* **ASSP-29:**147-154 (1981)

Real-Time Simulation of Adaptive Transform Coding

RICHARD V. COX, MEMBER, IEEE, AND RONALD E. CROCHIERE, SENIOR MEMBER, IEEE

***Abstract*—Adaptive transform coding (ATC) has recently been proposed as a technique for speech coding at bit rates in the range of 9.6-16 kbits/s. In this paper we report on two new developments: 1) the use of a homomorphic vocoder model for the "side-information" channel in ATC and 2) a real-time simulation of ATC on an array processing computer. It is shown that the choice of the homomorphic "side-information" model leads to a convenient form of the ATC algorithm for real-time block processing using array processing techniques. It is also shown that the log spectrum output of the homomorphic model is in a convenient form for input to both the bit assignment algorithm in ATC (which becomes a straightforward quantization operation) and the quantization of the transform coefficients (which may be done in the log domain). An array processor simulation of this form of the algorithm has been implemented and it serves as a highly useful and convenient tool for studying the ATC algorithm in real time in a Fortran programming environment. It has allowed us, for the first time, to perform actual telephone conversations over a transform coder. The quality of this ATC algorithm was found to be essentially equivalent to that of a previous version using an LPC vocoder model for the side information.**

I. Introduction

ADAPTIVE transform coding (ATC) has been shown to be an effective method of digitally encoding speech at low-bit rates typically in the range of 9.6-16 kbits/s [1]-[4]. Fig. 1 illustrates a basic block diagram of the algorithm. The input speech is blocked into frames of data (typically $N = 128$ to $N = 256$ samples long at 8 kHz) and transformed by an appropriate fast-transform algorithm. The transform coefficients $X(k)$, $k = 0, 1, 2, \cdots, N - 1$ are then adaptively quantized by a set of PCM quantizers whose step-sizes $\Delta(k)$ and number of bits $b(k)$ are dynamically varied from block to block to take advantage of known spectral properties of speech production and perception. The choice of bit assignment and step-size for the quantizers is determined through the aid of a separate "side-information" channel which models and parameterizes the local spectral characteristics of the speech in each block. This "side-information" is also quantized and transmitted to the receiver for use in decoding (about 2 kbits/s are needed for this).

The above algorithm requires a substantial amount of signal processing. In previous work it has been studied by nonreal-time simulation on general purpose computers. In this paper we report on recent real-time simulations of ATC using an array processing computer. This real-time capability has allowed us to study, more realistically, the behavior of the ATC algorithm for real telephone conversations as opposed to stored speech utterances.

Manuscript received June 30, 1980; revised October 14, 1980.
The authors are with the Department of Acoustics Research, Bell Laboratories, Murray Hill, NJ 07974.

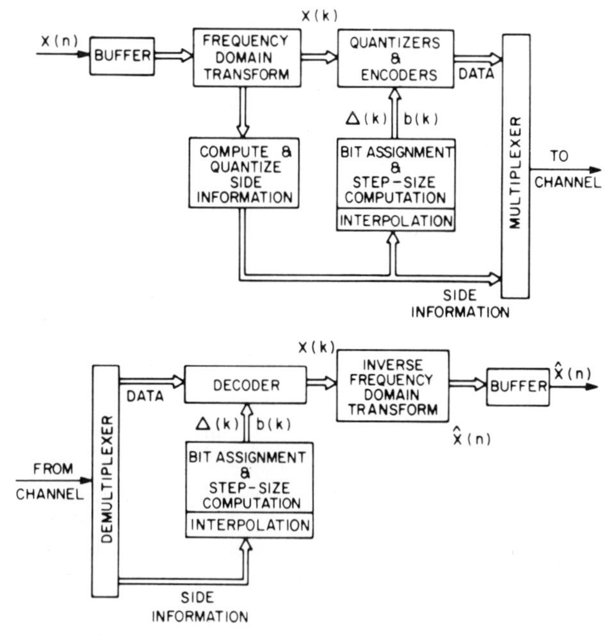

Fig. 1. Block diagram of adaptive transform coding.

A number of modifications have been made in the ATC algorithm over previous proposals [1], [4]. These modifications have made the algorithm more convenient for an array processing environment. In particular, we have studied the use of a homomorphic model for estimating and parameterizing the side information in ATC. This differs from the earlier proposal by Zelinski and Noll [1] in which spectral side information was obtained by locally averaging spectral coefficients into a smaller number of spectral magnitude components and then linearly interpolating between them (in the log domain) to obtain the spectral estimate. It also differs from the "speech-specific" model by Tribolet and Crochiere [3], [4] which uses an LPC and pitch model to specifically take advantage of known spectral properties of speech and tailor the performance of ATC for speechlike signals. The homomorphic model discussed here is also a "speech-specific" algorithm which has similar properties to that of the LPC and pitch model, however, it is more conveniently accommodated in the framework of array processing.

We will begin in Section II by discussing the general configuration, and the capabilities and limitations of a particular array processing computer. We will then discuss the implementation of the ATC algorithm within this framework. Section III considers the block transform analysis. Section IV presents details of the homomorphic model and discusses its properties. Section V discusses the implementation of the bit assignment and quantization routines and finally Section VI discusses results of real-time simulations.

II. Characteristics of the Array Processing Computer

The array processing computer used for this study is a Data General Eclipse AP130 computer which has an internal array processor [5]. The array processor (AP) has 2048 floating point words of high-speed (bipolar) memory which can be mapped into the main memory of the machine. Thus, the same memory can be accessed in the normal way by the Eclipse CPU or directly by the array processor. Because of this shared memory approach, the setup time is kept to a minimum.

The AP is basically a high-speed floating point vector processor which is controlled through Fortran calls from the Eclipse (it cannot be programmed directly by the user). It can perform specific functions on blocks of data stored within its high-speed floating point memory, such as adding or multiplying two arrays, taking the FFT of an array, and creating an array with the aid of a look-up table [5]. Combinations of these functions can be used to generate other functions. For example, a look-up table can be used to implement a fast logarithm. Operations involving integers (such as quantization) can be implemented with the aid of the Eclipse.

The array processor is inherently a block processing device whose speed is obtained in part from pipeline processing. As a result, stream processing operations (operations which proceed on a sample-by-sample basis using past values of the same vector as in a feedback loop) cannot be efficiently programmed on the AP through the Fortran callable routines. Fortunately, the homomorphic ATC model can be conveniently computed in a block processing manner on the AP and this implementation will be discussed in more detail in the next sections.

TABLE I
Execution Times for Parts of the Algorithm

SDFT Transform-Inverse Transform Pair			6.56 msec
$\log_2$ (for 257 point block)			1.78 msec
2^x (for 257 point block)			3.15 msec
Initial Transform (SDFT)		3.28	
Side Information Analysis			
Abs Value	.18		
$\log_2$	1.78		
$(\mathrm{SDFT})^{-1}$	3.28	6.73	
Pitch Extraction	.34		
Quantizing Side Info*	1.15		
Quantization Preparation			
SDFT (to get est. spectrum)	3.28		
Bit Assignment*	4.72	8.00	
Quantization			
Quant. in log domain*	3.23		
exponential	3.15	7.79	
Sign bits in linear domain*	1.41		
Inverse Transform		3.28	
Reading data from buffers/ Writing results to buffers		1.93	
Total Time		31.01 msec	

*These operations are not performed entirely in the array processor and the number of instruction performed in the Eclipse is a data dependent random variable.

III. Block Transform Analysis for ATC

In practice the frequency domain transform in the ATC algorithm in Fig. 1 is generally chosen to be the discrete cosine transform (DCT) [6] or a closely related symmetrical discrete Fourier transform (SDFT). This choice leads to a better signal-to-noise ratio performance for ATC [1] and a reduction in end effects in the blocks [4] over the choice of the more conventional discrete Fourier transform (DFT). In addition, the DCT or SDFT leads to a set of N real coefficients instead of $N/2$ complex coefficients as obtained with the DFT, where N is the size of the transform. Since an $N+1$ point SDFT can be conveniently computed from a $2N$ point DFT (which is available on the AP) in the manner described below, this was the form of the transform that was used for the real-time simulations.

The symmetrical DFT (SDFT) of an $N+1$ point sequence can be defined as follows. Given a block of $N+1$ samples $x(n)$, $n = 0, 1, 2, \cdots, N$, a $2N$ point block of samples, $\hat{x}(n)$, is generated such that

$$\hat{x}(n) = \begin{cases} x(n) & n = 0, 1, \cdots, N \\ x(2N-n) & n = N+1, N+2, \cdots, 2N-1. \end{cases} \tag{1}$$

The sequence $\hat{x}(n)$ represents a $2N$ point symmetrical sequence and its DFT is also a real, symmetrical, $2N$ point sequence (the imaginary parts are zero)

$$\hat{X}(k) = \sum_{n=0}^{2N-1} \hat{x}(n)\, e^{-j2\pi nk/2N}. \tag{2}$$

Then the SDFT of $x(n)$ can be defined as the first $N+1$ points of $\hat{X}(k)$, i.e.,

$$\begin{aligned} X(k) &= \text{SDFT of } x(n) \\ &= \hat{X}(k) \qquad k = 0, 1, 2, \cdots, N. \end{aligned} \tag{3}$$

Using the above approach a block of 257 speech samples (representing 32.125 ms of speech at an 8 kHz sampling rate) can be transformed with an SDFT transform in 3.28 ms on the AP or 10.2 percent of the total available block time. Table I lists the execution times for various parts of the algorithm and will be discussed later.

IV. Homomorphic Representation of the Side Information

The next operation that must be performed after the input transform is the spectral estimation for the side information (see Fig. 1). As stated earlier, this process is accomplished with a homomorphic model [7], [8].

Fig. 2 illustrates this basic model. The speech spectrum $X(k)$ is obtained from the SDFT. The log magnitude of $X(k)$ is computed and transformed with an $N+1$ point inverse SDFT. The resulting signal $c(n)$ is generally referred to as the cepstrum [7], [8]. However, in conventional homomorphic analysis the signal $X(k)$ is obtained from a properly windowed (e.g., a Hamming window) DFT analysis to obtain a good spectral estimate. In this case, due to the limitations of the ATC

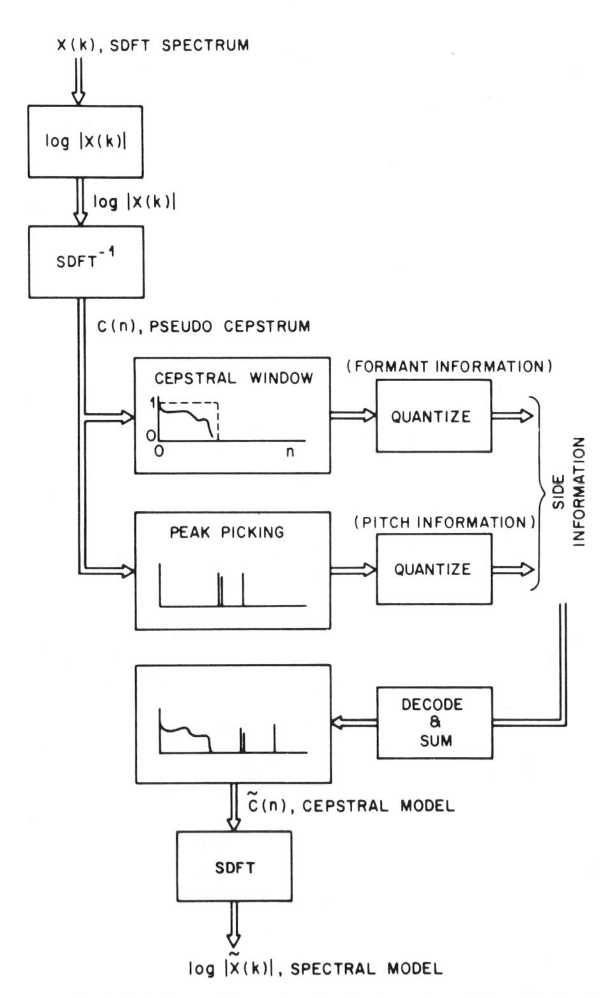

Fig. 2. Side information processing for homomorphic analysis.

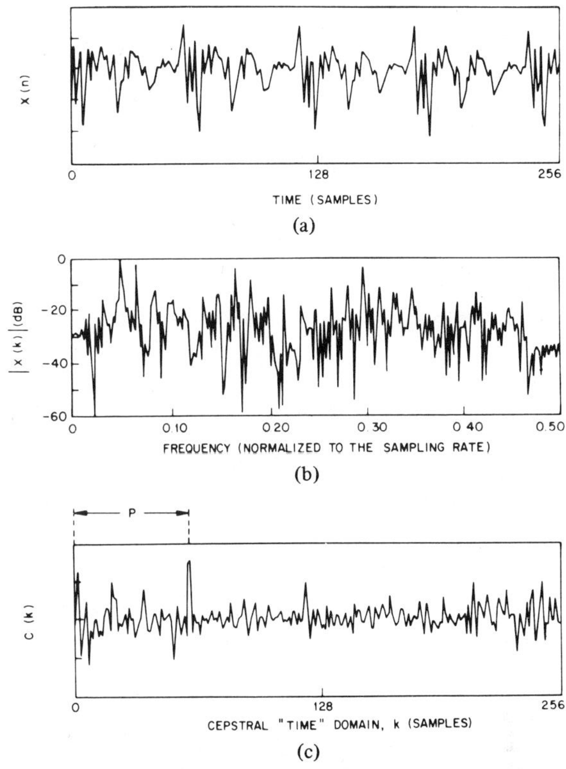

Fig. 3. Typical waveforms: (a) block of speech, (b) pseudospectrum, (c) pseudocepstrum.

framework, it is obtained from an SDFT analysis using rectangular windowing which does not give as good a spectral estimate as a properly windowed DFT approach. Therefore, in this paper we often refer to $X(k)$ as the "pseudospectrum" and to $c(n)$ as its "pseudocepstrum" to distinguish it from the more conventional procedure for homomorphic analysis.

From the pseudocepstrum the parameters can be extracted which model the formant structure (i.e. the smooth spectral envelope) and pitch structure of speech. The parameters associated with the formant structure are extracted with the aid of a cepstral window (see Fig. 2). The details of this process will be described in Section IV-A. The parameters associated with the pitch structure in speech are obtained from a "peak picking" process on the cepstrum (see Fig. 2) and this will be discussed in Section IV-B.

The formant parameters and pitch parameters are encoded for use as side information as seen in Fig. 1. Finally, in reconstructing the spectral model, the formant parameters and pitch parameters are combined, as shown at the bottom of Fig. 2, to obtain a model $\tilde{c}(n)$ of the cepstrum. The SDFT of this cepstral model then produces the desired log spectral, model log $|\tilde{X}(k)|$, which is used for bit assignment and step-size control in the ATC algorithm.

The bit assignment for the cepstral coefficients is done *a priori* as a function of the number of formant coefficients and the bit rate. Typically 10, 12, and 14 formant coefficients were quantized for rates of 9.6, 12 and 16 kbits/s, respectively. The quantization rates of the cepstral coefficients were 2.0, 2.7, and 2.9 kbits/s, respectively, for these rates. The quantization was done with fixed PCM quantizers based on statistics obtained for the cepstral coefficients of several male and female speakers. Throughout the rest of the paper only total rates are mentioned.

In spite of the nonideal analysis framework, which is necessary for ATC, the pseudocepstrum exhibits similar properties to that of the true cepstrum and it is suitable for the purposes of spectral modeling for the ATC algorithm. Due to the rectangular analysis window the pseudospectrum is much noisier and does not represent as refined a spectral estimate as a conventional DFT spectrum obtained using a smoothly windowed speech segment. This can be seen, for example, by comparing the pseudospectrum in Fig. 3(b) with the windowed DFT spectrum in Fig. 5 (solid line). However, as discussed in [4], the spectral shape of the pseudospectrum is closely related to that of the actual short-time spectrum.

A. Representation of the Formant Structure of Speech

The cepstrum (and the pseudocepstrum) exhibits the property that the information about the general shape (the smooth spectral envelope) of the spectrum is contained within the first 10 to 25 coefficients of the cepstrum, i.e., the "low-time" region of the cepstrum [7], [8]. This information represents the spectral shape of speech due to its formant structure (i.e., the vocal tract model). Thus, by windowing the first 10 to 25

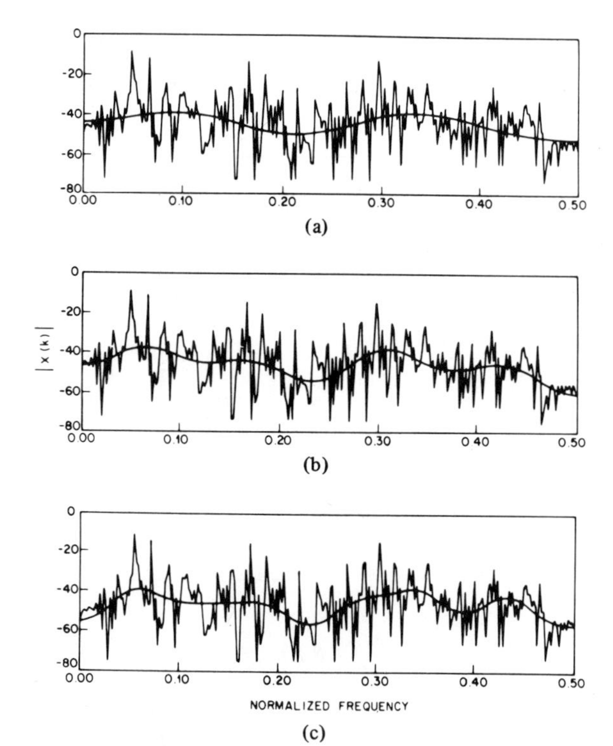

Fig. 4. Homomorphic spectral estimates: (a) from 6 cepstral coefficients, (b) from 10 cepstral coefficients, and (c) from 14 cepstral coefficients.

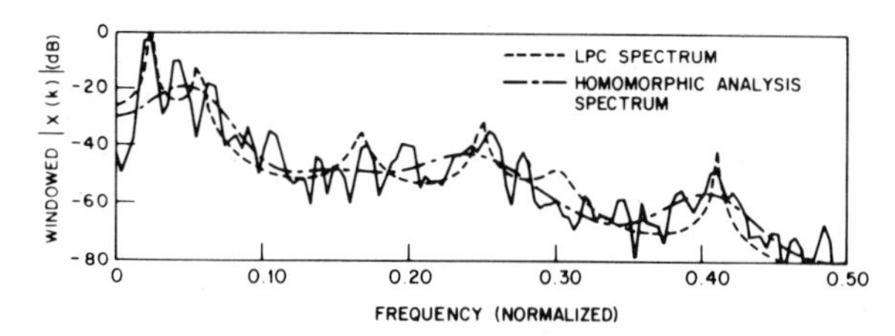

Fig. 5. LPC spectral estimate and homomorphic spectral estimate versus actual spectrum.

samples of the pseudocepstrum and quantizing these parameters (see Fig. 2) a spectral representation of the envelope of the 257-point pseudospectrum, $X(k)$, can be obtained. To retrieve this spectral envelope the pseudocepstral coefficients are decoded and padded with zeros. The SDFT of this sequence then gives the log magnitude envelope of the pseudospectrum $X(k)$.

Fig. 3 illustrates an example of the waveforms encountered in this analysis. Fig. 3(a) shows a section of a rectangularly windowed voiced speech segment. Fig. 3(b) shows the log magnitude of the SDFT spectrum (the pseudospectrum) and Fig. 3(c) shows the resulting pseudocepstrum.

Fig. 4(a)-(c) shows an SDFT spectrum and the smooth spectral envelopes obtained from the homomorphic model by retaining only the first 6, 10, and 14 pseudocepstral coefficients in Fig. 3(c), respectively. Generally, in simulations 10 to 14 pseudocepstral coefficients were used in the side information. This was sufficient to give a similar ATC performance to that of the LPC driven approach reported in [4].

Fig. 5 further illustrates the differences between the homomorphic model and the LPC model of spectral estimation. A block of 256 samples of speech was windowed by a Hamming window and analyzed with a DFT, a 10th-order LPC analysis, and a homomorphic analysis. The solid line represents the actual short-time DFT spectrum which the homomorphic and LPC models are trying to estimate. Note that this spectrum is composed of both a formant envelope and a pitch harmonic structure. The purpose of the LPC or homomorphic models are to extract the formant envelope. The pitch structure is modeled by the cepstral peak picking to be discussed later. Although the homomorphic and LPC models give different outputs, they give roughly equivalent mean-square-error performance in estimating the actual smooth component of the spectrum. As seen in Fig. 5 the LPC spectral estimate does a better job of estimation in the formant regions but a poorer job in the low-amplitude regions. This is a consequence, of course, of the all-pole nature of the LPC model. In terms of distributions in the bit assignment in ATC (see Section V), the LPC model tends to place more bits in the formant regions leaving the regions between the formants with fewer or no bits. The homomorphic model tends to distribute the bits more evenly across the frequency domain, giving less emphasis to the formants than the LPC model. Thus, each method has different strengths and weaknesses. In general, they both give roughly equivalent performance in the ATC algorithm for 9.6 kbits/s and up.

B. Representation of Pitch Structure in the Homomorphic Model

A second property exhibited by the homomorphic model of speech is that of the "pitch" structure or pseudoperiodicity of speech in voiced regions [7], [8]. This pseudoperiodicity manifests itself as a periodic pulse train in the cepstrum. If P represents the local pitch period then the pulses in the cepstrum for pseudoperiodic signals is clearly seen in Fig. 3(c). P, i.e., $n = mP$, $m = 0, 1, 2, \cdots$, and they decay rapidly with increasing values of n or m. This periodic property of the cepstrum for pseudoperiodic signals is clearly seen in Fig. 3c. In this example, P is approximately 60 samples, corresponding to a fundamental pitch frequency of 133 Hz.

In the homomorphic side information model this information about the harmonic pitch structure in the spectrum can therefore be obtained by measuring the location and the amplitude of the pulses in the cepstrum (i.e., "peak picking"). Since the amplitude of these pulses decay rapidly with increasing n, it is sufficient to measure the location and amplitude of only the first one or two pulses in the cepstrum in order to get a good estimate of the pitch harmonic structure in the spectral model. This information is then quantized and transmitted as side information.

Fig. 2 illustrates the entire processing for the side information. The pseudocepstrum is obtained by first taking the log of the absolute value of the pseudospectrum values and then taking the inverse SDFT. The pseudocepstrum is then analyzed to produce information on the smooth spectrum through the low-time coefficients as discussed above and the pitch information is found through the "peak-picking" procedure explained below. The remaining cepstral values are set to zero. After quantization, a complete spectral estimate is obtained

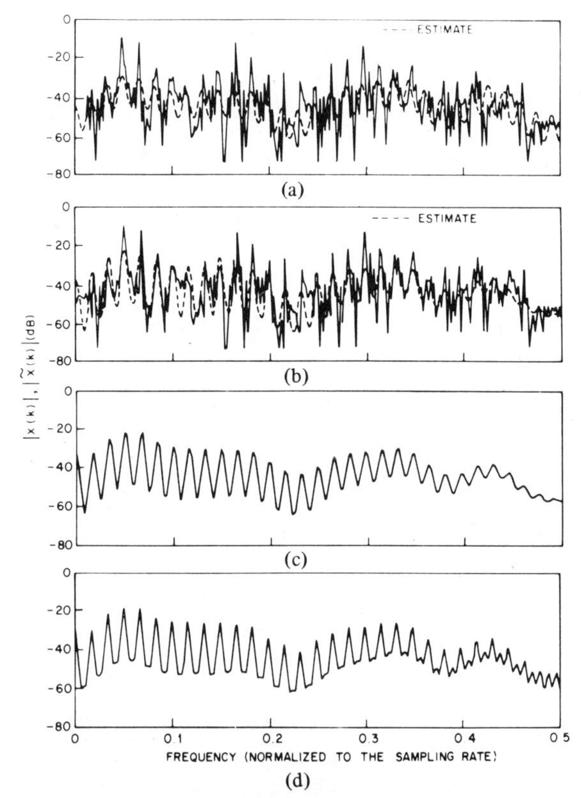

Fig. 6. Spectral estimates with pitch: (a) 1 peak model, (b) peak, 2 sample model, (c) 2 peak model, (d) 2 peak model with amplification.

by taking the SDFT of $\tilde{c}(n)$. This estimate is used for bit assignment and step-size computation.

Fig. 6 illustrates examples of the SDFT spectra for a block of speech and the spectral estimates obtained from the homomorphic model. An important requirement in obtaining a good estimate of the spectrum is that the information about the pitch pulses in the cepstrum be properly determined. In initial trials a single-pitch period, $\tilde{P}$, (where $\tilde{P}$ is an estimate of P) and pulse amplitude, A_1, were used by choosing the peak value of the cepstrum in the cepstral "time" range of 2 ms to 16 ms. Fig. 6(a) illustrates an example of a spectral estimate obtained in this way. The solid line corresponds to the original SDFT spectrum and the dashed line corresponds to the spectral estimate.

A considerable improvement to the above one pulse model was obtained by recognizing the fact that this main cepstral peak, $\tilde{P}$, was often more than one sample wide (i.e., pitch is not necessarily an integer number of samples). This improvement, illustrated in Fig. 6(b), was obtained by representing the main peak of the cepstrum as a pair of adjacent pulses (samples), the largest, A_1, and the next largest adjacent sample A_2. Thus, one location $\tilde{P}$ and two amplitudes are needed to be expressed in the side information for pitch. One more bit is needed to indicate whether A_2 is to the left or right of the main peak.

A further improvement in the pitch model was obtained with a two peak model by searching for cepstral peaks in the neighborhood of $\tilde{P}/2$ and $2\tilde{P}$. The location and amplitude, A_3, of the largest of these two pulses was then encoded and added to the side information. Note that it is important to search both in the neighborhood of $\tilde{P}/2$ and $2\tilde{P}$ since in some cases $\tilde{P}$ represents twice the pitch. This model, therefore, requires the transmission of two peak locations and three amplitudes. Fig. 6(c) illustrates an example of the spectral estimate obtained with this approach when the second peak was located near $2\tilde{P}$.

A final modification to the above pitch model was made by applying an artificial gain factor of approximately 1.5 to the pitch-pulse amplitudes A_1, A_2 and A_3 in the cepstral model. This helps to raise the peak amplitude of the pitch harmonics in the cepstral model by approximately 3.5 dB and provides a better spectral match to the peaks of the actual spectra. If this is not done, all of the pitch peaks in the spectral estimate tend to be uniformly too low. This is because homomorphic analysis averages in the log domain so that pitch peaks would ordinarily be lowered, just like the formant peaks for the homomorphic analysis are lower than the LPC formant peaks in Fig. 5. By introducing the artificial gain factor of 3.5 dB, the entire spectral estimate is raised by this amount so that the estimate matches the peaks in log spectrum rather than the average log spectral values. Fig. 6(d) represents the spectral estimate obtained with this final modification. Note that the pitch harmonics here are narrower and higher than those in Fig. 6(c) making them more like those of the actual spectrum in Fig. 6(a).

V. Bit Assignment and Quantization

The bit assignment routine determines the number of bits, $b(k)$, used by the ATC algorithm in quantizing each transform coefficient (see Fig. 1). This decision is based on the spectral estimate obtained from the side-information channel. In effect, the number of bits used to quantize each coefficient is proportional to the log spectral estimate, $\log|\tilde{X}(k)|$ [1]-[4]. Since this is exactly the form of the output for the homomorphic model, it is particularly convenient for implementation.

Let $b(k)$ denote the number of bits used for quantizing the kth coefficient. Then the bit assignment is determined as

$$b(k) = \lfloor \log_2^2 |\tilde{X}(k)| - D \rfloor^* \tag{4}$$

where $\log_2 |\tilde{X}(k)|$ is obtained directly from the side information channel, as discussed, and D is a constant. The operation $\lfloor \cdot \rfloor^*$ defined as

$$\lfloor u \rfloor^* = \begin{cases} 0, & \text{if } u < 0 \\ \text{greatest integer} \leqslant u, & \text{if } 0 \leqslant u < n_b \\ n_b, & \text{if } u \geqslant n_b \end{cases} \tag{5}$$

where n_b is the maximum number of bits allowed for quantizing any coefficient. Thus, it is seen that the bit assignment is simply a quantization operation which can be conveniently implemented in a block fashion with the aid of the AP.

The constant D in (4) must be chosen such that the bit assignment satisfies the condition.

$$B = \sum_{k=0}^{N} b(k) \tag{6}$$

where B is the total number of bits allowed in each block for quantizing the transform coefficients. The value of D is obtained by a two-pass operation [9]. The first estimate of D is determined as

$$D_1 = \frac{1}{N+1}\left[\sum_{k=0}^{N} \log_2 |\tilde{X}(k)| - B\right]. \tag{7}$$

This usually results in an estimate of D that is too high because there are many points for which the quantity $\log_2 |\tilde{X}(k)| - D$ is negative. If we define the set S^+ as the set of values of k for which this quantity is greater or equal to zero, i.e.,

$$S^+ = \{k \text{ such that } \log_2 |\tilde{X}(k)| - D_1 > 0\} \tag{8}$$

where $\{\cdot\}$ denotes "the set of" and define N^+ as the number of elements in S^+, then a second estimate of D is given by

$$D_2 = D_1 - \frac{1}{N^+}\left[\sum_{k\in S^+} (\log_2 |\tilde{X}(k)| - D_1) - B\right]. \tag{9}$$

The value of D_2 is usually close to the optimal value of D. At this point the assignment process is repeated by substituting D_2 for D in (4) to obtain the second estimate of the bit assignment. Values of k not belonging to the set S^+ are given assignments of zero bits. A heuristic rule is then used to add or subtract the few additional bits, such that the total bit assignment satisfies (6). The rule used is that when additional bits must be subtracted they should come from those frequencies receiving the most bits and when bits must be added they should go to the frequencies receiving only one bit.

The quantization of the transform coefficients can be implemented with either a linear or a logarithmic quantizer characteristic. Both methods were investigated and it was found that the logarithmic characteristic allowed the quantizer to span a larger dynamic range and avoid overflow in regions where the spectral estimate does not match the true spectrum very well (such as in very high Q formant regions). Since the quantity $\log_2 |X(k)|$ is already computed in the process of computing the side information, it can also readily be used for implementing the logarithmic quantizer characteristic as a linear quantizer in the log domain. In Fig. 1 $\Delta(k)$ represents the step-size for a linear quantizer in the linear frequency domain and it is proportional to $\tilde{X}(k)$. In effect it is the ratio $X(k)/\tilde{X}(k)$ that is being quantized. In the log domain this ratio is given by $\log_2 |X(k)| - \log_2 |\tilde{X}(k)|$ and this is precisely how the logarithmic quantizer characteristic can be implemented. The sign bit of the ratio must be sent separately. In practice it takes less processing time to implement the log quantizer characteristic than the linear quantizer characteristic (the sign is still obtained from $X(k)$).

Fig. 7 shows a plot of the quantity $\log_2 |X(k)| - \log_2 |\tilde{X}(k)|$. A difference of 2 in the log domain along the vertical axis corresponds to a factor of 4 in the linear domain (e.g., if the error is +2 the linear estimate is too small by a factor of 4). In a linear quantizer characteristic the input range of a 3-bit optimal Gaussian-Max [10] linear quantizer with a unit step-size is +1.22 to −1.78 on this logarithmic scale. If such a quantizer is used on the ratio $X(k)/\tilde{X}(k)$, many of the data points in Fig. 7 would be clipped, i.e., they would fall outside the range of the quantizer. Alternatively, if a logarithmic characteristic is

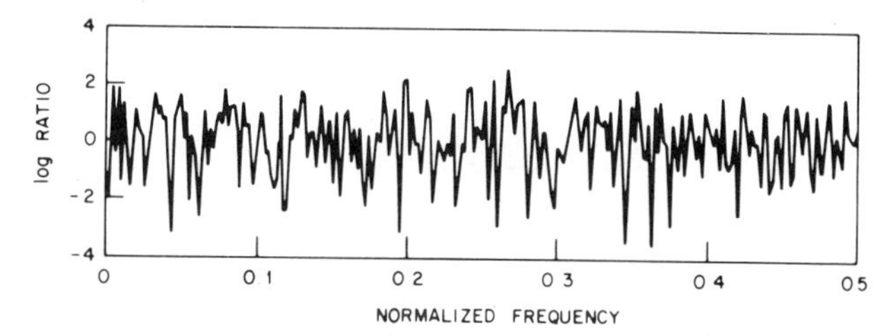

Fig. 7. Input to logarithmic quantizer.

used for a 3-bit quantizer the range is ±2.34 and fewer points are clipped. Raising the step-size by a factor of 1.5 expands the range to ±3.5 in the log domain, encompassing almost all of the input values.

VI. Array Processor Simulations of ATC

Fig. 8 gives a more detailed block diagram of the ATC algorithm that has been described in the above discussion. To review briefly, we see that a block of speech is transformed. The sign bits of the spectral coefficients are saved for later quantization and the base 2 logarithm is taken of the magnitudes. The side information is determined according to Fig. 2 from the pseudocepstrum to model the pitch and formant structure and to produce a log-spectral estimate. The bit assignment is determined from the log-spectral estimate according to the procedure discussed in Section V. The difference between the actual log-spectrum, $\log_2 |X(k)|$, and the estimate, $\log_2 |\tilde{X}(k)|$, is quantized and transmitted. This produces a quantized spectrum at the receiver which is then inverse transformed and buffered to produce the reconstructed speech. All of the major steps first introduced in Fig. 1 have been retained, although some have been slightly modified (such as the implementation of the quantizer in the log domain), while others are shown in more detail (such as the homomorphic side information model).

The above ATC algorithm has been implemented in real-time on the Eclipse AP130. The average computation time for a 257 point block of data is 32.8 ms and therefore, it can operate at a maximum (real-time) input sampling rate of 7.8 kHz. It has been used to demonstrate and study the performance of ATC over dialed-up telephone lines. Two parties can hold a conversation and control which user is coded and the bit rate of the ATC coder.

One artifact in the ATC algorithm is the frame rate noise, caused by the discontinuity between adjacent quantized blocks. This noise is perceived as a regular clicking sound (at low-bit rates) at the block rate. The use of block overlapping and tapering of the blocks with a trapezoidal window prior to quantization helps to subdue this noise to some extent [4]. In this work it was found that tapering of the blocks *after* quantization rather than before leads to a slightly better performance. Substantial overlap, however, (more than 16 samples) reduces the number of total bits allowed for each block and results in a corresponding loss of quality. It also reduces the maximum sampling rate that can be used in real-time simulations. With an overlap of 5 samples or more the noise is no longer perceptible except during sustained tones at low bit rates.

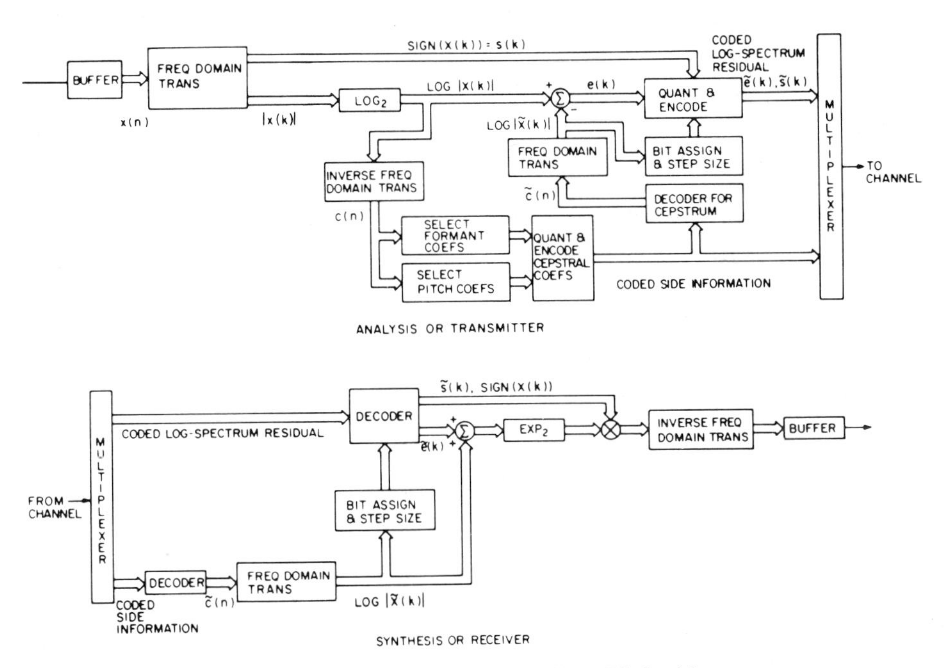

Fig. 8. Block diagram of homomorphic ATC algorithm.

At a bit rate of 16 kbits/s some slight frame rate noise is still perceptible on sustained voiced sounds and the effect is lessened (but still perceivable) when block overlapping was used. At 20 kbits/s the frame rate noise is eliminated with either method. At 9.6 kbits/s the frame-rate noise is more noticeable but the overall quality was still quite good. On standard test utterances the segmental SNR was 13.3 dB, 18.9 dB, and 21.2 dB for the bit rates of 9.6, 16, and 20 kbits/s.

In most situations block sizes of 257 samples were used; however, block sizes of 33, 65, or 129 may also be used. The lower block sizes require a proportionally larger amount of side information if it is transmitted for each block.

The use of the homomorphic analysis does not model the spectra of tones very well since it cannot follow high Q resonances as well as the LPC method (see Fig. 5). Consequently, pure tones tend to get clipped by this version of the ATC algorithm. Theoretically, however, the ATC algorithm should operate well on pure tones if a spectral model appropriate for pure tones is used for side information, i.e., the one currently used is appropriate for speech. This observation is mentioned because some consideration has been given to how well this algorithm would code modem signals. As yet no experiments have been made.

Table I lists the execution times of major parts of the algorithm. In addition to the times listed, there is an additional time for A/D and D/A buffer swapping which cannot be measured, but only inferred. For instance at an 8000 Hz sampling rate the algorithm frequently "runs out of time," even though a block is 32.125 ms and the total time accounted for in Table I is only 31.01 ms. (The reason for this is that some parts of the algorithm take a random amount of time.) Since not all blocks are processed in less than 32.125 ms (but not all blocks take more time either) it seems likely that the buffer swapping must consume about 1 ms. The buffers that are swapped each contain 4 blocks of data or 1028 points. The use of longer buffers would reduce this time further.

VII. Conclusions

In this paper we have considered the use of a homomorphic model for the side information channel of an adaptive transform coder and have implemented this form of the ATC algorithm in real-time on an array processing computer. The use of the homomorphic model leads to a speech-specific version of the ATC algorithm that is in a convenient form for a block-by-block array processing environment in which speed is obtained by means of a high degree of pipeline processing. It represents a good example of how we can modify the form of a signal processing algorithm to fit the characteristics of a particular class of machine architectures, namely, array processing. The resulting homomorphic vocoder driven ATC algorithm performs at least as well as its LPC predecessor [3].

The real-time capability and the Fortran programming environment have provided a highly useful and flexible means for studying and modifying the ATC algorithm in a realistic telephone environment. It has allowed us, for the first time, to conduct telephone conversations over the transform coder and experiment with the algorithm and its parameters in this manner.

Acknowledgment

The authors wish to thank D. Malah for his comments in the course of this work and R. Steele for his extensive editorial comments which have helped to improve this paper.

References

[1] R. Zelinski and P. Noll, "Adaptive transform coding of speech signals," *IEEE Trans. Acoust., Speech, Signal Processing*, vol. ASSP-25, pp. 299-309, Aug. 1977.

[2] ——, "Approaches to adaptive transform speech coding at low-bit

rates," *IEEE Trans. Acoust., Speech, Signal Processing*, vol. ASSP-27, pp. 89-95, Feb. 1979.

[3] R. E. Crochiere and J. M. Tribolet, "Frequency domain techniques for speech coding," *J. Acoust. Soc. Amer.*, vol. 66, pp. 1642-1646, Dec. 1979.

[4] J. M. Tribolet and R. E. Crochiere, "Frequency domain coding of speech," *IEEE Trans. Acoust., Speech, Signal Processing*, vol. ASSP-27, pp. 512-530, Oct. 1979.

[5] *Eclipse AP130 Array Processor Programmer's Reference*, 1st ed., Data General Corp., Sept. 1978.

[6] N. Ahmed and K. R. Rao, *Orthogonal Transforms for Digital Processing*. New York: Springer-Verlag, 1975.

[7] A. V. Oppenheim, "Speech analysis-synthesis system based on homomorphic filtering," *J. Acoust. Soc. Amer.*, vol. 45, pp. 459-462, Feb. 1969.

[8] A. V. Oppenheim and R. W. Schafer, "Homomorphic analysis of speech," *IEEE Trans. Audio Electroacoust.*, vol. AU-16, pp. 221-226, June 1968.

[9] R. Zelinski and P. Noll, "Adaptive block quantization of speech signals," Heinrich-Hertz Inst., Berlin, Germany, Tech. Rep. 181, 1975.

[10] J. Max, "Quantizing for minimum distortion," *IRE Trans. Inform. Theory*, vol. 6, pp. 16-21, Mar. 1960.

Performance of Transform and Subband Coding Systems Combined with Harmonic Scaling of Speech

DAVID MALAH, MEMBER, IEEE, RONALD E. CROCHIERE, SENIOR MEMBER, IEEE, AND RICHARD V. COX, MEMBER, IEEE

***Abstract*—In this study an approach for improving the performance of waveform coders, based on coding a frequency scaled speech signal, is examined and subjectively evaluated for specific subband and transform coding systems. The recently developed simple and efficient time-domain harmonic scaling (TDHS) algorithms are used to frequency scale the speech signal. The underlying frequency-domain model of the pitch-adaptive TDHS algorithms provides insight and guidelines for their use in this application, as outlined in this work. The subjective evaluation is based on an A-B comparison test involving 12 listeners and shows a meaningful improvement in quality for the waveform coders used at low bit rates. In particular, subband coding (SBC) combined with TDHS (SBC/HS) at 9.6 kbits/s was found to provide a quality equivalent to that of SBC alone at 16 kbits/s, i.e., a bit-rate advantage of about 7 kbits/s was realized. For the speech specific adaptive transform coder (ATC) used, the combined system (ATC/HS) achieves a bit-rate advantage of 4 kbits/s at 7.2 kbits/s. The SBC/HS system emerges as a particularly attractive method for speech encoding at the data rate of 9.6 kbits/s since its quality is comparable to that of ATC/HS (or SBC at 16 kbits/s). Yet, its complexity is lower than ATC and the system is amenable to real-time hardware implementation using current technology.**

I. Introduction

Waveform coding techniques attempt to reproduce encoded signals at lower bit rates than PCM encoding by utilizing the temporal and spectral properties of the signal. Speech-specific coders have achieved sizeable reductions in bit rate by incorporating in the coder design known properties of both short-time and long-time (pitch period) correlations which are related to spectral envelope (formants) and fine structure (pitch harmonics) of speech, respectively.

However, because they attempt to replicate the speech waveform, even the more complex forms of waveform coders, such as adaptive-predictive coders (APC) [1], [2] and adaptive transform coders (ATC) [3], [4], require bit rates typically at or above 9.6 kbits/s in order to have acceptable communication quality [5]. This range of bit rates is several times higher than the bit-rate range of vocoders which typically encode speech at 2.4 kbits/s and below [5]. Vocoders are based on a speech production model and achieve low bit rates by analyzing and then resynthesizing a speech signal which *sounds* like the original signal but does not necessarily replicate the original signal waveform. Because of this tight adherence to a speech production model, vocoders are found to be far less robust than waveform coders and lack naturalness in quality. While naturalness can be improved by adding more bits to better represent the excitation signal [6], the high sensitivity of vocoders to voicing decision errors, environmental conditions (background noise, simultaneous speakers), and channel errors make vocoders unattractive in many applications.

In this work we examine the performance of a speech encoding system which appears to meaningfully extend the range of waveform coder operation at the low bit-rate end, with only a modest loss of robustness. In the system under consideration the chosen waveform coder is used to encode a frequency scaled (compressed) speech signal which is rescaled (expanded) at the receiver—following waveform decoding. The frequency scaling operations are based on reducing (for compression) or increasing (for expansion) the interharmonic spectral gaps of the pitch by a factor of up to three using frequency shifting of the pitch harmonics. The actual scaling operations are done, however, in the time domain by means of the recently developed time-domain harmonic scaling (TDHS) algorithms [7], [8]. These algorithms are pitch-adaptive and perform a time-varying weighting of adjacent speech segments, with an appropriate weighting (window) function, in a way which assures continuity. Once the pitch period is known, the remaining operations required in the TDHS algorithms are typically only one multiplication and two additions per output sample. Furthermore, since the frequency compressed signal is decimated by a factor equal to the frequency compression factor, the computational load on the waveform coder is usually reduced by the same factor. A general block diagram which illustrates the way the waveform coders used in this study were combined with TDHS is given in Fig. 1.

The combination of a waveform coder with this particular method of frequency scaling can be viewed as an approach for exploiting the harmonic structure (pitch) of voiced speech signals in a different way than is currently done in waveform coders (e.g., a pitch prediction loop around the quantizer [5] or pitch dependent bit allocation [4]). Another point of view is to consider the combined system as a form of "soft vocoding," since unlike waveform coders the reconstructed

Manuscript received July 9, 1980; revised October 10, 1980.
D. Malah is with the Department of Acoustics Research, Bell Laboratories, Murray Hill, NJ 07974, on leave from the Department of Electrical Engineering, Technion–Israel Institute of Technology, Haifa, Israel.
R. E. Crochiere and R. V. Cox are with the Department of Acoustics Research, Bell Laboratories, Murray Hill, NJ 07974.

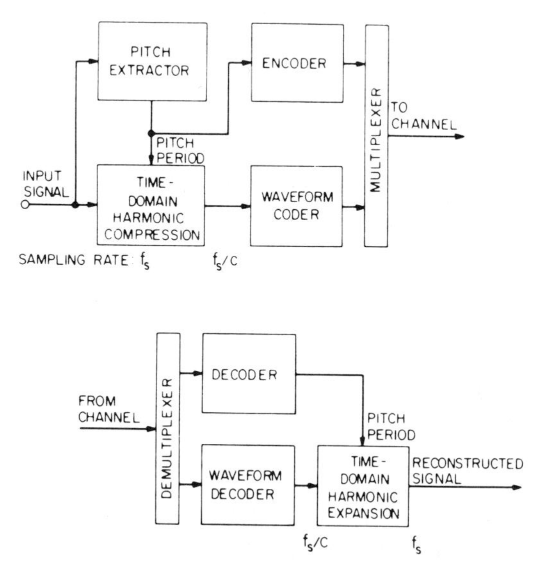

Fig. 1. General block diagram of combined TDHS–waveform coding system. C is the compression factor.

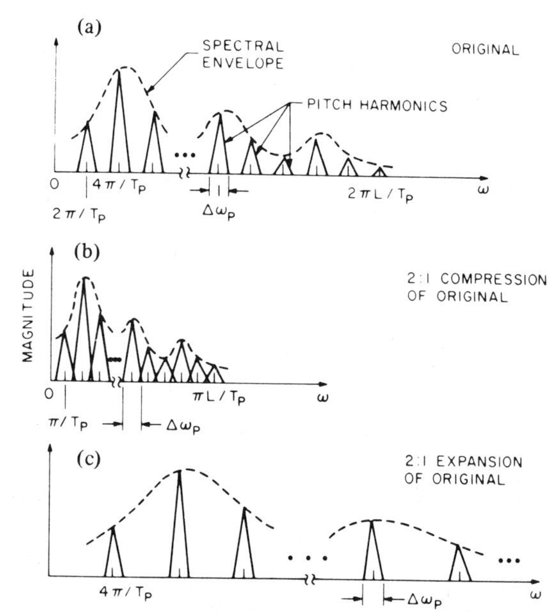

Fig. 2. Schematic spectral representation of (a) input voiced speech, (b) frequency compressed signal by pitch harmonic shifting, (c) frequency expanded signal. T_p is the pitch period duration and $\Delta\omega_p$ is the spectral width of a pitch harmonic.

signal may not replicate the input signal due to interharmonic aliasing caused by the frequency compression method used. However, unlike vocoders, the combined system is found to be robust and has a natural sounding quality. This robustness stems from the avoidance of voicing decisions (voiced-unvoiced), from the way the TDHS algorithms use pitch information, and the surprising ability of the system to encode speech signals of several simultaneous speakers (to be reported in the sequel).

Informal subjective tests suggest that by combining TDHS with waveform coders (particularly those which do not exploit pitch information), a reduction in bit rate by up to the scaling factor used (typically two) can be obtained. The performance advantage is found to be gained primarily at the low end of the bit-rate range of the coder, i.e., the performance "knee" (the bit rate below which the quality of the encoded signal starts falling rapidly) is pushed to lower bit rates. The more efficiently the waveform coder exploits pitch information the smaller is the expected improvement. However, it appears from the results of this work that even with a waveform coder which exploits pitch and with a scaling factor of two, the combined system achieves a meaningful improvement at the lower end of the bit-rate range of the waveform coder.

There is much interest to date in robust speech encoders which are capable of operating at or below 9.6 kbits/s with adequate quality to allow transmission of voice on digital data links [5]. For this reason, we have chosen in this study to examine the combination of TDHS with subband coding (SBC) [4], [9] and with adaptive transform coding (ATC) [3], [4]. The SBC technique efficiently exploits the formant structure in the speech spectrum, whereas the speech specific ATC technique takes advantage of both formant and pitch information in the way the bits are allocated. These coders offer high communications quality at 16 kbits/s and appear to have the potential of encoding speech at 9.6 kbits/s, or below, with adequate quality if combined with TDHS. Indeed, from informal listening and from an A-B comparison test, SBC combined with TDHS (SBC/HS) at 9.6 kbits/s was found to provide a quality equivalent to that of SBC alone at 16 kbits/s, i.e., a bit-rate advantage of about 7 kbits/s was realized. As could be expected, since pitch information is exploited by the particular ATC system used, the improvement gained by ATC/HS is smaller than for SBC/HS but still a 4 kbits/s bit-rate advantage was obtained at 7.2 kbits/s. The above results are in agreement with an earlier, less exhaustive, study in which a simple CVSD (continuously variable slope delta modulation) coder was combined with TDHS [10] and was found to be a useful approach for encoding at bit rates below the performance "knee" of the CVSD coder used.

In the following sections we briefly describe the TDHS algorithms and the waveform coders used in this study. We then discuss issues involved in combining TDHS with waveform coders and present results of subjective tests performed for comparing the various coding schemes.

II. Time-Domain Harmonic Scaling

The time-domain harmonic scaling (TDHS) algorithms [7], [8] provide simple and efficient means for frequency scaling of speech signals. The development of the algorithms is based on a frequency domain model of voiced speech as schematically shown in Fig. 2(a). The method used for frequency scale compression is to frequency shift the pitch harmonics to lower frequencies as shown in Fig. 2(b). The presence of relatively wide interharmonic gaps in the original spectral representation [Fig. 2(a)] allows for compression factors of two to three with tolerable interharmonic aliasing. Similarly, frequency scale expansion is achieved by shifting the pitch harmonics upwards as shown in Fig. 2(c). This approach for frequency scaling is different from the earlier phase-vocoder frequency division technique [11] which in addition to relocating the pitch harmonics also attempts to scale the width $\Delta\omega_p$ of each pitch "tooth" by the frequency scaling factor. The advantage of the frequency shifting approach is that it can be performed simply and efficiently in the time domain, if pitch information is known, as explained below.

In principal, the frequency shifting operations can be performed by using a filter bank analysis to isolate the different pitch harmonics and then modulate (frequency shift) each harmonic to its designated new location. A convenient mathematical framework for a uniform filter bank analysis, modification, and synthesis is provided by the short-time Fourier transform (STFT) [12]-[14]. In this framework the filter bank analysis can be performed by frequency shifting each desired frequency band centered at $\omega = \omega_k$ to baseband, using complex modulation, filtering the baseband signal with the prototype lowpass filter $h(t)$ of the filter bank, and then remodulating back to the appropriate frequency band. Hence, if no modification is performed, the reconstruction $y(t)$ of the input signal $x(t)$ is given by

$$y(t) = \sum_{k=-L}^{L} X(\omega_k, t)\, e^{j\omega_k t} \tag{1}$$

where $X(\omega_k, t)$ is the STFT of $x(t)$ evaluated at $\omega = \omega_k$ [11], i.e.,

$$X(\omega_k, t) = \int_{-\infty}^{\infty} x(\tau)\, h(t-\tau)\, e^{-j\omega_k \tau}\, d\tau \tag{2}$$

and for a uniform filter bank the center frequencies satisfy $\omega_k = k\Delta\omega$, $k = 0, \pm 1, \cdots, \pm L$ (where $\Delta\omega$ is the bandwidth of each band).

If the pitch period, T_p, is known, the center frequencies of the filter bank can be chosen to coincide with the pitch harmonics (i.e., $\Delta\omega = 2\pi/T_p$). In that case frequency scaling by a factor q ($q < 1$ for compression, $q > 1$ for expansion), by means of frequency shifting the pitch harmonics, results in an output signal $y^q(t)$ given by

$$y^q(t) = \sum_{k=-L}^{L} X(\omega_k, t)\, e^{jq\omega_k t}, \tag{3}$$

i.e., the kth pitch harmonic, which was originally located at ω_k, is shifted to $q\omega_k$.

Substituting (2) into (3), interchanging the order of summation and integration, and changing the variable $(t - \tau)$ to τ, one obtains [8]

$$y^q(t) = \int_{-\infty}^{\infty} x(t-\tau)\, h(\tau)\, K_N((q-1)\,t + \tau)\, d\tau \tag{4}$$

where $K_N(t)$ is a kernel function given by [8]

$$K_N(t) = \sum_{k=-L}^{L} e^{-j\omega_k t} = \frac{\sin(N\pi t/T_p)}{\sin(\pi t/T_p)} \tag{5}$$

with $N = 2L + 1$ and T_p being the pitch period of the input speech signal $x(t)$. This kernel function is periodic and has $N - 1$ zeroes in each period T_p. To be useful in a digital implementation (4) must be discretized. It turns out that if q is assumed to be rational ($q = \mu/\delta$, where μ and δ are relatively prime integers); the output signal is sampled at a rate corresponding to its bandwidth; and the kernel function $K_N(t)$ is sampled at its zeroes, the following discrete-time approximation to (4) results [8]

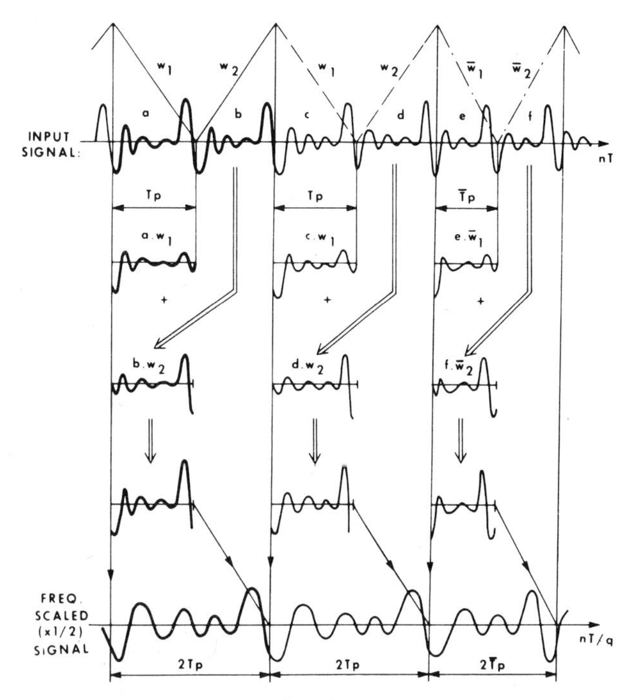

Fig. 3. Illustration of the time-domain operations for 2:1 compression ($q = \frac{1}{2}$) using a triangular window. T_p and $\overline{T}_p$ are two different pitch period durations.

$$\hat{y}^q(nT/q) = \sum_{l=-\infty}^{\infty} x(nT - lT_p)\, \hat{h}(lT_p - n(\mu - \delta)\, T'). \tag{6}$$

In (6), T is the Nyquist interval (or shorter) of the input signal $x(t)$ (so that $T_p = NT$), $T' = T/\mu$ (recall, $q = \mu/\delta$), and $\hat{h}(t) = h(t)\, T_p$.

Since speech signals are not stationary, the prototype filter impulse response (or window function) $h(t)$ is chosen to be of finite duration (FIR), avoiding the weighting of speech segments which are not within the same quasi-stationary interval. If we assume the finite duration of $h(t)$ to be mT_p, m an integer, then (6) appears to involve only m multiplications and $(m - 1)$ additions. However, due to constraints that $h(t)$ should satisfy [7], the computations can be rearranged into $(m - 1)$ multiplications and m additions.

Finally the TDHS algorithms presented in [7], [8] (one for compression and one for expansion) are obtained from (6) by using an FIR filter and letting $q = 1/C < 1$ for compression and $q = S > 1$ for expansion.

In the current work, as in most of our previous work with TDHS [7], [10], the simple triangular window (corresponding to $m = 2$) is used. Other suitable window functions can be applied [7], [8] but the triangular window is particularly convenient and simple to implement. It also provides adequate results with a scaling factor of two which is used in this work.

The way the compression algorithm is applied for 2:1 compression ($q = \frac{1}{2}$), using a triangular window, is shown in Fig. 3. It is seen that each two adjacent speech segments of pitch period duration are weighted and overlapped to provide one

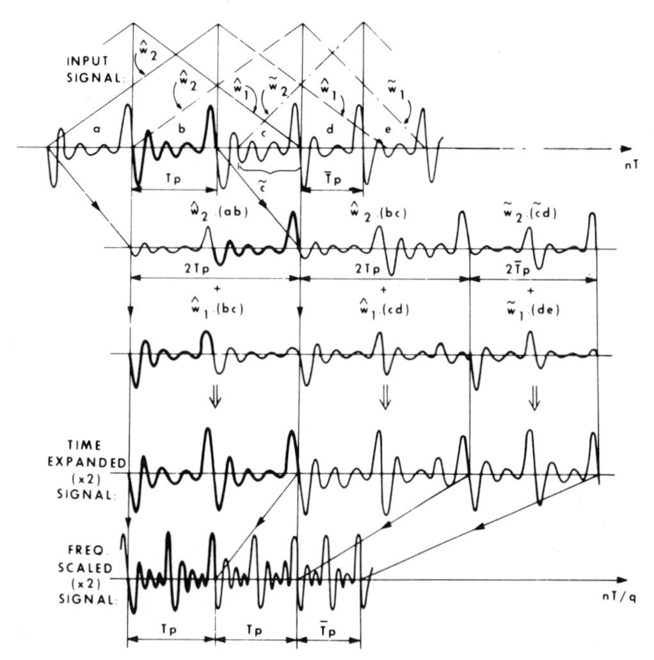

Fig. 4. Illustration of the time-domain operations for 2:1 expansion ($q = 2$) using a triangular window.

output segment of pitch period duration. If the output segments are concatenated and played out at the original sampling rate ($1/T$) a time compressed signal results. If, however, each segment is output at the decimated rate (i.e., with a sampling interval of $T/q = 2T$) a frequency compressed signal, which occupies the original time slot, results–as shown at the bottom of Fig. 3. Note that due to the way the input signal is weighted and due to the properties of the window, the last point in each output segment and the first point in the next segment are actually adjacent points in the original waveform. Hence, the output waveform suffers no discontinuities at segment boundaries. The procedure can, therefore, be viewed as an extension of earlier time-domain methods which are based on splicing the speech waveform, i.e., replacing each two speech segments with one segment. Such a crude approach corresponds to actually using a rectangular window instead of the triangular window used here. The deficiency of the rectangular window is quite clear from both time-domain (discontinuities) and frequency-domain (high sidelobes) considerations [7].

The expansion process for $q = 2$ is illustrated in Fig. 4, again using a triangular window. As before, adjacent speech segments are weighted to produce an output segment. However, this time for each input segment of pitch period duration, an output speech segment of two pitch periods duration is generated. This way a time-expanded signal is obtained if the original sampling rate is retained. If the output sampling interval is $T/q = T/2$, a frequency expanded signal is obtained. Obviously, if the expansion is applied to the compressed signal the reconstructed signal occupies the original frequency and time spans. Again, this procedure can be considered as an extension of the simple reiteration technique which corresponds to using a rectangular window instead of a better window function, such as the triangular window used here.

III. Subband Coding

Subband coding (SBC) is a waveform coding technique in which the speech band is partitioned into typically four to eight subbands by bandpass filters [5], [9], [15]-[17]. Each subband is effectively low pass-translated to *dc*, sampled at its Nyquist rate (twice the width of the subband), and then digitally encoded using adaptive PCM (APCM) as shown in Fig. 5. By carefully selecting the number of bits/sample used for encoding the subbands, each band can be preferentially coded to give a maximum overall perceptual quality with a minimum overall bit rate. The step-sizes in each band adapt independently in proportion to the rms speech level in their respective bands. In this way subband coding can take advantage of the properties of temporal nonstationarity, spectral formant structure, and auditory masking in speech production and perception.

The subband coder used in this experiment is based on the octave band approach of Barabell and Crochiere [16]. It utilizes quadrature mirror filters (QMF) in a "tree structure" to achieve an efficient filter bank framework [17]. The actual filter designs are selected from the family of designs given by Johnston [18]. The first QMF pair splits the initial speech band into two equally spaced bands, such that the aliasing or "leakage" of energy between bands is canceled in the reconstruction process. A second QMF pair in the "tree structure" then subdivides the lower band into two bands in a similar manner. This process is continued in the "tree structure" until the desired number of bands is obtained as shown in Fig. 6 for a four-band split. Tables I and II show the parameters of the resulting designs that are obtained from this approach for five-band and four-band systems at 16 and 9.6 kbits/s, respectively. The bit allocations for the combined systems will be detailed in a later section. Note that the initial sampling rates for the above designs are 6400 Hz and 5760 Hz, respectively. Since, typically, the input and output sampling rates are in the range of 8 to 10 kHz, digital decimation-interpolation techniques for rate conversion need to be applied.

IV. Adaptive Transform Coding

Adaptive transform coding of speech is a frequency domain technique in which a high resolution transform is applied to the speech signal on a block by block basis [3], [4]. The resulting transform coefficients (frequency components) are quantized using both step-size adaptation and dynamic bit assignment for each transform coefficient, i.e., the number of bits used to encode each transform coefficient is dynamically varied from block to block. The adaptive quantization is based on a smoothed spectral estimate of each speech block. This estimate is parametrized and encoded for transmission as "side information." A general block diagram of an ATC system is shown in Fig. 7. In this work a "speech-specific" coder was used similar to that of Tribolet and Crochiere [4]. It uses, however, the homomorphic model for parametrizing the smoothed spectral estimate of each block, as reported by Cox and Crochiere [19] instead of the LPC model used in [4]. Yet, as in [4], the spectral model also includes pitch information

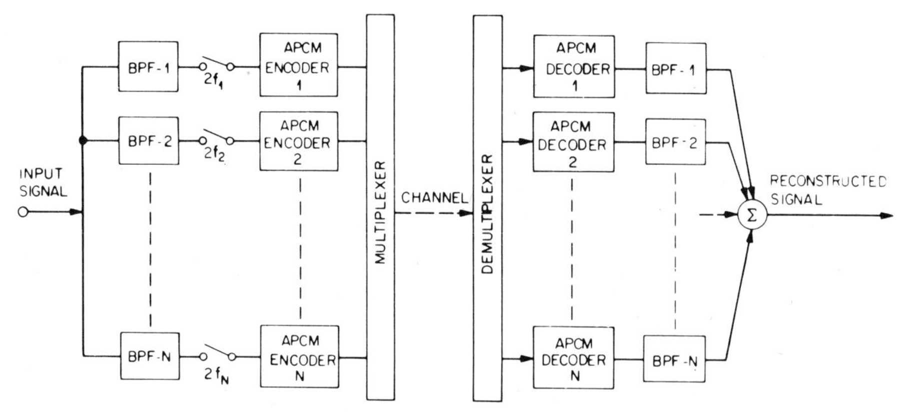

Fig. 5. General block diagram of a subband coding (SBC) system.

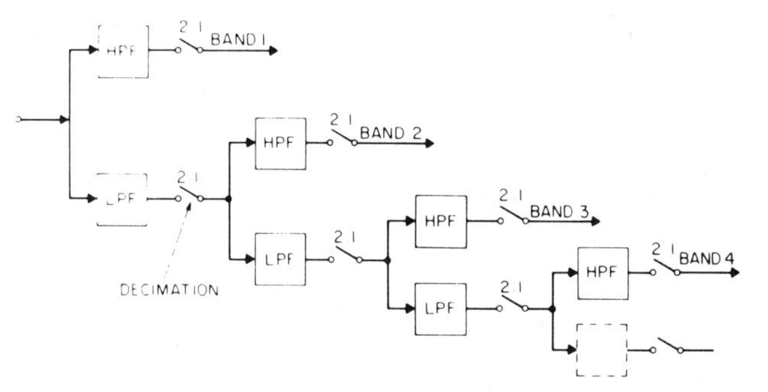

Fig. 6. Illustration of a four-band split using quadrature mirror filters (QMF).

TABLE I
FIVE-BAND SUBBAND CODER PARAMETERS FOR 16 kbits/s ENCODING*

Band No.	1	2	3	4	5
Freq. Range [Hz]	3200-1600	1600-800	800-400	400-200	200-100
No. of Taps in QMF Splitting Filter	32	16	16	16	8
Bit Allocation	2	2	4	5	5

*Input Sampling Rate: 6400 Hz

TABLE II
FOUR-BAND SUBBAND CODER PARAMETERS FOR 9.6 kbits/s ENCODING*

Band No.	1	2	3	4
Freq. Range [Hz]	2880-1440	1440-720	720-360	360-180
No. of Taps in QMF Splitting Filter	32	16	16	8
Bit Allocation	1-1/3	2	2	3

*Input Sampling Rate: 5760 Hz

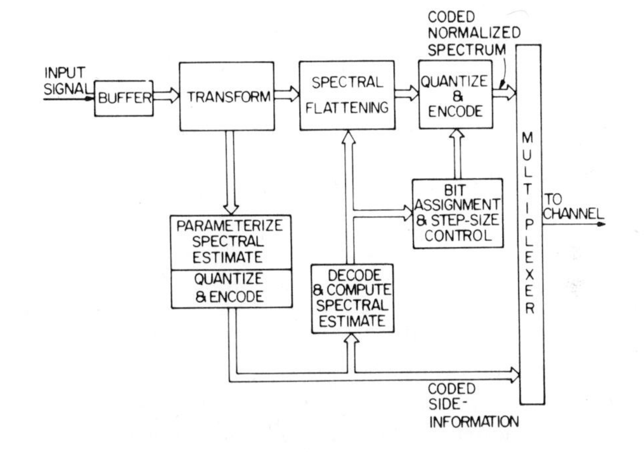

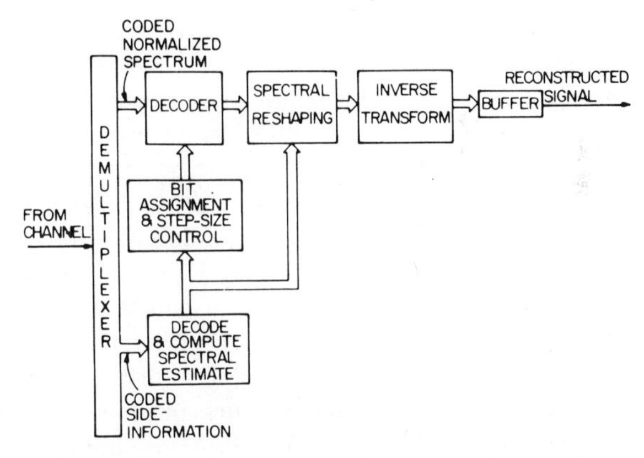

Fig. 7. General block diagram of an adaptive transform coding system (ATC).

which results in a more efficient bit assignment by assigning more bits to represent the spectral peaks at pitch harmonics.

Fig. 8 describes in more detail the system used to simulate adaptive transform coding of speech [19]. The transform used is a symmetric discrete Fourier transform, a close relative of the discrete cosine transform. It is obtained by forming a symmetric sequence of length $2M$ from a sequence of $M + 1$ time samples

$$y(n) = \begin{cases} x(n), & n = 0, 1, \cdots M \\ x(2M - n), & n = M + 1, \cdots, 2M - 1 \end{cases} \quad (7)$$

The $2M$ point DFT of this input sequence is taken via an FFT routine on an array processor. Since the input to the DFT is both real and symmetric, the output will also be both real and symmetric. In this way only $M + 1$ values are needed to represent the $2M$ values in the transform domain.

The overall operation of the coder proceeds as follows. A block of $M + 1 = 257$ samples of speech (at an 8 kHz sampling

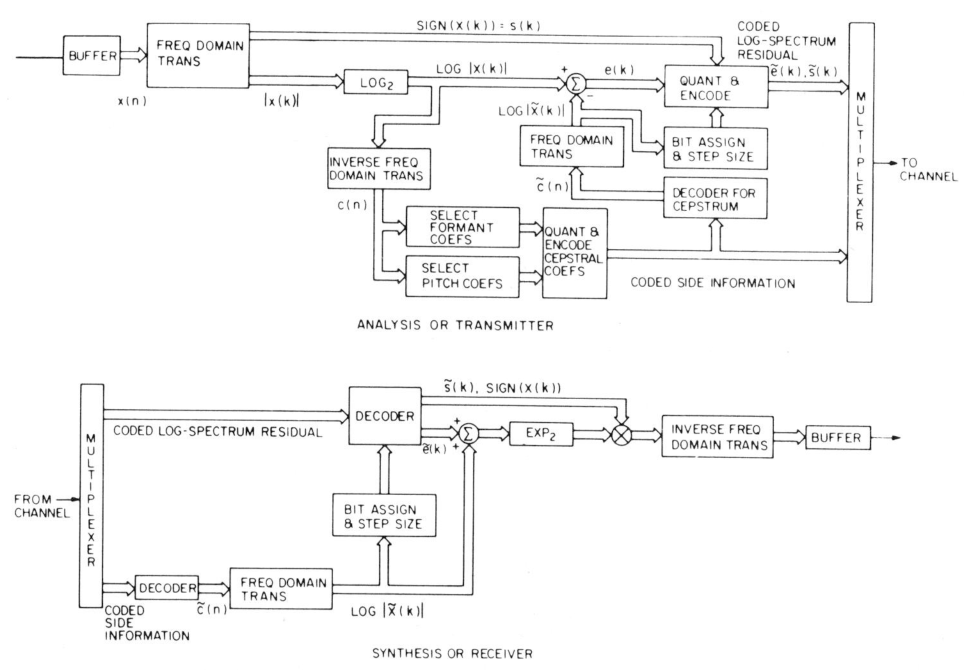

Fig. 8. Block diagram of the homomorphic-based ATC system used in the simulations with adaptive log-PCM quantization of the transform coefficients.

rate) is assembled in the input buffer. No analysis window is applied to the block; however, a trapezoidal synthesis window is used later at the output. Consequently, successive blocks have a small amount of overlap, typically about ten samples. The block is then transformed by the symmetric DFT (SDFT) described above yielding a real symmetric spectrum $X(k)$. The sign of each frequency domain sample $X(k)$ is saved for encoding and the base two logarithm of the magnitude of each sample is taken, yielding log $|X(k)|$. The log magnitude spectrum is used for log-PCM quantization (to be described later). It is also used to generate a "pseudocepstrum," $c(n)$, by inverse SDFT transformation. This pseudocepstrum can be used to model the combined pitch and formant structure of the speech segment with relatively few parameters. The first 8 to 14 cepstral coefficients are used to generate the smooth spectral envelope estimate and the peaks further from the origin in the cepstrum provide the pitch information. This information is quantized, coded, and transmitted as side information, and also decoded to provide a rough estimate of the cepstrum, $\tilde{c}(n)$. Taking the SDFT of $\tilde{c}(n)$ then gives an estimate of the log magnitude spectrum, log $|\tilde{X}(k)|$ of the speech segment. This estimate is used for the bit assignment in the ATC algorithm. It is also subtracted from the actual log magnitude spectrum $X(k)$ to provide a residual log magnitude spectrum $e(k)$. This residual is quantized and encoded according to the bit assignment (i.e., linear quantization of the difference, $e(k)$, in the log domain is equivalent to an adaptive step-size PCM of the transform coefficients using a quantizer with a logarithmic characteristic). At the receiver the quantized residual is decoded, the estimated log magnitude spectrum is added back and the frequency samples are exponentiated back to the linear frequency domain. Their signs are then reassociated and an inverse SDFT is taken.

TABLE III
BIT RATES FOR SIMULATED ATC (ALONE) SYSTEM

Overall Bit-Rate [kb/s]	Spectral-Residual Rate [kb/s]	Side-Information Rate [kb/s]	No. of Cepstral Coeff.
16	12.6	3.4	14
9.6	7	2.6	12
7.2	5	2.2	8

The resulting output is then windowed with a trapezoidal window and the overlap added to the previous output block.

The bit rates simulated (for ATC alone) and the breakdown of the overall bit rate into spectral residual and side information transmission is detailed in Table III. The corresponding information for the combined ATC/HS system is given in the next section.

V. COMBINING TDHS WITH WAVEFORM CODING

In the previous sections the TDHS algorithms and the waveform coding techniques were described. In this section we discuss issues involved in combining TDHS with waveform coding, with particular emphasis on the operation of the TDHS algorithms and the simulations performed.

The combination of TDHS with waveform coding can be done in two ways. The first is shown in Fig. 1, according to which the original pitch-period data is encoded and transmitted to the receiver. A second way is to extract the pitch information at the receiver from the decoded frequency-compressed signal. This approach was used in our earlier work which involved combining CVSD with TDHS [10] and it was chosen to keep the simplicity of CVSD transmission. However, since the number of pitch periods per unit time, in the frequency-compressed signal, is reduced by a factor equal to the scaling factor used (see Fig. 3), the task of pitch extraction is more difficult. This results in a somewhat higher degradation in the reconstructed signal. In this work, since relatively complex coders are used anyway, the configuration of Fig. 1 was used.

It is evident from the earlier discussion on the TDHS algorithms that the most fundamental and crucial operation is the pitch extraction. Many pitch extraction algorithms were reported in the literature [20] and most of them are probably adequate for this application. However, the choice of the most suitable one would typically depend on its amenability to hardware implementation, its robustness to environmental conditions, the computational load involved, etc. To help in the choice of a suitable pitch extractor the following observations are made.

1) A simple mathematical analysis performed in [7], for a stationary periodic signal, shows that even in the presence of a pitch-period error the reconstructed signal has the correct pitch if the relative error ϵ in the fundamental Frequency $F_p = 1/T_p$ is limited according to

$$\epsilon \triangleq \Delta F/F_p < 1/(2MC) \tag{8}$$

where M is the number of harmonics desired to be exactly reconstructed, ΔF is the error in estimating F_p, and C is the frequency compression factor.

Since the number of harmonics present in a given frequency band is dependent on the pitch period itself we have derived from (8) the following upper limit on the pitch-period error ΔT_p.

$$\Delta T_p < 1/(2F_M C) \tag{9}$$

where F_M is the highest frequency in the band for which the voiced speech signal is to be reconstructed exactly (theoretically). Typically, it is required that F_M be at least 1.5 to 2 kHz. With $C = 2$ this results in a permitted pitch-period error of 0.125 to 0.16 ms. Practically, however, even errors of twice as high are tolerated with a modest increase in the output signal degradation. This range of allowed pitch-period error usually complies with the performance of most known pitch extractors.

2) The difficult task of voiced-unvoiced decision is practically avoided since any random value (within a limited range) provided by the pitch extractor at unvoiced segments can be used without noticeable perceptual effects. This is found to be true in spite of the spectral distortion caused by the TDHS algorithms to unvoiced signals due to their noiselike characteristics.

3) The effect of a double pitch-period decision by the pitch extractor is far less detrimental than for vocoders. Actually, if it does not occur at voiced-unvoiced transitions its effect is quite small. This can be understood from the underlying frequency domain model since using a double pitch-period corresponds to performing the filter bank analysis with twice as many filters. This means that only every other filter contains a pitch harmonic. However, since the filters are now narrower the tolerated pitch measurement error is now also reduced.

In the simulations performed in the course of this study, a homomorphic pitch extractor based on a cepstral analysis [21] of the speech signal was used. This rather complex pitch extractor was applied because of its high performance and the availability of an array processor, on the computer system used for the simulations, which enables pitch extraction—in almost real time. In our earlier work [10] we have used a simpler time-domain autocorrelation type pitch extractor [22] which was found to be adequate for that application.

Since the simulations were not performed in real time, we have found it convenient to first prepare the pitch data of the tested utterance (up to six sentences) and then use these data for experimenting with the TDHS algorithms.

The pitch data file is prepared using a data window of 32 ms (256 samples) and a fixed update of 8 ms (64 samples at the 8 kHz sampling rate used). No smoothing of the raw pitch data was needed. Since the TDHS algorithms are pitch-adaptive it is extremely important that the pitch data pointer be carefully matched to the input signal pointer which is in itself pitch dependent. In particular, since the compressed signal has less pitch periods per time unit and may also be delayed this matching of pitch and signal pointers is particularly important at the reconstruction phase (expansion).

As stated earlier, the TDHS algorithms were applied in this work with the simple triangular window. If a scaling factor which is higher than two is attempted, the choice of more complex window functions [7] can be advantageous. Alternatively, the application of the TDHS algorithms in cascade using lower scaling factors in each stage (e.g. 1.5 and 2 for achieving a scaling factor of 3) was also found to be effective. The use of an adaptive window[1] as suggested in [8] improves somewhat the resulting signal at transitions and also can reduce the spectral distortion in unvoiced segments. However, when combined with waveform coding this improvement appears to be masked.

Although not part of this study, we have examined the effect of some environmental conditions on the TDHS algorithms performance. In particular, we have found that in spite of being a pitch-dependent system, the TDHS algorithms were able to perform exceptionally well on speech of several simultaneous speakers (the test performed used speech of three speakers). This result could perhaps be attributed to the tracking of the dominant speaker at each short-time interval, and to the masking properties of the ear. The TDHS system was also found to be relatively robust to noisy or degraded speech (e.g. room reverberations) and performed adequately, as long as the pitch detector did not break down. With a cepstral pitch detector the system was able to operate down to a signal-to-noise ratio of 0 dB. At high noise levels, however, structuring (coloration) of the noise (due to the pitch synchronous processing) can be perceived.

Another issue of interest is the effect of channel errors. Preliminary simulations indicate that the TDHS system is quite insensitive to channel errors in the pitch data up to a bit-error-rate of 10^{-2}. At higher error rates, error protection of the

[1] E.g., a trapezoidal window with variable slopes such that it becomes a triangular window at sustained voiced segments, whereas at unvoiced or transition (voiced-unvoiced) segment a rectangular window with a small amount of tapering is used.

TABLE IV
RATES AND BIT ALLOCATIONS FOR SIMULATED COMBINED SBC/HS SYSTEMS

SBC/HS Rate [kb/s]	SBC Rate [kb/s]	Pitch Transmission [kb/s]	Bit Allocation Bands: 1	2	3	4	5
16	31	0.5	5	5	5	5	5
9.6	18.2	0.5	2	3	5	5	5
7.2	13.2	0.6	2	2	4	5	-

TABLE V
BIT RATES FOR SIMULATED COMBINED ATC/HS SYSTEMS

ATC/HS Rate [kb/s]	ATC Rate [kb/s]	Additional Pitch Transmission [kb/s]	Spectral-Residual Rate [kb/s]	Side-Information Rate [kb/s]
16	31	0.5	27.1	3.9
9.6	18.2	0.5	14.8	3.4
7.2	·13.9	0.25	10.5	3.4
4.8	9.1	0.25	6.5	2.6

pitch data might be necessary. If the additional bits needed for pitch protection cannot be afforded, the reextraction of pitch at the receiver can be considered.

In the remainder of this section we will discuss the simulations of TDHS with the particular waveform coders used in this study, namely SBC and ATC, and the issue of exploiting pitch information within the waveform coder.

Again, since the simulations were not performed in real time it was convenient first to frequency scale (or time compress) the whole tested utterance, and then to encode the compressed signal by the waveform coders at different bit rates. Furthermore, since we used available waveform coding programs, it was most convenient to assume that the input sampling rate is the original 8 kHz and consider the input utterance to the waveform coders as a time compressed signal (see Fig. 3). Thus, to simulate, for example, the combined TDHS-SBC system at 9.6 kbits/s, the time-compressed (2 : 1) signal was encoded by the subband coder at the rate of 18.2 kbits/s which is equivalent to a 9.1 kbits/s transmission rate. Adding 500 kbits/s for the pitch information results in the overall desired bit-rate of 9.6 kbits/s. Table IV shows the different bit rates simulated with the combined SBC/HS system and the number of bits allocated to each band. (For SBC alone, the bit rates and bit allocations shown in Tables I and II were used.) Table V gives the corresponding information for the combined ATC/HS system.

Finally, the issue of exploiting pitch information within the waveform coder is discussed. Since TDHS already exploits pitch to reduce the redundancy present in speech signals, the effectiveness of using pitch information within the waveform coder itself is quite reduced. For systems which use a pitch prediction loop this can be explained by the fact that following the compression operation the correlation between adjacent pitch periods is reduced, since each pitch period in the compressed signal represents two pitch periods of the original signal. For systems which exploit the harmonic structure in the frequency domain (like ATC), the relative broadening of the pitch teeth reduces the gain obtained through the dynamic bit allocation (which assigns more bits at the pitch teeth and less bits at the interharmonic gaps). Yet, since TDHS requires pitch extraction due to the harmonic structure, the use of pitch information within the waveform coder can be useful if it is also found to be economic. In this study, with SBC and ATC, we found it to be uneconomical to use a pitch loop within the SBC due to its complexity [16] and the extremely small improvement. However, we continued to use pitch information within the ATC, since it is naturally embedded in the homomorphic model used.

VI. SUBJECTIVE EVALUATION OF THE COMBINED SYSTEMS

Due to the nature of the TDHS algorithms, which do not necessarily replicate the speech waveform, we could not use signal to noise measurements as performance indicators for the combined systems. To evaluate the performance of the combined systems an informal A-B comparison test for quality, similar to the one reported in [23], was performed. In the test, two groups of six listeners each compared the quality of the different coding systems by listening to pairs of sentences processed by the different systems. The preparation of the source material was done from six sentences (three sentences spoken by males and three sentences by females) and the material was presented to the listeners in a randomized order. Each of the 14 systems in the test was compared against all other systems, using both A-B and B-A comparisons (in random order). In all, each listener compared 182 pairs, and a total of 24 comparisons was done for each pair of coding systems. From the test results, the total number of "votes" given to each system (i.e., the number of times each system was preferred) was computed and used for rank ordering the different systems. The maximum number of votes that any coding system could get was 312. This number was used to present the results on a percentage basis. Since two different waveform coding systems were combined with TDHS, we found it useful to partition the test results and show separately the performance of the two different combined systems. Fig. 9 presents the results obtained for SBC alone in comparison to SBC combined with TDHS (denoted by SBC/HS), for bit rates of 16, 9.6, and 7.2 kbits/s (SBC alone at 7.2 kbits/s was not included due to its extremely low quality at this rate), and in comparison to the original signal and the reconstructed signal by TDHS only (with a scaling factor of two). The corresponding results for ATC and ATC/HS are shown in Fig. 10. Here the combined system was also operated at 4.8 kbits/s. To show how the two different systems relate to each other (subjectively) we present in Fig. 11 the overall test results. Before we turn to a discussion of the results we would like to mention that we have noted a certain variability in individual results due to the type of speaker (male or female). Since the test was balanced, we could extract the separate results for male and female speakers and these are presented in Figs. 12 and 13, respectively, for comparison with Fig. 11.

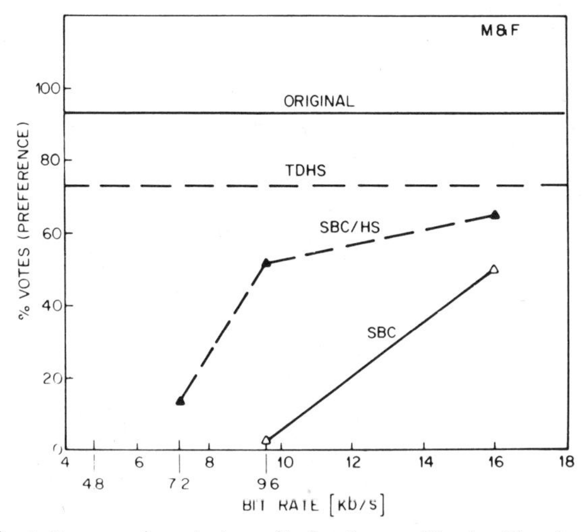

Fig. 9. A-B comparison test results for the combined subband coding system (SBC/HS) in comparison to SBC alone as well as to the original and TDHS (alone).

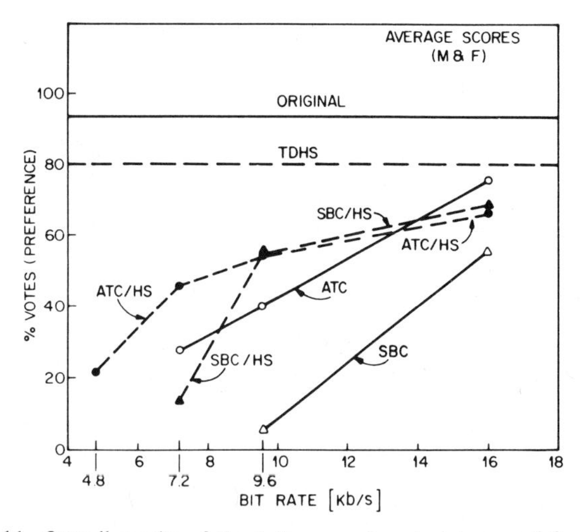

Fig. 11. Overall results of the A-B comparison test (averaged for male and female speakers).

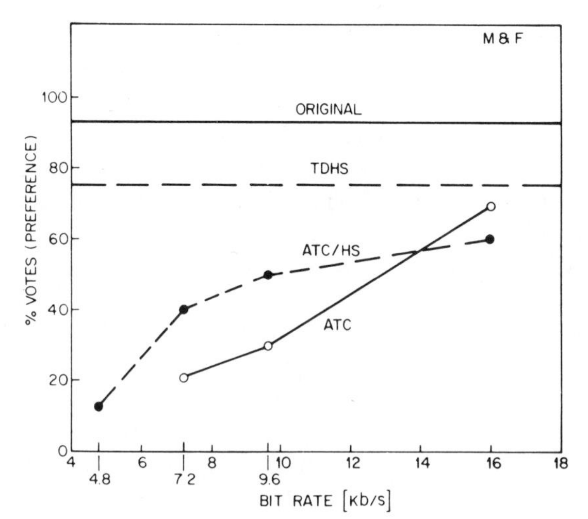

Fig. 10. A-B comparison test results for the combined ATC/HS system in comparison to ATC alone, as well as to the original and TDHS.

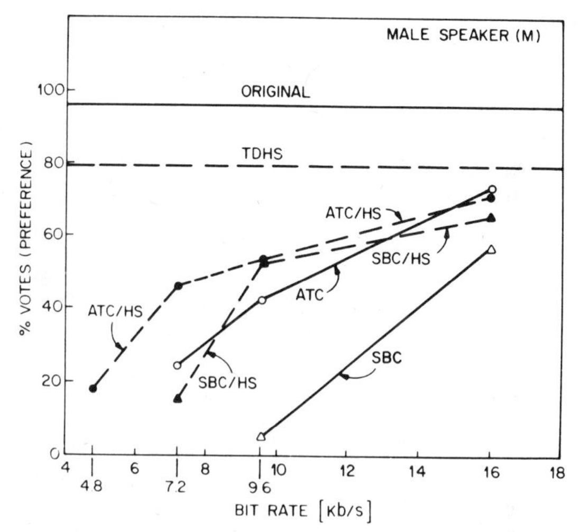

Fig. 12. Overall A-B comparison test results for male speaker.

A. Discussion of Results

We refer first to Fig. 9 which presents the results of the A-B comparison for SBC and SBC/HS. The rapid decrease in the quality of SBC alone, from 16 to 9.6 kbits/s, is quite evident. The combined SBC/HS system offers here a substantial improvement at 9.6 kbits/s, which corresponds to a bit-rate advantage of 7 kbits/s, i.e., the quality of the combined system at 9.6 kbits/s is equivalent to that of SBC alone at 16 kbits/s. Below 9.6 kbits/s the combined system degrades rapidly due to the rapid degradation of SBC below 16 kbits/s, and the added degradation by the TDHS operations which are not transparent. At 7.2 kbits/s the quality of the combined system is judged to be still acceptable for narrow band communication applications and the bit-rate advantage of the combined system at this rate is found to be 4 kbits/s.

For ATC, one observes from Fig. 10 that the improvement offered by the combined ATC/HS system is lower than for SBC, as the maximum bit-rate advantage is only 4 kbits/s (at 7.2 kbits/s), in comparison to a bit-rate advantage of 7 kbits/s for SBC/HS. From our earlier discussion, this is expected due to the fact that the ATC system simulated already exploits pitch information, leaving less redundancy to be removed by the TDHS system. Thus, the use of TDHS with a scaling factor of 2 results in a maximum improvement factor of 1.73 for SBC (at 9.6 kbits/s) and 1.55 for ATC (at 7.2 kbits/s). This appears to be a worthwhile improvement in light of the relatively small increase in the complexity of these systems.

Examining the overall test results in Fig. 11, one observes that the subjective qualities of SBC/HS and ATC/HS at 9.6 kbits/s and above are quite close to each other, whereas the complexity of SBC/HS is considerably lower than even ATC alone. As noted in the previous section, and shown in Figs. 12 and 13, the results for male and female speakers differ somewhat locally, but not globally, so that the above more general conclusions are not affected.

An additional interesting observation from the results for ATC is that at 16 kbits/s the combined system quality is actually slightly lower than the quality of ATC alone. This can be attributed to the relatively high quality of ATC at 16 kbits/s,

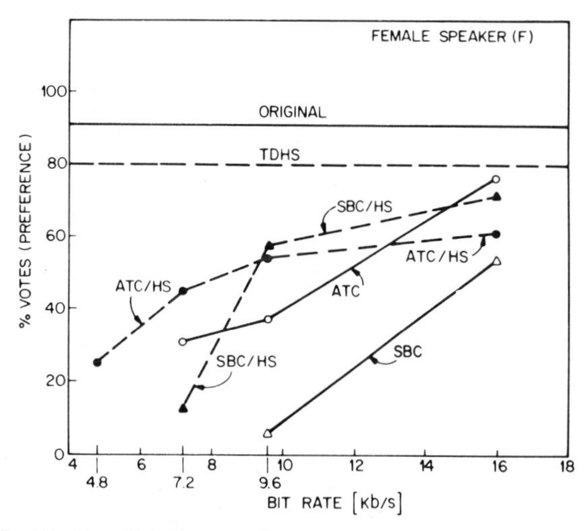

Fig. 13. Overall A-B comparison test results for female speaker.

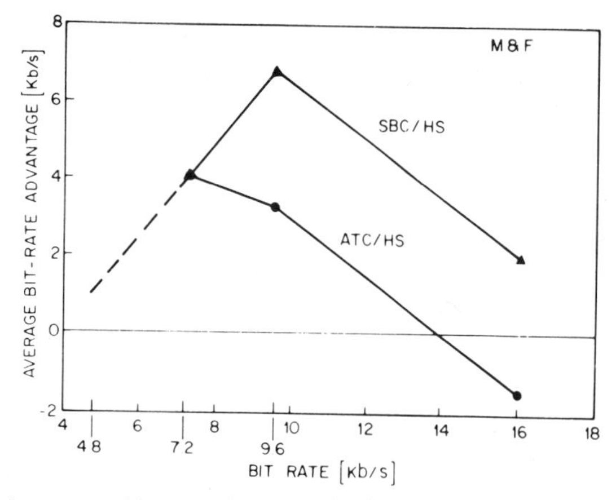

Fig. 14. Average bit-rate advantage obtained by the combined systems at different bit rates.

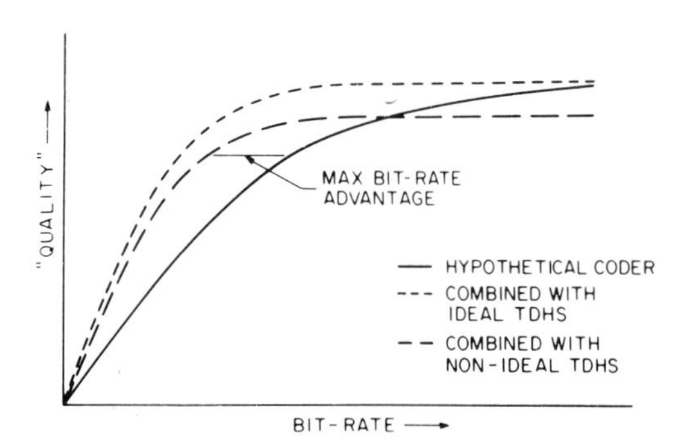

Fig. 15. Hypothetical coder characteristics showing the expected improvements by combining with TDHS.

and the nontransparency of the TDHS operations. For SBC, which has a lower quality at 16 kbits/s than ATC, there is some gain by using the combined SBC/HS system at this rate. As a summary the bit-rate advantages of the two systems at different bit rates is presented in Fig. 14.

The particular characteristic of the combined systems to improve the quality of the waveform coders mainly at rates which are below the performance knee of the coder, and to cause a degradation in quality at higher bit rates, is qualitatively explained by Fig. 15. In this figure the solid line is the performance curve of a hypothetical waveform coder. If an ideal TDHS system were available, one would expect the combined system to have the performance given by the short dashed-line, which is always above the solid line and is obtained by assuming a scale factor of two (each point on the original curve is moved to half the corresponding bit rate assuming no change in perceived quality). However, as is evident from the test results shown in the previous figures, the TDHS system introduces a certain degree of degradation (but is still judged to be of higher quality than ATC at 16 kbits/s). This causes the performance curve of the ideal combined system to be lowered, as shown by the long dashed-line in Fig. 15. It is clearly seen that the highest improvement is below the performance knee of the waveform coder and that for high bit rates one might actually obtain a loss in quality, as was found in our experimental results and discussed above.

VII. Conclusion

Waveform coding of the frequency scaled speech signal by means of time-domain harmonic scaling (TDHS) is shown to meaningfully improve the perceived performance of the waveform coding systems used—at their lower bit-rate end. In particular, bit-rate advantages of 7 and 4 kbits/s were obtained over subband coding (SBC) and adaptive transform coding (ATC), by the respective combined systems (SBC/HS and ATC/HS), at 9.6 and 7.2 kbits/s, respectively.

In spite of the "soft-vocoding" nature of the frequency scaling operations, the combined waveform coding-harmonic scaling system is almost as robust as the waveform coder used. The robustness is due to the avoidance of the voicing decisions and smoothing of the raw pitch-period data, the tolerance of the TDHS algorithms to limited pitch errors and to double pitch period decisions, and the ability to operate under different environmental conditions. In particular, it was found that the TDHS algorithms perform adequately on the speech of several simultaneous speakers.

The requirement for pitch extraction to implement the TDHS algorithms appears to increase the complexity of the system, but since the waveform coder operates on decimated samples, the overall computational load might even be reduced. However, the additional delay (of up to 100 ms) caused by the pitch extraction and harmonic scaling operations (transmitter and receiver) could be a limiting factor in some applications—but certainly not in many others.

The subjective evaluation which was performed by means of an A-B comparison test singles out the combined SBC/HS system at 9.6 kbits/s as having the highest perceived quality of all of the other systems examined at this rate. Since this system is much less complex than ATC, and is amenable to real-time hardware implementation using current technology, it provides an attractive system for speech coding at 9.6 kbits/s. The quality offered by this system is perceptually equivalent to that of 16 kbits/s SBC, which was judged in an earlier experiment [24] to be perceptually equivalent to 24 kbits/s ACPCM. An efficient implementation of this system could be based on the recently announced digital processing (DSP) chip by Bell Laboratories [25], [26]. Moreover, the SBC/HS system could also be attractive at 16 kbits/s since its quality is only somewhat lower than ATC at this rate but is sufficiently less complex.

The approach of combining waveform coding with TDHS is not limited to specific waveform coders. However, the improvement to be achieved is dependent on the extent that the pitch structure is exploited by the waveform coder used. It appears, however, that even waveform coders which do quite efficiently exploit the pitch structure, such as the speech-specific ATC used in this study, can still be meaningfully improved at low bit rates by combining them with TDHS.

REFERENCES

[1] B. S. Atal and M. R. Schroeder, "Adaptive predictive coding of speech signals," *Bell Syst. Tech. J.*, vol. 49, pp. 1973-1986, Oct. 1970.

[2] J. Makhoul and M. Berouti, "Predictive and residual encoding of speech," *J. Acoust. Soc. Amer.*, vol. 66, pp. 1633-1641, Dec. 1979.

[3] R. Zelinski and P. Noll, "Adaptive transform coding of speech signals," *IEEE Trans. Acoust., Speech, Signal Processing*, vol. ASSP-25, pp. 299-309, Aug. 1977.

[4] J. M. Tribolet and R. E. Crochiere, "Frequency domain coding of speech," *IEEE Trans. Acoust., Speech, Signal Processing*, vol. ASSP-27, pp. 512-530, Oct. 1979.

[5] J. L. Flanagan, *et al.*, "Speech coding," *IEEE Trans. Commun. Technol.*, vol. COM-27, pp. 710-746, Apr. 1979.

[6] B. S. Atal and N. David, "On synthesizing natural-sounding speech by linear prediction," in *Proc. 1979 IEEE Int. Conf. Acoust., Speech, Signal Processing*, 1979, pp. 44-47.

[7] D. Malah, "Time-domain algorithms for harmonic bandwidth reduction and time scaling of speech signals," *IEEE Trans. Acoust., Speech, Signal Processing*, vol. ASSP-27, pp. 121-133, Apr. 1979.

[8] —, "Harmonic scaling of speech signals by linear time-varying spectral modifications," to be published.

[9] R. E. Crochiere, S. A. Webber, and J. L. Flanagan, "Digital coding of speech in subbands," *Bell Syst. Tech. J.*, vol. 55, pp. 1069-1085, Oct. 1976.

[10] D. Malah, "Combined time-domain harmonic compression and CVSD for 7.2 kbit/s transmission of speech signals," in *Proc. 1980 IEEE Int. Conf. Acoust., Speech, Signal Processing*, Apr. 1980, pp. 504-507.

[11] J. L. Flanagan and R. M. Golden, "Phase vocoder," *Bell Syst. Tech. J.*, vol. 45, pp. 1493-1509, Nov. 1966.

[12] J. B. Allen and L. R. Rabiner, "A unified approach to short-time Fourier analysis and synthesis," *Proc. IEEE*, vol. 65, pp. 1558-1566, Nov. 1977.

[13] R. E. Crochiere, "A weighted overlap-add method of short-time Fourier analysis synthesis," *IEEE Trans. Acoust., Speech, Signal Processing*, vol. ASSP-28, pp. 99-102, Feb. 1980.

[14] M. R. Portnoff, "Time-frequency representation of digital signals and systems based on short-time Fourier analysis," *IEEE Trans. Acoust., Speech, Signal Processing*, vol. ASSP-26, pp. 55-69, Feb. 1980.

[15] R. E. Crochiere, "On the design of sub-band coders for low bit-rate speech communication," *Bell Syst. Tech. J.*, vol. 56, pp. 747-770, May-June 1977.

[16] A. J. Barabell and R. E. Crochiere, "Sub-band coder design incorporating quadrature filters and pitch prediction," in *Proc. 1979 IEEE Int. Conf. Acoust., Speech, Signal Processing*, Apr. 1979, pp. 530-533.

[17] D. Esteban and C. Galand, "Application of quadrature mirror filters to split band voice coding schemes," in *Proc. 1977 IEEE Int. Conf. Acoust., Speech, Signal Processing*, May 1977, pp. 191-195.

[18] J. D. Johnston, "A filter family designed for use in quadrature mirror filter banks," in *Proc. IEEE Int. Conf. Acoust., Speech, Signal Processing*, Apr. 1980, pp. 291-294.

[19] R. V. Cox and R. E. Crochiere, "Real-time simulation of adaptive transform coding," *IEEE Trans. Acoust., Speech, Signal Processing*, this issue, pp. xx-xx.

[20] L. R. Rabiner and R. W. Schafer, *Digital Processing of Speech Signals*. Englewood Cliffs, NJ: Prentice-Hall, 1978.

[21] A. M. Noll, "Cepstrum pitch determination," *J. Acoust. Soc. Amer.*, vol. 41, pp. 293-309, Feb. 1967.

[22] J. J. Dubnowski, R. W. Schafer, and L. R. Rabiner, "Real-time pitch detector," *IEEE Trans. Acoust., Speech, Signal Processing*, vol. ASSP-24, pp. 2-8, Feb. 1976.

[23] J. M. Tribolet and R. E. Crochiere, "A modified adaptive transform coding scheme with post-processing-enhancement," in *Proc. 1980 Int. Conf. Acoust., Speech, Signal Processing*, Apr. 1980, pp. 336-339.

[24] J. M. Tribolet, P. Noll, B. J. McDermott, and R. E. Crochiere, "A comparison of the performance of four low-bit-rate speech waveform coders," *Bell Syst. Tech. J.*, vol. 58, pp. 699-712, Mar. 1979.

[25] J. S. Thompson and J. R. Boddie, "An LSI digital signal processor," in *Proc. 1980 IEEE Int. Conf. Acoust., Speech, Signal Processing*, Apr. 1980, pp. 383-385.

[26] R. E. Crochiere, "Sub-band coding using the DSP," *Bell Syst. Tech. J.*, to be published.

HADAMARD-TRANSFORMATION TECHNIQUE OF SPEECH CODING: SOME FURTHER RESULTS

E. Frangoulis and L. F. Turner

Indexing terms: *Digital communication systems, Encoding, Transforms, Voice communication*

Abstract

The results of an extensive investigation of the properties of 64-point Hadamard transformed speech are presented. Detailed information is given about the probability density functions of the Hadamard coefficients, the average power-density spectrum in the Hadamard domain and the logical-autocorrelation function. The results indicate that good-quality speech can be reconstructed from 6 to 8 dominant Hadamard coefficients, but that the use of fewer coefficients is unlikely to lead to the reconstruction of speech of acceptable quality. The results of a preliminary series of listening tests are presented and these confirm conclusions drawn from the statistical properties of the transformed speech. It is shown that the number of bits needed for coefficient labelling constitutes a significant proportion of the total number of bits needed to represent Hadamard transformed speech. A technique is presented for reducing by more than 50% the number of labelling bits needed, and it is explained how, by using this technique, it should be possible to obtain good quality speech when using a transmission bit rate of 8 k bits/s.

List of principal symbols

C_i = Hadamard coefficient
x_i = data sample
H = Hadamard matrix
$L_S(k)$ = local, or short term, logical-autocorrelation function
$\hat{L}(k)$ = average value of logical-autocorrelation function
W = transmission bit rate in bits/s
L = number of dominant coefficients used to reconstruct speech signal
M = number of bits used to encode each coefficient
l = number of bits needed to label a coefficient
R = rate at which speech signal is sampled
Q = number of components in Walsh-Fourier spectrum
G_i = component of Walsh-Fourier spectrum
γ_i^2 = sequency spectral component
$a_s(n)$ = sal (n) coefficient
$a_c(n)$ = cal (n) coefficient

1 Introduction

In recent years, interest has grown in the possibility of using orthogonal transformations as a means of reducing the bit rate necessary for the transmission of speech signals. This interest stems mainly from the ease with which techniques of this kind can be implemented in practice.

One of the earliest works in which orthogonal transformations were applied to the processing of speech signals was that of Boesswetter,[1] who examined a 1·4 s section of German speech, and showed that if 2 ms samples of the speech were transformed using a 16-point Hadamard transformation, then reasonably intelligible speech could be reconstructed using only the two largest coefficients from the set of 16 Hadamard coefficients. This was then followed by the work of Campanella and Robinson.[2,3] In an early paper,[2] Campanella and Robinson performed a sequency analysis on the vowels a, e, i, o and u and then went on in a later paper[3] to show mathematically that, by using a 16-point Hadamard transformation, it should be possible, for a given signal/quantisation noise ratio, to transmit speech using 48·5 k bits/s, rather than 56 k bits/s as in conventional p.c.m. In a recent important paper, Shum *et al.*[4] have reported on the application of a 64-point Walsh-Hadamard transformation to speech compression, and have stated that their tentative results indicate that fair to good reproduction of speech can be obtained when using from 4 to 8 of the 64 Hamamard coefficients. In a practical system the complete set of coefficients would be determined and the dominant (largest) coefficients would be obtained and transmitted. The 4 to 8 coefficient representation corresponds to bit rates ranging from 6·5 k bits/s to 13 k bits/s.

The results of Shum *et al.*,[4] although only of a provisional nature, are indicative and sufficiently important to warant a more detailed examination of the question of speech compression by orthogonal transformations. This paper is a report on the results of an extensive and systematic investigation that has been conducted into the application of the Hadamard transformation to speech compression. In Section 2 of the paper the question of compression by use of a limited number of coefficients is examined, with particular attention being paid to the existence and time varying nature of dominant coefficients, and to the sensitivity of these coefficients to factors such as the sampling rate and changes in speaker. The quantitative results contained in this Section are new and are of considerable importance, since, if speech compression is to be achieved by using a limited number of coefficients, then it is essential that the variation and sensitivity of these coefficents to various parameters be clearly understood. In Section 3 the question of the subjective quality of speech reconstructed from dominant coefficients is considered briefly, and some comments are made as to how the speech quality varies with variations in the number of coefficients used, the sampling rate and the number of bits used to represent each coefficient. The comments made in Section 3 concerning speech quality should only be considered as indicative, since an extensive series of subjective tests is necessary to obtain a quantitative assessment of speech quality. In the fourth Section of the paper, the nature and occurrence of the dominant coefficients is considered, and it is shown that the dominant coefficients are always members of a fixed subset of the 64 Hadamard coefficients, and that by making use of this fact it is possible to reduce the necessary transmission bit rate without affecting speech quality. In Section 5 of the paper, some initial results concerning window widths are presented, and it is argued that, if the results that have been obtained so far can be shown to be generally true, then it should be possible to obtain acceptable speech quality when using transmission bit rates of approximately 6 k bits/s.

2 Some properties of Hadamard-transformed speech signals

2.1 Introduction

In this Section, some of the results of an extensive study of the properties of Hadamard transformed speech are presented. The results, although only a selection from a very large number of results, are totally representative of the general pattern of behaviour that has emerged from the tests, and conclusions drawn from them are of general validity.

The specific properties that are considered in this Section are as follows:

(a) the probability density function (p.d.f.) of the Hadamard coefficients for the 64-point transformation
(b) the average power-density function in the Hadamard domain
(c) the variation of the logical autocorrelation function in terms of the number of coefficients.

These properties are studied with respect to various parameters, such as the length of the speech section transformed, the sampling rate, speaker variation and sentence variation.

Paper 7971 E, first received 9th December 1976 and in revised form 12th May 1977

Mr. Frangoulis and Dr. Turner are with the Department of Electrical Engineering, Imperial College of Science Technology, South Kensington, London SW7 2BT, England

2.2 Probability density function of the Hadamard coefficients for the 64-point transformation

In all the tests reported in this paper the speech signals were passed through low-pass filter of cutoff frequency 3·4 kHz and then sampled. The sampled values were then analogue-digital converted using a 10-bit/sample convertor and the digitised samples were then processed in a PDP-15 computer. After being processed, the digitised sampled speech was then reconverted to analogue form using a 10-bit digital-analogue convertor and an associated low-pass filter.

In applying the Hadamard transformation technique to speech signals, the n Hadamard coefficients $C_1, C_2, \ldots, C_n$ are obtained by taking blocks of $n(=r^2)$ data samples $x_1, \ldots, x_n$ and then performing the following matrix multiplication:

$$\begin{bmatrix} C_1 & \cdots & C_r \\ \vdots & & \vdots \\ C_{r^2-r+1} & \cdots & C_n \end{bmatrix} = [H] \begin{bmatrix} x_1 & \cdots & x_r \\ \vdots & & \vdots \\ x_{r^2-r+1} & \cdots & x_n \end{bmatrix} [H]$$

where $[H]$ is an $r \times r$ Hadamard matrix.

In the first series of tests long sections* of speech, with the lengths ranging from parts of a sentence to a number of sentences, were Hadamard transformed and the p.d.f.s associated with the 64 $(= n)$ Hadamard coefficients were obtained. A selection from the results are shown in Figs. 1 to 9. It is not possible in these Figures to show separately each of the 64 coefficients. What is done, where appropriate, is to display the coefficient p.d.f.s in bounded sets. Figs. 1 and 2 show the coefficient p.d.f.s associated with a complete sentence spoken by both male and female speakers. These two Figures are perhaps the most important of the first set of nine Figures, since they appear to indicate the existence of a small set of dominant (large valued) coefficients. The Figures indicate that the dominant coefficients exist irrespective of whether the speaker is male or female. Figs. 3 and 4, which relate to parts of a sentence, also indicate the existence of dominant coefficients. Although it is not clear from Figs. 1 to 4, the dominant set of coefficients may change, and in fact do change, from speaker to speaker and from one part of a sentence to another. This is a point that is taken up in more detail later in the paper. Figs. 5 to 8 are essentially a repeat of Figs. 1 to 4, but with the difference that attention is focused on isolated dominant coefficients. Figs. 5 and 6 show how the p.d.f. of a dominant coefficient varies as the length of speech considered is changed from a whole sentence to part of a sentence. Also, within each of the two Figures, a comparison is given as to how the p.d.f. of the coefficient varies from the male to the female speaker. Figs. 7 and 8 show how the p.d.f. of a dominant coefficient varies from sentence to sentence. Figs. 1 to 8 indicate that dominant coefficients tend to have higher average values for female speakers than for male speakers, which suggests that it may be necessary to use a larger average number of binary digits to represent female speech. Fig. 9 shows how the p.d.f. of a dominant coefficient varies as a function of the rate at which the speech signal is sampled. It should be noted from Fig. 9 that the dominant coefficient tends to have a higher average value for lower sampling rates, and, thus, this again indicates that an increased number of bits/dominant-coefficient may be necessary at lower sampling rates.

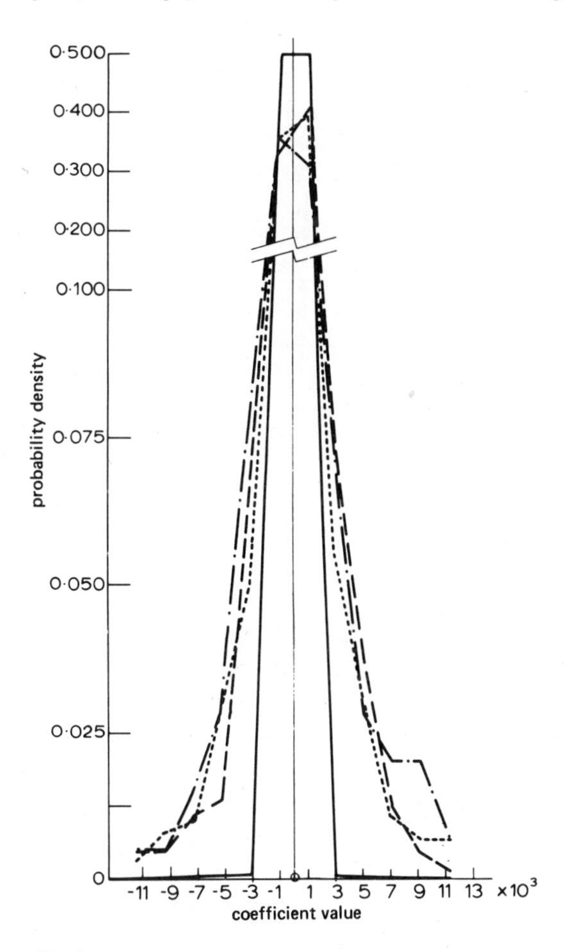

Fig. 1
Probability density functions of Hadamard coefficients

Complete sentence spoken by male, sampling rate 8 k samples/s

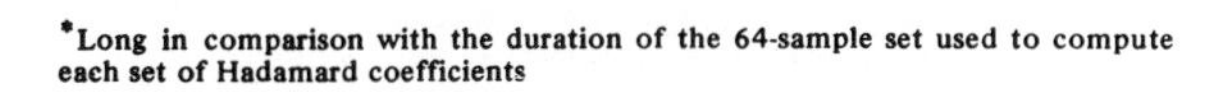

---- bounds on set of 56 nondominant coefficients
-·-·- bounds on set of 8 dominant coefficients

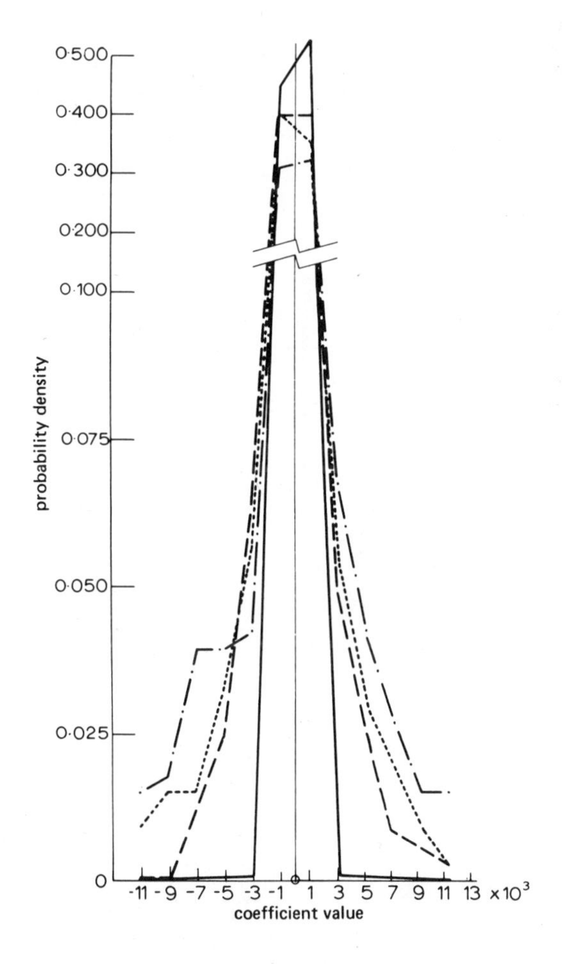

Fig. 2
Probability density functions of Hadamard coefficients

Complete sentence spoken by female, sampling rate 8 k samples/s

---- bounds of set on 56 nondominant coefficients
-·-·- bounds on set of 8 dominant coefficients

*Long in comparison with the duration of the 64-sample set used to compute each set of Hadamard coefficients

2.3 Power-density functions for 64-point Hadamard transformed speech

Although the p.d.f.s of the Hadamard coefficients indicate the existence of certain large valued dominant Hadamard coefficients, this information provides little real evidence as to how accurately speech waveforms can be represented when using a few dominant coefficients. In order to obtain some quantitative information relating to this matter, speech signals were examined in terms of their average power-density spectrum in the Hadamard domain. The results of this examination are summarised in Figs. 10 to 12 and in Table 1. Fig. 10, which was obtained using a complete sentence spoken by a male speaker, shows how, as a function of sampling rate, the power-density spectrum in the Hadamard domain is distributed among the Hadamard coefficients. Fig. 11 shows the way in which the Hadamard power-density spectrum changes from male to female speakers, and Fig. 12 shows how, for a sentence spoken by a male, the percentage of the total power contained in 5 and 8 dominant coefficients varies with increasing sampling rate. Table 1 provides important general information about the percentage of the total waveform power that is contained in 5 and 8 dominant coefficients for various sentences spoken by both male and female speakers.

The results of this Section indicate that, for both male and female speakers, between 60% and 90% of the total average power is contained in the 6 to 8 dominant coefficients when using a sampling rate of 8 k samples/s. The results also indicate that there is approximately a 10% reduction in the power contained in the dominant coefficients as the sampling rate is reduced from 10·8 k samples/s to 6·1 k samples/s. In addition, it is clear that for female speakers the dominant coefficients contain a smaller percentage of the total power than for male speakers. The results contained in Table 1 indicate that the average power contained in 5 dominant coefficients is between 20% and 25% less than that contained in 8 dominant coefficients, and this thus casts considerable doubt on the feasibility of obtaining good-quality speech reconstruction from 5, or fewer, coefficients.

2.4 Logical autocorrelation function for 64-point Hadamard transformed speech

Further insight into the acceptability of speech reconstructed from a limited number of Hadamard coefficients can be gained by considering the normal (arithmetic) autocorrelation function of the reconstructed speech, or, alternatively, by considering the so-called[5] logical autocorrelation function.

The local, or short term, logical autocorrelation function of a sample of speech is defined[5, 6] to be

$$L_s(k) = \frac{1}{N}\sum_{j=0}^{N-1} x(j \oplus k)x(j)$$

where $k = 0, 1, 2, \ldots, N-1$ and $j \oplus k$ denotes the bit by bit modulo-2 addition of the integers j and k expressed in binary notation, and $N = 2^n$ is the length of the sequence. The logical autocorrelation function $L(k)$ is defined to be the expected (or average) value of the local logical autocorrelation function, and it can be estimated by its sample

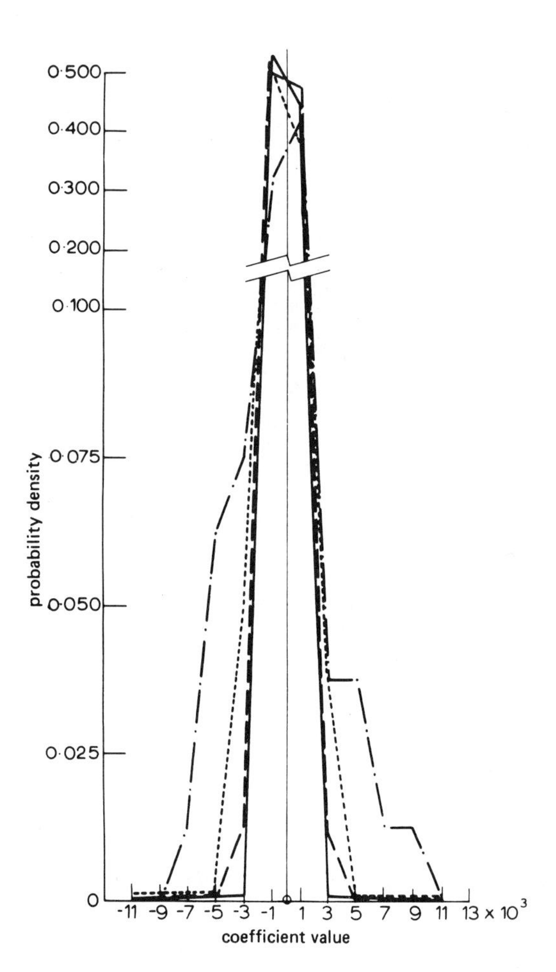

Fig. 3
Probability density functions of Hadamard coefficients

Part sentence spoken by male, sampling rate 8 k samples/s

---- bounds on set of 56 nondominant coefficients
-·-·- bounds on set of 8 dominant coefficients

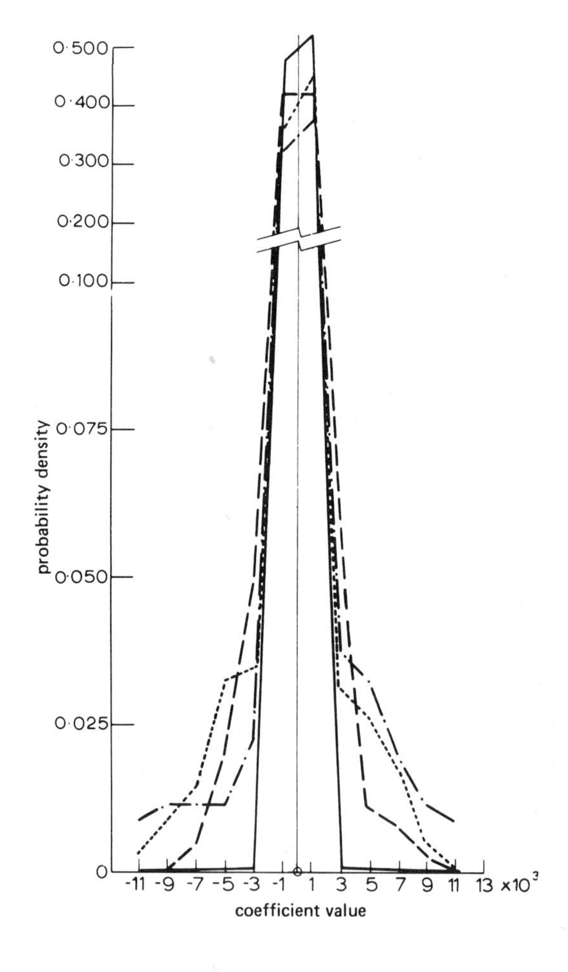

Fig. 4
Probability density functions of Hadamard coefficients

Part sentence spoken by female, sampling rate 8 k samples/s

---- bounds on set of 56 nondominant coefficients
-·-·- bounds on set of 8 dominant coefficients

Table 1
PERCENTAGE OF TOTAL POWER CONTAINED IN 5 AND 8 DOMINANT COEFFICIENTS

Sentence Number	Percentage of total power contained in 8 dominant coefficients		Percentage of total power contained in 5 dominant coefficients	
	male	female	male	female
1	62	59	41	26
2	67	63	44	43
3	73	67	48	46
4	75	68	50	46
5	88	84	69	64

mean $\hat{L}(k)$, where

$$\hat{L}(k) = \frac{1}{M} \sum_{s=1}^{M} L_s(k)$$

where $k = 0, 1, \ldots, N-1$. The logical autocorrelation function is considerably easier to compute than the normal arithmetic autocorrelation function, and it has the property[5, 6] that the arithmetic autocorrelation function can be derived from it by a relatively simple non-singular linear transformation.

Figs. 13 and 14 show some of the results of a study and comparison of the logical autocorrelation function of speech and logical autocorrelation function of the corresponding speech signals reconstructed from dominant Hadamard coefficients. Figs. 13 to 15 show the logical autocorrelation functions for complete sentences spoken by a male speaker when the speech is reconstructed from 8, 7 and 5 dominant

Table 2
SPEECH QUALITY AGAINST SAMPLING RATE, BITS/COEFFICIENT AND NUMBER OF COEFFICIENTS

Sampling rate	Number of coefficients used	Number of bits/coefficient	Resultant transmission bit rate	Speech quality
k sample/s			k bits/s	
10·8	8	7	17·55	very good
	6	7	13·16	very good
	6	6	13·14	very good
10·0	8	7	16·25	very good
	6	7	12·18	very good
	6	6	11·24	very good
9·0	8	7	14·62	very good
	6	7	10·96	very good
	6	6	10·12	very good
8·0	8	7	13·00	very good
	6	7	9·74	very good
	6	6	8·99	good
7·0	8	7	11·37	good
	6	7	8·53	good
	6	6	7·87	average
6·1	8	7	9·91	poor
	6	7	7·43	poor
	6	6	6·85	poor

Sentences used in all tests were taken from the Harvard list number 1. A tape recording of the results is available from the authors on request.
Very good = very little different from original speech and of low noise
Good = highly intelligible but noticeably noisy
Average = less acceptable than 'good', intelligible but noisy
Poor = of low intelligibility and very noisy

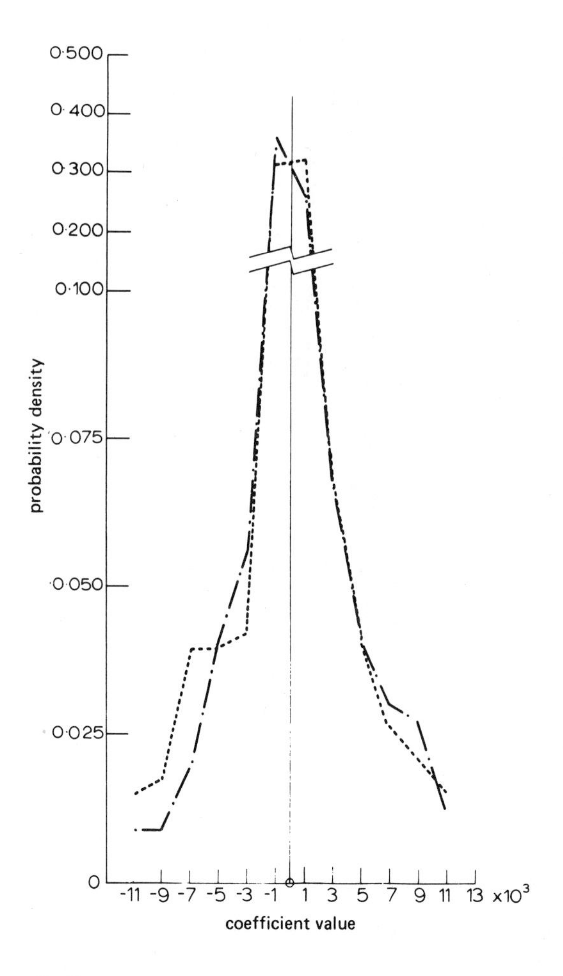

Fig. 5
Probability density function of dominant Hadamard coefficient

Complete sentence; sampling rate 8 k samples/s
------ female speaker — — — male speaker

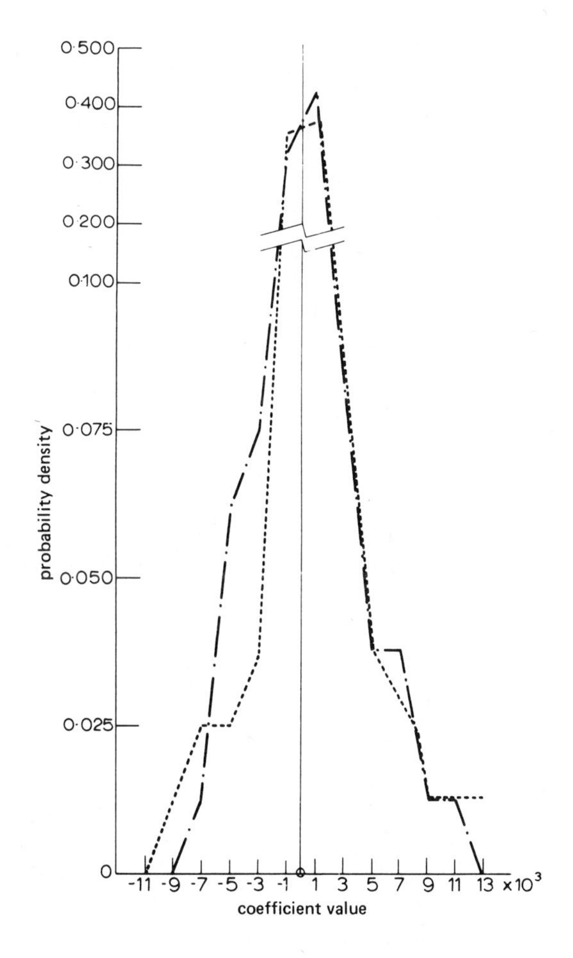

Fig. 6
Probability density function of dominant Hadamard coefficient

Part sentence; sampling 8 k samples/s
------ female speaker —·— male speaker

coefficients, respectively, and Figs. 16 to 18 show the corresponding functions for the same sentences spoken by a female speaker. Examination of these Figures shows that the extent of the agreement between the true logical autocorrelation function and that associated with speech reconstructed from 7 or 8 dominant Hadamard coefficients is quite remarkable. This is a feature that was found to be true for all sentences examined in the current investigation. The extent and generality of the agreement is such that it adds much weight to the idea of using 7 or 8 dominant Hadamard coefficients to represent speech that can be reconstructed from a limited number of the 64 dominant coefficients is not likely to be so satisfactory.

3 Quality of speech reconstructed from limited number of Hadamard coefficients

In addition to performing a detailed investigation of the statistical properties of Hadamard transformed speech signals, an initial series of tests were performed to gain some 'feel' for the quality of speech that can be reconstructed from a limited number of the 64 Hadamard coefficients. The tests were of a provisional nature and a detailed and extensive series of subjective tests are to be performed to obtain a quantitative assessment of speech quality and to obtain a comparison of the effectiveness of the technique as compared with other methods of speech coding. In the tests,† speech quality was considered, though somewhat superficially, as a function of various factors such as sampling rate, number of coefficients used, and the number of bits used to encode each coefficient.

The results of the tests are shown in Table 2, with the transmission rates W quoted in the table being derived using the expression

$$W = L\frac{(m+l)}{64}R \qquad (1)$$

where L is the number of coefficients used in reconstructing the signal, m is the number of bits used to encode each coefficient, l is the number of bits needed to indicate to the receiver which of the 64 coefficients a particular received coefficient is, and R is the sampling rate.

From Table 2 it can be seen that when using 8 coefficients and transmission bit rates in excess of 13 k bits/s the speech quality is very little different from the original speech and is virtually noise free. Speech of this quality is designated as 'very good'. When using 8 coefficients and a sampling rate of 7 k samples/s the speech is highly intelligible, but slightly more noisy than at higher bit rates. When the sampling rate is lowered to 6·1 k samples/s, with a corresponding transmission bit rate of 9·91 k bits/s, the speech is very noisy and is of low intelligibility. In the light of the information contained in Fig. 12, this fall-off in performance when using a sampling rate of 6·1 k samples/s is to be expected. When using 6 coefficients, and 7 bits/coefficient, the quality of the reconstructed speech was found to be very little different from that obtained when using 8 coefficients. When using 6 coefficients, and 6 bits/coefficients, the speech quality was found to be slightly more noisy than when using 7 bits to encode each coefficient. However, despite this slight increase in noise level, the speech remains highly intelligible for transmission bit rates of 8·99 k bit/s, and greater.

4 Possibility of further bit-rate reductions

The results presented in the previous Sections indicate that, when using a 64-point Hadamard transformation, it is unlikely that acceptable quality speech can be obtained when using fewer than 6 coefficients, and fewer than 6 bits to encode each of the coefficients. In view of these facts the question arises as to whether it is at all possible to reduce the transmission bit rates below those values quoted in Table 2, and still maintain acceptable speech quality.

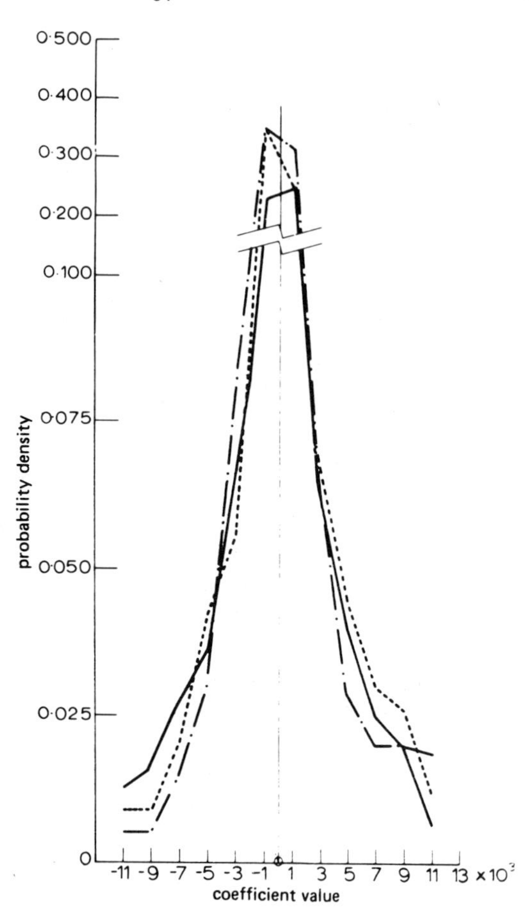

Fig. 7
Probability density function of dominant coefficient for two separate sentences and the combination of the sentences

Sentences spoken by male; sampling rate 8 k samples/s
— · — sentence 1
——— sentence 2
- - - - - - sentence 1 and 2

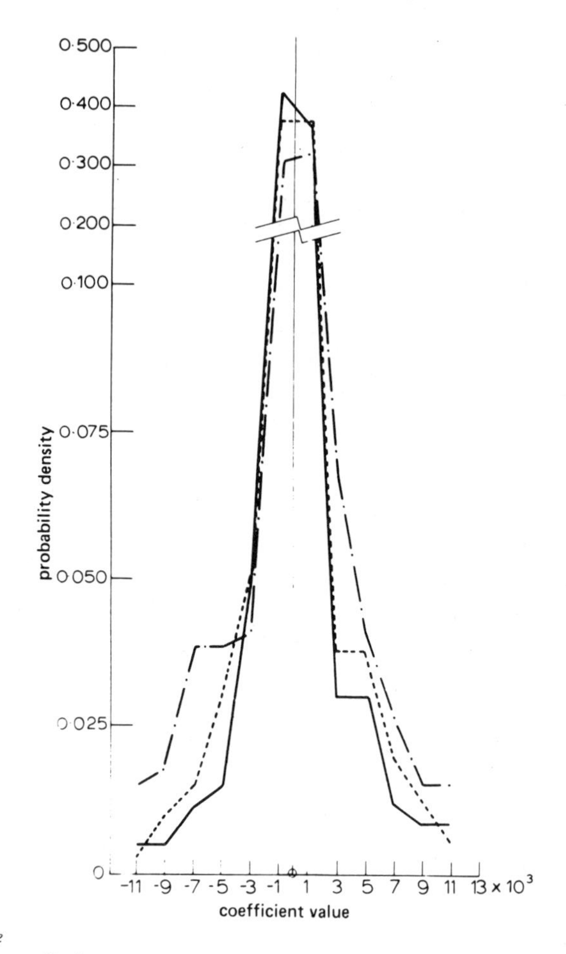

Fig. 8
Probability density function of dominant coefficient for two separate sentences and the combination of the sentences

Sentences spoken by female; sampling rate 8 k samples/s
— · — sentence 1
——— sentence 2
- - - - - - sentences 1 and 2

†The sentences used in all tests were taken from the Harvard list number 1.

For the 64-point Hadamard transformation, the transmission bit rate is given by eqn. 1, from which it is clear that, if L, m and R are fixed, then the only way in which the transmission bit rate can be reduced is by reducing the number of bits l used to indicate to the receiver which of the possible 64 coefficients a particular transmitted coefficient is. It is necessary to provide this information since the coefficients which dominate change as the speech varies. In the work of Shum *et al.*,[4] and in obtaining the bit rates quoted in Table 2, it was assumed that to provide this information it was necessary to use 6 labelling bits per coefficient. This means that approximately half of the transmission bit rate is being used to provide the necessary labelling information.

In an attempt to determine whether it is possible to reduce the number of labelling bits, an investigation was carried out, with attention being focused on the question of the occurrence of the dominant coefficients.

Examination of the occurrence of the dominant coefficients reveals that the 6 or 8 dominant coefficients are always members of the following group of 32 coefficients:

1, 4, 9, 11, 13, 15, 17, 20, 21, 25, 27, 29, 31, 33, 35, 37, 39, 40, 41, 43, 45, 47, 49, 51, 53, 55, 57, 59, 60, 61, 63, 64

This means that the number of labelling bits can be reduced by 1 bit/coefficient without the speech quality being affected in any way. Although this saving is of value, it is not a major saving. A second aspect of the investigation, involving a consideration of the Walsh-Fourier spectrum of speech, has revealed, however, that it is possible to reduce the average number of labelling bits by more than 50%.

The Walsh-Fourier spectrum of a signal has been considered previously[2] and it has been shown that, for an N-point transformation, the Walsh-Fourier spectrum consists of $Q = 1 + \log_2 N$ components $G_0, G_1, \ldots, G_{Q-1}$, and that the groups are comprised sequency spectral components γ_n^2 as follows:

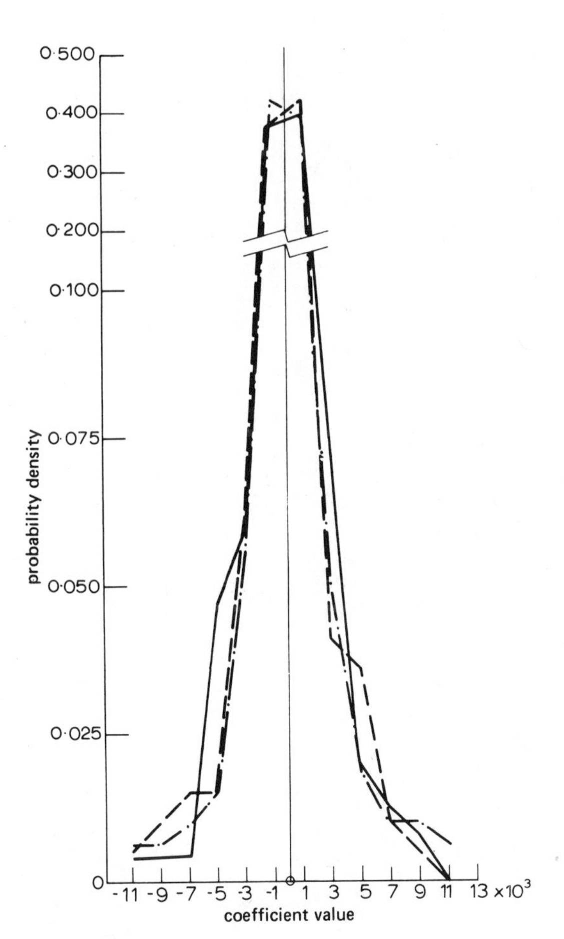

Fig. 9
Variation of probability density function of Hadamard coefficients with sampling rate

Complete sentence spoken by male
— · — 6 k samples/s
— — — 8 k samples/s
——— 10 k samples/s

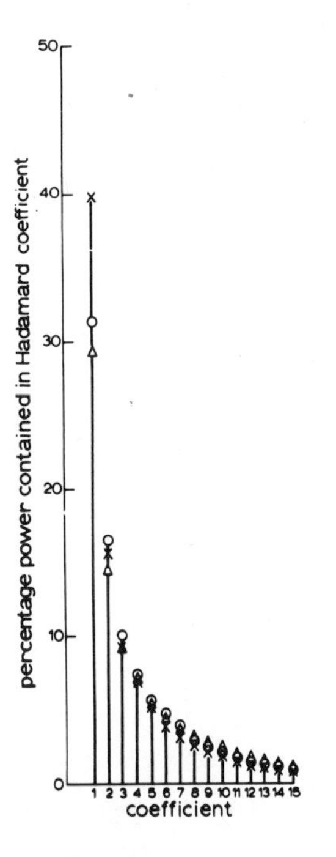

Fig. 10
Power density against Hadamard coefficient

Complete sentence spoken by male; sampling rate 8 k samples/s
X 10 k samples/s
o 8 k samples/s
Δ 6 k samples/s

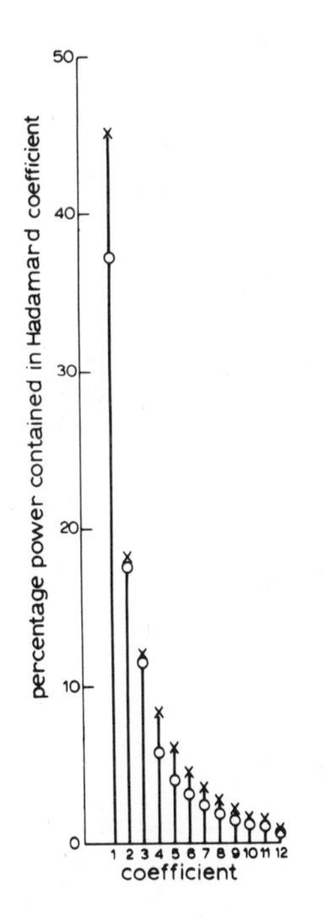

Fig. 11
Variation of power density from male to female speaker

Complete sentence; sampling rate 8 k samples/s
X male speaker
o female speaker

$$
\left.\begin{aligned}
G_0 &= a_0^2 \\
G_1 &= \gamma_1^2 + \gamma_3^2 + \gamma_5^2 + \ldots \\
G_2 &= \gamma_2^2 + \gamma_6^2 + \gamma_{10}^2 + \ldots \\
G_3 &= \gamma_4^2 + \gamma_{12}^2 + \gamma_{20}^2 + \ldots \\
G_4 &= \gamma_8^2 + \gamma_{24}^2 + \gamma_{40}^2 + \ldots \\
&\vdots \\
G_{\log_2 N} &= a_s^2(N/2)
\end{aligned}\right\} \qquad (2)
$$

where $\gamma_n^2 = a_c^2(n) + a_s^2(n)$ and where $a_s(n)$ and $a_c(n)$ are the sal(n) and cal(n) coefficients, respectively, obtained by Walsh transforming the signal. In the case of the 64-point transformation, the set of eqns. 2 becomes

$$
\left.\begin{aligned}
G_0 &= a_0^2 \\
G_1 &= \gamma_1^2 + \gamma_3^2 + \gamma_5^2 + \ldots + \gamma_{31}^2 = \sum_{i=1}^{i=N/4} \gamma_{2i-1}^2 \\
G_2 &= \gamma_2^2 + \gamma_6^2 + \gamma_{10}^2 + \ldots + \gamma_{30}^2 \\
G_3 &= \gamma_4^2 + \gamma_{12}^2 + \gamma_{20}^2 + \gamma_{28}^2 \\
G_4 &= \gamma_8^2 + \gamma_{24}^2 \\
G_5 &= \gamma_{16}^2 \\
G_{\log_2 N} &= G_{\log_2 64} = G_6 = a_s^2(32)
\end{aligned}\right\} \qquad (3)
$$

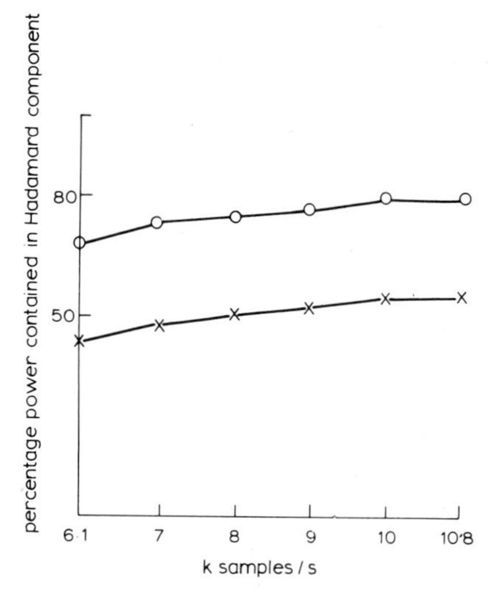

Fig. 12
Variation of percentage power contained in limited number of coefficients against sampling rate

Complete sentence spoken by male

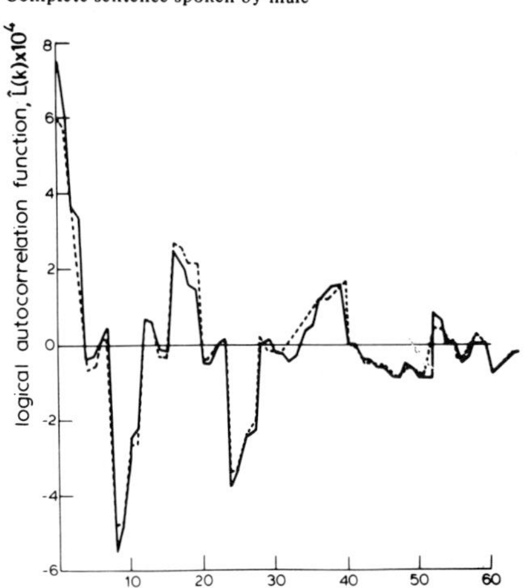

Fig. 13
Logical autocorrelation function

Complete sentence spoken by male; sampling rate 8 k samples/s
- - - - - - logical autocorrelation function reconstructed from 8 dominant coefficients
——— logical autocorrelation function of original sentence

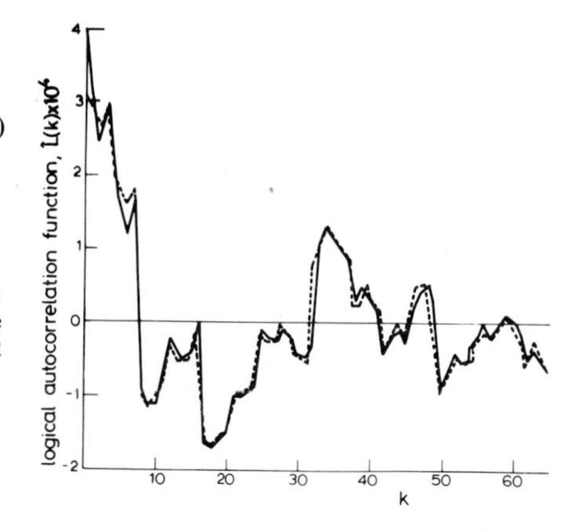

Fig. 14
Logical autocorrelation function

Complete spoken by male; sampling rate 8 k samples/s
- - - - - - logical autocorrelation function reconstructed from 7 dominant coefficients
——— logical autocorrelation function of original sentence

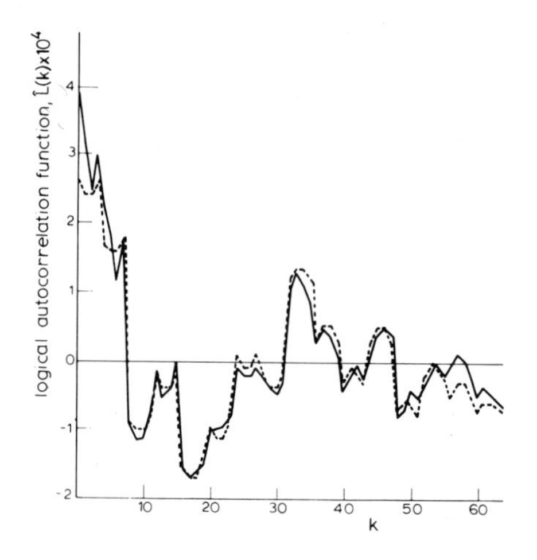

Fig. 15
Logical autocorrelation function

Complete sentence spoken by male; sampling rate 8 k samples/s
- - - - - - logical autocorrelation function reconstructed from 5 dominant coefficients
——— logical autocorrelation function of original sentence

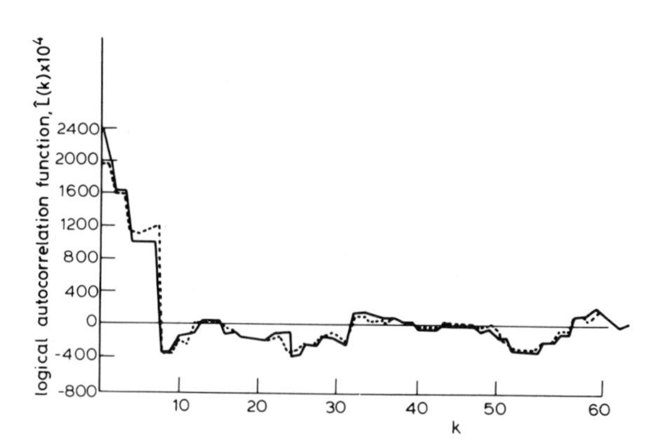

Fig. 16
Logical autocorrelation function

Complete sentence spoken by female; sampling rate 8 k samples/s
- - - - - - logical autocorrelation function reconstructed from 8 dominant coefficients
——— logical autocorrelation function of original sentence

An examination of the relationship between the dominant Hadamard coefficients and the Walsh-Fourier spectrum has shown that of the 8 dominant coefficients, 7 tend to predominate, and that one of the 7 coefficients occurs in each of the Walsh-Fourier components, and further that the Walsh-Fourier component is determined almost totally by the dominant coefficient. The occurrence of one dominant coefficient in each of the Walsh-Fourier components is highly significant from the point of view of reducing the number of labelling bits. The number of labelling bits can be reduced, since, if it is arranged so that the first dominant coefficient to be transmitted is that associated with G_0, the second dominant coefficient to be transmitted is that associated with G_1 etc., then, in order for the receiver to know exactly which Hadamard coefficient is dominating each G_i term, it is only necessary to supply a number of bits as indicated in Table 3. This is highly significant, since it means that only 15 bits have to be used to label 7 dominant coefficients, rather than the 42 ($=6 \times 7$) bits that would be necessary when using the labelling scheme adopted in connection with Table 2. Expressed in terms of transmission bit rates this means that, if with a 64-point transformation a sampling rate of 8 k samples/s is used, and 7 dominant coefficients are used to represent the speech signal, then, with the simplified labelling scheme, the transmission bit rate is 8 k bit/s. Clearly, this constitutes a significant bit-rate reduction when compared with the 11·375 k bit/s transmission rate that would be necessary when using the labelling scheme associated with Table 2.

5 16-point and 512-point Hadamard transformed speech signals

In addition to considering the 64-point Hadamard transformed speech, some initial consideration has also been given to 16-point and 512-point Hadamard transformed speech. An analysis of a number of sentences has shown that, in the case of the 16-point transformation, excellent quality speech can be obtained if the 16 Hadamard coefficients are encoded in the manner indicated in Table 4. This encoding corresponds to a transmission rate of 16 k bits/s. Furthermore, an initial examination of the coefficient p.d.f.s and the average power spectrum in the Hadamard domain, indicates that it may be possible to obtain good-quality speech reconstruction when using only 3 dominant coefficients from the set of 16 Hadamard coefficients. The results suggest that 2 coefficients may not be sufficient for the reconstruction of good quality speech, and this thus suggests that the question of intelligibility touched upon by Boesswetter[1] should be examined carefully.

In the case of the 512-point transformation, a study of the average power spectrum in the Hadamard domain suggest that the use of 30 dominant coefficients from the set of 512 coefficients may be sufficient to reconstruct good quality speech. Some initial calculations, based on a 7 k sample/s sampling rate, the use of 6 bits/coefficient and the use of 9 labelling bits/coefficient, indicate that if 30 coefficients are sufficient, then it should be possible to obtain good quality speech with a transmision bit rate of 6·15 k bits/s.

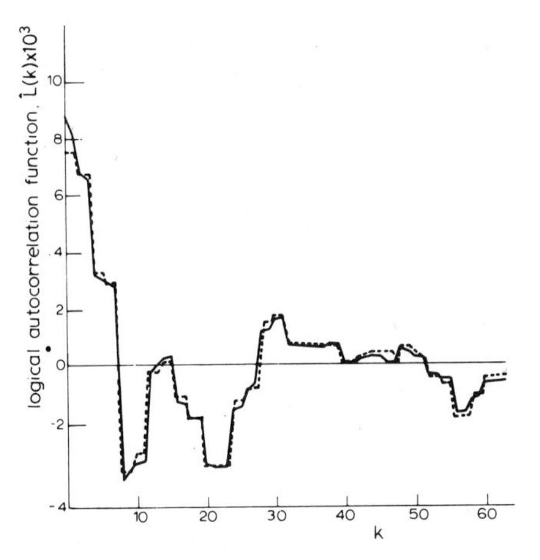

Fig. 17
Logical autocorrelation function

Complete sentence spoken by female; sampling rate 8 k samples/s
------ logical autocorrelation function reconstructed from 7 dominant coefficients
——— logical autocorrelation function of original sentence

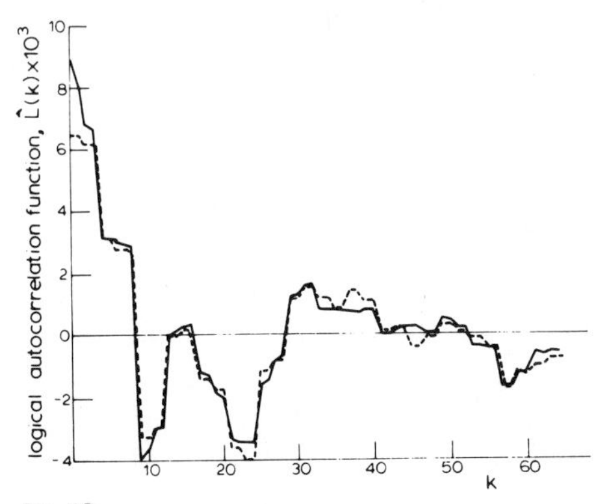

Fig. 18
Logical autocorrelation function

Complete spoken by female; sampling rate 8 k samples/s
------ logical autocorrelation function reconstructed from 5 dominant coefficients
——— logical autocorrelation function of original sentence

6 Conclusions

In the paper the results of an extensive analysis of the properties of 64-point Hadamard transformed speech are presented. Detailed information is given in Section 2 about the p.d.f.s of the Hadamard coefficients, the average power-density spectrum in the Hadamard domain, and the logical autocorrelation function of speech reconstructed from a limited number of Hadamard coefficients. The results contained in Section 2 indicate that good quality speech can be reconstructed from 6 to 8 dominant (large valued) Hadamard coefficients, and the initial listening tests reported in Section 3 confirm this. The indications are that the quality of the speech that can be reconstructed from 5, or fewer, Hadamard coefficients will not be acceptable. In Section 4 of the paper it is explained that, at first sight, it appears that approximately 50% of the transmission bit rate is needed for the sending of labelling information. However, the results of an investigation are presented, and from these results it is argued that the necessary number of labelling bits can be reduced by more than 50%. When using the modified-labelling technique it should be possible to obtain good quality speech when using a bit rate of 8 k bits/s.

Table 3
NUMBER OF BITS NEEDED TO IDENTIFY DOMINANT HADAMARD COEFFICIENTS ASSOCIATED WITH WALSH-FOURIER COMPONENT

Walsh-Fourier component	Number of bits needed to identify dominant Hadamard coefficients associated with Walsh-Fourier component
G_0	0
G_1	5
G_2	4
G_3	3
G_4	2
G_5	1
G_6	0

Table 4
SCHEME FOR ENCODING THE COEFFICIENTS OF THE 16-POINT HADAMARD TRANSFORMATION

Coefficient Number	1	2	3	4	5	6	7	8	9	10	11	12	13	14	15	16
Number of bits/coefficient	5	1	2	1	2	1	2	1	5	1	2	1	4	1	2	1

7 References

1 BOESSWETTER, C.: 'Analog sequency analysis and synthesis of voice signals'. Proceedings of the symposium on applied Walsh functions. Washington, D.C., 1970, pp. 220–229

2 CAMPANELLA, S.J., and ROBINSON, G.S.: 'Digital sequency decomposition of voice signals'. Proceedings of the symposium on applied Walsh functions, Washington, DC, 1970, pp. 230–237

3 CAMPANELLA, S.J., and ROBINSON, G.S.: 'A comparison of orthogonal transformations for digital speech processing', *IEEE Trans.*, 1971, **COM-19**, pp. 1045–49

4 SHUM. Y.Y., ELLIOTT, A.R., and BROWN, O.W.: 'Speech processing with Walsh-Hadamard transforms', *ibid.*, 1973, **AU-21**, pp. 174–179

5 LOPRESTI, P.V., and SURI, H.L.: 'A fast algorithm for the estimation of autocorrelation functions', *ibid.*, 1974, **ASSP-22**, pp. 449–453

6 ROBINSON, G.S.: 'Logical convolution and discrete Walsh and Fourier power spectra', *ibid.*, 1972, **AU-20**, pp. 271–280

Editor's Comments on Papers 38 Through 42

IMAGE CODING

Wintz (Paper 38), in a tutorial review paper, describes the philosophy of transform image coding and outlines the work that had been done up to that time (1972). While several researchers have developed the techniques for transform processing, it appears that the paper by Pratt, Kane, and Andrews (1969) on Hadamard transform image coding has stimulated the investigation of other discrete transforms for applications in image coding. The advantage of transform coding is that the image energy tends to be concentrated in low-frequency coefficients that need to be finely quantized while the rest can either be coarsely quantized or discarded altogether without incurring any appreciable error. The deleted coefficients can also be extrapolated at the receiver. Both the selection of transform coefficients and their quantization (bit allocation) can be a priori or adaptive.

Other advantages of transform coding are that it is less sensitive to picture statistics and less vulnerable to channel noise compared to predictive coding. However, in general, it is more complex to implement compared to the predictive schemes. Very efficient algorithms and specialized architecture have made the implementation of transform coding quite practical (Jalali and Rao, 1982; Belt, Keele, and Murray, 1977; Bessette and Schaming, 1980; Chan and Whiteman, 1983; Chen, 1981; Peters and Kanters, 1983; Ward and Stanier, 1983; Ohira, Hayakawa, and Matsumoto, 1978; Lynch and Reis, 1976*a*, 1976*b*; Papers

2, 21, 39, and 40). Based on transform coding, hardware has been built for commercial and defense applications in digital image coding for transmission and storage.

Transform coding has been applied to single frames (Ploysongsang and Rao, 1982; Camana, 1979), sequence of frames (Paper 41; Natarajan and Ahmed, 1977; Kamangar and Rao, 1981), monochrome, and color images (Chen and Smith, 1977). It has also been applied to hybrid coding, which is a combination of transform/predictive coding (Paper 42). It has been applied to subblocks of images and to full frames (Pearlman and Jakatdar, 1981). Also, motion compensation (Jain and Jain, 1981; Netravali and Stuller, 1979; Hein, 1982) has been added to transform coding. Such an intensive and varied activity implies enormous growth and potential for image processing by transform techniques.

REFERENCES

Belt, R. A., R. V. Keelé, and G. G. Murray, 1977, Digital TV Microprocessor System, *IEEE Natl. Telecommun. Conf., Los Angeles, Proc.* **10:**6-1-6-6.

Bessette, O. E., and W. B. Schaming, 1980, *A Two Dimensional Discrete Cosine Transform Video Bandwidth Compression System,* IEEE National Aerospace Electronics Conference (NAECON) Proceedings 1-7.

Camana, R., 1979, Video-bandwidth Compression: A Study on Tradeoffs, *IEEE Spectrum* **16:**24–29.

Chan, L. C., and P. Whiteman, 1983, Hardware Constrained Hybrid Coding of Video Imagery, *IEEE Aerosp. Electr. Syst. Trans.* **AES-19:**71–83.

Chen, W. H., 1981, Scene Adaptive Coder, *Int. Conf. Commun. Proc.* **22:**5.1–5.6.

Chen, W. H., and C. H. Smith, 1977, Adaptive Coding of Monochrome and Color Images, *IEEE Trans. Commun.* **COM-25:**1285–1292.

Hein, D. N., 1982, *Video Data Compression Using Motion Compensation,* IEEE MIDCON, Dallas, Tex., pp. 1-9.

Jain, J. R., and A. K. Jain, 1981, Displacement Measurement and Its Application in Interframe Image Coding, *IEEE Commun. Trans.* **COM-29:**1799–1808.

Jalali, A., and K. R. Rao, 1982, A High Speed FDCT Processor for Real Time Processing of NTSC Color TV Signal, *IEEE Electromagn. Compat. Trans.* **EMC-24:**278–286.

Kamangar, F. A., and K. R. Rao, 1981, Interfield Hybrid Coding of Component Color Television Signals, *IEEE Commun. Trans.* **COM-29:**1740–1753.

Lynch, R. T., and J. J. Reis, 1976*a,* Class of Transform Digital Processors for Compression of Multidimensional Data, U.S. Patent 3981443.

Lynch, R. T., and J. J. Reis, 1976*b, Haar Transform Image Coding,* IEEE National Telecommunications Conference, Dallas, Tx., pp. 44.3-1–44.3-5.

Natarajan, T. R., and N. Ahmed, 1977, Interframe Transform Coding of Monochrome Pictures, *IEEE Commun. Trans.* **COM-25:**1323–1329.

Netravali, A. N., and J. A. Stuller, 1979, Motion Compensated Transform Coding, *Bell Syst. Tech. J.* **58:**1703–1718.

Ohira, T., M. Hayakawa, and K. Matsumoto, 1978, Orthogonal Transform Coding System for NTSC Color Television Signals, *IEEE Commun. Trans.* **COM-26:**1454–1463.

Pearlman, W. A., and P. Jakatdar, 1981, *Hybrid DFT/DPCM Interframe Image Quantization,* IEEE International Conference on Acoustics, Speech, and Signal Processing, Atlanta, Ga., pp. 1121–1124.

Peters, J. H., and J. T. Kanters, 1983, *Hadamard Transform of Composite Video for Consumer Recording,* Picture Coding Symposium Proceedings, University of California, Davis, pp. 32-33.

Ploysongsang, A., and K. R. Rao, 1982, DCT/DPCM Processing of NTSC Composite Video Signal, *IEEE Commun. Trans.* **COM-30:**541-549.

Pratt, W. K., J. Kane, and H. C. Andrews, 1969, Hadamard Transform Image Coding, *IEEE Proc.* **57:**58-68.

Ward, J. S., and B. J. Stanier, 1983, Fast Discrete Cosine Transform Algorithm for Systolic Arrays, *Electron. Lett.* **19:**58-60.

Transform Picture Coding

PAUL A. WINTZ, MEMBER, IEEE

***Abstract*—Picture coding by first dividing the picture into subpictures and then performing a linear transformation on each subpicture and quantizing and coding the resulting coefficients is introduced from a heuristic point of view. Various transformation, quantization, and coding strategies are discussed. A survey of all known applications of these techniques to monochromatic image coding is presented along with a summary of the dependence of performance on the basic system parameters and some conclusions.**

I. Introduction

Bits and Pieces

CONSIDER a digitized image consisting of an $N \times N$ array of pels (picture elements) x_{ij} $i, j = 1, 2, \cdots, N$ each of which is quantized to one of 2^K gray levels $1, 2, \cdots, 2^K$. Any such array can be encoded by a sequence of N^2 K-bit code words that specify the gray levels of the N^2 pels, i.e., PCM encoding [1]. PCM requires K bit/pel or KN^2 bit/picture to code any of the 2^{KN^2} possible pictures. For $K=8$ and $N=256$ we have $N^2 = 65\,536$ pels, $KN^2 = 524\,288$ bit/picture, 8 bit/pel, and $2^{524\,288} \approx 10^{158\,000}$ possible pictures. The cameraman picture presented in Fig. 1 was reconstructed from a 256×256 array of 8-bit pels.

Fig. 1. The cameraman was reconstructed from a 256×256 array of 8-bit pels.

We can conceive of a more efficient encoder that has stored a unique code word for each of the $2^{524\,288}$ pictures. If the Huffman code is used to assign short code words to the more likely pictures and long code words to the least likely pictures the average number of bits required per picture would be close to the entropy

$$- \sum_{i=1}^{2^{524\,288}} p_i \log p_i$$

where p_i is the probability of the ith picture. Two practical considerations prohibit this approach: We do not know the probabilities p_i $i = 1, 2, \cdots, 2^{524\,288}$ and the storage and speed requirements for implementing the code book look-up procedure are beyond present-day technology.

One possibility that comes to mind is to block code the picture by first partitioning the picture into a number of $n \times n$ arrays of subpictures with $n \ll N$ as illustrated in Fig. 2, and then use the code book look-up procedure for each subpicture. However, a few quick calculations show that the code book is still too large. For example, the number of 4×4 arrays of pels with each pel quantized to 2^8 gray levels is $2^{128} \approx 10^{40}$.

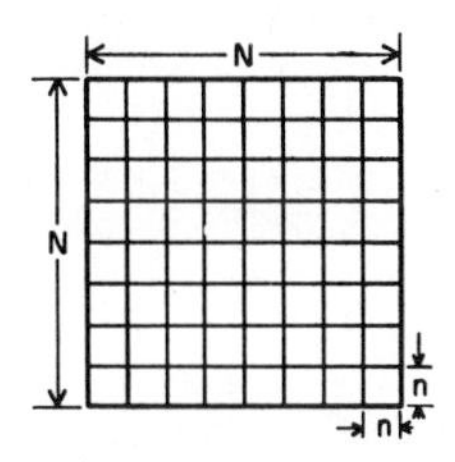

Fig. 2. Partitioning of the $N \times N$ picture into $(Nn)^2$ $n \times n$ subpictures.

Picture Structure

Pictorial data contain significant structure. Structure is a departure from randomness. Efficient source encoding can be accomplished by first determining the structure of the data and then developing encoding algorithms that are efficient for data having that structure. The more structure inherent in the data the greater the efficiency that can be achieved by matching the encoder to the data structure. On the other hand, if an encoder designed for a particular data structure is used to encode data having a different structure, the performance is degraded proportional to the amount of mismatch between the design structure and the actual structure.

Pictorial data are not homogenous—different regions of a picture contain different structures. Nonadaptive encoders are matched to the average data structure. Adaptive encoders are matched to the local data structure by first determining the local structure and then processing the local data with an algorithm that is efficient for that structure.

The structure inherent in pictorial data is not well understood. Pictures of natural objects have some structure due to the structure inherent in the universe, e.g., the shape of the earth, the direction of gravitational forces, etc. Man-made objects tend to have more structure. A few simple kinds of structure in pictures are more or less obvious. Pictures consist of a number of areas of nearly constant brightness. Statistics on the numbers of areas, their brightness, sizes, etc., have been collected by Nishikawa, Massa, and Mott-Smith [2], Gattis and Wintz [3], and others. See [3] for a literature survey.[1] A definite structure also exists in the boundaries between the areas of nearly constant brightness. These boundaries are

Manuscript received September 20, 1971; revised February 25, 1972.
The author is with the School of Electrical Engineering, Purdue University, Lafayette, Ind. 47907.

[1] For a more general survey see L. C. Wilkins and P. A. Wintz, "Bibliography on data compression, pictures properties, and picture coding," *IEEE Trans. Inform. Theory*, vol. IT-17, pp. 180–197, Mar. 1971.

usually smooth lines as illustrated in [4, figs. 6(a) and 7(a)]. Coding strategies that take these structures into account include the contour coding techniques developed at MIT [4] and Purdue [3], [5].

Statistical Coding

One manifestation of picture structure is in the picture statistics. Encoding techniques that match to the data statistics are referred to as *statistical coding* [6], [7]. Although Schreiber [8] measured a few third-order statistics, only first- and second-order moments can be thoroughly measured. Furthermore, techniques for matching to moments of higher order than first and second have not been developed. Unfortunately, means, covariances, and first-order probability density functions are very gross measures of picture structure. This can be demonstrated by measuring them for a set of pictures and then programming a computer to generate sample pictures having these same statistics [9]. The resulting pictures do not contain birds and bees, etc., but look like random noise. We conclude that pictures contain significantly more structure than can be accounted for by first- and second-order moments.

Psychovisual Coding

The sensitivity of the human visual system to errors in the reconstructed picture depends on the frequency spectrum of the error, the gray level, and amount of detail in the picture in the vicinity of the error, etc. Hence it is possible to increase the efficiency of the coder by allowing distortions that do not degrade subjective quality (how the picture looks to a human observer). Picture coders that take the characteristics of the human visual system into account are called *psychovisual coders* [6] or *psychophysical* coders [7]. Schreiber [7] gives a good summary of the properties of human vision and presents some coding techniques that take them into account.

Transform Coding

Transform coding is a method for accomplishing some aspects of both statistical and psychovisual coding. Transform coders perform a sequence of two operations, the first of which is based on statistical considerations and the second on psychovisual considerations. The first operation is a linear transformation that transforms the set of statistically dependent pels into a set of "more independent" coefficients. The second operation is to individually quantize and code each of the coefficients. The number of bits required to code the coefficients depends on the number of quantizer levels which is dictated by the sensitivity of the human vision to the subjective effect of the quantization error.

II. Linear Transformations

Coordinate System for Pictures

Let us interpret an $n \times n$ subpicture $n \leq N$ as a point in an n^2-dimensional coordinate system where each of the n^2 coordinates corresponds to one of the n^2 pels; the value of each coordinate is the gray level of the corresponding pel.

Consider a simplified example. $n=1$ is too simple while $n=2$ requires a four-dimensional space that is difficult to visualize. Hence we consider a 1×2 array (a subpicture consisting of two adjacent pels). Let $K=3$ so that each of the two pels can take on any of $2^3=8$ gray levels. Then each of the $2^{2 \cdot 3}=64$ possible 1×2 arrays can be represented by one of the 64 points in the two-dimensional space illustrated in Fig. 3(a). Since adjacent pels are likely to have nearly the same gray

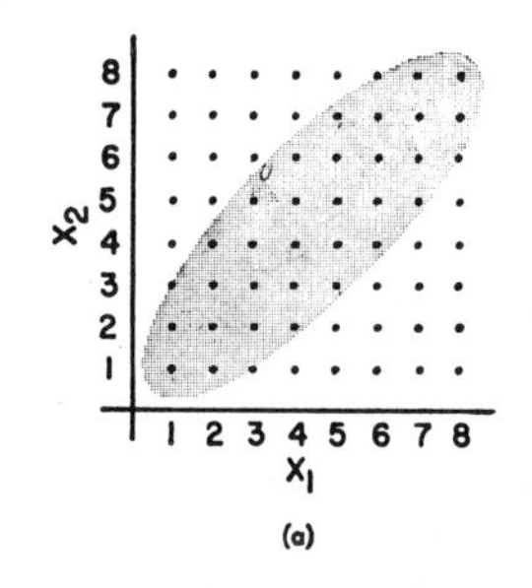

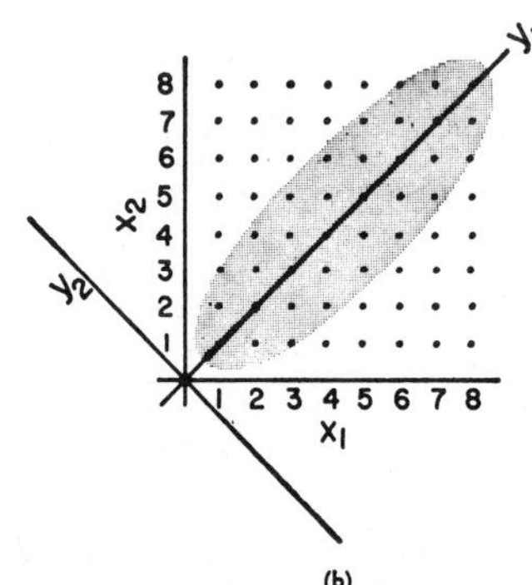

Fig. 3. (a) Coordinate system for the gray levels of two adjacent pels x_1 and x_2. The most likely subpictures are the ones in the shaded area. (b) New coordinate system.

level the most likely subpictures are the ones in the vicinity of $x_1=x_2$, i.e., those in the shaded area.

Now rotate the coordinate system as illustrated in Fig. 3(b). In the new coordinate system the more likely subpictures are not in the vicinity of $y_1=y_2$ but are lined up with the y_1 axis. Hence the variables y_1 and y_2 are "more independent" than were x_1 and x_2, e.g., y_2 is likely to be small independent of the value of y_1. Rotating the coordinate system also rearranged the variances. The total is the same, $\sigma_{y_1}^2+\sigma_{y_2}^2=\sigma_{x_1}^2+\sigma_{x_2}^2$, but whereas both pels had the same variance, $\sigma_{x_1}^2=\sigma_{x_2}^2$, more of the variance is now in the first coefficient, $\sigma_{y_1}^2>\sigma_{y_2}^2$. Finally, note that given the coefficients y_1 and y_2 we can perform the inverse rotation to obtain the pels x_1 and x_2.

The same procedure can be used for an $n \times n$ array of pels each of which is quantized to one of 2^K gray levels. In this case an n^2-dimensional coordinate system is required with each coordinate labeled with the values $1, 2, \cdots, 2^K$. Each of the 2^{n^2K} points corresponds to one of the 2^{n^2K} possible pictures.

One-Dimensional Transformations

Rotating the n^2-dimensional coordinate system of the n^2 pels corresponds to arranging the n^2 pels into the n^2 vector $\boldsymbol{x}=[x_1, x_2, \cdots, x_{n^2}]^t$ and performing the linear transformation

$$\boldsymbol{y} = \boldsymbol{a}\boldsymbol{x} \tag{1}$$

where $\boldsymbol{a}$ is an $n^2 \times n^2$ unitary matrix with elements a_{ki} $k, i=1, 2, \cdots, n^2$ that determines the rotation and $\boldsymbol{y}=[y_1, y_2, \cdots, y_{n^2}]^t$ is an n^2 vector whose elements give the values of the coefficients on the rotated coordinates. The inverse rotation is accomplished by the inverse transformation

$$\boldsymbol{x} = \boldsymbol{a}^t\boldsymbol{y} \tag{2}$$

where $\boldsymbol{a}^t$ is the transpose of $\boldsymbol{a}$. (For unitary matrices $\boldsymbol{a}^{-1}=\boldsymbol{a}^t$.)

According to (1) each coefficient y_k is a linear combination of all of the pels

$$y_k = \sum_{i=1}^{n^2} a_{ki} x_i, \qquad k = 1, 2, \cdots, n^2 \tag{3}$$

and similarly (2) gives each pel as a linear combination of all the coefficients

$$x_i = \sum_{k=1}^{n^2} a_{ik} x_k, \qquad i = 1, 2, \cdots, n^2. \tag{4}$$

The best reversible transformation, linear or otherwise, would be one that results in independent variables y. This transformation cannot be determined for two reasons. First, it evidently depends on very detailed statistics—the joint probably density function of the n^2 pels—which have not been deduced from basic physical laws and cannot be measured. Second, even if the joint density function of the n^2 pels was known, the problem of determining a reversible transformation that results in independent coefficients is unsolved. The closest we can get with linear transformations to a transformation that produces independent coefficients is the one that produces uncorrelated coefficients. The resulting coefficients are uncorrelated but not necessarily independent.

The transformation matrix $\boldsymbol{a}$ that produces uncorrelated coefficients can be computed from the covariance matrix of the pels

$$\boldsymbol{C}_x = E\{(\boldsymbol{x} - E\{\boldsymbol{x}\})(\boldsymbol{x} - E\{\boldsymbol{x}\})^t\}. \tag{5}$$

The columns of $\boldsymbol{a}$ are the normalized eigenvectors of $\boldsymbol{C}_x$, i.e., the n^2 column vector solutions $\boldsymbol{\phi}$ to the matrix equation

$$\boldsymbol{C}_x \boldsymbol{\phi} = \lambda \boldsymbol{\phi}. \tag{6}$$

The covariance matrix of the coefficients is then given by

$$\boldsymbol{C}_y = \begin{bmatrix} \lambda_1 & 0 & \cdot & \cdot & \cdot & 0 \\ 0 & \lambda_2 & \cdot & \cdot & \cdot & 0 \\ \cdot & \cdot & \cdot & & & \cdot \\ \cdot & \cdot & & \cdot & & \cdot \\ \cdot & \cdot & & & \cdot & \cdot \\ 0 & 0 & \cdot & \cdot & \cdot & \lambda_{n^2} \end{bmatrix} \tag{7}$$

where $\lambda_1, \lambda_2, \cdots, \lambda_{n^2}$ are the eigenvalues of (6). This transformation is known as the *eigenvector transformation* or the *Hotelling transformation* [10].

Two-Dimensional Transformations

It is also possible to arrange the $n \times n$ array of pels into an $n \times n$ matrix $\boldsymbol{X}$ with elements x_{ij} $i, j = 1, 2, \cdots, n$

$$\boldsymbol{X} = \begin{bmatrix} x_{11} & x_{12} & \cdot & \cdot & \cdot & x_{1n} \\ x_{21} & x_{22} & \cdot & \cdot & \cdot & x_{2n} \\ \cdot & \cdot & \cdot & & & \cdot \\ \cdot & \cdot & & \cdot & & \cdot \\ \cdot & \cdot & & & \cdot & \cdot \\ x_{n1} & x_{n2} & \cdot & \cdot & \cdot & x_{nn} \end{bmatrix} \tag{8}$$

and then transform this $n \times n$ matrix into another $n \times n$ matrix $\boldsymbol{Y}$ with elements y_{kl} $k, l = 1, 2, \cdots, n$

$$\boldsymbol{Y} = \begin{bmatrix} y_{11} & y_{12} & \cdot & \cdot & \cdot & y_{1n} \\ y_{21} & y_{22} & \cdot & \cdot & \cdot & y_{2n} \\ \cdot & \cdot & \cdot & & & \cdot \\ \cdot & \cdot & & \cdot & & \cdot \\ \cdot & \cdot & & & \cdot & \cdot \\ y_{n1} & y_{n2} & \cdot & \cdot & \cdot & y_{nn} \end{bmatrix} \tag{9}$$

by premultiplying $\boldsymbol{X}$ by the fourth-order point tensor $\boldsymbol{A}$ with elements a_{klij} $k, l, i, j = 1, 2 \cdots, n$. Each coefficient is now given by

$$y_{kl} = \sum_{i=1}^{n} \sum_{j=1}^{n} a_{klij} x_{ij}, \qquad k, l = 1, 2, \cdots, k. \tag{10}$$

Similarly, the inverse transformation gives each pel as a linear combination of the coefficients

$$x_{ij} = \sum_{k=1}^{n} \sum_{l=1}^{n} a_{ijkl} y_{kl}, \qquad i, j = 1, 2, \cdots, n. \tag{11}$$

Clearly, both the one-dimensional transformation (3) and the two-dimensional transformation (10) result in a set of n^2 coefficients each of which is a linear combination of the n^2 pels. Hence the difference in these two transformations is simply a matter of notation.

A number of two-dimensional linear transformations other than the eigenvector transformation have been proposed for picture coding. These include the Fourier transformation

$$a_{klij} = \frac{1}{n} \exp\left[-2\pi\sqrt{-1}\,(ki + lj)/n\right] \tag{12}$$

and the Hadamard transformation

$$a_{klij} = \frac{1}{n}(-1)^{b(k,l,i,j)} \tag{13}$$

where

$$b(k, l, i, j) = \sum_{h=0}^{\log_2 n - 1} [b_h(k) b_h(l) + b_h(i) b_h(j)]$$

$b_h(\cdot)$ is the hth bit in the binary representation of $(\cdot)$, and n is a power of 2. Both are members of a class of Kroneckered matrix transformations [11] that have $2n^2 \log_2 n^2$ degrees of freedom and can, therefore, be implemented with $2n^2 \log_2 n^2$ computer operations as opposed to the n^4 operations required by the Hotelling transformation. The Fourier transformation requires $2n^2 \log_2 n^2$ multiplications and a like number of additions whereas the Hadamard transformation requires only $2n^2 \log_2 n^2$ additions since all of the entries (13) are $+1$ or -1 except for the normalizing constant n. The coefficients produced by both of these transformations are usually more dependent than those produced by the Hotelling transformation. Andrews [12] and Pearl [13] have investigated "distances" between the Hotelling, Fourier, Hadamard, and Haar transformations.

Basis Pictures

Another interpretation of the two-dimensional transformation of the $n \times n$ array of pels $\boldsymbol{X}$ into the $n \times n$ array of coefficients $\boldsymbol{Y}$ is possible. Let us write (11) in the form

$$X = \sum_{k=1}^{n} \sum_{l=1}^{n} y_{kl} \boldsymbol{a}_{kl} \qquad (14)$$

and interpret this as a series expansion of the $n \times n$ picture $\boldsymbol{X}$ onto the n^2 $n \times n$ basis pictures

$$\boldsymbol{a}_{kl} = \begin{bmatrix} a_{kl11} & a_{kl12} & \cdot & \cdot & \cdot & a_{kl1n} \\ a_{kl21} & a_{kl22} & \cdot & \cdot & \cdot & a_{kl2n} \\ \cdot & \cdot & \cdot & & & \cdot \\ \cdot & \cdot & & \cdot & & \cdot \\ \cdot & \cdot & & & \cdot & \cdot \\ a_{kln1} & a_{kln2} & \cdot & \cdot & \cdot & a_{klnn} \end{bmatrix}, \quad k, l, = 1, 2, \cdots, n \qquad (15)$$

with the y_{kl} $k, l = 1, 2, \cdots, n$ the coefficients of the expansion. Hence (14) gives the picture $\boldsymbol{X}$ as a weighted sum of the basis pictures $\boldsymbol{a}_{kl}$. The weights y_{kl} are given by (10) which can be written in the form

$$y_{kl} = \boldsymbol{a}_{kl} \boldsymbol{X}. \qquad (16)$$

The weight y_{kl} given to basis picture $\boldsymbol{a}_{kl}$ in the sum (14) can be interpreted as the amount of correlation between the picture $\boldsymbol{X}$ and the basis picture $\boldsymbol{a}_{kl}$.

The 256 16×16 basis pictures for the Hotelling and Hadamard transformations for $n = 16$ are presented in Fig. 4. The Hadamard basis pictures are not picture dependent whereas the Hotelling basis pictures are the "eigenpictures" for the cameraman.[2]

We have already noted that if the set of basis pictures is chosen such that the coefficients are "more independent" than the pels, then the variances of the coefficients are unequal. Therefore, we can index the basis pictures such that the terms in the sum (14) are ordered according to the variances of the coefficients so that successive terms contribute proportionally less and less, on the average, to the total. Indeed, for some choices of basis pictures the coefficients become insignificant after the first, say η, terms so that an approximate representation for the picture can be obtained by truncating the series after the first η terms, i.e.,

$$X = \sum_{k=1}^{n} \sum_{l=1}^{n} y_{kl} \boldsymbol{a}_{kl} \approx \sum_{k=1}^{\eta}{}' \sum_{l=1} y_{kl} \boldsymbol{a}_{kl} = X'. \qquad (17)$$

The mean-square approximation error between the original picture $\boldsymbol{X}$ and the approximate picture $\boldsymbol{X}'$ is given by

$$\begin{aligned} \epsilon_a^2 &= E\{\|X - X'\|^2\} \\ &= E\left\{\left\| \sum_{k=1}^{n} \sum_{l=1}^{n} y_{kl} \boldsymbol{a}_{kl} - \sum_{k=1}^{\eta}{}' \sum_{l=1} y_{kl} \boldsymbol{a}_{kl} \right\|^2\right\} \\ &= E\left\{\left\| \sum_{\eta+1}^{k=n} \sum^{l=n} y_{kl} \boldsymbol{a}_{kl} \right\|^2\right\} \\ &= \sum_{\eta+1}^{k=n} \sum^{l=n} \sigma_{y_{kl}}^2. \end{aligned} \qquad (18)$$

The last step follows because the basis pictures are orthonormal. Equation (18) states that the mean-square approximation error is given by the sum of variances of the discarded coefficients. Equation (16) states that the y_{kl} and, therefore, their variances depend on the basis pictures.

[2] Computing the actual Hotelling basis pictures would have required inverting a 256×256 matrix. Therefore, the basis pictures presented in Fig. 4(b) and used in the sequel were obtained by approximating the elements in the covariance matrix $C_x(\Delta_h, \Delta_v)$ with the elements exp $(-0.125|\Delta_h| - 0.249|\Delta_v|)$ which is a good approximation for the cameraman [63], [68].

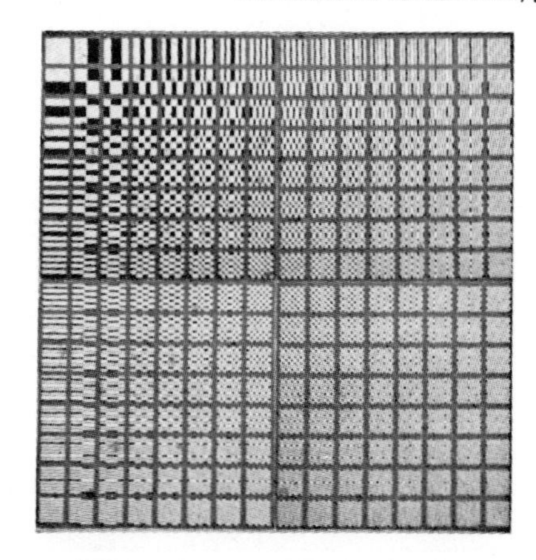

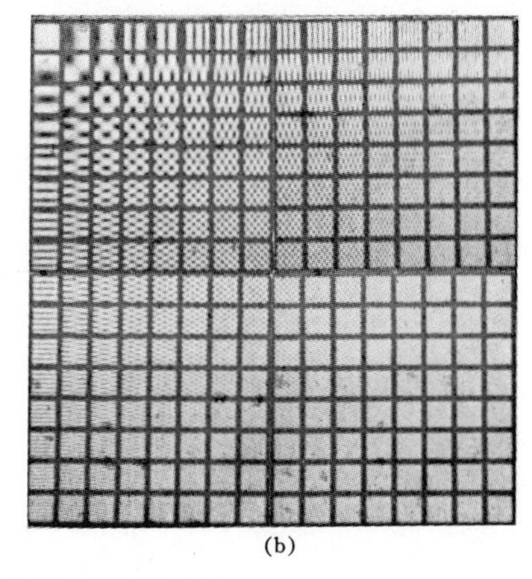

(b)

Fig. 4. (a) The 256 16×16 sequency ordered basis pictures for the Hadamard transformation. (b) The 256 16×16 Hotelling basis pictures for the cameraman picture of Fig. 1.

We now pose the following problem: What set of basis pictures minimizes the mean-square error (18) by packing the most variance into the first η coefficients? The solution to this problem is the same as the solution to the seemingly unrelated problem of determining the set of basis pictures that produce uncorrelated coefficients. The Hotelling transformation: a) produces uncorrelated coefficients; b) minimizes the mean-square approximation error; and c) packs the maximum amount of variance into the first η coordinates (for any η).

The cameraman picture was divided into 16×16 subpictures and each subpicture expanded in a Hotelling, Fourier, and Hadamard series expansion. The sample variances of the coefficients, presented in Fig. 5(a), can be interpreted as generalized power spectra of the pictures. The Fourier variances give the usual power spectrum. Note from Fig. 5(b) that all three transformations are approximately equally efficient in packing the variances into lower order coefficients.

Retaining the η coefficients with the largest variances corresponds to dividing the domain of the coefficients (9) into two zones one of which contains the retained coefficients and the other the discarded coefficients. Hence we can view the series truncation process (17) as a "zonal filtering" or "masking" in the transform domain. This procedure is sometimes referred to as "zonal sampling." Zonal sampling is a nonadap-

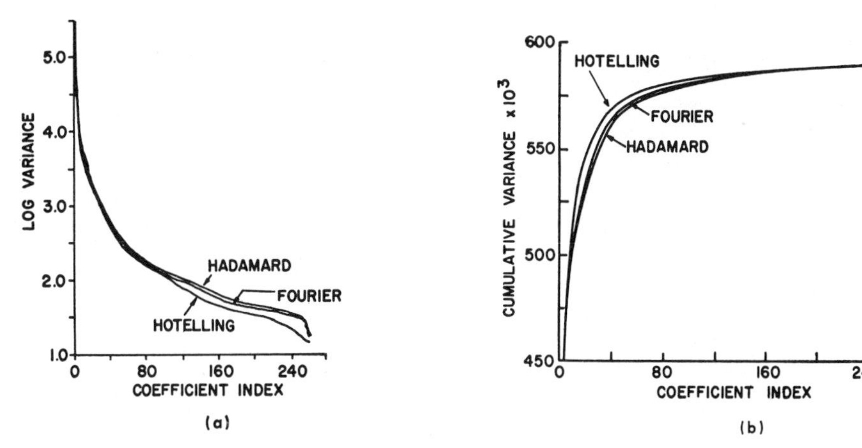

Fig. 5. (a) The sample variances $\sigma_{kl}^2 k$, $l=1, 2, \cdots, 16$ of the 256 coefficients y_{kl} averaged over the 256 subpictures of the cameraman and ordered according to their magnitudes. (b) Cumulative sums of the variances.

Fig. 6. Reconstructed pictures X' obtained by retaining the first $\eta=128$ of the $n^2=256$ terms in the sum (17) for each of the 256 16×16 subpictures of the 256×256 cameraman. (a) Hotelling transformation, $\epsilon_a^2=0.34$ percent: (b) Fourier transformation, $\epsilon_a^2=0.45$ percent. (c) Hadamard transformation, $\epsilon_a^2=0.49$ percent.

Fig. 7. Reconstructed pictures X' obtained by retaining the first $\eta=64$ of the $n^2=256$ terms in the sum (17) for each of the 256 16×16 subpictures of the 256×256 cameraman. (a) Hotelling transformation, $\epsilon_a^2=0.86$ percent. (b) Fourier transformation, $\epsilon_a^2=1.06$ percent. (c) Hadamard transformation, $\epsilon_a^2=1.14$ percent.

tive technique for retaining those terms in the sum (17) that, on the average, have the largest energy. Another alternative would be to first evaluate all n^2 coefficients (16) and then retain only those coefficients that exceed a preset threshold. This is an adaptive technique that retains only those coefficients that are large for the particular picture being processed and is sometimes referred to as "threshold sampling." When threshold sampling is used certain "bookkeeping information" that specifies which coefficients have been retained must also be coded.

Some approximations to the cameraman picture are presented in Figs. 6 and 7 along with the mean-square errors

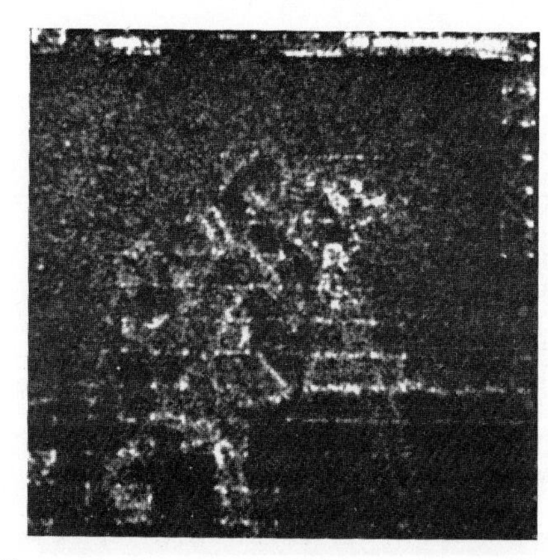

Fig. 8. The magnitude error between the original picture X of Fig. 1 and the picture X' of Fig. 6(b) scaled by eight, i.e., $8|X-X'|$. Black corresponds to zero error and white to an error of $256/8=32$ gray levels. The error was scaled by 8 to make it more visible.

between them and the original picture. These pictures were obtained by dividing the picture into 16×16 subpictures, representing each subpicture with its expansion onto the 16×16 basis pictures, and truncating the expansion after η terms (zonal sampling). Approximately one half of the terms ($\eta=128$) can be discarded with no visible degradation in picture quality although some mean-square error (18) is incurred. Truncating the expansion prior to the 128th term results in a loss of resolution due to the low-pass filtering effect. That is, from the basis pictures presented in Fig. 4 we note that increasing index corresponds to increasing spatial frequency content. Hence truncating the expansion (17) at η terms corresponds to discarding all picture energy at frequencies higher than those corresponding to the first η basis pictures. Also note that for small η the subpicture edges are visible. Fig. 8 shows the error between the reconstructed picture of Fig. 6(b) and the original picture of Fig. 1. Note that the largest errors occur in the high detail (high-frequency) parts of the picture and at the block edges.

Finally, the reader is referred to Landau and Slepian [44] who describe a somewhat different picture coding philosophy based on the basis picture concept. Whereas we seek basis pictures that result in independent coefficients (basis pictures matched to the picture statistics) Landau and Slepian seek a set of basis pictures matched to the area properties of pictures. They argue that the Hadamard basis are such a set.

Historical Notes

Hotelling [10] was the first to derive and publish the transformation that transforms discrete variables into uncorrelated coefficients. He referred to this technique as *the method of principal components*. His paper gives considerable insight into the method and is worth reading. Hotelling's transformation was rediscovered by Kramer and Mathews [14] and Huang and Schultheiss [15].

The analogous transformation for transforming continuous data into a set of uncorrelated coefficients was discovered by Karhunen [16] and Loéve [17] and is called the Karhunen–Loéve expansion. (See Selin [18] for an excellent discussion.) The result that the Karhunen–Loéve expansion minimizes the mean-square truncation error was first published by Koschman [19] and rediscovered by Brown [20]. (See also Totty [21].)

Transform picture coding evolved as a natural extension to pictorial data of the basic transform coding techniques developed for one-dimensional data such as speech, electrocardiograms, etc. Since pictorial data are sometimes modeled as a Markov process we list a few references concerned with the transform coding of one-dimensional Markov data [12]–[15], [22]–[35].

III. Quantizing the Coefficients

After the $n\times n$ array of pels has been transformed into $\eta\leq n^2$ coefficients each coefficient must be quantized and coded.

Since the variances of the coefficients vary widely as illustrated in Fig. 5(a) it would be inefficient to use the same quantizer for all coefficients. That is, if the quantizer output levels are adjusted to span the range of the coefficient with the largest variance, then the coefficients with much smaller variances would fall in a much smaller range with the result that most of the quantizer levels are not used. This effect can be negated by first scaling each coefficient by the inverse of its standard deviation to form the normalized coefficients $\gamma_{kl}=y_{kl}/\sigma_{kl}$ all of which have unit variance and can be efficiently quantized with the same quantizer. The decoder must, of course, scale each γ_{kl} by σ_{kl} to obtain the y_{kl}.

Using the same quantizer for each normalized coefficient γ_{kl} and the natural or gray code to assign equal length code words to all quantizer output levels results in each coefficient requiring the same number of bits. Since the coefficients with the larger variances contribute significantly more to the reconstructed picture (17), on the average, than the coefficients with the smaller variances, it appears that the total distortion due to quantizing the coefficients could be lessened by alloting more quantization levels and/or bits to the coefficients with the larger variances and proportionally fewer to the coefficients with the smaller variances.

Furthermore, recall that each coefficient corresponds to a particular spatial frequency band. Since the sensitivity of the human visual system to distortion is dependent on the frequency of the distortions, it appears that better subjective quality could be obtained by alloting more quantization levels and/or bits to those coefficients corresponding to the frequencies to which the eye is most sensitive.

Quantizing for Minimum Mean-Square Error

The total mean-square quantization error is usually defined as

$$\epsilon_q^2 = E\left\{\sum_{k=1}^{n}\sum_{l=1}^{n}(y_{kl}-\hat{y}_{kl})^2\right\} \tag{19}$$

where $\hat{y}_{kl}$ is the quantized value of y_{kl}. Equation (19) depends on the joint probability density function of y_{kl} and $\hat{y}_{kl}$. Since each coefficient is a linear combination of n^2 pels the central limit theorem indicates that the distributions of the y_{kl}'s tend toward Gaussian since some of the pels are more or less independent. Indeed, histograms for the coefficients for various transformations have been constructed and found to be roughly "bell-shaped." This effect becomes more pronounced with increasing index.

Panter and Dite [36] and Max [37] investigated quantization strategies that minimize the mean-square error $E\{(y_{kl}-\hat{y}_{kl})^2\}$ of a single coefficient. They found that if the probability density function of y_{kl} is uniform a uniform quantizer (uniformly space output levels) is optimum. For other distributions the mean-square error can be decreased by using a nonuniform quantizer with the spacing between output levels decreased in regions of high probability and increased in regions of low probability. For the Gaussian distribution the nonuniform quantizer can be 20–30 percent more efficient than the uniform quantizer [37].

Nonuniform quantization is usually accomplished by companding [38]. This involves passing y_{kl} through a nonlinearity called a compressor and then into a uniform quantizer. The combined effect is identical to that of a nonuniform quantizer. The decoder must then pass $\hat{y}_{kl}$ through the inverse nonlinearity called an expandor.

Gish and Pierce [39] and Wood [40] showed that the uniform quantizer minimizes the entropy of the quantizer output for a large class of distortion measures independent of the probability distribution of the input provided the quantization is done sufficiently fine. This indicates that uniform quantizers are inherently more efficient than nonuniform quantizers for a single coefficient. Sophisticated coding schemes, e.g., the Huffman code, are usually required to achieve this efficiency.

Quantization strategies for minimizing the total mean-square error (19) have been investigated by Huang and Schultheiss [15] and Hayes and Bobilin [41]. Huang and Schultheiss determined the optimum allocation of a total of M bits to the n^2 coefficients. They found that the number of bits m_{kl} used to code coefficient y_{kl} should be proportional to $\log \sigma_{kl}^2$. They give an algorithm for computing the m_{kl} $k, l=1, 2, \cdots, n$ such that (19) is minimized for a given $M=\sum_{k=1}^{n}\sum_{l=1}^{n} m_{kl}$ and set of variances σ_{kl}^2 $k, l=1, 2, \cdots, n$. Ready and Wintz [74] give an algorithm for computing the m_{kl} $k, l=1, 2, \cdots, n$ and M such that (19) is minimized for a given n^2 and set of variances σ_{kl}^2 $k, l=1, 2, \cdots, n$. This technique is called *block quantization*. It is significantly more efficient than using the same number of bits $m_{kl}=M/n^2$ for all coefficients. Its disadvantage lies in the problems inherent in handling binary words of unequal lengths.

Hayes and Bobilin suggested a different approach. They use the same uniform quantizer for all of the coefficients. The number of quantizer output levels and the spacings are determined to achieve adequate quantization of the coefficient with the largest variance. The coefficients with smaller variances will then tend to fall near the center of the quantizer range. Hence the probabilities associated with the quantizer outputs, averaged over all coefficients, are large near the center of the range and small at the ends. This set of probabilities is used to generate a Huffman code. This quantization strategy is significantly more efficient than block quantization. Its disadvantage is the complexity required to implement the Huffman code.

Quantizing for Best Subjective Quality

Experiments have been performed to determine the bit assignments m_{kl} for the block quantizer that yield reconstructed pictures having the best subjective quality. In general, the results indicate bit assignments quite close to those given by the mean-square error criterion although in some cases the subjective quality was slightly improved by assigning more bits to the coefficients with the larger variances and fewer to the coefficients with the smaller variances than suggested by the rule $m_{kl} \sim \log \sigma_{kl}^2$. This is probably due to the fact that the sensitivity of the human visual system to distortion as well as the coefficient variances is inversely proportional to spatial frequency.

Experiments have also been performed to optimize Hayes and Bobilin's method relative to subjective quality by finding the set of numbers u_{kl} $k, l=1, 2, \cdots, n$ such that if coefficient y_{kl} is first scaled by u_{kl} and then quantized with the same uniform quantizer, an approximately equal amount of subjective distortion is generated by each coefficient [66].

Effect of Quantization Error

The effect of quantizing the coefficients on the reconstructed picture is illustrated in Figs. 9 and 10. The cameraman was divided into 16×16 subpictures and each subpicture expanded onto the Fourier basis pictures. The expansions were truncated at $\eta=128$ terms and the 128 retained coefficients quantized by each of the three methods discussed in the preceding paragraphs. Fig. 11 shows the error between the pictures of Figs. 9(b) and 6(b).

IV. Discussion

Transform Coding Parameters

Transform coding performance depends on a number of parameters: 1) the transformation; 2) the quantization strategy; 3) the subpicture size; 4) the subpicture shape.

Transformation: The best transformation from both a mean-square error and subjective quality viewpoint is the Hotelling transformation, but it is closely followed by the Fourier transformation which is closely followed by the Hadamard transformation. Each is seprated by 0.1 or 0.2 bit/pel for $n=8$ or 16. For $n=4$ the performances are essentially the same.

Quantization Strategy: Both mean-square error and subjective quality are quite sensitive to the efficiency with which bits are used to code the coefficients. The most simple strategy is to form the normalized coefficients $\gamma_{kl}=y_{kl}/\sigma_{kl}$ and use the same quantizer (number of bits) for each coefficient. If η coefficients are retained and m bits used to code each coefficient then a total of $m\eta/n^2$ bit/pel are required. For good quality reproductions half the coefficients must be retained and at least 7 bit per coefficient must be used. Hence $m\eta/n^2 \approx m(n^2/2)/n^2=m/2=3.5$ bit/pel are required. If the normalized coefficients are quantized with the same 7- or 8-bit quantizer, but only the $m_{kl} \sim \log \sigma_{y_{kl}}$ most significant bits retained (block quantization) the same quality pictures can be obtained with a savings of about 1 bit/pel. Sometimes a further 0.1 bit/pel can be saved by choosing the bit assignments to give the best subjective quality. A further saving of 0.2 to 0.3 bit/pel can be achieved by quantizing the coefficients with the same 9-bit quantizer and using a Huffman code to assign code words of unequal lengths to the quantizer output levels.

Subpicture Size: Mean-square error performance should improve with increasing n since the number of correlations taken into account increases with n. However, most pictures contain significant correlations between pels for only about 20 adjacent pels although this number is strongly dependent on the amount of detail in the picture. (See [68, fig. 2].) Hence a point of diminishing returns is reached and $n>16$ is not warranted. Even smaller n, say $n=8$, does not significantly increase the error.

This argument does not appear to apply when subjective quality is the criterion of goodness. The subjective quality appears to be essentially independent of n for $n \geq 4$. Since the

(a) (b) (c)

Fig. 9. Reconstructed pictures obtained by quantizing the $\eta=128$ coefficients of the picture in Fig. 6(b) using an average of 2 bit/pel (4 bit per retained coefficient). (a) All 128 normalized coefficients y_{kl}/σ_{kl} quantized to 16 levels, $\epsilon_q^2=2.09$ percent. (b) The 128 coefficients block quantized, i.e., the number of quantization levels for coefficient $y_{kl} \sim \sigma_{kl}$, $\epsilon_q^2=0.78$ percent. (c) The 128 coefficients quantized with the same 746-level quantizer with a Huffman code, $\epsilon_q^2=0.02$ percent.

(a) (b) (c)

Fig. 10. Reconstructed pictures obtained by quantizing the $\eta=128$ coefficients of the picture of Fig. 6(b) using an average of 1 bit/pel (2 bit per retained coefficient). (a) All 128 normalized coefficients y_{kl}/σ_{kl} quantized to 4 levels, $\epsilon_q^2=8.68$ percent. (b) The 128 coefficients block quantized, i.e., the number of quantization levels for coefficient $y_{kl} \sim \sigma_{kl}$, $\epsilon_q^2=2.21$ percent. (c) The 128 coefficients quantized with the same 158 level quantizer with a Huffman code, $\epsilon_q^2=0.26$ percent.

number of computations per pel is proportional to n, 4×4 is a reasonable choice for subpicture size.

An optimum subpicture size is expected for the adaptive techniques since too large a subpicture size, say $n=N$, does not allow adaptation to the local picture characteristics while too small a size, say $n=1$, does not allow pel correlations to be taken into account. It appears that the best size is between $n=4$ and $n=8$ with $n=4$, 6, 8 yielding essentially the same performance relative to both mean-square error and subjective picture quality.

Subpicture Shape: Transforming two-dimensional arrays ($n\times n$ arrays) of pels yields better performance than one-dimensional ($1\times n$) arrays of pels but the gain is surprisingly small—about 0.2 bit/pel [42], [66], [71]. However, a larger n is required for one-dimensional arrays. Whereas two-dimensional arrays of size 4×4 are reasonable, one-dimensional arrays of 1×16 arrays are reasonable.

Adaptive Techniques

Transform coders can be made to adapt to local picture structure by allowing a number of modes of operation and, for each subpicture, choosing the mode that is most efficient for that subpicture. Bookkeeping information that indicates which mode was used must be coded along with the subpicture. In general, increasing the number of modes decreases the number of bits required to code the subpicture pels, but increases the number of bits required to code the bookkeeping information.

Since the Hotelling transformation is matched to the subpicture statistics one is tempted to use different transformations for subpictures with different statistics. However, if we segregate the cameraman subpictures with different statistics into different groups and compute the eigenpictures for each group we would find them strikingly similar. Hence using different transformations for different subpictures is not generally warranted. On the other hand, once the transformation is chosen (whether it be Hotelling, Fourier, Hadamard, or another) significant efficiencies can be achieved by adapting to the coefficients generated for each subpicture. A number of schemes for accomplishing this have been reported, but most are variations of the following:

Method 1: Compute all n^2 coefficients (ordered according to their variances) and determine the smallest η for which

$$\sum_{k=1}^{\eta}\sum_{l=1} |y_{kl}|^2 \Big/ \sum_{k=1}^{n}\sum_{l=1}^{n} |y_{kl}|^2$$

Fig. 11. The magnitude error between the pictures of Fig. 6(b) and Fig. 9(b) scaled by eight. Black corresponds to zero error and white to an error of 256/8 = 32 gray levels. The error was scaled by 8 to make it more visible.

(a)

(b)

Fig. 12. Reconstructed pictures obtained after dividing the picture into 16×16 subpictures, expanding each subpicture into a Fourier series expansion and retaining the first $\eta = 128$ terms, block quantizing the coefficients using 2 bit/pel, and making random bit error in the coefficients. (a) Bit error rate = 10^{-3}. (b) Bit error rate = 10^{-2}.

exceeds a predetermined threshold (like 0.99). Code the first η coefficients and code the number η. Making the threshold dependent on the average subpicture brightness y_{11} improves the subjective quality because of the Weber–Fechner law.

Coding method	bits/pel
PCM	8
DPCM	3
DPCM, 2-dimensional, adaptive	2.3
Transform, 1-dimensional	2.3
Transform, 2-dimensional	2
Transform, 2-dimensional, adaptive	1

Fig. 13. Comparison of coding methods on the basis of the number of bits per pel required to achieve approximately the same quality as 7-bit PCM for moderate detail pictures such as the cameraman.

Method 2: Compute all n^2 coefficients (ordered according to their variances) and retain all that exceed a predetermined threshold. Code the retained coefficients and code which subset of coefficients was retained. As in method 1 the threshold can depend on the average brightness.

Note that method 1 wastes bits by coding all the small coefficients contained within the first η coefficients, but requires only a few bits to code the bookkeeping information. Method 2 codes only the large coefficients, but requires many more bits to code the bookkeeping information. Surprisingly, both of these schemes, as well as most variations of them, achieve almost identical performances. They all require about 1 bit/pel less than the nonadaptive techniques to achieve the same mean-square error or subjective quality.

Picture Complexity

The numbers given in the preceding subsections are for pictures containing a moderate amount of detail such as the camerman. For pictures with less detail, such as head and shoulder portraits, the same mean-square error and subjective quality can be achieved with about 0.5 fewer bit/pel. More detailed pictures such as aerial photographs require about 0.5 more bit/pel for the same quality.

Effect of Bit Errors in the Coefficients

Since the decoder reconstructs the pels from linear combinations of the coefficients, an error in a coefficient leads to errors in all the pels reconstructed from it. If the complete picture is transformed as a unit, i.e., $n = N$, then one or more errors in the coefficients result in some error in all the reconstructed pels [92], [93]. If the $N \times N$ picture is first divided into $n \times n$ subpictures and each subpicture coded independently, then only the subpictures with errors are affected.

We obtained the pictures presented in Fig. 12 by starting with the picture of Fig. 9(b) and making random bit errors in the coefficients at error rates of 10^{-2} and 10^{-3}, respectively. The structure of the blocks containing errors can be explained by recalling that each subpicture is a weighted sum of basis pictures as in (17). Hence changing the value of a coefficient results in the corresponding basis picture receiving the wrong weight in the reconstruction process.

V. Conclusions

Fig. 13 lists a number of picture coding techniques and their approximate performances. Since adaptive two-dimensional DPCM is capable of approximately the same performance as nonadaptive transform coding while being easier to implement it appears that nonadaptive transform coding has no general utility. Adaptive transform coding appears to be the only interframe technique capable of rates approaching 1 bit/pel. Habibi [43] contains a more detailed comparison of DPCM and transform techniques relative to performance, complexity, and picture-to-picture variations.

The adaptive transform techniques appear to do a good job of matching to first- and second-order picture statistics. Significant improvements in performance by developing strategies that achieve a better match to first- and second-order moments are not likely.

Pictures contain significant amounts of structure that cannot be accounted for with means, covariances, and first-order probability density functions. Significant advances in picture coding performance are possible, but must be based on a better understanding of picture structure.

Appendix

Survey of Contributions to Transform Picture Coding

Bell Laboratories [44]

Landau and Slepian used the Hadamard transform with 4×4 subpictures and block quantization. The number of bits used to code the coefficients were chosen on the basis of subjective tests. Good quality pictures were obtained at 2 bit/pel. The paper contains an excellent philosophical discussion.

Boeing [45]

Claire, Farber, and Green used the Hadamard transform with 16×16, 64×64, and 256×256 subpictures of a 256×256 aerial photograph. All coefficients for all subpictures were computed first, and then threshold sampling applied. (Different numbers of coefficients were retained for different subpictures.) All retained coefficients were quantized with the same 8-bit quantized.

Centre National D'Etudes des Telecommunications [46]–[50]

Schwartz, Marano, Poncin, and Fino used the Hadamard, Fourier, and Haar transformations with block quantization of the coefficients. Both zonal and threshold sampling were investigated. Hadamard and Haar coding were shown to be subjectively equivalent.

COMSAT [51]

Robinson compared the Hotelling, Fourier, and Hadamard transformations on 16×16 subpictures of a 512×512 picture (moonscape). Zonal sampling and block quantization was used with the bit assignment chosen on the basis of the coefficient variances. The picture was coded with 7, 4, and 2 bit/pel and compared to PCM coding of the same picture with respect to both subjective quality and the improvement obtained in the signal-to-quantizing-noise ratio. The paper includes an extensive list of references on various orthogonal transforms and their application to speech and image processing.

McDonnell Douglas [52], [53]

Kennedy, Clark, and Parkyn describe some experiments with the Hadamard transformation and threshold sampling. The effects of quantization and the bookkeeping bits required to specify which coefficients are retained were neglected. They also expanded the picture $\boldsymbol{X}$ in terms of the eigenvalues and eigenvectors of $\boldsymbol{XX}^t$ and $\boldsymbol{X}^t\boldsymbol{X}$.

MIT [54]–[58]

Huang and Woods divided 256×256 pictures into $n\times n$ subpictures with $n=4$, 8, 16, 256 and used the Hadamard transform, zonal sampling, and block quantization. They also found the structure required of a covariance matrix in order that it can be diagonalized by the Hadamard transformation. For small block sizes ($n=4$) the covariance matrices of pictures are approximately of this form, but for $n=8$, 16, etc., they diverge considerably.

Anderson and Huang developed an adaptive technique based on the Fourier transform and threshold sampling. $n\times n$ blocks with $n=8$, 16, 32, and 256 were investigated. They also used a windowing technique to combat the edge effect between the subpictures. The average picture energy over the $n\times n$ subpicture was first computed and this number used to determine a set of three thresholds which are then used to determine whether the magnitudes and phases of the coefficients are quantized to 2, 3, 4, or 5 bit. 7 bit were used to code the average gray level y_{11} and 10 bit to specify the largest coefficient aside from y_{11}. A run length (Huffman) code was used to code the location of the retained coefficients. The same technique was used with one-dimensional blocks each of which consists of one row of the 256×256 array of pels. 1.25 bit/pel are required to achieve negligible visible distortion with the two-dimensional technique.

Purdue University [59]–[74]

Habibi and Wintz divided 256×256 pictures into 16×16 arrays and used the Hadamard, Fourier, and Hotelling transformations and zonal sampling. Block quantization was used with the bit assignments optimized relative to both the mean-square error and subjective quality criterion. The performances of the four transformations are compared at 0.5, 1.0, and 2.0 bit/pel.

Tasto and Wintz developed an adaptive technique that takes into account the sensitivity of the human visual system to both gray level and spatial frequency. The 256×256 pictures were divided into 6×6 arrays, transformed with the Hotelling transformation, and classified as one of three types. The number of coefficients retained and the quantization strategy used to code them depends on the classification. Both block quantization and the Hayes and Bobilin method were optimized relative to mean-square error and subjective quality. Some relationships between mean-square error and subjective quality are presented. 0.75–1.5 bit/pel depending on the picture gray level and spatial frequency content are required for negligible distortion.

Proctor and Wintz divided the picture into 6×6 arrays of subpictures and used the Hotelling and Hadamard transformations. The system was made adaptive by truncating the expansion after η terms with η determined by computing the fraction of retained energy

$$\sum_{k=1}^{\eta}\sum_{l=1} y_{kl}^2 \Big/ \sum_{k=1}^{n}\sum_{l=1}^{n} y_{kl}^2$$

and comparing it to a threshold that depends on the average gray level of the subpicture. Three quantization strategies were tried. The effect of bit errors was demonstrated.

Ready and Wintz used these techniques to code aircraft and spacecraft multispectral scanner data.

Teledyne Brown Engineering [75], [76]

Bowyer used the Hadamard transformation and threshold sampling on a 32 gray-level test pattern. All retained coefficients were quantized with a 2^7 level quantizer and transmitted over a noisy channel. A BCH code was used for error control.

Tokyo Institute of Technology and KDD Research Laboratory [77]–[83]

Enomoto and Shibata performed experiments with 1×8 and 2×4 arrays of pels arranged into 8 vectors and transformed by the one-dimensional Hadamard transformation and some variations of it. Block quantization was used. They also constructed hardware for coding TV 2.5 to 3 bit/pel resulted in picture quality comparable to that obtained with 6-bit PCM coding of the video signal. They also suggested that color TV could be coded with an additional 0.75 bit/pel by increasing the number of bits assigned to the sixth and seventh coefficients since these correspond to the frequencies containing the color signal in the composite color TV signal.

TRW [84]

Agarwal and Stephens implemented a hardware Hadamard transform coder using 16×16 subpictures and threshold sampling. All coefficients exceeding the threshold were quantized with a 6-bit quantizer. A run length code was used to specify which coefficients exceeded the threshold.

University of Southern California [85]–[99]

Pratt, Andrews *et al.* used the Hotelling, Hadamard, and Fourier transformations and both zonal and threshold sampling with 256×256 pictures and $n=16$ and 256. An equal number of bits used to code all retained coefficients which were scaled before being quantized. "Bookkeeping information" that indicates which coefficients have been retained was included by using a run length code to indicate the sequence of retained coefficients. They also investigated the effect of bit errors in the coefficients on the quality of the reconstructed pictures and performed experiments using BCH codes for error control. They also applied these techniques to color pictures.

Acknowledgment

The author wishes to thank H. Andrews, D. Slepian, and R. Totty for their critical comments on the original manuscript.

References

[1] T. S. Huang, "PCM picture transmission," *IEEE Spectrum*, vol. 2, pp. 57–63, Dec. 1965.

[2] A. Nishikawa, R. J. Massa, and J. C. Mott-Smith, "Area properties of television pictures," *IEEE Trans. Inform. Theory*, vol. IT-11, pp. 348–352, July 1965.

[3] J. L. Gattis and P. A. Wintz, "Automated techniques for data analysis and transmission," School of Electrical Engineering, Purdue University, Lafayette, Ind., Tech. Rep. TR-EE 71-37, Aug. 1971.

[4] D. N. Graham, "Image transmission by two-dimensional contour coding," *Proc. IEEE* (*Special Issue on Redundancy Reduction*), vol. 55, pp. 336–346, Mar. 1967.

[5] L. C. Wilkins and P. A. Wintz, "Studies on data compression, Part I: Picture coding by contours, Part II: Error analysis of run-length codes," School of Electrical Engineering, Purdue University, Lafayette, Ind., Tech. Rep. TR-EE 70-17, Sept. 1970.

[6] T. S. Huang, "Digital picture coding," in *Proc. Nat. Electron. Conf.*, pp. 793–797, 1966.

[7] W. F. Schreiber, "Picture coding," *Proc. IEEE* (*Special Issue on Redundancy Reduction*), vol. 55, pp. 320–330, Mar. 1967.

[8] ——, "The measurement of third order probability distributions of television signals," *IRE Trans. Inform. Theory*, vol. IT-2, pp. 94–105, Sept. 1956.

[9] N. A. Broste, "Digital generation of random sequences with specified autocorrelation and probability density functions," U. S. Army Missile Command, Redstone Arsenal, Ala., Rep. RE-TR-70-5, Mar. 1970.

[10] H. Hotelling, "Analysis of a complex of statistical variables into principal components," *J. Educ. Psychology*, vol. 24, pp. 417–441, 498–520, 1933.

[11] H. C. Andrews and K. L. Caspari, "Degrees of freedom and modular structure in matrix multiplication," *IEEE Trans. Comput.*, vol. C-20, pp. 133–141, Feb. 1971.

[12] H. C. Andrews, "Some unitary transformations in pattern recognition and image processing," in *IFIP Congress '71* (Yugoslavia, Aug. 1971).

[13] J. Pearl, "Basis-restricted transformations and performance measures for spectral representations," in *Proc. 1971 Hawaiian Conf.*; also, *IEEE Trans. Inform. Theory* (Corresp.), vol. IT-17, pp. 751–752, Nov. 1971.

[14] H. P. Kramer and M. V. Mathews, "A linear coding for transmitting a set of correlated signals," *IRE Trans. Inform. Theory*, vol. IT-2, pp. 41–46, Sept. 1956.

[15] J. J. Y. Huang and P. M. Schultheiss, "Block quantization of correlated Gaussian random variables," *IEEE Trans. Commun. Syst.*, vol. CS-11, pp. 289–296, Sept. 1963.

[16] H. Karhunen, "Über lineare Methoden in der Wahrscheinlich-Keitsrechnung," *Ann. Acad. Sci. Fenn.*, Ser. A.I. 37, Helsinki, 1947. (An English translation is available as "On linear methods in probability theory" (I. Selin transl.), The RAND Corp., Doc. T-131, Aug. 11, 1960.)

[17] M. Loéve, "Fonctions aléatoires de seconde ordre," in P. Lévy, *Processus Stochastiques et Mouvement Brownien.* Paris, France: Hermann, 1948.

[18] I. Selin, *Detection Theory.* Princeton, N. J.: Princeton Univ. Press, 1965.

[19] A. Koschman, "On the filtering of nonstationary time series," in *Proc. 1954 Nat. Electron. Conf.*, p. 126.

[20] J. L. Brown, Jr., "Mean-square truncation error in series expansions of random functions," *J. SIAM*, vol. 8, pp. 18–32, Mar. 1960.

[21] R. E. Totty and J. C. Hancock, "On optimum finite-dimensional signal representation," in *1963 Proc. 1st Ann. Allerton Conf. on Circuits and System Theory.*

[22] J. Pearl, "Walsh processing of random signals," University of California at Los Angeles, Los Angeles, Calif.

[23] W. R. Crawther and C. M. Rader, "Efficient coding of vocoder channel signals using linear transformation," *Proc. IEEE* (Lett.), vol. 54, pp. 1594–1595, Nov. 1966.

[24] L. M. Goodman, "A binary linear transformation for redundancy reduction," *Proc. IEEE* (Lett.) (*Special Issue on Redundancy Reduction*), vol. 55, pp. 467–468, Mar. 1967.

[25] P. A. Wintz and A. J. Kurtenbach, "Waveform error control in PCM telemetry," *IEEE Trans. Inform. Theory*, vol. IT-14, pp. 650–661, Sept. 1968.

[26] C. J. Palermo, R. V. Palermo, and H. Horwitz, "The use of data omission for redundancy removal," in *1965 Rec. Int. Space Electron. and Telemetry Symp.*, pp. (11)D1–(11)D16.

[27] A. J. Kurtenbach and P. A. Wintz, "Data compression for second order processes," in *1968 Proc. Nat. Telemetering Conf.*

[28] C. E. Shannon, "Coding theorems for a discrete source with a fidelity criterion," in *1959 IRE Nat. Conv. Rec.*, pt. 4, pp. 142–163.

[29] T. J. Goblick, Jr., and J. L. Hoslinger, "Analog source digitation: A comparison of theory and practice," *IEEE Trans. Inform. Theory* (Corresp.), vol IT-13, pp. 323–326, Ar. 1967.

[30] L. M. Goodman and P. R. Drouihet, Jr., "Asymptotically optimum pre-emphasis and de-emphasis networks for sampling and quantizing," *Proc. IEEE* (Lett.), vol. 54, pp. 795–796, May 1966.

[31] S. Watanabe, "The Loéve-Karhunen expansion as a means of information compression for classification of continuous signals," IBM Watson Res. Cen., Yorktown Heights, N. Y., Tech. Rep. AMRL-TR-65-114.

[32] T. Berger, *Rate Distortion Theory.* Englewood Cliffs, N. J.: Prentice-Hall, 1971.

[33] E. Donchin, "A multivariate approach to the analysis of average evoked potentials," *IEEE Trans. Biomed. Eng.*, vol. BME-13, pp. 131–139, July 1966.

[34] C. K. Stidd, "The use of eigenvectors for climatic estimates," *J. Appl. Meteorology*, vol. 6, pp. 255–264, Apr. 1967.

[35] J. Pearl, "On the distance between representations," presented at Symp. Picture Coding, North Carolina State Univ., Raleigh, N. C., Sept. 1970.

[36] P. F. Panter and W. Dite, "Quantization distortion in pulse-count modulation with nonuniform spacing of levels," *Proc. IRE*, vol. 39, pp. 44–48, Jan. 1951.

[37] J. Max, "Quantizing for minimum distortion," *IRE Trans. Inform. Theory*, vol. IT-6, pp. 7–12, Mar. 1960.

[38] B. Smith, "Instantaneous companding of quantized signals," *Bell Syst. Tech. J.*, vol. 36, pp. 653–709, May 1957.

[39] H. Gish and J. N. Pierce, "Asymptotically efficient quantizing," *IEEE Trans. Inform. Theory*, vol. IT-14, pp. 676–683, Sept. 1968.

[40] R. C. Wood, "On optimum quantization," *IEEE Trans. Inform. Theory*, vol. IT-15, pp. 248–252, Mar. 1969.

[41] J. F. Hayes and R. Bobilin, "Efficient waveform encoding," School of Electrical Engineering, Purdue University, Lafayette, Ind., Tech. Rep. TR-EE 69-4, Feb. 1969.

[42] D. J. Sakrison and V. R. Algazi, "Comparison of line-by-line and two-dimensional encoding of random images," *IEEE Trans. Inform. Theory*, vol. IT-17, pp. 386–398, July 1971.

[43] A. Habibi, "Comparison of *n*th order DPCM encoder with linear transformations and block quantization techhiques," *IEEE Trans.*

Commun. Technol., vol. COM-19 (pt. 1), pp. 948–956, Dec. 1971.

[44] H. J. Landau and D. Slepian, "Some computer experiments in picture processing for bandwidth reduction," *Bell Syst. Tech. J.*, vol. 50, pp. 1525–1540, May–June 1971.

[45] E. J. Claire, S. M. Farber, and R. R. Green, "Practical techniques for transform data compression/image coding," in *Proc. 1971 Applications of Walsh Function Symp.*, pp. 2–6 (Washington, D. C., Apr. 1971).

[46] P. Marano and P. Y. Schwartz, "Compression d'information sur la transformee de Fourier d'une image," *Onde Elec.*, vol. 50, pp. 908–919, Dec. 1970.

[47] P. Y. Schwartz, "Analysis of Fourier compression on images" (note technique interne est.TDS/423), presented at Symp. Picture Coding, Raleigh, N. C., Sept. 1970.

[48] ——, "Analyse de la compression d'information sur la transformee de Fourier d'une image," *Ann. Telecommun.*, Mar.–Apr. 1971.

[49] J. Poncin, "Utilisation de la transformation de Hadamard pour le codage et la compression de sigaux d'images," *Ann. Telecommun.*, July–Aug. 1971.

[50] B. Fino, "Traitement d'images nar transformations orthogonales," Note Tech. ITD/TTI/52.

[51] G. S. Ribonson, "Orthogonal transform feasibility study," prepared for NASA/MSC by COMSAT, Contract no. NAS 9-11240, Suppl. Reps. 1 (Aug. 1971), 2 (Sept. 1971), 3 (Oct. 1971), and Final Rep. (Oct. 1971).

[52] J. D. Kennedy, S. J. Clark, and W. A. Parkyn, Jr., "Digital imagery data compression techniques," McDonnell Douglas Corp., Huntington Beach, Calif., Rep. MDC G0402, Jan. 1970.

[53] J. D. Kennedy, "Walsh function imagery analysis," in *Proc. 1971 Applications of Walsh Functions Symp.* (Washington, D. C., Apr. 1971), pp. 7–10.

[54] J. W. Woods, "Video bandwidth compression by linear transformation," MIT Research Laboratory of Electronics, Quarterly Progress Rep. 91, Oct. 15, 1968, pp. 219–224.

[55] T. S. Huang and J. W. Woods, "Picture bandwidth compression by block quantization," presented at the 1969 Int. Symp. Inform. Theory, Ellenville, N. Y.

[56] J. W. Woods and T. S. Huang, "Picture bandwidth compression by linear transformation and block quantization," presented at the 1969 Symp. Picture Bandwidth Compression, MIT, Cambridge, Mass.

[57] G. B. Anderson and T. S. Huang, "Picture bandwidth compression by pieceiwse Fourier transformation," in *Proc. Purdue Centennial Year Symp. Inform. Processing*, 1969.

[58] ——, "Piecewise Fourier transformation for picture bandwidth compression," *IEEE Trans. Commun. Technol.*, vol. COM-19, pp. 133–140, Apr. 1971.

[59] J. E. Essman and P. A. Wintz, "Redundancy reduction of pictorial information," presented at the Midwestern Simulation Council Meeting on Advanced Mathematical Methods in Simulation, Oakland University, 1968.

[60] G. G. Apple and P. A. Wintz, "Experimental PCM system employing Karhunen–Loéve sampling," presented at the 1969 Int. Symp. Inform. Theory, Ellenville, N. Y., Jan. 1969.

[61] A. Habibi and P. A. Wintz, "Optimum linear transformations for encoding 2-dimensional data," presented at the Symp. Picture Bandwidth Compression, MIT, Apr. 1969.

[62] ——, "Optimum linear transformations for encoding 2-dimensional data," School of Electrical Engineering, Purdue University, Lafayette, Ind., Tech. Rep. TR-EE 69-15, May 1969.

[63] ——, "Linear transformations for encoding 2-dimensional sources," School of Electrical Engineering, Purdue University, Lafayette, Ind., Tech. Rep. TR-EE 70-2, Jan. 1970.

[64] M. Tasto and P. A. Wintz, "An adaptive picture bandwidth compression system using pattern recognition techniques," in *Proc. 3rd Hawaii Int. Conf. on System Sciences*, pp. 513–515, Jan. 1970.

[65] A. Habibi and P. A. Wintz, "Picture coding by linear transformation," presented at the 1970 IEEE Int. Symp. Inform. Theory, Noordwijk, The Netherlands, June 1970.

[66] M. Tasto and P. A. Wintz, "Picture bandwidth compression by adaptive block quantization," School of Electrical Engineering, Purdue University, Lafayette, Ind., Tech. Rep. TR-EE 70-14, June 1970.

[67] M. Tasto, A. Habibi, and P. A. Wintz, "Adaptive and nonadaptive image coding by linear transformation and block quantization," presented at the Symp. Picture Coding, North Carolina State Univ., Raleigh, N. C., Sept. 10, 1970.

[68] A. Habibi and P. A. Wintz, "Image coding by linear transformations and block quantization," *IEEE Trans. Commun. Technol.*, vol. COM-19, pp. 50–60, Feb. 1971.

[69] M. Tasto and P. A. Wintz, "Image coding by adaptive block quantization," *IEEE Trans. on Commun. Technol.*, vol. COM-19, pp. 957–971, Dec. 1971.

[70] C. W. Proctor and P. A. Wintz, "Picture bandwidth reduction for noisy channels," School of Electrical Engineering, Purdue Univ., Lafayette, Ind., Tech. Rep. TR-EE 71-30, Aug. 1971.

[71] M. Tasto and P. A. Wintz, "A bound on the rate-distortion function and application to images," *IEEE Trans. Inform. Theory*, vol. IT-18, pp. 150–159, Jan. 1972.

[72] P. J. Ready, P. A. Wintz, and D. A. Landgrebe, "A linear transformation for data compression and feature selection in multispectral imagery," Lab. for Applications of Remote Sensing, Purdue Univ., Lafayette, Ind., Info. Note 072071, July 1971.

[73] P. J. Ready, P. A. Wintz, S. J. Whitsitt, and D. A. Landgrebe, "Effects of compression and random noise on multispectral data," in *Proc. 7th Int. Symp. on Remote Sensing of Environment* (Univ. of Michigan, May 1971).

[74] P. J. Ready and P. A. Wintz, "Multispectral data compression thru transform coding and block quantization," School of Electrical Engineering, Purdue Univ., Lafayette, Ind., Tech. Rep. TR-EE 72-2, Jan. 1972.

[75] D. E. Bowyer, "Rate calculations for compressed Hadamard-transformed data," Teledyne Brown Engineering, Huntsville, Ala., Summary Rep. MSD-INT-1338, June 1971.

[76] ——, "Walsh functions, Hadamard matrices and data compression," presented at 1971 Walsh Function Symp.

[77] H. Enomoto and K. Shibata, "Features of Hadamard transformed television signal," presented at 1965 Nat. Conf. IECE Jap., Paper 881.

[78] ——, "Television signal coding method by orthogonal transformations," presented at 1966 Joint Conv. IECE Jap., Paper 1430.

[79] ——, "Television signal coding method by orthogonal transformation," presented to 6th Research Group on Television Signal Transmission. The Institute of Television Engineers of Japan (May 1968).

[80] S. Inoue and K. Shibata, "Quantizing noise of orthogonal transform TV PCM system," presented at 1968 Nat. Conf. IECE Jap., Paper 1310.

[81] H. Enomoto, K. Shibata *et al.*, "Experiments on television PCM system by orthogonal transformation," presented at 1969 Joint Conv. IECE Jap., Paper 2219.

[82] K. Shibata, T. Ohira *et al.*, "PCM terminal equipment for bandwidth compression of television signal," presented at 1969 Joint Conv. IECE Jap., Paper 2619.

[83] H. Enomoto and K. Shibata, "Orthogonal transform coding system for television signals," *Television, J. Inst. TV Eng. of Japan*, vol. 24, pp. 99–108, Feb. 1970; also in *Proc. 1971 Applications of Walsh Functions Symp.* (Washington, D. C., April 1971), pp. 11–17.

[84] V. K. Agarwal and T. J. Stephens, Jr., "On-board data processor (picture bandwidth compression)," Data Systems Lab., IRAD Rep. R 733.3-32 Feb. 1970.

[85] W. K. Pratt and H. C. Andrews, "Fourier transform coding of images," in *Proc. Hawaii Int. Conf. on System Sciences*, Jan. 1968.

[86] ——, "Television bandwidth reduction by Fourier image coding," presented at 103rd Society of Motion Picture and Television Eng. Conf., 1968.

[87] ——, "Television bandwidth reduction by encoding spatial frequencies," *J. Motion Picture and Television Eng.*, pp. 1279–1281, Dec. 1968.

[88] ——, "Two dimensional transform coding for images," presented at 1969 IEEE Int. Symp. on Inform. Theory, Jan. 1969.

[89] ——, "Transformation coding for noise immunity and bandwidth reduction," in *Proc. 2nd Hawaii Int. Conf. on System Sci.*, Jan. 1969.

[90] W. K. Pratt, J. Kane, and H. C. Andrews, "Hadamard transform image coding," *Proc. IEEE*, vol. 57, pp. 58–68, Jan. 1969.

[91] W. K. Pratt and H. C. Andrews, "Application of Fourier-Hadamard transformation to bandwidth compression," presented at MIT Symp. on Picture Bandwidth, Apr. 1969.

[92] H. C. Andrews and W. K. Pratt, "Transform image coding," presented at Polytech. Inst. of Brooklyn Int. Symp. on Computer Processing in Communications, Apr. 1969.

[93] W. K. Pratt and H. C. Andrews, "Transform image coding," University of Southern California, Los Angeles, Calif. Final Rep. USCEE 387, Mar. 1970.

[94] H. C. Andrews and W. K. Pratt, "Digital image transform processing," presented at Symp. Applications of Walsh Functions, Apr. 1970.

[95] W. K. Pratt, "Karhunen-Loéve transform coding of images," presented at 1970 IEEE Int. Symp. Inform. Theory, June 1970.

[96] ——, "Application of transform coding to color images," presented at Symp. Picture Coding, North Carolina State Univ., Raleigh, N. C., Sept. 1970.

[97] ——, "A comparison of digital image transforms," presented at University of Missouri at Rolla—Mervin J. Kelly Communications Conf., Rolla, Mo., Sept. 1970.

[98] H. C. Andrews, "Fourier and Hadamard image transform channel error tolerance," in *Proc. UMR—Mervin J. Kelly Communications Conf.*, Rolla, Mo., pp. 10-4-1–10-4-6, Oct. 1970.

[99] W. K. Pratt, "Spatial transform coding of color images," *IEEE Trans. Commun. Technol.*, vol. COM-19, pp. 980–991, Dec. 1971.

39

Reprinted from pages 10.6.1-10.6.5 of *National Telecommunications Conference, Birmingham, Alabama, December 3-6,* Institute of Electrical and Electronics Engineers, New York, 1978

ADAPTIVE ORTHOGONAL TRANSFORM CODING SYSTEM FOR NTSC COLOR TELEVISION SIGNALS

T. Ohira, M. Hayakawa, K. Matsumoto

OKI Electric Industry Co., Ltd.

1-9-6, Kounan, Minato-ku,

Tokyo, 108 JAPAN

Abstract

Adaptive orthogonal transform coding system (ADP-OTC) which performs two-dimensional 32nd-order transformation in real-time for NTSC color television signals has been developed. This system equips three kinds of bit assignments and the optimum one among those is adaptively selected block by block by detecting the feature of input television signal. This adaptive technique is expected to improve the system performance compared with conventional orthogonal transform coding system (OTC).

According to an experiment, ADP-OTC showed some improvement on the special kinds of pictures such as containing slant edge components, high frequency vertical components and high saturated color components etc.

1. Introduction

Although the most television signals are, at present, transmitted by analog basis such as FM, digital television transmission is considered to be advantageous from the standpoint of the occupancy of channel capacity particularly when a high efficient encoding scheme is employed. Among various types of encoding schemes which have been investigating, orthogonal transform coding is one of the representative techniques.

We developed an orthogonal transform coding system (OTC) which was designed to perform three types of transformations by Hadamard matrix and two slant matrices in order to examine the performance of codings from the standpoint of hardware realization and presented the paper at ICC'77.[1] The system uses two-dimensional 32nd-order (4 lines x 8 samples) transform coding which provides highly efficient compression and relatively compact hardware.

OTC, in general, performs bit-assignment using the fixed code compression characteristics when the output of orthogonal transformation is compressed. Well-balanced bit-assignment is made so that the possible degradation in picture quality due to compression becomes minimum at time of subjective evaluation, focusing on the pictures frequently appeared. In other words, the OTC is not provided with the enough bits to be sufficiently rare pictures such as delicate picture, picture with high saturated color component etc. Namely, the picture quality of the rare picture reconstructed at decoder side is inferior to those of common pictures. Although adaptive coding methods have been investigating in the field of bandwidth compression techniques [2] - [4]. We have developed an adaptive orthogonal transform coding system (ADP-OTC) in order to eliminate above demerits of conventional type, producing equal picture quality for input picture of any kind with different code compression characteristics being provided.

This paper describes an experimental result based on ADP-OTC for NTSC color television signals and evaluates picture quality based on this scheme.

2. Adaptive Orthogonal Transform Coding

A block diagram of ADP-OTC is shown in Fig.1. The system adopts a two-dimensional 32nd-order slant matrix, whose lowest order is shown below:

$$MH1_8 = \frac{1}{\sqrt{8}} \begin{bmatrix} & (1 & 1 & 1 & 1 & 1 & 1 & 1 & 1 \\ \sqrt{7}/\sqrt{3} & (1 & 5/7 & 3/7 & 1/7 & -1/7 & -3/7 & -5/7 & -1 \\ 3/\sqrt{5} & (1 & 1/3 & -1/3 & -1 & -1 & -1/3 & 1/3 & 1 \\ 17/\sqrt{105} & (7/17 & -1/17 & -9/17 & -1 & 1 & 9/17 & 1/17 & -7/17 \\ & (1 & -1 & -1 & 1 & 1 & -1 & -1 & 1 \\ & (1 & -1 & -1 & 1 & -1 & 1 & 1 & -1 \\ 3/\sqrt{5} & (1/3 & -1 & 1 & -1/3 & -1/3 & 1 & -1 & 1/3 \\ 3/\sqrt{5} & (1/3 & -1 & 1 & -1/3 & 1/3 & -1 & 1 & -1/3 \end{bmatrix} \quad (1)$$

The 32nd-order matrix is expanded from the lowest order matrix by the following alogorithm.

$$[H_{2n}] = \frac{1}{2} \begin{bmatrix} Hn & Hn \\ Hn & -Hn \end{bmatrix} \quad (2)$$

The operation at the transmitter side is the following. After input signal is sampled and encoded into 8 bits/sample by A/D converter, a block is constructed taking 8 samples from each of 4 adjacent lines by block composition circuit. With execution of 32nd-order slant transformation, mode control circuit detects the nature of input signal block by block, and adaptively selects the best bit-assignment which is determined by bit-reduction circuit. Each bit-reduction circuit compresses original input signal into information bits equivalent to transmission bit rate. After passing through single-error-correcting Hamming code encoder, compressed codes in parallel form are converted to serial stream and transmitted. The signal flow at the receiver side is reverse processing of the transmitter side.

Table 1 shows performance specifications of the system.

3. Adaptive Processing Section

The nature of input signal is analized at every transformation so that adaptive selection of optimum bit-assignment becomes possible.

The available number of modes for code compression shall be determined in light of system performance, size of hardware, number of bits assigned for discrimination code, etc. Maximum two modes of code compression are available when only one bit is assigned as discrimination code, whereas maximum four modes are available with two bits of discrimination code. When larger number of modes are employed, adaptability for variety of pictures can be increased. However, it inevitably increases the number of bits assigned for discrimination code, which shall result in sacrificing the bits assigned for picture information making the size of hardware larger at same time. After synthetic consideration, ADP-OTC employed two bits for discrimination code and three modes for code compression. As is shown in Fig.1, one mode is selected among three modes of code compression by mode control and mode select switch.

The three modes are determined to cope with higher frequency component, picture with high saturation point of color component, etc. which cannot be covered satisfactorily with OTC.

Fig.2 shows mode control algorithm of ADP-OTC.

Among the outputs after orthogonal transformation, "H17, H24, H25, H32" are arranged as group A while "H20, H21, H28, H29" are group B or group C and "H10, H11, H14, H15" are group D. The outputs of these three combinations and four groups being monitored, the optimum mode can be selected. For example, when any one of the outputs of group A among H17, H24, H25 and H32 exceeds 1/16 of maximum amplitude, it is judges as "1". On the other hand, it is judges as "0", if the output does not exceed the 1/16 of the maximum amplitude. In the same manner, the output of group B is judged as either "1" or "0" with threshold level at 1/8 of maximum amplitude. The outputs of group C and group D are also judged in the same way, but with different threshold levels.

The bit-assignment of mode 1 is to cope with the picture of slant component. Being compared with mode 2 or mode 3, more bits are assigned to group D (H10, H11, H14, H15) which is to restore slant edge components.

Mode 2 assigns more bits to H20, H21, H28 and H29 to cope with the picture with high saturation point of color component.

Mode 3 allocates more bits to H17, H24, H25 and H32 to cope with high frequency vertical component.

Among three modes, top priority is given to mode 1 to be selected. When the output of group D is "1" in Fig.2, mode 1 is always selected. Next priority is given to output of group B, and mode 2 is selected when the output is "1". The output of group A has higher priority than that of group C. Accordingly, mode 3 is selected when the output of both group A and group C are "1". As an exception, mode 3 is selected when all the outputs of groups A, B, C and D are "0".

Selecting the optimum mode among above three depending on the nature of input picture signal, ADP-OTC assures better reproduction of any type of pictures, compared with OTC.

4. Experimental Result

Mode control algorithm was experimentally decided by the subjective evaluation of picture quality.

Table 2 shows the optimum bit-assignment, supposing the transmission bit rate is 32 Mb/s.

The outputs of H1 through H8, which are common to all three modes are used to restore DC component and lower frequency component of picture information, while other outputs starting from H9 can be selected in accordance with the nature of input pictures. The best mode among three is selected at every block in accordance with the mode control algorithm shown in Fig.2.

Photo 1 shows the processed pictures obtained by ADP-OTC. As ADP-OTC is based on an intraframe coding technique, it provides the same picture quality for a moving picture and a still picture.

When we evaluated the picture quality of OTC, we could, strictly speaking, notice a little zigzag degradation on a slanting edge component and a little background noise in large flat areas of a still picture for the 32 Mb/s transmission rate.

Comparing the processed pictures by ADP-OTC with OTC, adaptive technique showed some improvement on the picture quality such as containing slant components and high frequency vertical components etc. But the quality was almost same for general kind of picture.

Since the visual resolution of human observers for a moving picture is not as good as it is a still picture, the degradation in moving pictures was unnoticed.

Photo 1 (c) - (e) show the effect of transmission errors on the system. As can be seen from Photo 1 (c), with the bit rate of 10^{-5}, no degradation can be seen in comparison with the noiseless case. The system uses a single-error-correcting Hamming code which corrects one of the most significant 11 bits in a block, so the system can greatly reduce the effect of transmission errors. Photo 1 (d), with a bit error rate of 10^{-4}, shows the occurence of a few block errors. When the bit error rate becomes 10^{-3}, block errors increases tolerably as in Photo 1 (e). The error effect of orthogonal transform coding does not propagate to the following block, and the picture with a bit error rate of 10^{-3} is still recognizable.

5. An Example of Application

Satellite communication links which are expensive to use are one of the most suitable application to bandwidth compression system for TV signals. Particularly, fixed rate coding such as ADP-OTC would be advantageous for a transmission channel like satellite links whose error probability is not small.

Fig. 3 shows an example of application to satellite communication links. Supposing the transmission capacity is 64 Mb/s, it transmits two channels of color TV signal with fine quality of pictures, or of monochrome TV signal with excellent quality by the 32 Mb/s transmission rate.

6. Conclusion

This paper has described an ADP-OTC which executes real-time processings for NTSC color TV signals. The system equips three kinds of bit assignments and the optimum one among those is adaptively selected block by block by detecting the feature of input television signal.

Adaptive technique of the system showed some improvement on the picture quality such as containing slant edge components, high frequency vertical components and high saturated color components etc. in comparison with processed pictures by OTC.

References

[1] T. Ohira, M. Hayakawa and K. Matsumoto, "Orthogonal transform coding system for NTSC color television signal", Proceedings of 1977 International Conference on Communications, Chicago, vol.1, pp.86-90, June 1977.

[2] A. Habibi, "Survey of adaptive image coding techniques", IEEE Trans. on COM., vol.COM-25 No.11, pp.1275-1284, Nov. 1977.

[3] W.H. Chen and C.H. Smith, "Adaptive coding of monochrome and color images", IEEE Trans. on COM., vol. COM-25, No.11, pp.1285-1292, Nov. 1977.

[4] A. Netravali, B. Prasada and F.W. Mounts, "Adaptive Hadamard transform coding of pictures", Proceedings of 1976 National Telecommunications Conference, Dallas, Nov. 1976.

1. Transformation	
Bandwidth of input signal:	DC-4.5MHz
Number of code bits:	8 bits/sample
Sampling frequency:	9.6923MHz
Transform matrix:	slant matrix (MH1)
Order of matrix:	32nd (4 lines x 8 samples)
Number of bits for transformation:	10 bits/sample
Quantizing: linear:	Dynamic range 1/64∿63/64
nonlinear:	Given by ROM
Number of control mode:	3 modes
Control algorithm:	Featrue detection of transformed output per block
2. Transmission	
Framing:	Sequential arrangement from H1 to H32
Synchronizing:	1 bit shift by frame pulse (1010 ...... pattern)
Average number of bits:	3.03 bits/sample
Transmission bit rate:	31.5 Mb/s
Compression ratio:	0.379
Error protection:	(15,11) single-error-correcting Hamming code.

Table 1. Performance Specifications of ADP-OTC

V\H	1	2	3	4	5	6	7	8
1	8	6	5	4	0	0	0	0
2	6	4	4	3	3	2	0	0
3	6	4	4	3	0	2	0	0
4	6	4	0	3	4	6	6	4

MODE 1

V\H	1	2	3	4	5	6	7	8
1	8	6	5	5	4	0	0	0
2	6	4	0	0	0	2	2	0
3	6	4	0	0	0	4	4	0
4	6	4	0	5	5	6	6	5

MODE 2

V\H	1	2	3	4	5	6	7	8
1	8	6	5	5	5	6	6	5
2	6	4	3	2	2	2	0	0
3	6	4	0	0	0	0	0	0
4	6	4	0	0	3	4	5	0

MODE 3

TABLE 2. BIT ASSIGNMENTS OF ADP-OTC FOR 32 Mb/s TRANSMISSION RATE

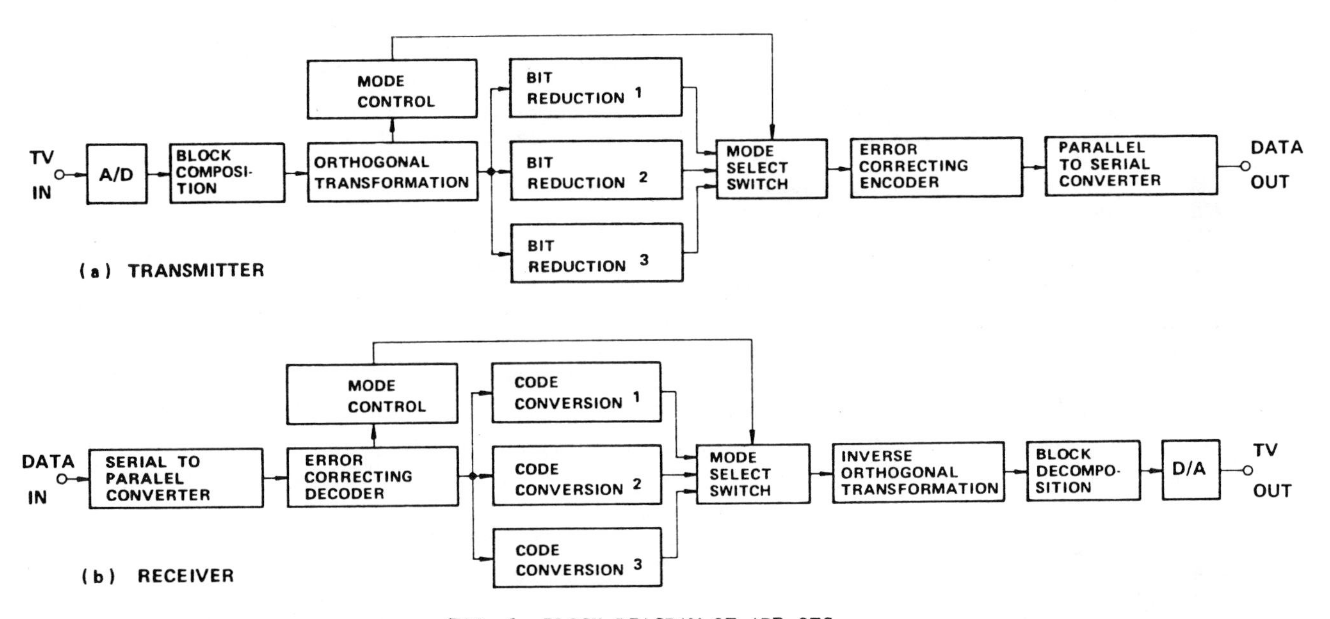

FIG. 1 BLOCK DIAGRAM OF ADP-OTC

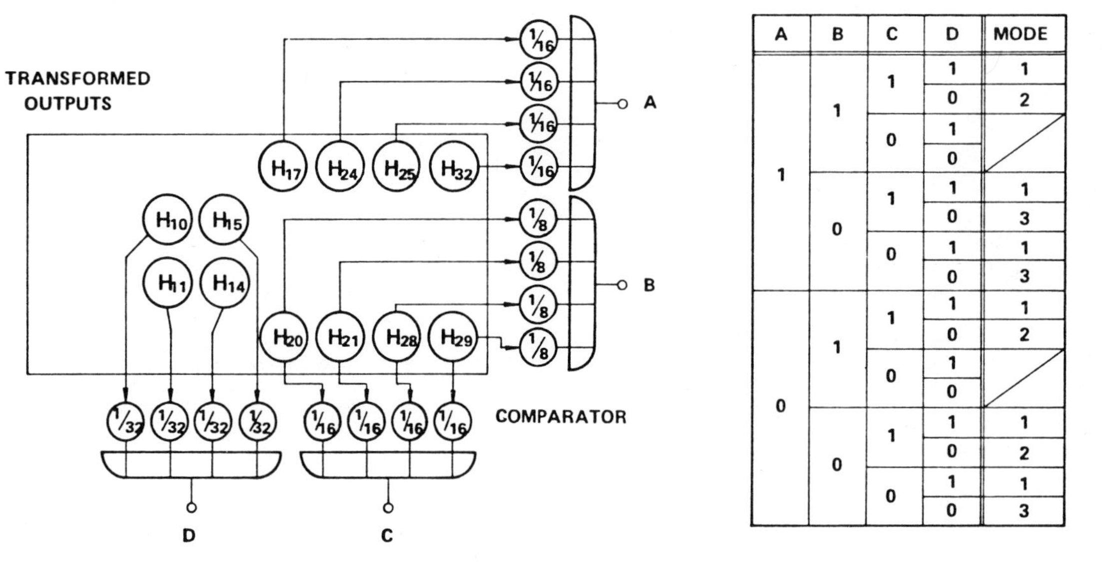

A	B	C	D	MODE
1	1	1	1	1
			0	2
		0	1	
			0	
	0	1	1	1
			0	3
		0	1	1
			0	3
0	1	1	1	1
			0	2
		0	1	
			0	
	0	1	1	1
			0	2
		0	1	1
			0	3

FIG. 2 MODE CONTROL ALGORITHM

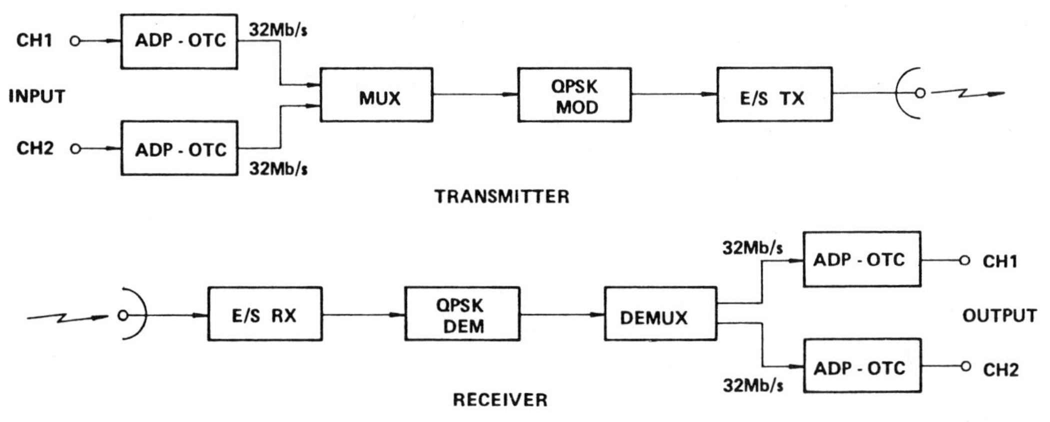

FIG. 3 EXAMPLE OF APPLICATION TO SATELLITE COMMUNICATION LINK

(a) 32 Mb/s transmission rate

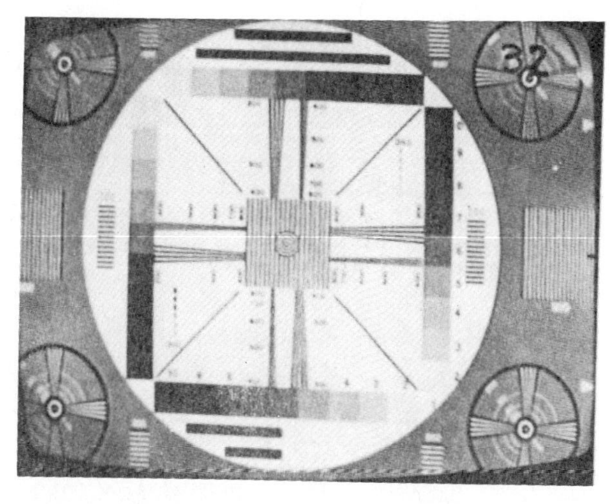

(b) 32 Mb/s transmission rate

(c) BER = 10^{-5} at 32 Mb/s

(d) BER = 10^{-4} at 32 Mb/s

(e) BER = 10^{-3} at 32 Mb/s

Photo 1. Processed Pictures

Reprinted from pages 58-62 of *24th Midwest Symposium on Circuits and Systems Proceedings*, S. Karni, ed., University of New Mexico, Albuquerque, June 29-30, 1981

A REAL-TIME COLOR VIDEO COMPRESSOR*

S. C. Knauer and D. N. Hein

ABSTRACT

A real-time video compressor has been constructed at the NASA-Ames Research Center. This compressor uses adaptive intraframe Hadamard transform coding on the Y, I, and Q color components. It is able to transmit color video at a rate of 16 Megabits per second, or it may be used to study different data rates and compression schemes through the use of a programmable quantizer.

1. INTRODUCTION

Video compression is the reduction of the data rate needed to transmit digital video information. The use of video compressors provides significant economic advantages when transmitting video images through satellite channels. By transmitting several low rate video images through one satellite channel the cost per user can be reduced by sharing the cost of the channel among all the users. One major use of video compression would be in the area of video teleconferencing.

One method of bandwidth reduction is source data compression, which is the removal of redundant information in the television signal, and the consequent reduction in the bit rate used to send the signal. Information that the eye cannot see, or that the camera and scene will not generate, can be isolated and discarded. This paper describes the construction and testing of a real-time intraframe digital television encoder and decoder used to transmit NTSC (National Television Systems Committee) standard color video at 16 Mega-bits per second (Mbps). This compressor was developed and constructed at the NASA-Ames Research Center. The unit uses a two-dimensional 8x8 Walsh-Hadamard transform (WHT) and adaptive quantization to remove the redundancy within each frame, producing a picture quality suitable for teleconferencing.

2. INTRAFRAME TRANSFORM COMPRESSION

Transform compression works by taking the correlated image data and translating it into a form in which the resulting data is much less correlated and in a more compact form [1]. Each transformed value is then quantized to a certain number of bits to achieve the desired results. Quantizing has the effect of taking a number and mapping it into one of a finite number of values. As an example, we could take an 8 bit number and quantize it to 5 bits by simply throwing away the 3 least significant bits. In doing so we are throwing away information which cannot be recovered. This type of quantizing is known as uniform quantization because the number is sliced up into uniformly sized bins. In general, non-uniform quantizers are used which have variable size bins [2].

The WHT is one of the most widely used transforms for video compression [3-5]. Since the WHT transform matrix consists only of plus and minus ones, it can be implemented by additions and subtractions and does not require any multiplications. For this reason the WHT is very appealing for use in hardware implementations. The basis vectors of the WHT are depicted in Figure 1. Each square represents one of

*Part of this work was performed under NASA contract NAS -10481.

the 64 8x8 WHT vectors. The white areas represent additions of data samples and the black areas represent subtractions. For example, the H01 vector is the sum of the right half of an 8x8 subpicture minus the sum of the left half.

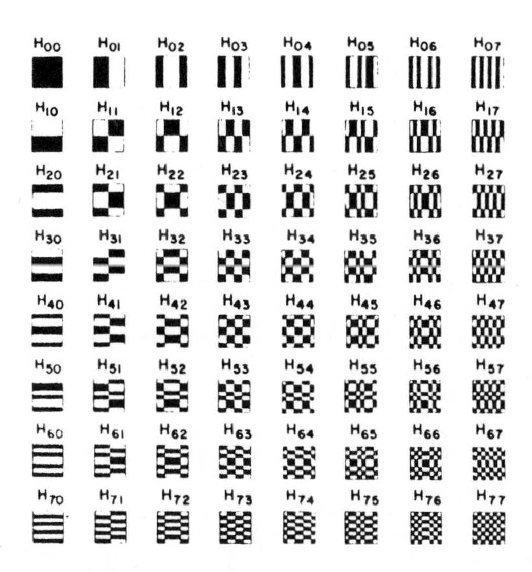

Figure 1. 8x8 WHT basis vectors

Figure 2 shows the basic structure of a transform compressor. The video is digitized and then processed by the two-dimensional transformer, which works on small blocks of data at a time. The transformed data is then quantized and converted to a one bit wide serial stream, which is transmitted over some type of communications link.

At the receiver, representative values are assigned to the quantized transform video signals. An inverse transform is performed to obtain a facsimile of the original video image. The quality of this facsimile image depends on the rate available for transmitting it. For better quality more rate is needed. If lesser quality is acceptable then less rate will be required.

Each picture element in a color image is made up of a combination of three primary colors. For television displays these primary colors are red, green, and blue. When analog video images are transmitted they are represented by another set of primaries known as Y, I, and Q. We can convert from one primary set to another through linear combinations of the color components. In the YIQ format the Y component is the luminance of the image and the I and Q components contain the chrominance information. The advantage of using this format is that the human eye has much less spatial resolution for chrominance information than it does for luminance information. This allows us to reduce the rate used to transmit the color information [6].

3. IMPLEMENTATION

Block diagrams of the compression hardware constructed at Ames are given in Figures 3 and 4. An analog tristimulus color signal is fed into the encoder as separate red, green, and blue signals. A separate sync line is also required for timing purposes. These signals are then transformed to the NTSC Y, I, and Q components by an analog circuit and are sampled and digitized at 8 Mega-samples per second and 8 bits of resolution. The I and Q signals are averaged and subsampled horizontally and vertically by a factor of 4 to achieve an effective sampling rate of 1/2 Mega-samples per second for each component.

An 8 point WHT is first done in the horizontal dimension on the luminance component. Then, eight lines of the one-dimensional transformed video data are reformatted by scanning down the columns rather than across the lines. This has the effect of partitioning the image into small 8x8 blocks. A vertical 8 point WHT is then performed on the reformatted data to produce the 8x8 two-dimensional transformed coefficients. At the same time, the two chrominance components have independent 2x2 WHTs performed on them.

Once the transformed values have been obtained we can then perform the actual data compression. This reduction in data rate is possible because, by transforming the video data, we have condensed much of the signal information into a small fraction of the data samples. Several of the high frequency vectors may be eliminated altogether. We take advantage of this fact by using 8 of these "throw away" vectors to insert the values of the 4 I and 4 Q vectors. This merging is done in the color multiplexor. The color multiplexor combines the 4 I, 4 Q, and 56 of the 64 Y vectors to form one set of 64 vectors which may be processsed further to reduce the rate even more. This additional step of rate reduction is known as quantizing. A quantizer assigns a different number of bits to each of the vectors dependent on the amount of information that vector contains.

The quantizer used in this system is known as an adaptive quantizer. This differs from an unadaptive quantizer in the following way. An unadaptive quantizer uses only one set of quantizers to code a block

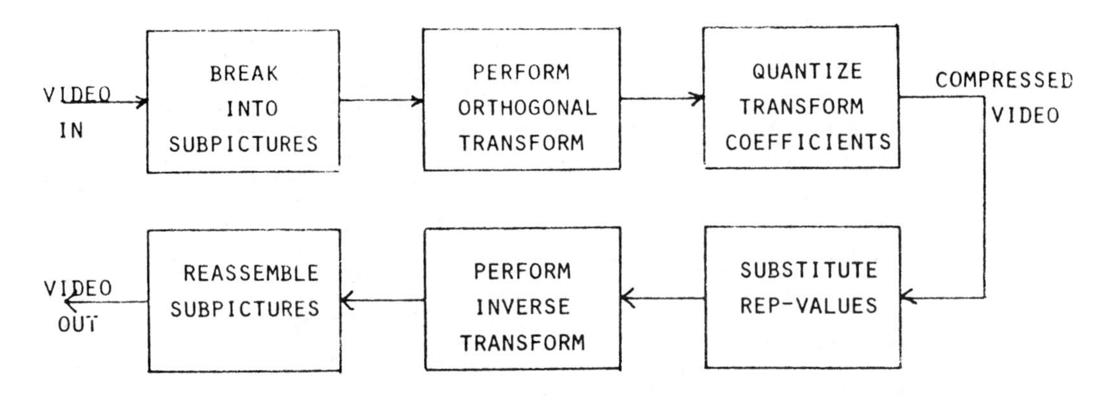

Figure 2. General block diagram of a transform compressor

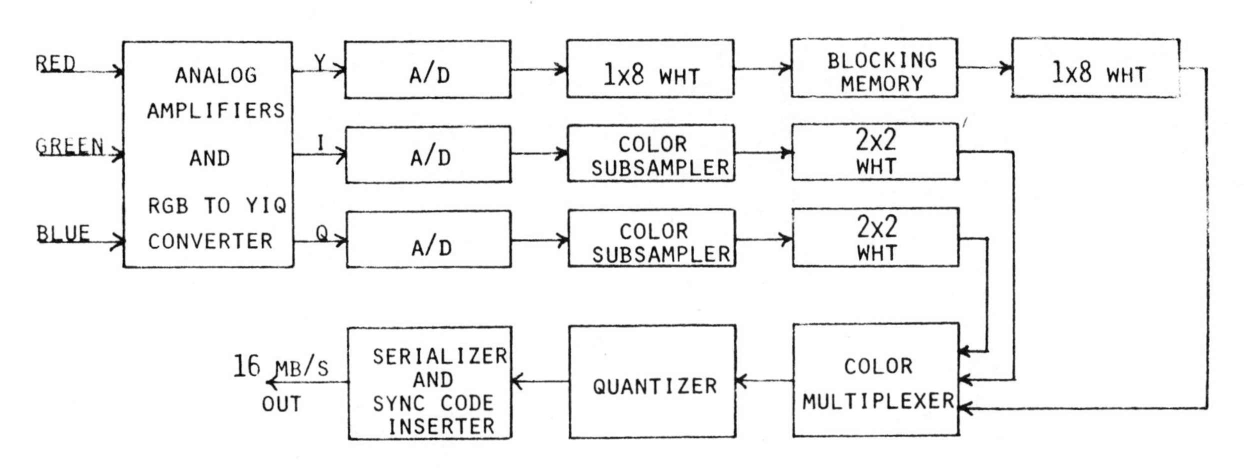

Figure 3. 16 Mbps transform compression encoder

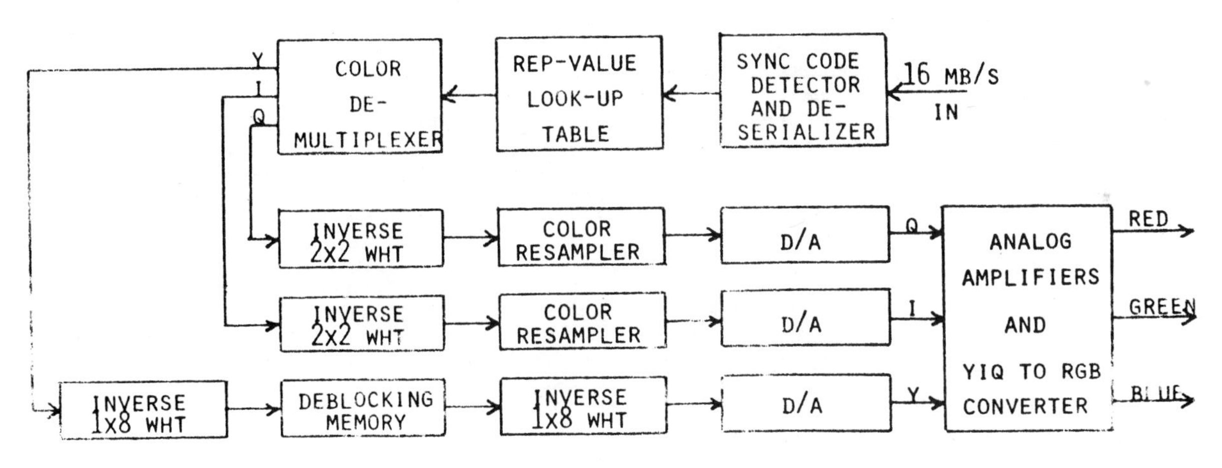

Figure 4. 16 Mbps transform compression decoder

of data. That is, any one vector will have one, and only one, quantizer assigned to it. An adaptive quantizer, on the other hand, may use any of several different quantizers assigned to a vector, depending on the type of data block that is being coded. The adaptive quantizer used for coding the 16 Mbps data rate has four sets of quantizers.

The quantizers consist of look-up tables that are implemented in hardware by high speed memories. The quantizers are actually made up of two tables. They are the quantizer selector tables and the quantizer tables themselves. There is one set of quantizer selector tables for each mode used. Each quantizer selector table consists of 64 entries, where each entry defines the number of bits to be used for a particular vector, and the actual quantizer that will be used to code that vector. The quantizer tables contain the actual mapping of the 8 bit vector coefficient to its quantized value.

The 2 bit mode value and the 64 coded vector values are fed into the encoder rate buffer where they are packed together prior to being transmitted. This produces a 16 Mbps serial data stream of compressed video. A 32 bit syncronization word is added to this serial data stream to indicate the beginning of the video frame. The serial bit stream, along with a 16 MHz clock, are meant to feed a quadra-phase shift key (QPSK) modem which in turn may feed into a satellite, or terrestrial, microwave communication link.

At the receiver, essentially the reverse of the encoding process is done to reconstruct the video data. The serial data is unpacked and representative values are substituted for the quantizer codes generated by the encoder. Inverse transforms are then performed on the data which is converted back to red, green, and blue signals after passing through the D/A converters. The outputs of the decoder may then be used to directly drive an RGB monitor for display. Color burst and burst gate are also generated and are used to drive an NTSC encoder, giving a standard NTSC composite output.

5. QUANTIZATIONS

The task of determining quantization parameters to be used by the encoder and decoder can be broken down into two steps: First, determine for each adaptive mode the vector coefficients to be retained and quantized (the rest are set to zero). Second, determine the number of bits in each mode, to be spent on each coefficient and determine the quantizer range (the maximum coefficient value the quantizer is expected to handle).

The bit assignments for the 16 Mbps quantization are shown in Table 1. Each number indicates how many bits are used to code the corresponding WHT vector value. The same bit assignments were used for all four quantizer modes, even though this doesn't necessarily have to be true in general. The total number of bits assigned adds up to 126, which averages out to 2 bits per sample after the 2 mode bits are included.

```
8 5 4 4 3 3 3 3        6 4
5 4 3 3 2 2 2 2        4 0
4 3 2 2 0 0 0 0     I-Component
4 3 2 2 0 0 0 0
3 2 0 0 0 0 0 0        6 4
3 2 0 0 0 0 0 0        4 0
3 2 0 0 0 0 0 0     Q-Component
3 2 0 0 0 0 0 0
  Y-Component
```

Table 1. 16 Mbps bit assignment

Subjective evaluation was used to verify the choice of parameters for a large class of video inputs and it was necessary to compare several sets of different parameters for each class. To accomplish this in a reasonable amount of time, it was necessary to use a quantizer whose parameters could be changed in real-time.

A programmable quantizer was built for the purpose of determining quantization parameters experimentally. This unit combines the functions of the encoder quantizers and the decoder rep-value look-up tables. The programmable quantizer is connected to the encoder in place of the standard encoder quantizer input and connected to the decoder in place of the representative value lookup table output. The programmable quantizer uses RAMs instead of ROMs to store the look-up tables. This allows the contents of the tables to be easily modified. This configuration turns the compressor into a programmable real-time simulator. The RAMs are loaded through a high speed data link connected to a SEL 32/77 computer. An interactive design program was developed on the SEL in conjunction with the programmable quantizer. This allows one to develop and test quantizer designs in an efficient manner using real-time video data.

In our real-time simulations we concentrated on coding rates of 24, 20, 16, and 12 Mbps. Four mode adaptive quantizers were developed for each of these rates and we were able to evaluate the performance of the video compressor relative to these four rates. Major emphasis was put on the development of a 16 Mbps quantization, since this is the rate for which the serialization hardware was designed.

6. CONCLUDING REMARKS

It was found that the quality of the 24 Mbps quantization was excellent and was nearly identical to the uncompressed video. The 20 Mbps quantization also performed quite well with some noticeable distortion occurring along edges. At this rate the resultant video is still quite acceptable. The 16 Mbps rate begins to introduce substantial artifacts which are displeasing for some sources of video. The two major artifacts introduced are sawtooth edges and an increase of noise in the video. However, we feel that this performance is still acceptable for video conferencing use.

At the 12 Mbps rate the video quality is poor. The random noise in the image is increased and contouring is quite visible. Edges appear very jagged because of a 2:1 subsampling done in the horizontal direction. Because of the coarse quantizations used on the chrominance part of the signal, errors in the color are apparent. However, this rate may still be acceptable for video conferencing purposes.

In comparing the higher data rates to the lower rates it becomes evident that most of the degradation appears in the areas containing a larger amount of detail, such as along edges. The mode detector can successfully detect those areas with high detail and special quantizers with wide dynamic ranges are used to code the resulting vectors. However, it is impossible to give a good representation of high detail data at the lower rates. On the other hand, the lower rate quantizations work almost as well as the high rate quantizations in areas of very low detail. The fundamental problem with using a fixed rate system such as ours, which allocates the same rate to each sub-block in the image, is that video data has an inherent non-stationarity in the information content from one area of an image to another. The solution is to allocate different data rates to blocks with varying information content. This variable rate has to be buffered and smoothed out to produce a constant rate.

We have done some simulations of variable rate systems, but have not constructed any real-time hardware of this type. However, we do plan to build more real-time hardware in the near future, and are considering variable rate systems and also interframe systems that utilize the redundancy in time. We are very interested in an interframe system known as conditional replenishment. In this system only the changing parts of an image are transmitted. This system has the potential of very low data rates for images with little motion in them.

7. REFERENCES

[1] H. F. Harmuth, "Transmission of Information by Orthogonal Functions," Springer-Verlag, pp. 22-33,42.

[2] J. Max, "Quantizing for Minimum Distortion," IRE Trans. Inform. Theory, Vol. 14-6, No. 1, pp. 7-12, Mar. 1960.

[3] W. K. Pratt, J. Kane, and H. C. Andrews, Hadamard Transform Image Coding, Proc. IEEE, Vol 57, No. 1, Jan. 1969.

[4] S. C. Knauer, "Real-Time Video Compression Algorithm for Hadamard Transform Processing," IEEE Trans. Electromagnetic Compatibility, Vol. EMC-18, No. 1, pp. 28-36, Feb. 1976.

[5] H. W. Jones, "A Real-Time Adaptive Hadamard Transform Video Compressor," Proc. of the SPIE, Vol. 87, pp. 2-9, Aug. 1976.

[6] W. Chen and C. H. Smith, "Adaptive Coding of Monochrome and Color Images," IEEE Trans. on Comm., Vol. COM-25, No. 11, pp. 1285-1292, Nov. 1977.

Reprinted with permission from *IEEE Commun. Trans.* **COM-25:**1329–1338 (1977)

Interframe Cosine Transform Image Coding

JOHN A. ROESE, WILLIAM K. PRATT, SENIOR MEMBER, IEEE, AND GUNER S. ROBINSON, MEMBER, IEEE

***Abstract*—Two-dimensional transform coding and hybrid transform/DPCM coding techniques have been investigated extensively for image coding. This paper presents a theoretical and experimental extension of these techniques to the coding of sequences of correlated image frames. Two coding methods are analyzed: three-dimensional cosine transform coding, and two-dimensional cosine transform coding within an image frame combined with DPCM coding between frames. Theoretical performance estimates are developed for the coding of Markovian image sources. Simulation results are presented for transmission over error-free and binary symmetric channels.**

1. INTRODUCTION

DURING the past decade there has been extensive research and development directed toward two-dimensional image coding systems based upon transform and linear predictive coding techniques [1-5]. These coding methods utilize the spatial redundancy within an image field to achieve efficient quantization and coding. In this paper the concept is extended to the coding of sequences of correlated image frames in order to exploit the temporal redundancy of television imagery.

Manuscript received November 30, 1976; revised April 30, 1977. This work was supported by the Advanced Research Projects Agency of the Department of Defense and monitored by the Wright Patterson Air Force Base under Contract F-33615-76-C-1203 and by the Naval Undersea Center, San Diego, CA, under Contract N00123-75-1192. This paper incorporates certain results which were presented at the SPIE Technical Sessions on "Efficient Transmission of Pictorial Information" and "Advances in Image Transmission Techniques" in 1975 and 1976, respectively.

J. A. Roese is with the Naval Ocean Systems Center, San Diego, CA 92152.

W. K. Pratt is with the Image Processing Institute, University of Southern California, Los Angeles, CA 90007.

G. S. Robinson is with the Northrop Research and Technology Center, Hawthorne, CA 90250.

2. INTERFRAME TRANSFORM CODING

The basic concept of three-dimensional transform coding is illustrated in figure 1 [6]. A sequence of image frames $F(j, k, l)$ undergoes a three-dimensional transform in blocks of $J \times K \times L$ pixels according to the general formula

$$\mathcal{F}(u, v, w) = \sum_{j=0}^{J-1} \sum_{k=0}^{K-1} \sum_{l=0}^{L-1} F(j, k, l) A(u, v, w; j, k, l) \quad (1)$$

to produce a sequence of transform planes $\mathcal{F}(u, v, w)$ where $A(u, v, w; j, k, l)$ represents a unitary transform kernel, and (J, K, L) denote the row, column, and interplane indices, respectively. The transform coefficients are then quantized and coded for transmission. Zonal sampling or zonal coding quantization and coding strategies are usually employed. In zonal sampling, transform coefficients with the greatest expected energy are selected for transmission, and the remainder are discarded. Zonal coding entails the quantization of transform coefficients with the number of quantization levels chosen to minimize the total mean square quantization

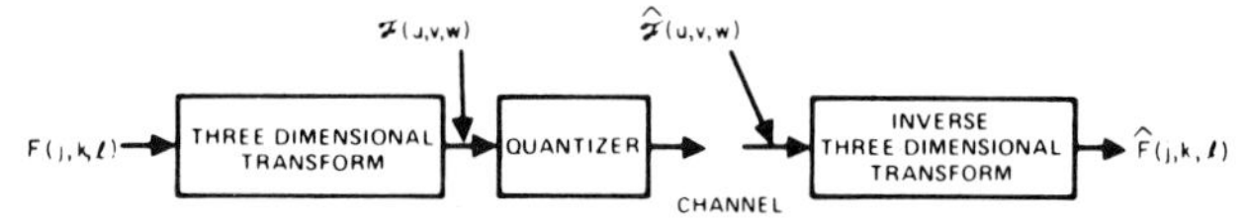

Fig. 1. Three-dimensional transform coder.

error. At the receiving unit the quantized transform coefficients $\hat{F}(u, v, w)$ are inverse transformed to produce the reconstructed frame sequence

$$\hat{F}(j, k, l) = \sum_{u=0}^{J-1} \sum_{v=0}^{K-1} \sum_{w=0}^{L-1} \hat{F}(u, v, w) A^{-1}(u, v, w; j, k, l) \tag{2}$$

where $A^{-1}(u, v, w; j, k, l)$ is the inverse unitary transform kernel.

Many different types of unitary transforms, including the Fourier, sine, cosine, Hadamard, slant, and Karhunen-Loeve transforms, have been investigated for transform image coding. One of the most attractive from the standpoint of coding performance and implementation simplicity is the discrete cosine transform defined by the separable kernel

$$A(u, v, w; j, k, l) = C(u, j)C(v, k)C(w, l) \tag{3a}$$

where

$$C(u, j) = \frac{1}{\sqrt{J}} \qquad u = 0 \tag{3b}$$

$$C(u, j) = \sqrt{\frac{2}{J}} \cos\left[\frac{(2j+1)u\pi}{2J}\right] \qquad u > 0. \tag{3c}$$

It has been shown that the theoretical mean square error coding performance of the cosine transform is nearly equivalent to that obtained by the optimal Karhunen-Loeve transform for Markovian image sources [7]. Furthermore, the cosine transform can be efficiently computed by transversal filtering methods [8]. For these reasons the cosine transform has been adopted for the analysis of interframe coding techniques.

A bandwidth reduction is achieved in transform coding by quantizing and coding transform coefficients according to their expected energy based upon a statistical model of coefficient variances. The conventional model assumes pixels to be zero mean samples of a separable Markov process. The correlation matrix along each image column is given by

$$K_{f_j} = \sigma_j^2 \begin{bmatrix} 1 & \rho_c & \rho_c^2 & \cdots & \rho_c^{J-1} \\ \rho_c & 1 & \rho_c & \cdots & \rho_c^{J-2} \\ \cdot & \cdot & \cdot & & \cdot \\ \cdot & \cdot & \cdot & & \cdot \\ \cdot & \cdot & \cdot & & \cdot \\ \cdot & \cdot & \cdot & & \cdot \\ \rho_c^{J-1} & \cdot & \cdot & \cdots & 1 \end{bmatrix} \tag{4}$$

where σ_j^2 represents the pixel variance and ρ_c is the adjacent pixel correlation. Similar expressions exist for the row and temporal correlation matrices. As a result of the separability of the cosine transform the covariance matrices of the coefficients can be obtained from

$$K_{f_u} = CK_{f_j}C^T \tag{5a}$$

$$K_{f_v} = CK_{f_k}C^T \tag{5b}$$

$$K_{f_w} = CK_{f_l}C^T \tag{5c}$$

where C represents the cosine transform kernel $C(j, u)$ in matrix form and K_{f_u}, etc. are the transform coefficient column, row, and temporal covariance matrices. Column, row, and temporal variance vectors v_{f_u}, etc. can then be obtained by extracting diagonal elements of the corresponding covariance matrices. Finally, a three-dimensional variance array is formed by the product

$$V_F(u, v, w) = v_{f_u}(u) v_{f_v}(v) v_{f_w}(w). \tag{6}$$

This array models the variance of each transform at coordinate (u, v, w).

For the zonal sampling form of transform coding the coefficient selection process can be conveniently represented by a three-dimensional transform domain selection operator $S(u, v, w)$. The reconstructed image can then be expressed as

$$\hat{F}(j, k, l) = \sum_{u=0}^{J-1} \sum_{v=0}^{K-1} \sum_{w=0}^{L-1} F(u, v, w) S(u, v, w) \cdot A^{-1}(u, v, w; j, k, l) \tag{7}$$

where $S(u, v, w)$ assumes the value of unity for selected transform coefficients and zero for discarded coefficients. Figure 2 illustrates coefficient selection. The resultant mean square coding error for zonal sampling is given by [9]

$$\mathcal{E} = E\{(F - \hat{F})^2\} = \sum_{u=0}^{J-1} \sum_{v=0}^{K-1} \sum_{w=0}^{L-1} E\{F^2(u, v, w)\} \cdot [1 - S(u, v, w)]. \tag{8}$$

The optimal coding strategy is to order the transform coefficients on the basis of largest variance as specified by eq. (6). Figure 3 contains an evaluation of the zonal sampling mean square error for an adjacent pixel correlation factor of $\rho = 0.95$ for a 32:1 sample reduction as a function of spatial block size and frame size. The curve for $L = 1$ represents the performance of a two-dimensional zonal sampling transform coder.

In the zonal coding process the number of quantization levels assigned to each coefficient is commonly set proportional to the logarithm of the coefficient variance; the coefficient quantization and decision levels are nonlinearly spaced to minimize the mean square quantization error according to a probability density model for the coefficients [9]. Those coefficients receiving no quantization levels are discarded as in the zonal sampling process. The resulting mean square error expression for a constant length binary code of each

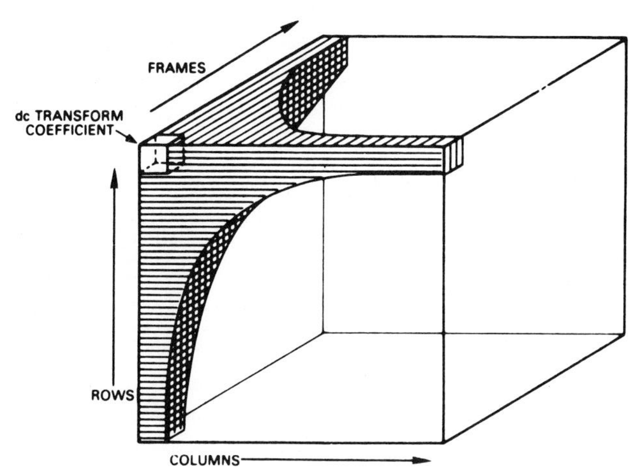

Fig. 2. Three-dimensional array of selected transform coefficients.

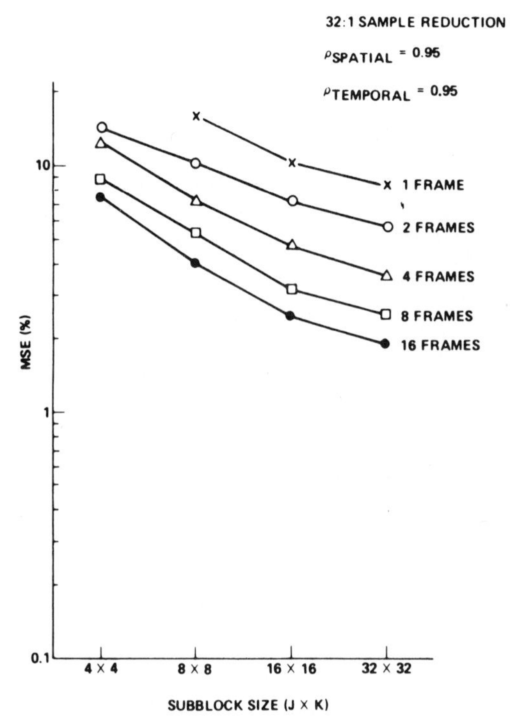

Fig. 3. Theoretical performance evaluation for three-dimensional cosine transform coder with zonal sampling for Markov process data source.

coefficient $b(u, v, w)$ is given by

$$E = \sum_{u=0}^{J-1} \sum_{v=0}^{K-1} \sum_{w=0}^{L-1} \left[E\{F^2(u, v, w)\} - \sum_{n=1}^{2^{b(u,v,w)}} R_n^2(u, v, w) P[R_n(u, v, w)] \right] \tag{9}$$

where $R_n(u, v, w)$ represents the n-th quantization level of a coefficient, $P(\cdot)$ is the occupancy probability of the n-th quantization level, and $b(u, v, w)$ denotes the number of bits assigned to a coefficient. The probability density of the "d.c." coefficient $F(0, 0, 0)$ is usually modeled as a Rayleigh density, while the remaining "a.c." coefficients are modeled

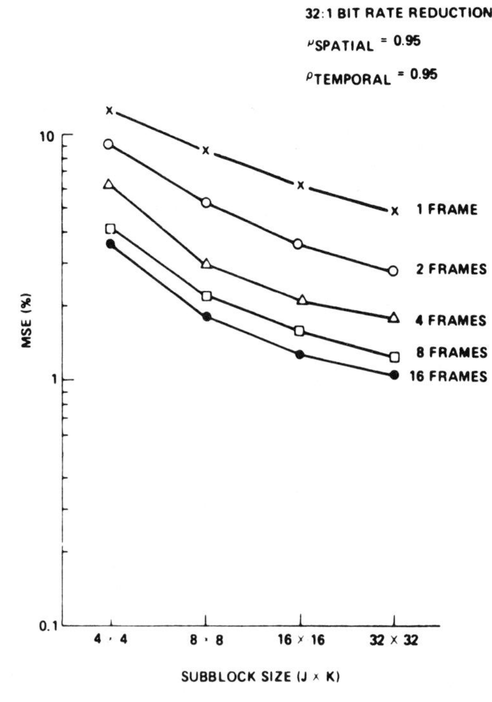

Fig. 4. Theoretical performance evaluation for three-dimensional cosine transform coder with zonal coding for Markov process data source.

as Gaussian densities with variances specified by eq. (6). Figure 4 illustrates the mean square error zonal coding performance for a separable Markov process image source with $\rho = 0.95$ and a 32:1 bit rate reduction, i.e. coding from 8 to 0.25 bits/pixel. The relative efficiency between zonal sampling and zonal coding is evident from a comparison of figures 3 and 4.

3. INTERFRAME HYBRID TRANSFORM/DPCM CODING

In 1974 Habibi introduced the concept of hybrid transform/DPCM image coding in which a unitary transform is taken along an image row and differential pulse code modulation (DPCM) coding is performed on the transform coefficients along each column [10]. This concept can be easily extended to three-dimensional image sources [10-13].

A block diagram of the basic interframe hybrid coder is shown in figure 5. In this coding system a two-dimensional unitary transform is performed on each partition or spatial subblock of the image. One of the bank of parallel DPCM linear predictive coders is then applied to each set of transform coefficients in the temporal direction. The resulting sequences of transform coefficient differences are quantized and coded for transmission. Image reconstruction occurs at the receiver where the transform coefficient differences are decoded and a replica of each transmitted image is reconstructed by a two-dimensional inverse transformation.

The three-dimensional array of elements obtained by a two-dimensional cosine transform in $J \times K$ pixel blocks may be expressed in general form as

$$H(u, v, l) = \sum_{j=0}^{J-1} \sum_{k=0}^{K-1} F(j, k, l) C(u, j) C(v, k) \tag{10}$$

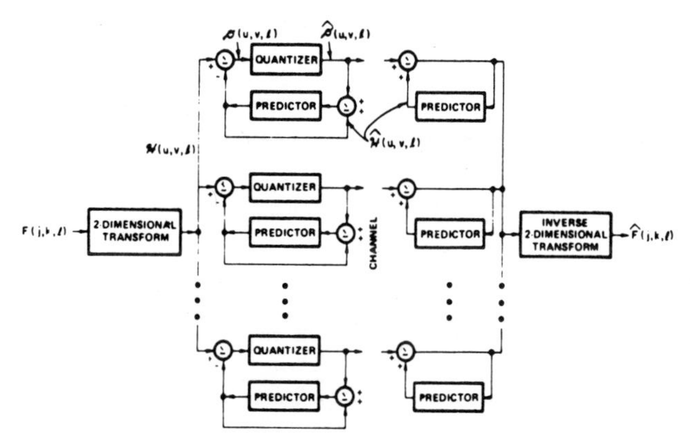

Fig. 5. Hybrid two-dimensional transform/DPCM coder.

where $C(u, j)$ represents the cosine transform kernel. At each spatial frequency (u, v) the DPCM coder quantizes and codes the difference signal

$$\mathcal{D}(u, v, l) = H(u, v, l) - \hat{H}(u, v, l) \quad (11)$$

where $\hat{H}(u, v, l)$ denotes the predicted value of $H(u, v, l)$. The difference signal $\mathcal{D}(u, v, l)$ is coded using a zonal coding strategy similar to that of transform coding in which the number of code bits assigned to each difference signal is set proportional to the logarithm of its variance [4]

$$V_H(u, v) = v_f(u)v_f(v)[1 - \rho_T{}^2] \quad (12)$$

where ρ_T represents the temporal correlation factor. Quantization levels are usually set according to a Laplacian model of each difference signal. The mean square error expression is given by

$$E = \sum_{u=0}^{J-1} \sum_{v=0}^{K-1} \left[V_H(u, v) - \sum_{n=1}^{2^{b(u,v)}} R_n{}^2(u, v)P(R_n(u, v)) \right] \quad (13)$$

where $R_n(u, v)$ represents the n-th reconstruction level of the quantized difference signal and $b(u, v)$ denotes the number of bits assigned at each spatial frequency coordinate. Figure 6 contains a plot of mean square error versus block size for a hybrid cosine transform/DPCM coder for a Markov process source with $\rho = 0.95$. A comparison of the theoretical performance of the hybrid and three-dimensional transform coders is shown in figure 7.

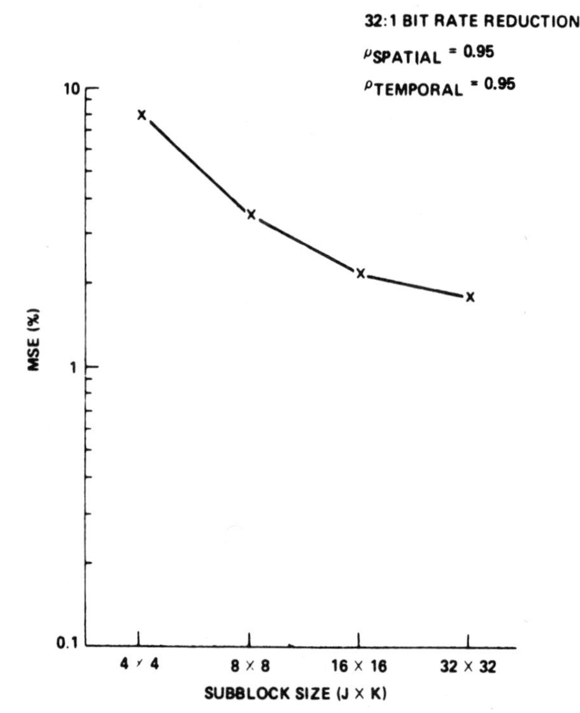

Fig. 6. Theoretical performance evaluation for hybrid two-dimensional transform/DPCM coder with zonal coding for Markov process data source.

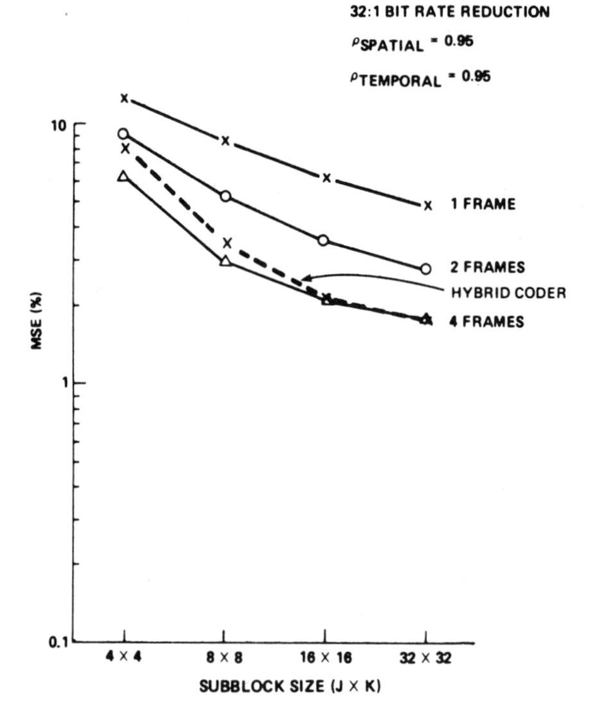

Fig. 7. Theoretical performance comparison of three-dimensional cosine transform coder and hybrid two-dimensional transform/DPCM coder for Markov process data source.

4. ADAPTIVE INTERFRAME CODING

Significant improvement in coding performance can be obtained by spatially adapting the hybrid/DPCM interframe coder to use calculated temporal statistical measures of the transform coefficient temporal difference signal $H(u, v, l)$ in each subblock. The statistical measures utilized are the mean and correlation for each temporal sequence of transform coefficients defined as

$$\bar{H}(u, v) = \frac{1}{L} \sum_{l=0}^{L-1} H(u, v, l) \quad (14a)$$

and

$$\rho(u, v) = \frac{\frac{1}{L-1} \sum_{l=1}^{L-1} H(u, v, l) H(u, v, l-1)}{\frac{1}{L} \sum_{l=0}^{L-1} H^2(u, v, l)}. \tag{14b}$$

The variance of the temporal coefficient difference is then estimated as

$$\sigma_D{}^2(u, v) = \frac{1}{L-1} \sum_{l=1}^{L-1} \tilde{H}_D{}^2(u, v, l) - \left[\frac{1}{L-1} \sum_{l=1}^{L-1} \tilde{H}_D(u, v, l)\right]^2 \tag{15a}$$

where

$$\tilde{H}_D(u, v, l) = H(u, v, l) - \rho(u, v) H(u, v, l-1) \tag{15b}$$

for $l = 1, 2, \cdots, L-1$. For each sequence of transform coefficients the temporal correlation $\rho(u, v)$ is the gain coefficient in the DPCM predictor feedback loop. The transform coefficient mean $\bar{H}(u, v)$ provides biasing to achieve a zero-mean input sequence of transform coefficients. The computed transform coefficient difference variances $\sigma_D{}^2(u, v)$ are used to generate the two-dimensional subblock bit assignment. Thus, the resulting bit assignment, predictor feedback loop gain coefficients, and associated biasing and scaling factors are, in general, different for different subblocks. Although the subblock bit assignment may vary between subblocks, the total number of bits available for coding each subblock is constrained to be equal. Local adaptation to the measured statistics of each subblock will normally produce improved coding results when compared with non-adaptive implementations. However, adaptation does result in increased coder complexity.

In the adaptive coding process described above the receiver must have available the three statistical parameters $\bar{H}(u, v)$, $\rho(u, v)$, and $\sigma_D{}^2(u, v)$ at each coefficient index (u, v). One approach is to form these statistics at the transmitter for L stored frames, and then quantize, code, and transmit each statistic to the receiver. The disadvantages of this approach are the data overhead and the effects of quantization error introduced in the transmission of each statistic. A better approach is to form the statistics jointly at the transmitter and receiver utilizing the coefficient feedback predicted value $\hat{H}(u, v, l)$ in eqs. (14) and (15). This scheme requires no channel overhead since the transmitter and receiver coefficient predictions are identical in the absence of channel errors. Furthermore, quantization error of the statistics is reduced by the DPCM feedback process. With either approach it is necessary to occasionally re-transmit the feedback prediction $\hat{H}(u, v, l)$ to correct for accumulated channel error effects.

The concept of temporal/spatial adaptation described above for the hybrid transform/DPCM interframe coding is not directly applicable to interframe three-dimensional transform coding since temporal averages are meaningless for a three dimensional coder. There are, however, a number of other means of adaptively computing subcube bit assignments for a three-dimensional transform coder based upon coefficient energy distribution, object motion, or edge structure occuring within a subcube. Such methods have not been explored in this paper.

5. EXPERIMENTAL RESULTS

An extensive set of computer simulations has been performed to experimentally evaluate the performance of interframe transform coders. The hybrid coders considered are parametric in many variables including choice of separable transforms, zonal sampling or zonal coding with various quantization strategies, presence or absence of channel error, spatial subblock size, and average pixel bit rate. In the evaluation of coder performance levels the normalized mean square error (NMSE) and signal-to-noise ratio criteria (SNR) are employed in conjunction with subjective visual evaluations. These error criteria are defined as

$$\text{NMSE} = \frac{\sum_{j=0}^{J-1} \sum_{k=0}^{K-1} [F(j, k, l) - \hat{F}(j, k, l)]^2}{\sum_{j=0}^{J-1} \sum_{k=0}^{K-1} [F(j, k, l)]^2} \tag{16}$$

$$\text{SNR} = -10 \log_{10} \frac{\frac{1}{JK} \sum_{j=0}^{J-1} \sum_{k=0}^{K-1} [F(j, k, l) - \hat{F}(j, k, l)]^2}{Y_M{}^2} \tag{17}$$

where F and $\hat{F}$ are the original and coded images, and Y_M represents the maximum luminance value of F, typically 255. The source data for the computer simulations consist of two sets of 16 images digitized from sequential frames of 24 frame-per-second motion pictures. One sequence, called "Walter," contains images from a fixed position camera of a moving subject engaged in conversation. The other contains images of a chemical plant photographed from an airplane in a flyby trajectory. Each frame is digitized at a spatial resolution of 256 X 256 pixels with each pixel amplitude linearly quantized to 256 levels. Photographs of the sixteenth frame of each data sequence are presented in figure 8.

Figure 9 contains photographs of image reconstructions of the sixteenth frame of an image sequence for three-dimensional transform coding in 16 X 16 X 16 pixel cubes. Coding is performed at average rates of 1.0, 0.5, 0.25, and 0.1 bits/pixel/frame. Normalized mean square error measurements are indicated in the figure. Subjectively, coding errors are not noticeable for coding down to 0.5 bits. Beyond this value some blotchiness becomes apparent.

Photographs of the image reconstruction of the sixteenth frame of the sequence for hybrid transform/DPCM coding system at average coding rates of 0.1, 0.25, 0.5, and 1.0 bits/pixel are shown in figures 10 and 11 for non-adaptive and adaptive coding, respectively. In these experiments the first of the sixteen frames is available at both the transmitter and

(a)

(b)

Fig. 8. Sixteenth frame of original sequences of test images. (a) Walter. (b) Chemical plant.

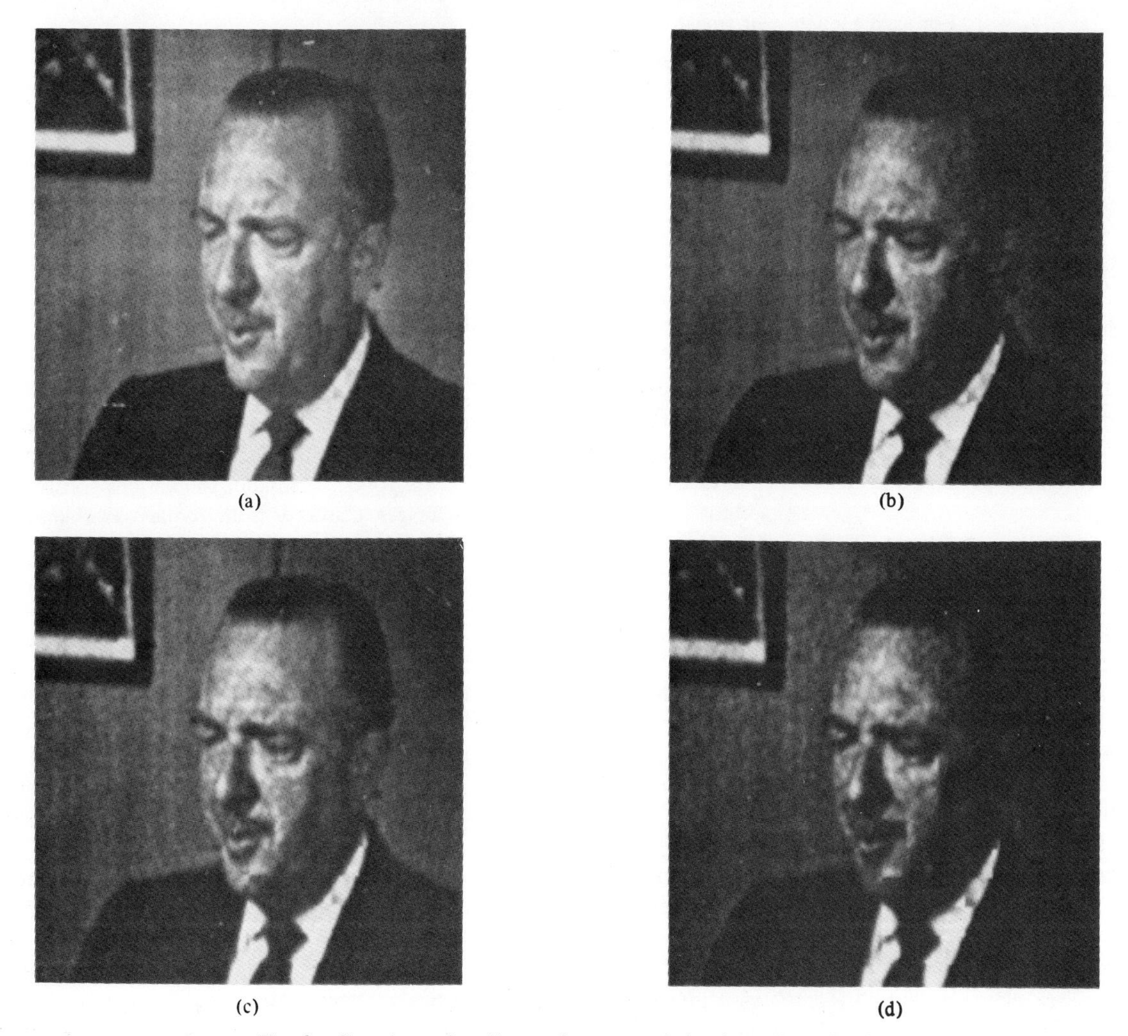

(a) (b) (c) (d)

Fig. 9. Experimental coding performance of the three dimensional cosine transform coder. (a) 1.0 bits/pixel/frame, NMSE = 0.163%. (b) 0.5 bits/pixel/frame, NMSE = 0.299%. (c) 0.25 bits/pixel/frame, NMSE = 0.491%. (d) 0.1 bits/pixel/frame, NMSE = 0.882%.

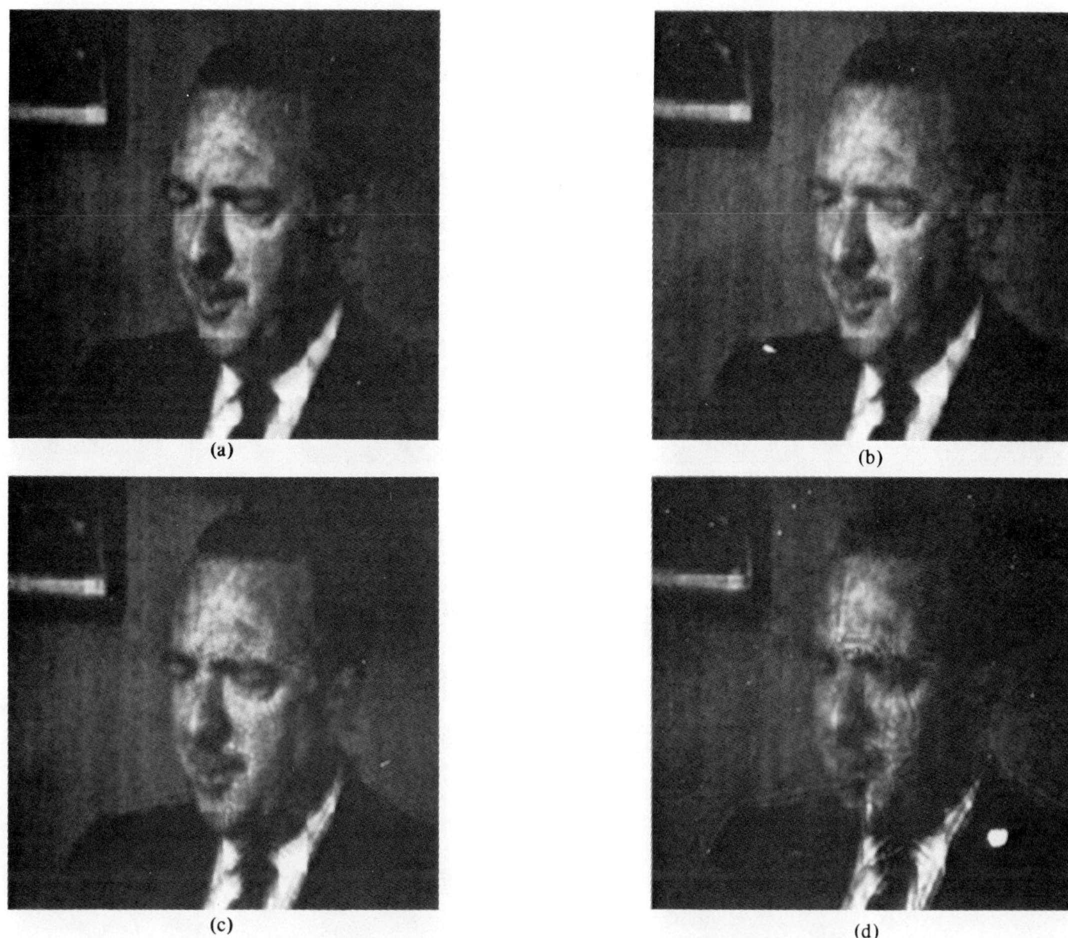

Fig. 10. Experimental coding performance of the hybrid transform/DPCM coder without adaptive coding. (a) 1.0 bits/pixel/frame, NMSE = 1.026%. (b) 0.5 bits/pixel/frame, NMSE = 1.227%. (c) 0.25 bits/pixel/frame, NMSE = 1.601%. (d) 0.1 bits/pixel/frame, NMSE = 2.634%.

receiver as an initial condition. The transform coefficient statistics in the adaptive coder simulation have been generated from the common feedback predicted coefficient $\hat{H}(u, v, l)$ available at both the transmitter and receiver, and therefore, there is no channel overhead for this form of adaptive coding. Visually, no image degradation can be seen for bit rates as low as 0.5 bits/pixel/frame. Some artifacting effects due to the 16 X 16 subblock partitioning of the images is apparent at the 0.25 bits/pixel/frame rate. Also, regions outlining the subject's head begin to show degradation at this bit rate because of head motion and the relatively few bits assigned to transmit high frequency coefficients. The observed image degradations are similar in nature, but more pronounced at the 0.1 bits/pixel/frame bit rate. The sixteen reconstructed frames produced by the coder have been displayed on a television monitor for real time subjective analysis. No error build-up has been perceived in the display.

Figure 12 illustrates NMSE and SNR measures as a function of frame number for the adaptive hybrid transform/DPCM coder implementation at average pixel bit rates from 0.1 to 1.0 bits/pixel/frame with 16 X 16 subblocks. These graphs indicate that, even at the lowest bit rate, stability in coder performance is achieved within the first eight frames. Perfomance stability occurs much earlier in the frame sequence at the higher bit rates. In these simulations the first frame is assumed available at the receiver, and consequently, zero NMSE is obtained for the first frame.

Performance of the adaptive hybrid transform/DPCM coder in the presence of noise has been investigated by computer simulation of a binary symmetric channel transmission system. The channel operates on each binary digit independently, changing each digit from 0 to 1 or from 1 to 0 with probability P and leaving the digit unchanged with probability $1-P$. At the receiver the encoded picture is reconstructed from the string of binary digits, including errors, transmitted across the channel.

Photographs showing the visual effects of channel noise on the sixteenth frame of the data base are given in figure 13. The results illustrated are for average pixel bit rates of 1.0 and 0.25 bits/pixel, and P of 10^{-2} and 10^{-3}. Simulations have

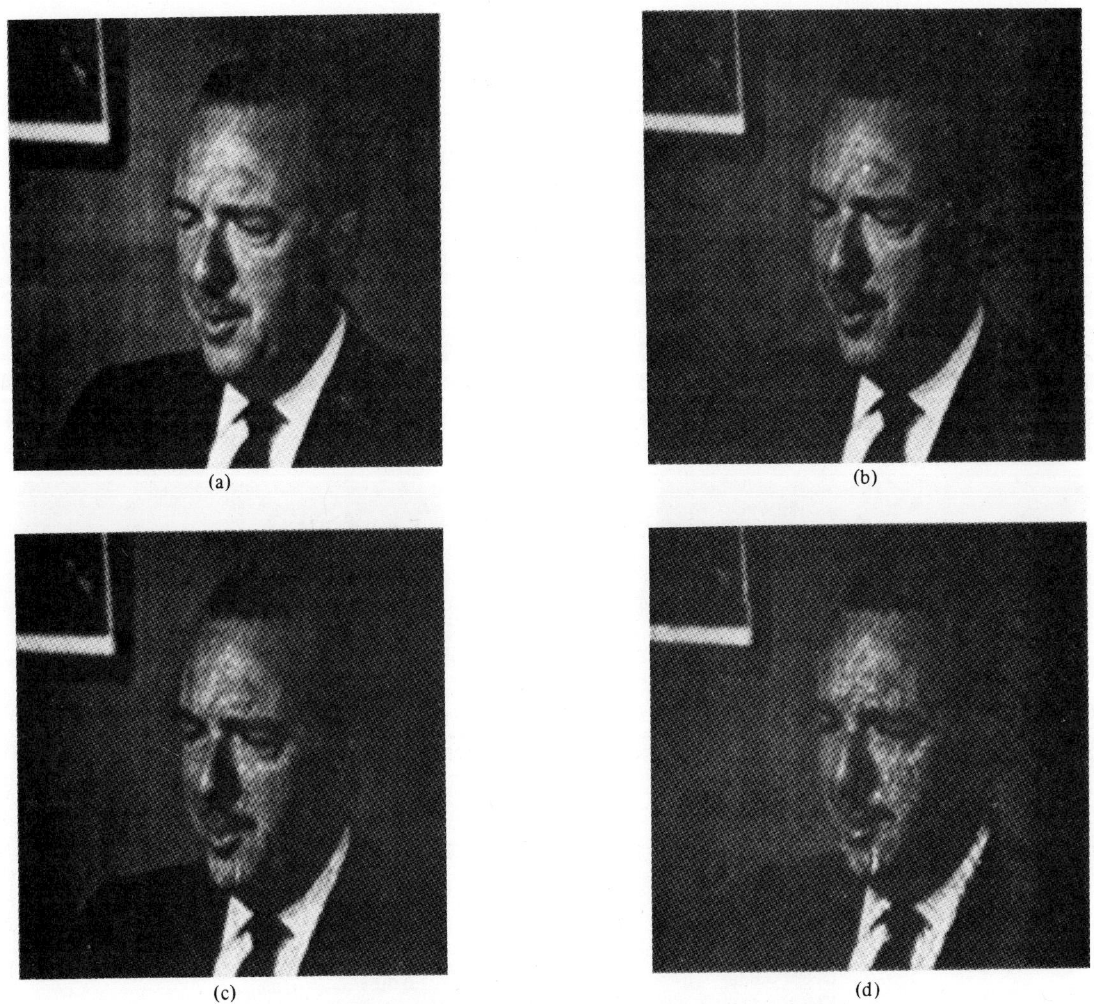

Fig. 11. Experimental coding performance of the hybrid transform/DPCM coder with adaptive coding. (a) 1.0 bits/pixel/frame, NMSE = 0.022%. (b) 0.5 bits/pixel/frame, NMSE = 0.067%. (c) 0.25 bits/pixel/frame NMSE = 0.186%. (d) 0.1 bits/pixel/frame, NMSE = 0.625%.

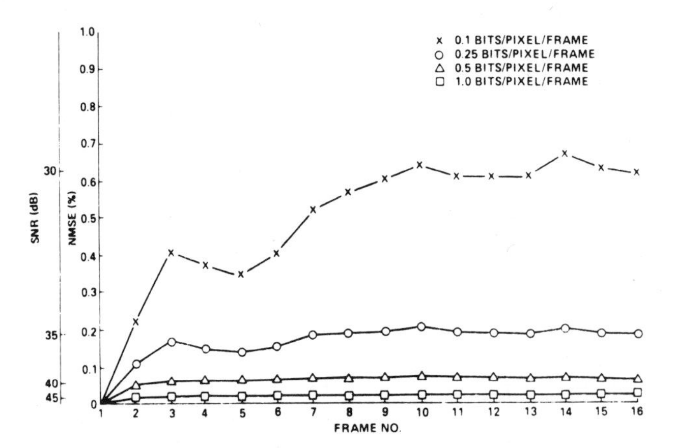

Fig. 12. Coding performance as a function of frame number for the hybrid two-dimensional transform/DPCM coder with adaptive coding.

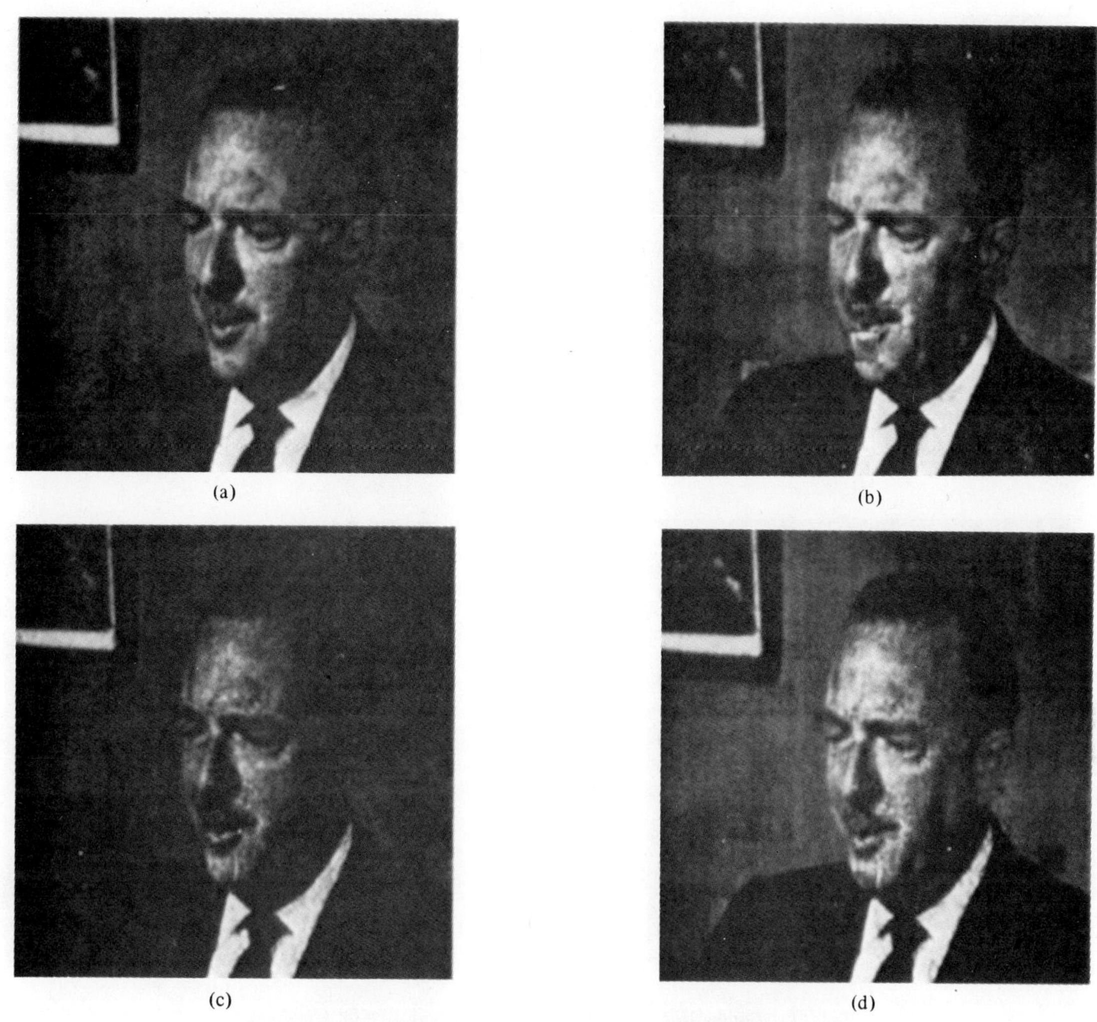

Fig. 13. Experimental coding performance of the hybrid transform/DPCM coder with adaptive coding and with channel error. (a) 1.0 bits/pixel/frame, $P = 10^{-3}$, NMSE = 0.043%. (b) 1.0 bits/pixel/frame, $P = 10^{-2}$, NMSE = 0.282%. (c) 0.25 bits/pixel/frame, $P = 10^{-3}$, NMSE = 0.222%. (d) 0.25 bits/pixel/frame, $P = 10^{-2}$, NMSE = 0.450%.

also been run with a channel error probability of 10^{-4}. Coding results obtained with $P = 10^{-4}$ are essentially indistinguishable from the case with $P = 0$ and, therefore, are not included.

Figure 14 compares the coding results of intraframe and interframe transform/DPCM coders for the two image data bases. The average pixel bit rates used give subjectively equivalent image reconstructions although the nature of the image degradations differs for the two coders. For the "Walter" data base, exploitation of temporal correlation results in an 8:1 reduction in average pixel bit rate. Comparison of coding performance using the "chemical plant" data base shows an average pixel bit rate improvement of 4:1. Since performance of the hybrid interframe coder is dependent on temporal correlation, reduced levels of performance are to be anticipated for image sequences distorted by the effects of camera motion.

6. SUMMARY

Based on the theoretical and experimental results obtained, it has been demonstrated that exploitation of temporal as well as spatial correlations is a viable technique for coding digital image sequences. In addition, theoretical performance levels have been predicted for the interframe transform and hybrid coders. Supplementary studies of alternative interframe coder operational modes and image motion compensation effects are described in reference [13].

ACKNOWLEDGMENT

The authors wish to acknowledge the contributions of Dr. Ali Habibi of TRW Systems in the initial stages of research leading to this paper while Dr. Habibi was a staff member of the University of Southern California Image Processing Institute. Also, the authors are grateful to Dr.

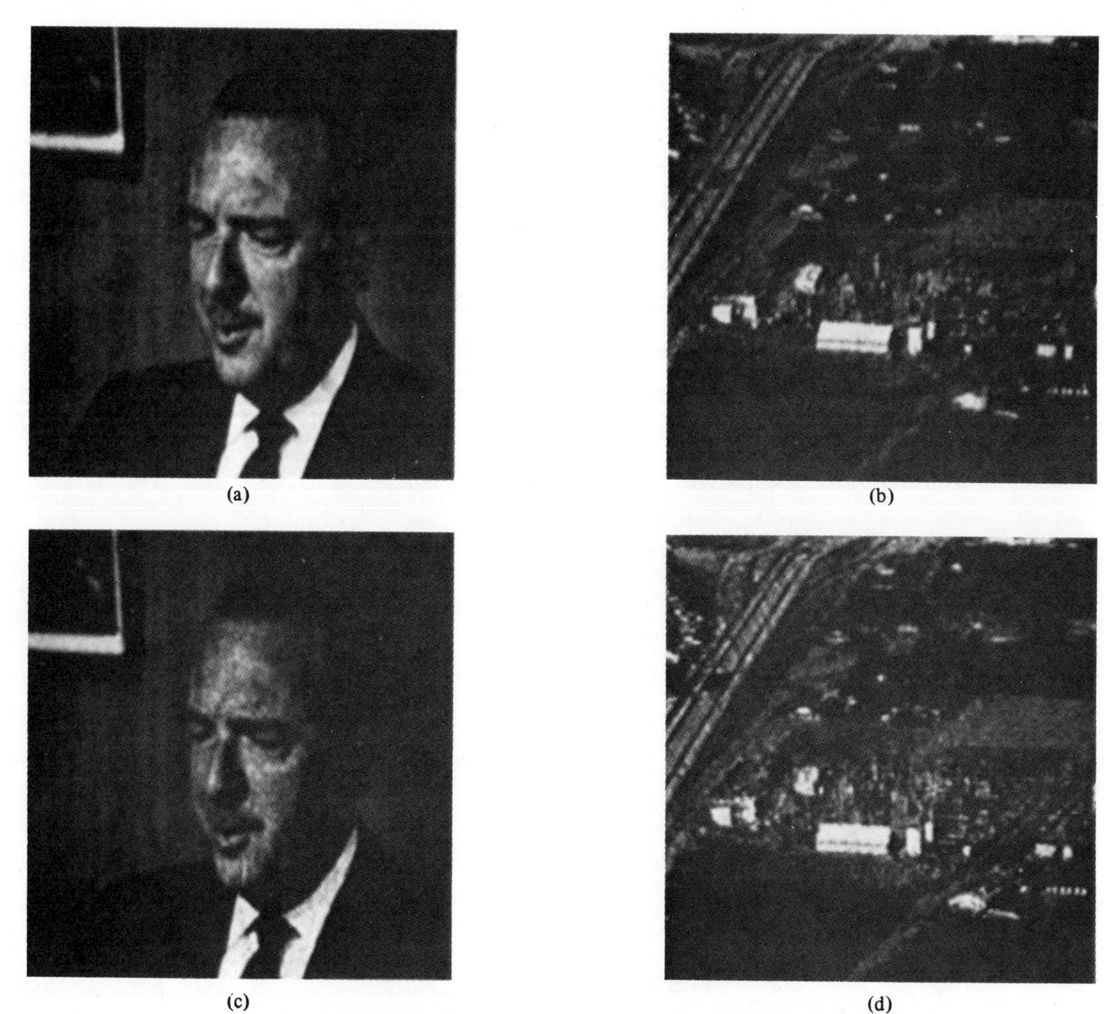

Fig. 14. Comparison of hybrid intraframe and hybrid interframe cosine transform/DPCM coders with adaptive coding. (a) Hybrid intraframe coder, 2.0 bits/pixel. (b) Hybrid intraframe coder, 0.8 bits/pixel. (c) Hybrid interframe coder, 0.25 bits/pixel/frame. (d) Hybrid interframe coder, 0.2 bits/pixel/frame.

Clifford Reader of Ford Aerospace Corporation of Palo Alto, California for his assistance in photographic recording of images.

REFERENCES

1. A. Habibi and G. S. Robinson, "A Survey of Digital Picture Coding," *IEEE Computer*, Vol. 7, No. 5, May 1974, pp. 22-34.
2. Special Issue on Redundancy Reduction, *Proceedings IEEE*, Vol. 55, No. 3, March 1967.
3. Special Issue on Digital Picture Processing, *Proceedings IEEE*, Vol. 6, No. 7, July 1972.
4. A. Habibi and P.A. Wintz, "Image Coding by Linear Transformation and Block Quantization," *IEEE Transactions on Communication Technology*, Vol. COM-19, February 1971, pp. 50-62.
5. A. Habibi, "Comparison of *n*-th Order DPCM Encoder with Linear Transformations and Block Quantization Techniques," *IEEE Transactions on Communication Technology*, Vol. COM-19, No. 6, December 1971, pp. 948-956.
6. J. A. Roese and W. K. Pratt, "Theoretical Performance Models for Interframe Transform and Hybrid Transform/DPCM Coders," *Proceedings SPIE*, Vol. 87, August 1976, pp. 172-179.
7. N. Ahmed, T. Natarajan, and K. R. Rao, "On Image Processing and a Discrete Cosine Transform," *IEEE Transactions on Computers*, Vol. C-23, No. 1, January 1974, pp. 90-93.
8. R. W. Means, H. J. Whitehouse, and J. M. Speiser, "Television Encoding Using a Hybrid Discrete Cosine Transform and a Differential Pulse Code Modulator in Real Time," *Proceedings National Telecommunications Conference*, San Diego, California, December 1974, pp. 61-66.
9. W. K. Pratt, *Digital Image Processing*, Wiley, New York, 1977.
10. A. Habibi, "Hybrid Coding of Pictorial Data," *IEEE Transactions on Communications*, Vol. COM-22, No. 5, May 1974, pp. 614-624.
11. J. A. Roese, et. al., "Interframe Transform Coding and Predictive Coding Methods," *Proceedings 1975 International Conference on Communications*, San Francisco, Vol. 2, June 1975, pp. 17-21.
12. J. A. Roese and G. S. Robinson, "Combined Spatial and Temporal Coding of Digital Image Sequences," *Proceedings SPIE*, Vol. 66, August 1975, pp. 172-180.
13. J. A. Roese. "Interframe Coding of Digital Images Using Transform and Hybrid Transform/Predictive Techniques," Naval Undersea Center, TP 534, June 1976.

Hybrid Coding of Pictorial Data

ALI HABIBI, MEMBER, IEEE

Abstract—**Two hybrid coding systems utilizing a cascade of a unitary transformation and differential pulse code modulators (DPCM) systems are proposed. Both systems encode the transformed data by a bank of DPCM systems. The first system uses a one-dimensional transform of the data where the second one employs two-dimensional transformations. Theoretical results for Markov data and experimental results for a typical picture are presented for Hadamard, Fourier, cosine, slant, and the Karhunen-Loeve transformations. The visual effects of channel error and also the impact of noisy channel on the performance of the hybrid system, measured in terms of the signal-to-noise ratio of the encoder, is examined and the performance of this system is compared to the performances of the two-dimensional DPCM and the standard two-dimensional transform encoders.**

I. INTRODUCTION

THE search for efficient techniques of transmitting pictorial data over digital communication channels has led various researchers to a common approach to the problem. Briefly, this approach is processing the correlated data (images) to generate a set of uncorrelated or as nearly uncorrelated as possible set of signals which in turn are quantized using a memoryless quantizer. The quantized signal is then encoded using either fixed or variable length code words and is transmitted over a digital channel. This is the general approach taken in designing differential pulse code modulators (DPCM) and the techniques that use unitary transformation and block quantization as well as many other techniques developed in recent literature [1]–[6]. Both DPCM and transform coding techniques have been used with some success in coding pictorial data. A study of both these systems has indicated that each technique has some attractive characteristics and some limitations. The transform coding systems achieve superior coding performance at lower bit rates; they distribute the coding degradation in a manner less objectionable to a human viewer and show less sensitivity to data statistics (picture to picture variation) and are less vulnerable to channel noise. On the other hand, DPCM systems, when designed to take advantage of spatial correlations of the data, achieve a better coding performance at a higher bit rate, the equipment complexity and the delay due to coding operation is minimal, and the system does not require the large memory needed in the transform coding systems. Perhaps the most desirable characteristic of this system is the ease of design and the speed of the operation that has made the use of DPCM systems in real time coding of television signals possible. The limitations of this system are the sensitivity of the well designed two-dimensional DPCM systems to picture statistics and the propagation of the channel error on the transmitted picture.

In this paper, two coding systems that use a cascade of unitary transformations and a bank of DPCM systems are proposed. These systems combine the attractive features of both transform coding and the DPCM systems, thus achieving good coding capabilities without many of the limitations of each system. The first system proposed here exploits the correlation of the data in the horizontal direction by taking a one-dimensional transform of each line of the picture, then it operates on each column of the transformed data using a bank of DPCM systems. The DPCM systems quantize the signal in the transform domain where it takes advantage of the vertical correlation of the transformed data to reduce the coding error. The unitary transformation involved is a one-dimensional transformation of individual lines of the pictorial data, thus the equipment complexity and the number of computational operations is considerably less than what is involved in a two-dimensional transformation. The visual effect of channel error on the encoded pictures and its effect on the performance of the system measured in terms of the signal-to-noise ratio of the encoder using a binary symmetric channel is examined. The system shows little degradation at small or moderate levels of channel noise, however its performance is degraded significantly at high levels of channel error. The system is particularly attractive in the sense that the principle can be expanded to utilize interframe coding of television signals. Such a coding system would start by taking a two-dimensional transformation of each frame of the television signal, then it would encode the transformed signal in the temporal direction by a number of parallel DPCM encoders, thus exploiting the correlation of data in temporal as well as spatial directions.

The second system proposed uses a two-dimensional unitary transformation on the pictorial data divided into small blocks. The elements of each block in the transformed domain are ordered in a one-dimensional array and are coded by a bank of DPCM systems. A small block size reduces the number of arithmetic operations needed to obtain the transformed data, but it reduces the efficiency of the transformed system since the elements of various blocks remain correlated in the transformed domain. How-

Paper approved by the Associate Editor for Communication Theory of the IEEE Communications Society for publication after presentation at the 1973 Image Coding Symposium, Los Angeles, Calif. Manuscript received May 14, 1973; revised October 24, 1973. This research was supported by the Naval Undersea Center, San Diego, Calif., under Contract N00123-73-C-1507 and by the Advanced Research Projects Agency of the Department of Defense and was monitored by the Air Force Eastern Test Range under Contract F08606-72-C-0008.

The author is with the Department of Electrical Engineering, University of Southern California, Los Angeles, Calif. 90007.

ever, DPCM coding of the elements in the transformed domain takes advantage of this correlation and improves the performance of the system.

II. ONE-DIMENSIONAL TRANSFORMATION AND DPCM CODING

In the system proposed here the pictorial data is scanned to form N lines with appropriate vertical resolution, then each line is sampled at a Nyquist rate. The sampled image is then divided into arrays of M by N picture elements $u(x,y)$, where x and y index the rows and the columns in each individual array such that the number of samples in a line of image is an integer multiple of M. One-dimensional unitary transformation of the data and its inverse are modeled by the set of equations

$$u_i(y) = \sum_{x=1}^{M} u(x,y)\Phi_i(x) \qquad i = 1,2,\cdots,M \quad y = 1,2,\cdots,N \tag{1}$$

$$u(x,y) = \sum_{i=1}^{M} u_i(y)\Phi_i(x) \tag{2}$$

where $\Phi_i(x)$ is a set of M orthonormal basis vectors. The correlation of the transformed samples $u_i(y)$ and $u_i(y+\tau)$ is given by

$$C_i(\tau) = \sum_{x=1}^{M}\sum_{\hat{x}=1}^{M} R(x,\hat{x},y,y+\tau)\Phi_i(x)\Phi_i(\hat{x}) \tag{3}$$

where $R(x,\hat{x},y,\hat{y})$ is the spatial autocovariance of the data.

Note that this equation indicates (1) the correlation of samples in each column of the transformed array is directly proportional to the correlation of sampled image in vertical direction, and (2) the correlation of samples in various columns of the transformed array is different. Thus, a number of different DPCM systems should be used to encode each column of the transformed data. The block diagram of the proposed system is shown in Fig. 1. A replica of the original image $u^*(x,y)$ is formed by inverse transforming the coded samples, i.e.,

$$u^*(x,y) = \sum_{i=1}^{n} v_i(y)\Phi_i(x), \quad n \le M. \tag{4}$$

The mean square value of coding error is

$$\epsilon^2 = E\left\{\frac{1}{MN}\sum_{y=1}^{N}\sum_{i=1}^{M}[u(x,y) - u^*(x,y)]^2\right\}. \tag{5}$$

Using (1)–(4) and assuming that $q_i(y)$, the quantization error encountered in the ith DPCM system, is uncorrelated with $u_i(y)$, the coding error ϵ^2 is

$$\epsilon^2 = R(0,0,0,0) - \sum_{i=1}^{n} C_i(0) + E\left\{\frac{1}{MN}\sum_{y=1}^{N}\sum_{x=1}^{n}[u_i(y) - v_i(y)]^2\right\} \tag{6}$$

where the first two terms are introduced because of using n (rather than M) DPCM systems. Study of DPCM systems has shown that [1]

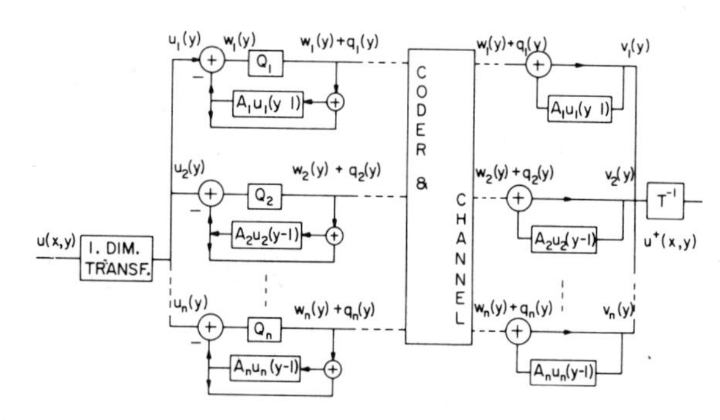

Fig. 1. Block diagram of the hybrid system using a cascade of one-dimensional transformations and a bank of DPCM systems to encode pictorial data.

$$E\{[u_i(y) - v_i(y)]^2\} = Eq_i^2(y) = K(m_i)e_i^2 \tag{7}$$

where e_i^2 is the variance of the differential signal in the ith DPCM and $K(m_i)$ is the quantization error of a variate with a unity variance in a quantizer with $(2)^{m_i}$ levels.

Analysis of Lloyd–Max quantizer has shown that $K(m_i)$ can be approximated fairly accurately by

$$K(m_i) \cong b \exp(-am_i) \tag{8}$$

where the best fit for a Gaussian variate is obtained using $a = 0.5 \ln 10$ and $b = 1.0$ [4], [5], [7]. Study of other quantization techniques has indicated similar results for various probability density functions [8]. Fig. 2 shows the functional form of $K(m_i)$ and the accuracy of approximation (8) using $a = 0.5 \ln 10$, $b = 2.0$ for a random variable with a double-sided exponential probability density function using an instantaneous companding quantizer [9].

From published results [1], [6] the variance of the differential signal in a DPCM system with an mth order linear predictor is

$$e_i^2 = C_i(0) - \sum_{j=1}^{m} A_{ij}C_i(j) \tag{9}$$

where A_{ij} are related to $C_i(j)$ by m algebraic equations.

Substituting (7) and (8) in (6), ϵ^2 is

$$\epsilon^2 = R(0,0,0,0) - \sum_{i=1}^{n} C_i(0) + \frac{b}{M}\sum_{i=1}^{n}\exp(-am_i)e_i^2 \tag{10}$$

where the error is defined in terms of n and m_i, $i = 1,\cdots,n$. Treating m_i as continuous variables and minimizing ϵ^2 with a constraint

$$\sum_{i=1}^{n} m_i = M_b$$

will give

$$m_i = \frac{M_b}{n} + \frac{1}{a}\left[\ln e_i^2 - \frac{1}{n}\sum_{i=1}^{n}\ln e_i^2\right] \tag{11}$$

where n is chosen such that ϵ^2 is minimum[1] and the quantizer in the ith DPCM system will have $(2)^{m_i}$ levels. m_i, as obtained from (11), is modified as discussed in [4]

[1] Ready and Wintz minimize ϵ^2 with respect to both n and m_i. It gives the same end result requiring fewer computations [10].

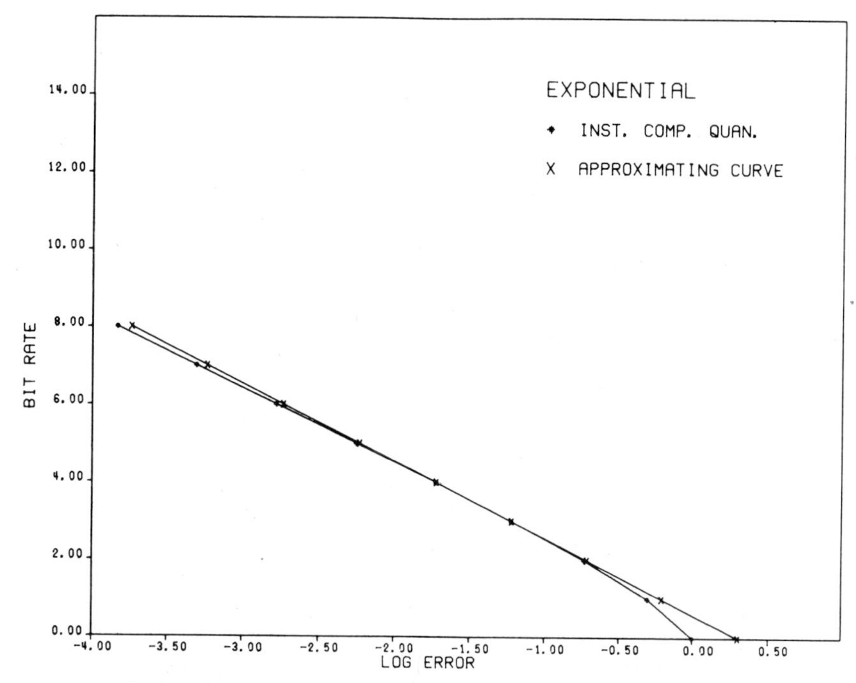

Fig. 2. Bit rate versus the logarithm of the quantization error and its linear approximation for a variate with a double sided exponential probability density function using an instantaneous-companding quantizer.

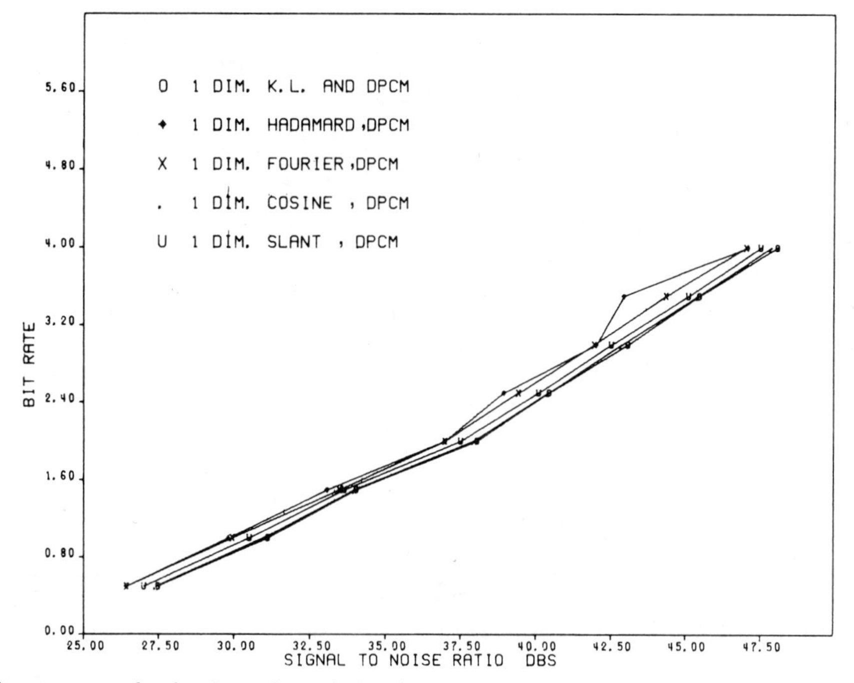

Fig. 3. Bit rate versus the signal-to-noise ratio for the proposed one-dimensional hybrid systems for the discrete random field.

and [5]. Note that expressions (10) and (11) are similar to those obtained for coding the transformed data by memoryless quantizers, the difference being that here the variance of the differential signal rather than the variance of the transformed data is used.

Fig. 3 shows the theoretical value of the coding error as given by (1) in terms of the peak to peak signal to rms noise ratio for a discrete random field with an autocovariance

$$R(x,\hat{x},y,\hat{y}) = \exp(-\alpha \mid x - \hat{x} \mid - \beta \mid y - \hat{y} \mid) \quad (12)$$

using Karhunen–Loeve, Hadamard, discrete Fourier, discrete cosine [11], and slant [12] transformations for $\alpha = 0.0545$, $\beta = 0.128$. These curves are obtained using a bank of n DPCM encoders with one-element predictors. The quantizers in the DPCM systems are of instantaneous-

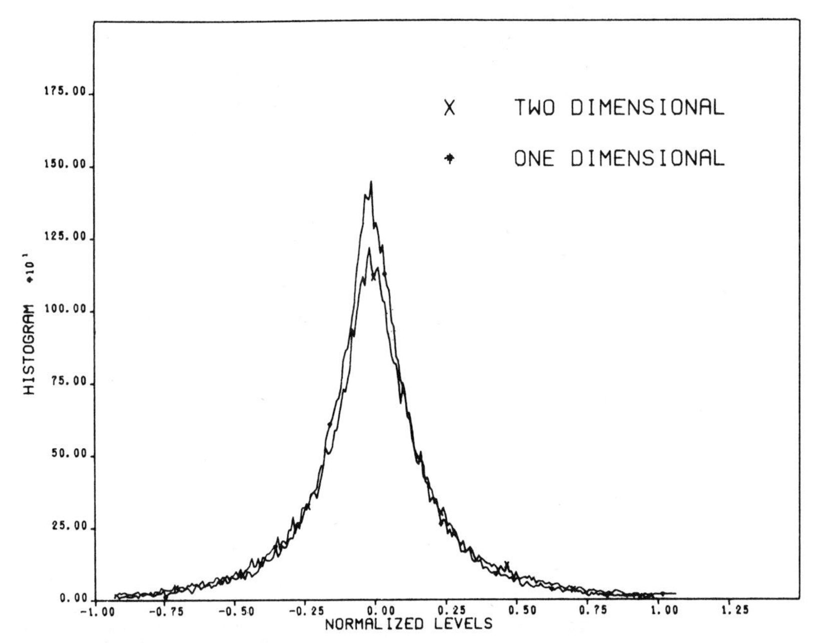

Fig. 4. Histogram of the differential signals in the DPCM systems for one-dimensional ($M = 16$) and two-dimensional (4×4) Hadamard transformations.

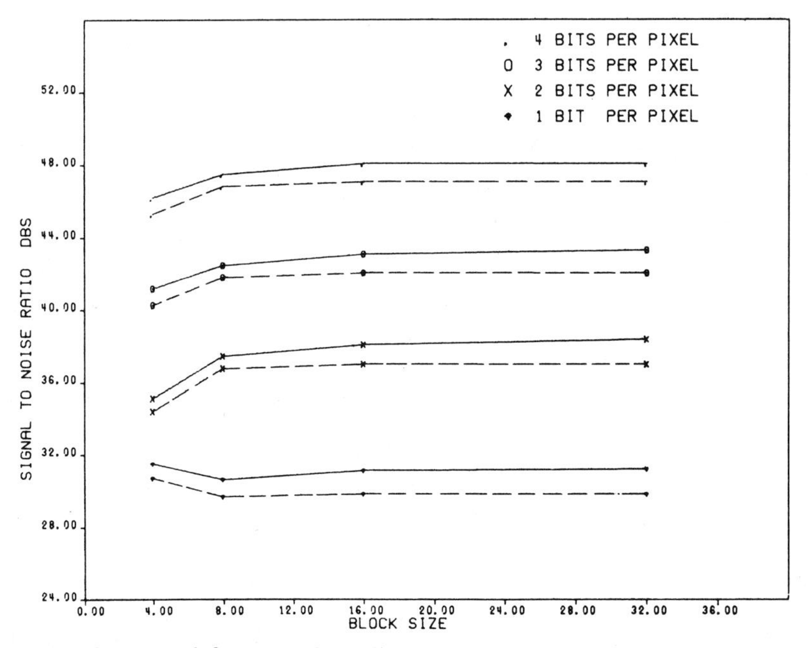

Fig. 5. Theoretical performance of the proposed one-dimensional encoders versus block size M. The solid and the dashed lines refer to the Karhunen–Loeve and the Hadamard transforms, respectively.

companding type and are designed for the probability density functions of the differential signals in the DPCM systems. Experimental results (see Fig. 4) indicate that the probability density functions of the differential signals are double-sided exponential functions.

Fig. 5 shows that increasing the block size M improves the theoretical performance of the proposed one-dimensional systems, however the improvement becomes negligible for values of M larger than 8. This makes this coding system less sensitive to the block size than the standard transform coding systems which use memoryless quantizers.

III. TWO-DIMENSIONAL TRANSFORMATION AND DPCM CODING

The most severe limitation of the two-dimensional transform coding technique is the large number of operations needed for coding pictorial data and the resulting equipment complexity. This limitation becomes less significant when the picture is divided into smaller block sizes then encoded, but this also limits the efficiency of the system since the elements of various blocks remain correlated in the transformed domain. A coding system utilizing two-dimensional transformation and a bank of DPCM systems would exploit this correlation, thus improving the coding efficiency of the system.

In this system the sampled image $u(\cdot,\cdot)$ of N by N elements is divided into arrays of M by M elements. The location of each array in the image is indexed by subscripts (k,l). A two-dimensional unitary transformation of each block of image is obtained and is ordered to form a one-dimensional array U_i, $i = 1,2,\cdots,M^2$ where $U_i(k,l)$ refers to the ith member of this array which is obtained by transforming the (k,l)th block of the data. This transformation and its inverse are modeled as

$$U_i(k,l) = \sum_{y=1}^{M} \sum_{x=1}^{M} u(kM + x, lM + y)\Phi_i(x,y) \tag{13}$$

$$u(kM + x, lM + y) = \sum_{i=1}^{M^2} U_i(k,l)\Phi_i(x,y) \tag{14}$$

where $\Phi_i(x,y)$ are a set of orthonormal basis matrices. The elements of $U_i(k,l)$ arrays for various values of i, $(i = 1,2,\cdots,M^2)$ are correlated, thus could be coded by M^2 DPCM systems. The analysis of this problem is similar to one discussed in Section II. The coding error and the bit assignment rules are

$$\epsilon^2 = R(0,0) - \sum_{i=1}^{n} C_i(0) + \frac{b}{M} \sum_{i=1}^{n} \exp(-am_i)e_i^2 \tag{15}$$

$$m_i = \frac{M_b}{n} + \frac{1}{a}\left[\ln e_i^2 - \frac{1}{n}\sum_{i=1}^{n} \ln e_i^2\right] \tag{16}$$

where

$$C_i(\tau) = \sum_{\hat{y}=1}^{M} \sum_{y=1}^{M} \sum_{x=1}^{M} \sum_{\hat{x}=1}^{M} R(x - M\tau - \hat{x}, y - \hat{y}) \cdot \Phi_i(x,y)\Phi_i(\hat{x},\hat{y}) \tag{17}$$

and e_i^2 is the variance of the differential signal in ith DPCM system

$$e_i^2 = C_i(0) - \sum_{j=1}^{m} A_{ij}C_i(j). \tag{18}$$

Note that in (16), $R(\cdot,\cdot)$ is the autocovariance of the image $u(\cdot,\cdot)$ and it is assumed that $u(\cdot,\cdot)$ is a covariance stationary random field.

The last term in (15) indicates the improvements due to the use of DPCM systems. Replacing the DPCM systems by memoryless quantizers will give a similar expression for ϵ^2 where e_i^2 is replaced by $C_i(0)$. Thus, the coding improvement due to the use of DPCM encoders depends on the relative values of e_i^2 and $C_i(0)$ which in turn depend on $R(\cdot,\cdot)$ and the block size M. In addition, note that the coding errors of the two-dimensional transform encoders are obtained from (15) using $C_i(0)$ for e_i^2 and $b = 1$, since it is generally assumed that the transformed data possesses a normal distribution and according to (8), $b = 1$. This assumption is not accurate for small block sizes and this partially accounts for the difference between the experimental and the theoretical performance of the standard two-dimensional transform encoders [4]. In the hybrid systems ϵ^2 is evaluated for $b = 2$ which gives a larger theoretical value for the coding error, however experimental results indicate that the differential signal in the hybrid systems possesses an exponential density regardless of the block size (see Fig. 4). This results in a better agreement between the theoretical and the experimental results in the hybrid system.

Fig. 6 shows the performance of the two-dimensional system for the random field that was considered in Section II. Here a block size of 4 by 4 is used, the bit assignment is according to (16), and the DPCM encoders are identical to those employed for the one-dimensional system. Fig. 7 shows the performance of the two-dimensional systems for various block sizes M. The results are similar to those observed for the one-dimensional system.

IV. EXPERIMENTAL RESULTS AND THE NOISY CHANNEL

The coding systems discussed in Sections II and III were simulated on a digital computer and were used to encode the picture shown in Fig. 9(a). This picture is composed of 256 by 256 picture elements, each element quantized to 64 levels. Values of α and β for (12) were approximated from Fig. 9(a) and are 0.0545 and 0.128, respectively. The experimental results for the hybrid system using one-dimensional transformations are obtained using a block size of 16 and for the hybrid system using two-dimensional transformations for a block size of 4 by 4. The results of the proposed systems for the cascade of one-dimensional Hadamard, Fourier, and Karhunen–Loeve transformations with the DPCM systems as well as the cascade of two-dimensional Hadamard transformation and the DPCM systems are plotted in Fig. 8. This figure also shows the performance of a simple third-order DPCM system discussed in [6] and the performance of the encoder using two-dimensional Hadamard transform and a block quantizer discussed in [4], which is included here for comparison. The two-dimensional Hadamard system uses a block size of 16 by 16 and gives a better signal-to-noise ratio than the two-dimensional hybrid system at low bit rates, however the two-dimensional hybrid system performs better at high bit rates. Note that the two-dimensional hybrid encoder uses a block size of 4 by 4 and its performance would have been inferior to the performance of the two-dimensional transform encoder at all bit rates

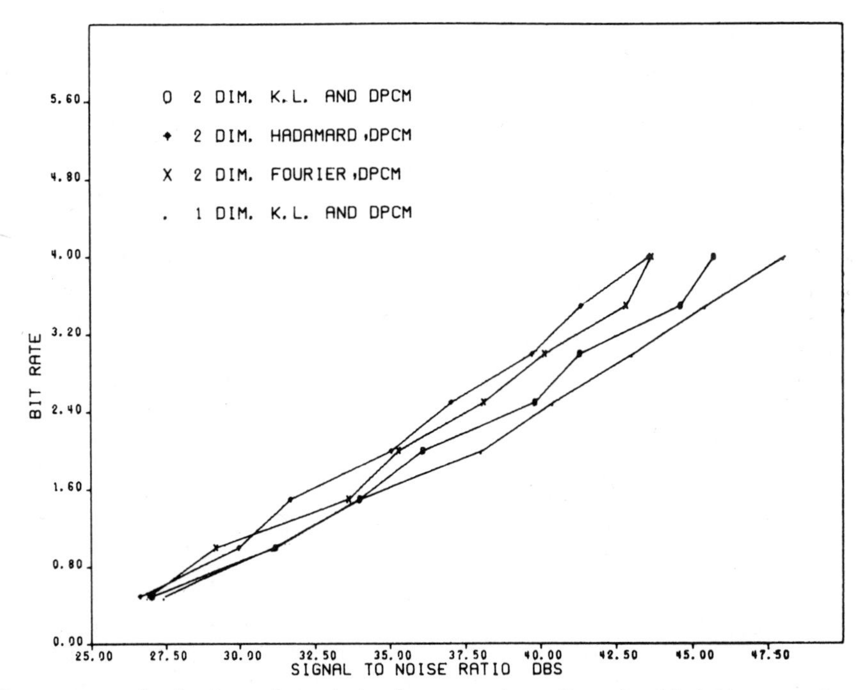

Fig. 6. Bit rate versus the signal-to-noise ratio for the proposed two-dimensional hybrid systems for the discrete random field. Performance of the hybrid system using one-dimensional Karhunen–Loeve transform and DPCM is included for comparison.

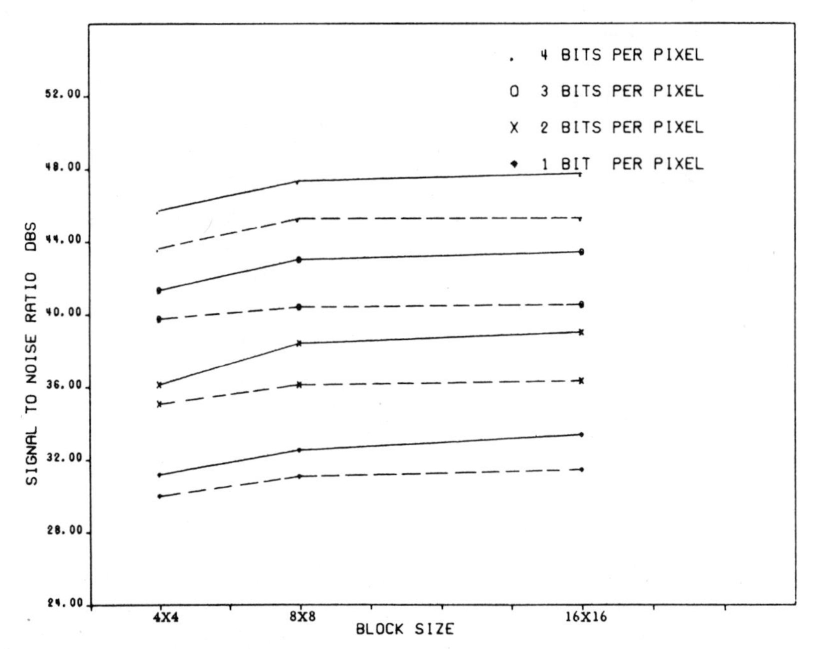

Fig. 7. Theoretical performance of the proposed two-dimensional encoders versus block size $M \times M$. The solid and the dashed lines refer to the Karhunen–Loeve and the Hadamard transforms, respectively.

if it were not a hybrid system. The one-dimensional hybrid systems perform better than the two-dimensional and also the simple DPCM system. The performance of the two-dimensional hybrid system is similar to the performance of the simple two-dimensional system in the sense that it performs better than the DPCM system at low bit rate, however it is worse at a higher bit rate. The encoded pictures corresponding to the one-dimensional and the two-dimensional hybrid systems as well as the system using two-dimensional Hadamard transform are shown in Figs. 9–11 for one and two bits per pixel. Good coding results are obtained at a bit rate of two bits/pixel where

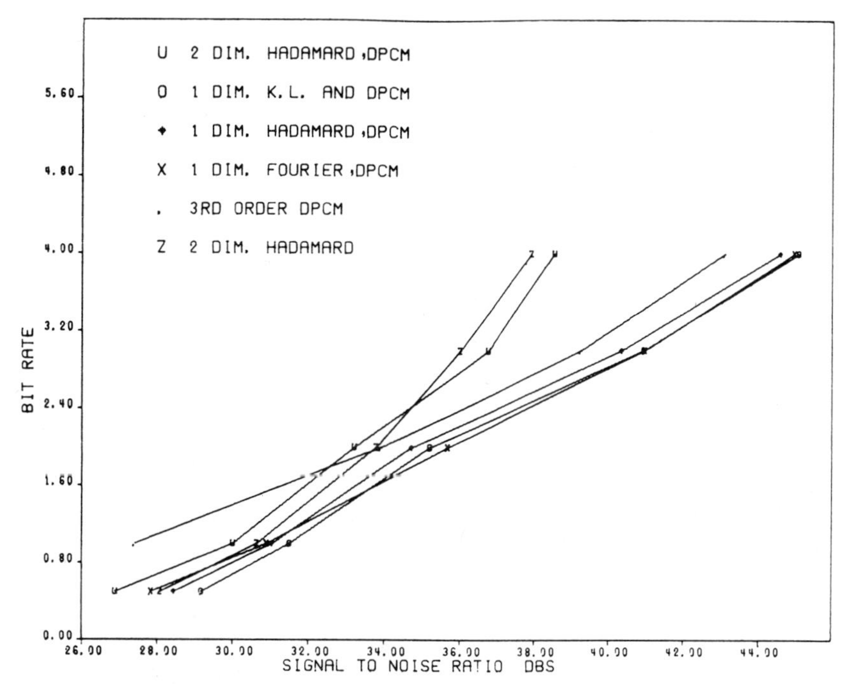

Fig. 8. Experimental results comparing the performance of the proposed hybrid systems using the picture in Fig. 9(a). The performance of a third-order simple DPCM and the encoder using two-dimensional Hadamard transform and a block quantizer is included for comparison.

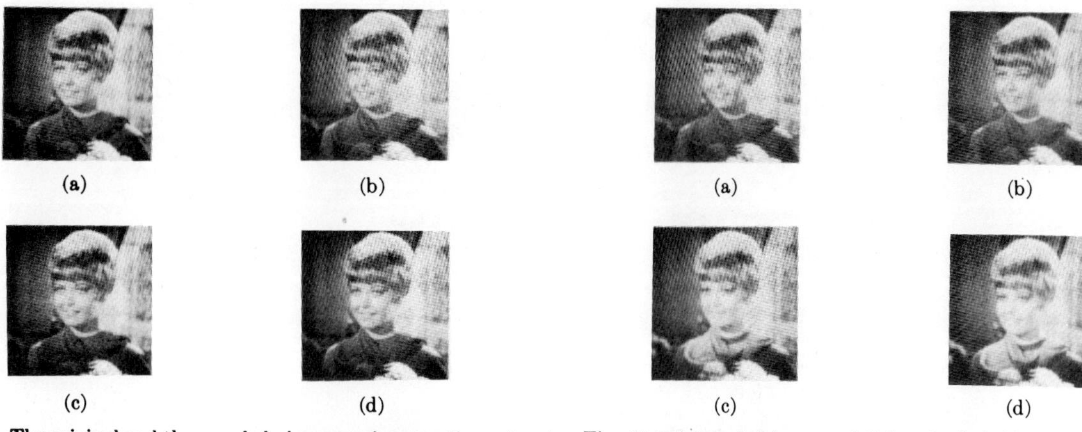

Fig. 9. The original and the encoded pictures using one-dimensional transformations and DPCM systems, 1 bit/pixel, $M = 16$. a) Original, b) Karhunen-Loeve, c) Fourier, d) Hadamard.

Fig. 11. (a) and (b) are coded by the hybrid system using two-dimensional Hadamard transform (4 by 4) and DPCM. (c) and (d) are coded by the system using two-dimensional Hadamard transform (16 by 16) and a block quantizer. a) 1 bit/pixel two-dimensional hybrid, b) 2 bits/pixel two-dimensional hybrid, c) 1 bit/pixel two-dimensional Hadamard, d) 2 bits/pixel two-dimensional Hadamard.

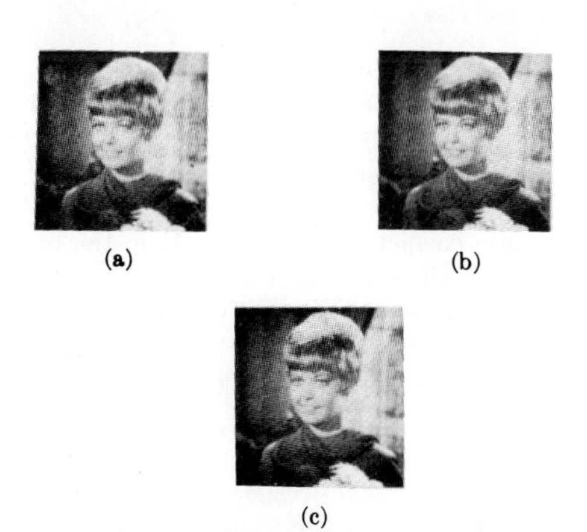

Fig. 10. The encoded pictures using the cascade of one-dimensional transformations and DPCM systems, 2 bits/pixel, $M = 16$. a) Karhunen–Loeve, b) Fourier, c) Hadamard.

the pictures encoded at one bit/pixel are acceptable; though some degradation is noticeable. We also note that the pictures encoded by the hybrid systems are less degraded than their counterparts encoded by the two-dimensional transform coder.

To study the effect of channel error in the performance of the hybrid encoders, a binary symmetric channel was simulated on a digital computer to link the transmitter with the receiver. This channel would operate on each binary digit independently changing each digit from 0 to 1 or from 1 to 0 with probability p and leaving the digit unchanged with probability $1 - p$. Then the same receiver discussed in Section II is used to reconstruct the encoded picture from the string of binary digits at the output of the

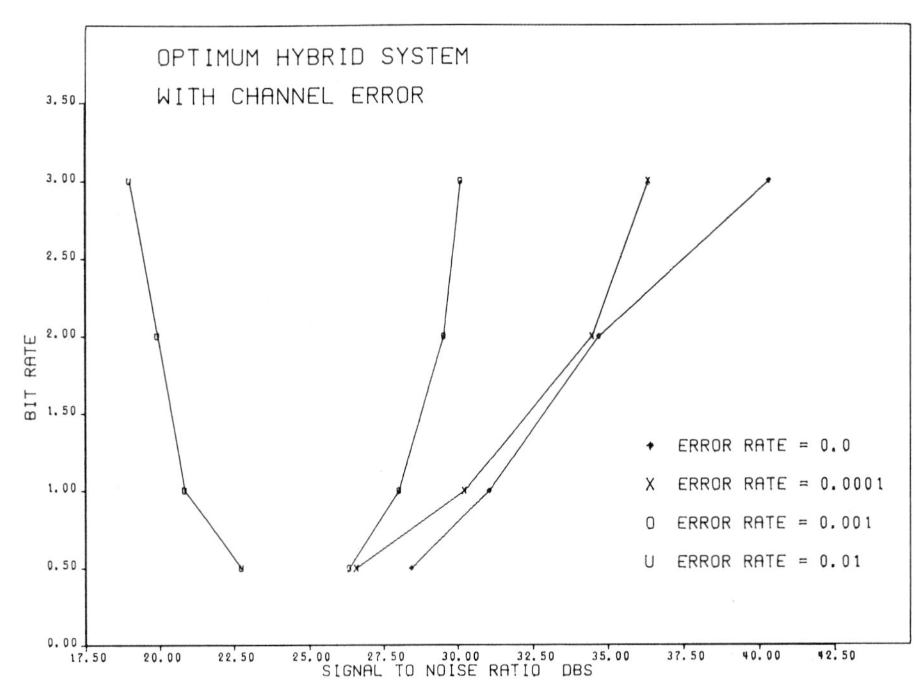

Fig. 12. Performance (experimental results) of optimum hybrid system (A_i's have correct value) using one-dimensional Hadamard transform and DPCM encoders for a noisy channel.

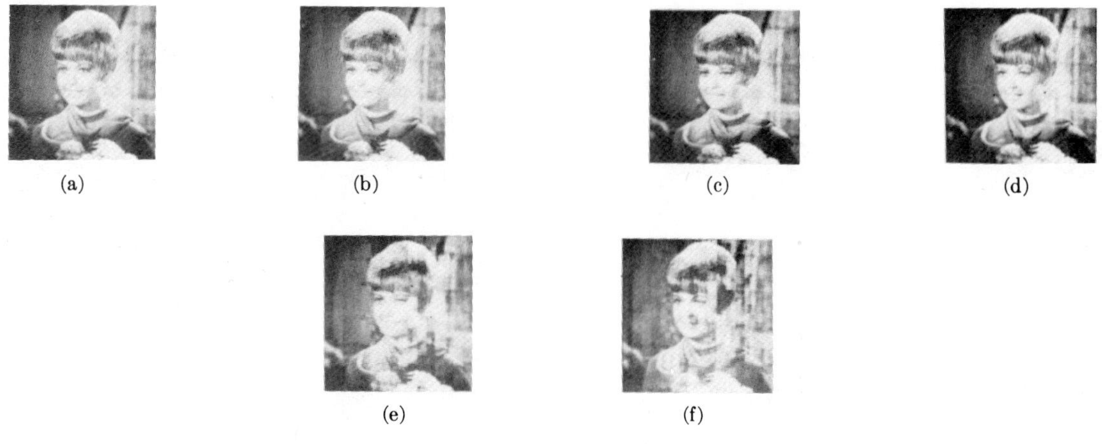

Fig. 13. Coded by the hybrid system using one-dimensional Hadamard transform and DPCM encoders employing a noisy channel. a) 1 bit, $P = 10^{-4}$, b) 1 bit, $P = 10^{-3}$, c) 1 bit, $P = 10^{-2}$, d) 2 bits, $P = 10^{-4}$, e) 2 bits, $P = 10^{-3}$, f) 2 bits, $P = 10^{-2}$.

channel. Fig. 12 shows the degradation of the one-dimensional hybrid system using Hadamard transformation and the DPCM encoders, in terms of signal-to-noise ratio, for bit-error probabilities of $p = 10^{-2}$, $p = 10^{-3}$, and $p = 10^{-4}$. It is observed that the performance of the hybrid system is affected very little for a channel noise corresponding to a bit-error probability of 10^{-4}, however it increases significantly at bit-error probabilities of 10^{-3} and 10^{-2}. Fig. 13 shows the encoded pictures using the above hybrid system at bit rates of one and two bits per pixel for bit-error probabilities of 10^{-2}, 10^{-3}, and 10^{-4}. These figures show that aside from the exceptionally high bit-error probability of 10^{-2}, the degradation of the encoded pictures because of noisy channel is not very significant. Visible degradations are in the form of blocks of 16 pixel width (the block length of one-dimensional transformations), which is due to the error in the output of the DPCM system that encodes the lower components of the transformed signal. The propagation of the error in the vertical direction is caused by the DPCM encoders.

The hybrid system we have investigated is one that uses the optimum coefficients $A_1, A_2, \cdots, A_n$ in the predictors employed in both the receiver and the transmitter of the system (see Fig. 1). These coefficients depend upon the statistics of the particular picture that we are encoding. To make the encoder independent of the signal statistics, a common value can be used for A_1–A_n. This makes the bank of DPCM systems similar to the differential encoders that have been used in practice [3]. Fig. 14 shows the performance of the hybrid system discussed above using a common value of 0.90 for A_i, $i = 1, \cdots, n$ at various bit rates and bit-error probabilities. This figure also shows

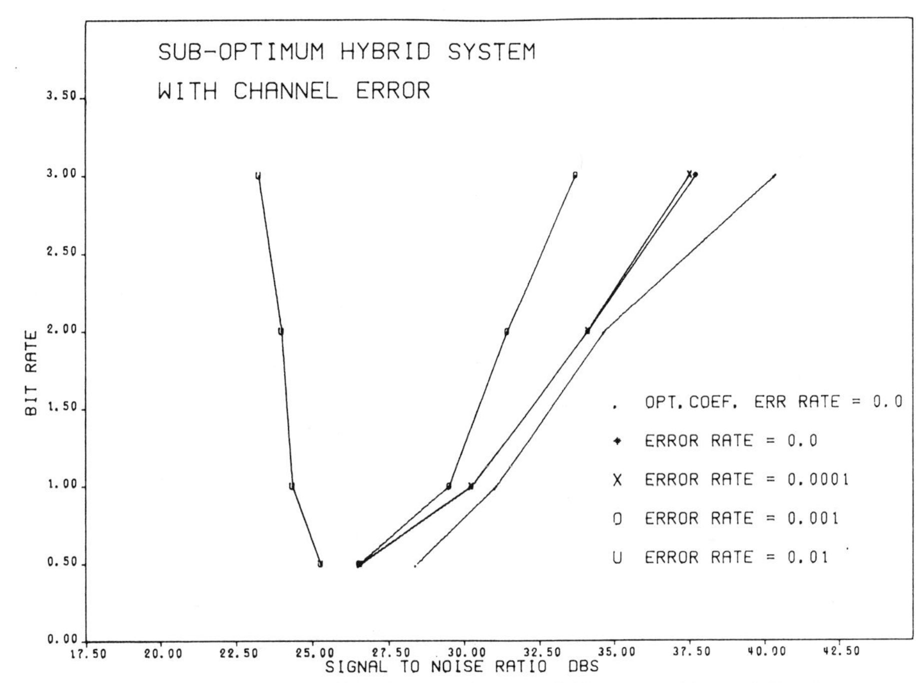

Fig. 14. Performance (experimental results) of suboptimum hybrid system ($A_i = 0.9$) using one-dimensional Hadamard transform and DPCM encoder for a noisy channel. Performance of optimum hybrid encoder is included for comparison.

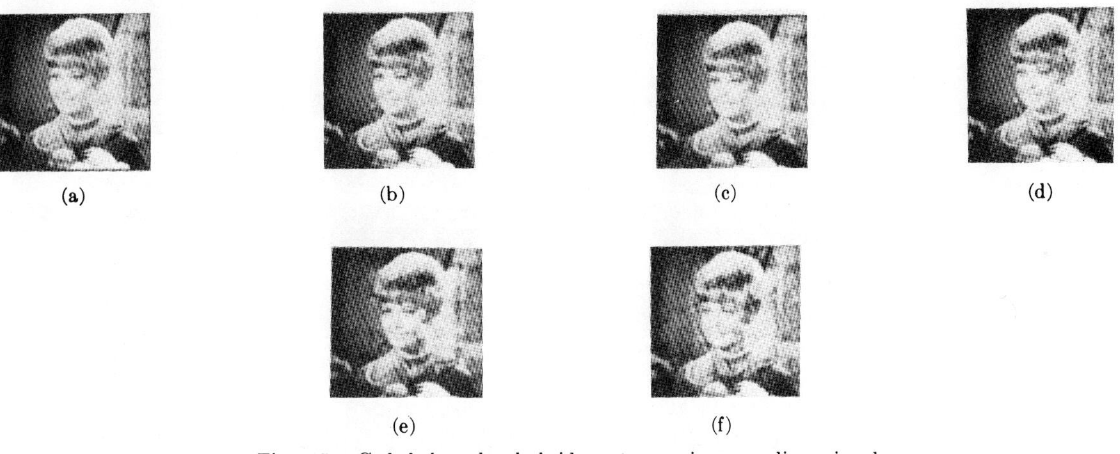

(a) (b) (c) (d)

(e) (f)

Fig. 15. Coded by the hybrid system using one-dimensional Hadamard transform and simplified DPCM encoders ($A_i = 0.9$) employing a noisy channel. a) 1 bit, $P = 10^{-4}$, b) 1 bit, $P = 10^{-3}$, c) 1 bit, $P = 10^{-2}$, d) 2 bits, $P = 10^{-4}$, e) 2 bits, $P = 10^{-3}$, f) 2 bits, $P = 10^{-2}$.

the performance of the same encoder using optimum coefficients and a noiseless channel. The degradation due to making these coefficients 0.9 is small and it is almost the same at various bit rates for a noiseless channel. The behavior of this system for various bit-error probabilities is actually somewhat better than the performance of the encoder using optimum coefficients shown in Fig. 12. This is because 0.9 is smaller than the optimum values of most Ai's and this causes a shorter propagation of the channel error in the encoder.

Encoded pictures using this suboptimum encoder for one and two bits per pixel at various bit-error probabilities are shown in Fig. 15. Comparison of Figs. 13 and 15 shows that there is no noticeable difference in the performance of the optimum and suboptimum encoder.

To compare the performance of the hybrid system with the encoder using two-dimensional transform and block quantization for a noisy channel, the binary symmetric channel discussed above is simulated in the system that uses two-dimensional Hadamard transform and a block quantizer for a block size of 16 by 16. Fig. 16 shows the performance of this system in terms of signal-to-noise ratio at various bit rates for bit-error probabilities of 10^{-4}, 10^{-3}, and 10^{-2}. Pictures encoded by this system using one and

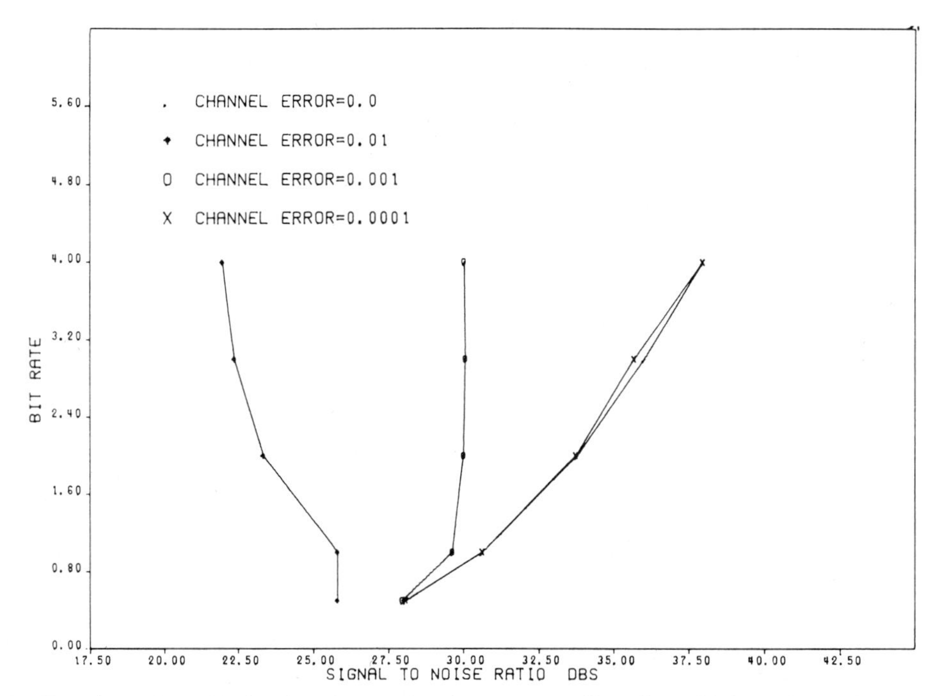

Fig. 16. Experimental results showing the encoder that uses two-dimensional Hadamard transform (16 × 16) and a block quantizer for a noisy channel.

two bits per pixel and a bit-error probability of 10^{-3} and 10^{-2} are shown in Fig. 17. Comparison of these pictures with those in Figs. 11 and 13 indicates that the subjective quality of the resultant pictures is affected more because of channel noise when coded by the hybrid system than it is using the two-dimensional transform technique, though the hybrid system produces pictures with a better overall quality.

V. CONCLUSIONS

The hybrid system using the two-dimensional transformations and the DPCM encoders is essentially the standard two-dimensional transform coding system where the encoding efficiency is improved by exploiting the interblock correlation in the data. The system is attractive since a small block size could be used reducing the computational complexity without affecting the performance of the system.

The hybrid system that uses one-dimensional transformation and DPCM systems attempts to uncorrelate the data in one spatial direction by the transform technique and in the other spatial direction by the DPCM encoders. The quantization and the channel error are introduced in the transform domain, where it is less objectionable to a human viewer. The fact that the system uses a one-dimensional transformation, it is not sensitive to the block size, and allows parallel operation on the data, makes the system attractive from an implementational viewpoint. The significant result is the improved performance of the system as indicated by both theoretical and experimental results that surpasses the performance of both the DPCM and the two-dimensional transform coding systems and the fact that the performance of the encoder is not impaired significantly for small or moderate levels of channel noise.

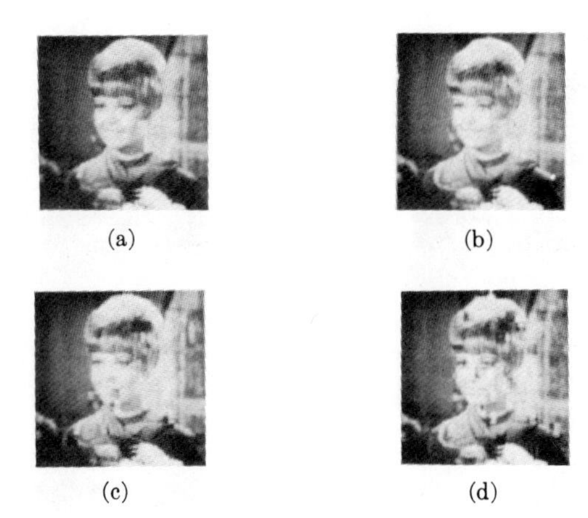

Fig. 17. Coded by the system using two-dimensional Hadamard transform (16 by 16) and a block quantizer with a noisy channel. a) 1 bit/pixel, $P = 10^{-3}$, b) 1 bit/pixel, $P = 10^{-2}$, c) 2 bits/pixel, $P = 10^{-3}$, d) 2 bits/pixel, $P = 10^{-2}$.

ACKNOWLEDGMENT

The author wishes to thank Dr. Wen-hsiung Chen of the University of Southern California for his efforts in simulating the noisy channel in the system.

REFERENCES

[1] J. B. O'Neal. Jr., "Delta modulation quantizing noise—Analytical and computer simulation results for Gaussian and television input signals," *Bell Syst. Tech. J.*, vol. 45, pp. 117–142, Jan. 1966.

[2] W. K. Pratt, J. Kane, and H. C. Andrews, "Hadamard transform image coding," *Proc. IEEE*, vol. 57, pp. 58–68, Jan. 1969.

[3] J. O. Limb, C. B. Rubinstein, and K. A. Walsh, "Digital coding of color picturephone signals by element-differential quantiza-

tion," *IEEE Trans. Commun. Technol.*, vol. COM-19, pp. 992–1006, Dec. 1971.

[4] A. Habibi and P. A. Wintz, "Image coding by linear transformations and block quantization," *IEEE Trans. Commun. Technol.*, vol. COM-19, pp. 50–62, Feb. 1971.

[5] J. J. Y. Huang and P. M. Schultheiss, "Block quantization of correlated Gaussian random variables," *IEEE Trans. Commun. Syst.*, vol. CS-11, pp. 289–296, Sept. 1963.

[6] A. Habibi, "Comparison of *n*th-order DPCM encoder with linear transformations and block quantization techniques," *IEEE Trans. Commun. Technol.*, vol. COM-19, pp. 948–956, Dec. 1971.

[7] P. A. Wintz and A. J. Kurtenbach, "Waveform error control in PCM telemetry," *IEEE Trans. Inform. Theory*, vol. IT-14, pp. 650–661, Sept. 1968.

[8] A. Habibi, "Performance of zero-memory quantizers using rate distortion criteria," to be published.

[9] B. Smith, "Instantaneous companding of quantized signals," *Bell Syst. Tech. J.*, vol. 36, pp. 44–48, Jan. 1951.

[10] P. J. Ready and P. A. Wintz, "Multispectral data compression through transform coding and block quantization," School of Elec. Eng., Purdue Univ., Lafayette, Ind., Tech. Rep. TR-EE 72-29, May 1972.

[11] N. Ahmed, T. Natarajan, and K. R. Rao, "Discrete cosine transform," *IEEE Trans. on Comput.*, vol. C-23, pp. 90–93, Jan. 1974.

[12] W. K. Pratt, L. R. Welch, and W. Chen, "Slant transforms for image coding." Proc. Symp. Application of Walsh Functions, Mar. 1972.

AUTHOR CITATION INDEX

SUBJECT INDEX

About the Editor

PROFESSOR RAO received the B. E. degree from the University of Madras in 1952, the M.S.E.E. and M.S.N.E. degrees from the University of Florida, Gainesville in 1959 and 1960 respectively, and the Ph.D. degree in electrical engineering from the University of New Mexico, Albuquerque in 1966. Since 1966 he has been with the University of Texas at Arlington, where he is currently a professor of electrical engineering. He has published extensively in reviewed technical journals in the areas of discrete transforms and digital image coding. With two other researchers, he introduced the discrete cosine transform in 1975, which has since become very popular in digital signal processing. He has organized and conducted short courses and conferences on thermoelectric energy conversion since 1970. Dr. Rao is coauthor of two books: *Orthogonal Transforms for Digital Signal Processing* (Springer-Verlag, 1975) and *Fast Transforms: Algorithms, Analyses and Applications* (Academic Press, 1982).